CAMBRIDGE MONOGRAPHS ON MATHEMATICAL PHYSICS

General editors: P. V. Landshoff, D. W. Sciama, S. Weinberg

SUPERSTRING THEORY

Volume 1: Introduction

D0222994

TO OUR PARENTS

SUPERSTRING THEORY

Volume 1

Introduction

MICHAEL B. GREEN

Queen Mary College, University of London

JOHN H. SCHWARZ

California Institute of Technology

EDWARD WITTEN

Princeton University

CAMBRIDGE
UNIVERSITY PRESS

CAMBRIDGE UNIVERSITY PRESS
Cambridge, New York, Melbourne, Madrid, Cape Town, Singapore, São Paulo

Cambridge University Press
The Edinburgh Building, Cambridge CB2 2RU, UK

Published in the United States of America by Cambridge University Press, New York

www.cambridge.org
Information on this title: www.cambridge.org/9780521323840

First published 1987
Reprinted 1987 (thrice)
First paperback edition with corrections 1988
Reprinted 1988, 1992, 1994, 1995, 1996, 1998, 1999, 2001, 2002

A catalogue record for this publication is available from the British Library

ISBN-13 978-0-521-35752-4 paperback
ISBN-10 0-521-35752-7 paperback

Transferred to digital printing 2005

Contents

Preface

Recent years have brought a revival of work on string theory, which has been a source of fascination since its origins nearly twenty years ago. There seems to be a widely perceived need for a systematic, pedagogical exposition of the present state of knowledge about string theory. We hope that this book will help to meet this need. To give a comprehensive account of such a vast topic as string theory would scarcely be possible, even in two volumes with the length to which these have grown. Indeed, we have had to omit many important subjects, while treating others only sketchily. String field theory is omitted entirely (though the subject of chapter 11 is closely related to light-cone string field theory). Conformal field theory is not developed systematically, though much of the background material needed to understand recent papers on this subject is presented in chapter 3 and elsewhere. Our discussion of string propagation in background fields is limited to the bosonic theory, and multiloop diagrams are discussed only in very general and elementary terms. The omissions reflect a combination of human frailty and an attempt to keep the combined length of the two volumes from creeping too much over 1000 pages.

We hope that these two volumes will be useful for a wide range of readers, ranging from those who are motivated mainly by curiosity to those who actually wish to do research on string theory. The first volume, which requires as background only a moderate knowledge of particle physics and quantum field theory, gives a detailed introduction to the basic ideas of string theory. This volume is intended to be self-contained. The second volume delves into a number of more advanced topics, including a study of one-loop amplitudes, the low-energy effective field theory, and anomalies. There is also a substantial amount of mathematical background on differential and algebraic geometry, as well as their possible application to phenomenology.

We feel that the the two volumes should be suitable for use as textbooks in an advanced graduate-level course. The amount of material is probably more than can be covered in a one-year course. This should provide the instructor the luxury of emphasizing those topics he or she finds especially important while omitting others. Despite our best efforts, it is inevitable that a substantial number of misprints, notational inconsistencies and other errors have survived. We will be grateful if they are brought to our attention so that we can correct them in future editions.

We have benefitted greatly from the assistance of several people whom we are pleased to be able to acknowledge here. Kyle Gary worked with skill and diligence in typing substantial portions of the manuscript, as well as figuring out how to implement the formatting requirements of Cambridge University Press in TEX, the type-setting system that we have used. Marc Goroff brought his wealth of knowledge about computing systems to help solve a myriad of problems that arose in the course of this work. We also received help with computing systems from Paul Kyberd and Vadim Kaplunovsky. Patricia Moyle Schwarz put together the index and made useful comments on the manuscript. Harvey Newman set up communications links that enabled us to transfer files between Pasadena, Princeton and London. Judith Wallrich helped to compile the bibliography. Useful criticisms and comments on the text were offered by Čedomir Crnković, Chiara Nappi, Ryan Rohm and Larry Romans.

We would like to dedicate this book to our parents.

1986   Michael B. Green
John H. Schwarz
Edward Witten

1. Introduction

1.1 The Early Days of Dual Models

In 1900, in the course of trying to fit to experimental data, Planck wrote down his celebrated formula for black body radiation. It does not usually happen in physics that an experimental curve is directly related to the fundamentals of a theory; normally they are related by a more or less intricate chain of calculations. But black body radiation was a lucky exception to this rule. In fitting to experimental curves, Planck wrote down a formula that directly led, as we all know, to the concept of the quantum.

In the 1960s, one of the mysteries in strong interaction physics was the enormous proliferation of strongly interacting particles or hadrons. Hadronic resonances seemed to exist with rather high spin, the mass squared of the lightest particle of spin J being roughly $m^2 = J/\alpha'$, where $\alpha' \sim 1(\text{GeV})^{-2}$ is a constant that became known as the Regge slope. Such behavior was tested up to about $J = 11/2$, and it seemed conceivable that it might continue indefinitely. One reason that the proliferation of strongly interacting particles was surprising was that the behavior of the weak and electromagnetic interactions was quite different; there are, comparatively speaking, just a few low mass particles known that do not have strong interactions.

The resonances were so numerous that it was not plausible that they were all fundamental. In any case consistent theories of fundamental particles of high spin were not known to exist. Consistent (renormalizable) quantum field theories seemed to be limited to spins zero, one-half, and one, the known examples being abelian gauge theories and scalar and Yukawa theories. That limitation on the possible spins in consistent quantum field theory still seems valid today, though now we would include Yang–Mills theory in the list of consistent theories for spin one. The apparent limitation of consistent quantum field theories to low spin was compatible with the existence of a successful field-theory description of the electromagnetic interactions, in which the basic particles have spin one half and spin one, and was compatible at least with attempts (which

1

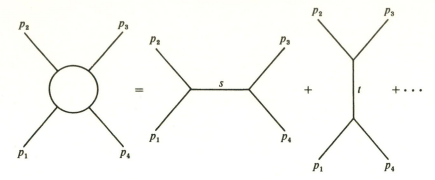

Figure 1.1. An elastic scattering process with incoming particles with momenta p_1, p_2 and outgoing particles with momenta $-p_3, -p_4$ (we adopt the convention that the labels refer to incoming momenta). Both s- and t-channel diagrams are indicated. In field theory the amplitude is constructed as a sum of s-channel and t-channel diagrams.

in time succeeded) at field theories of the weak interactions. But a similar approach to strong interactions did not appear promising.

A related puzzle about strong interactions concerned the high-energy behavior of the scattering amplitudes. Consider an elastic scattering process with incoming spinless particles of momenta p_1, p_2 and outgoing particles of momenta p_3, p_4. We adopt a metric with signature $\{-++\ldots+\}$, so that the mass squared of a particle is $m^2 = -p^2$. The conventional Mandelstam variables are defined as

$$s = -(p_1 + p_2)^2, \qquad t = -(p_2 + p_3)^2, \qquad u = -(p_1 + p_3)^2. \qquad (1.1.1)$$

They obey the one identity $s+t+u = \sum m_i^2$. We assume that the external states in fig. 1.1 are particles such as pions that transform in the adjoint representation of the flavor group, which for three flavors is $SU(3)$ or $U(3)$. The flavor quantum numbers of the ith external meson are specified by picking a flavor matrix λ_i. We will discuss a term in the scattering amplitude proportional to the group-theory factor $\text{tr}(\lambda_1 \lambda_2 \lambda_3 \lambda_4)$. Since this group-theory factor is invariant under the cyclic permutation $1234 \to 2341$, Bose statistics require that the corresponding amplitude should be cyclically symmetric under $p_1 p_2 p_3 p_4 \to p_2 p_3 p_4 p_1$. In terms of Mandelstam variables, this permutation of momenta amounts to $s \leftrightarrow t$, which is the symmetry we will require for the amplitude $A(s, t)$.

In quantum field theory, the leading nontrivial contributions to the scattering amplitude come from the tree diagrams of fig. 1.1. The basic reason that it is difficult to construct sensible quantum field theories of particles of high spin is that tree diagrams with the exchange of high spin particles have bad high-energy behavior. Asymptotically, they exceed unitarity bounds. Consider, for instance, the t-channel diagram. Denote the external particles in fig. 1.1 as ϕ and the exchanged particle as σ. If σ has spin zero fig. 1.1 may involve a simple $\phi^*\phi\sigma$ interaction; the amplitude is then simply $A(s,t) = -g^2/(t-M^2)$ with g being the coupling constant and M the mass of the σ particle. This amplitude vanishes for $t \to \infty$, this being one aspect of the excellent high-energy behavior of the cubic scalar interaction we are discussing.

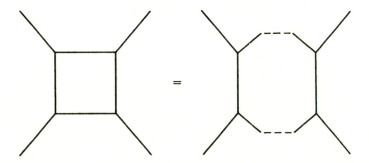

Figure 1.2. A one-loop diagram can be made by sewing together two tree diagrams, as indicated here.

Suppose instead that the sigma particle is a spin J field $\sigma_{\mu_1\mu_2...\mu_J}$. For such a field, the cubic coupling in fig. 1.1 must then be something like $\phi^* \overset{\leftrightarrow}{\partial}_{\mu_1} \overset{\leftrightarrow}{\partial}_{\mu_2} \ldots \overset{\leftrightarrow}{\partial}_{\mu_J}\phi \cdot \sigma^{\mu_1\mu_2...\mu_J}$. In fig. 1.1 there are now $2J$ factors of momenta. If the external particles are scalars then the contribution to the scattering amplitude of the exchange in the t channel of this spin J particle has the form

$$A_J(s,t) = -\frac{g^2(-s)^J}{t-M^2} \tag{1.1.2}$$

at high energies.[*] The behavior of this amplitude is therefore worse and worse (more and more divergent) for larger and larger J. An objective criterion for what is a 'bad' amplitude is to ask what will happen when we sew together amplitudes like that of (1.1.2) to make loops, as in fig. 1.2.

---

[*] This is the behavior of the tree-level scattering amplitude in the asymptotic region of large s, fixed t. The s^J behavior is easily found by contracting the momenta that appear in the interaction vertices in fig. 1.1. The exact formula (for moderate s) is more complicated, involving a Legendre polynomial $P_J(\cos\theta_t)$ (θ_t is the center-of-mass scattering angle in the t channel). We prefer to write only the high-energy behavior, which is transparent and adequate for our purposes.

The one-loop integrand in n dimensions is roughly $\int d^n p \, A^2/(p^2)^2$, with A being the tree amplitude of (1.1.2). In four dimensions such a loop diagram is convergent for $J < 1$, has a potentially renormalizable logarithmic divergence for $J = 1$, and has a nasty unrenormalizable divergence for $J > 1$.

There are strongly interacting particles of various mass and spin that might be exchanged in the t channel, so we must think of a t-channel amplitude of the general form

$$A(s,t) = -\sum_J \frac{g_J^2 (-s)^J}{t - M_J^2}, \qquad (1.1.3)$$

where now we allow for the possibility that the couplings g_J and masses M_J of the exchanged particles may depend on J (and perhaps on other quantum numbers that we do not indicate). Of course, one might take the point of view that the strong interactions are so strong that a Born-like approximation as in (1.1.3) is hopeless. But let us be optimists and see how well we can do. What is the high-energy behavior of the sum in (1.1.3)? If this is a *finite* sum, the high-energy behavior is simply determined by the hadron of largest J that contributes in (1.1.3). This is very different from what is observed in nature; the actual high-energy behavior of hadron scattering amplitudes is much softer than the behavior of any individual term in (1.1.3). (In fact, Regge asymptotic behavior of the type described in §1.1.2. is a reasonable approximation to experiment.) On the other hand, it is not reasonable to think of (1.1.3) as a finite sum. There certainly does not seem to be any such thing as a 'hadron of highest spin'. With (1.1.3) viewed as an infinite sum, it is certainly conceivable that the whole sum might have a high-energy behavior better than the behavior of any individual term in the series, just as the function e^{-x} is smaller for $x \to \infty$ than any individual term in its power series expansion $e^{-x} = \sum_{n=0}^{\infty} (-x)^n/n!$

Regarding (1.1.3) as an infinite sum has another consequence. In a physical process such as the elastic scattering of pions, we expect the t-channel poles that appear in (1.1.3), but we also expect s-channel resonances or in other words poles in the amplitude at certain values of s. In fact, the cyclic symmetry that we discussed earlier requires that the coefficient of $\mathrm{tr}(\lambda_1 \lambda_2 \lambda_3 \lambda_4)$ in the scattering amplitude have both s- and t-channel poles or neither. A *finite* sum (1.1.3) defines an amplitude $A(s,t)$ that has no s-channel poles; for fixed t, (1.1.3) manifestly defines an entire function of s, as long as there are only a finite number of terms in the sum. It is precisely for this reason that the perturbative expansion of ordinary quantum field theories satisfies crossing symmetry by including both s- and t-channel diagrams. In the case of an infinite sum, things are different. Though each term in (1.1.3) is an entire function of s, the infinite sum might diverge at some finite values of s, giving poles in the s channel. Thus, once we accept the fact that (1.1.3) is essentially an *infinite* series, it is no longer obvious that s-channel terms must be included separately; they may be already implicit in (1.1.3).

Similar remarks could be made if we took as our starting point resonant scattering or in other words contributions to scattering amplitudes with s-channel poles. We would then construct an amplitude analogous to (1.1.3) but with s-channel poles rather than t-channel poles:

$$A'(s,t) = -\sum_J \frac{g_J^2(-t)^J}{s - M_J^2}. \tag{1.1.4}$$

Symmetry under cyclic permutation of the external momenta requires that the same masses and couplings appear in (1.1.4) as in (1.1.3). Studying (1.1.4) we would again observe that a finite sum of the type in (1.1.4) inevitably has a high-energy behavior much worse than the observed behavior of hadrons, but this is not inevitably true for an infinite sum of this type. Furthermore, a finite sum (1.1.4) would certainly define (for fixed s) an entire function of t, but this might not be true for an infinite sum.

Pursuing these thoughts still further, one might imagine that if the couplings g_J and masses M_J are cunningly chosen, then the s-channel and t-channel amplitudes $A(s,t)$ and $A'(s,t)$ might be equal. In this case, the entire amplitude could be written as a sum over only s-channel poles, as in (1.1.4), *or* as a sum over only t-channel poles, as in (1.1.3). This would be a sharp contrast to the field-theory situation in which one ordinarily needs a sum over *both s- and t-channel poles*.

Equality of the s- and t-channel amplitudes was advocated around 1968 by Dolen, Horn and Schmid, who argued, on the basis of an approximate evaluation of (1.1.3) and (1.1.4) (carried out with the help of experimental data), that the equality $A(s,t) = A'(s,t)$ was indeed approximately obeyed for small values of s and t. This was called the 'duality' hypothesis, the hypothesis that s- and t-channel diagrams give alternative or 'dual' descriptions of the same physics. Is duality an approximation or a principle? At first sight it looks well nigh impossible to choose the resonance masses and couplings to obey exactly the duality relation $A(s,t) = A'(s,t)$. However, a way of doing this was found by Veneziano in 1968. Veneziano simply postulated a formula for the scattering amplitude, namely

$$A(s,t) = \frac{\Gamma(-\alpha(s))\Gamma(-\alpha(t))}{\Gamma(-\alpha(s) - \alpha(t))}. \tag{1.1.5}$$

Here Γ is the Euler gamma function,

$$\Gamma(u) = \int_0^\infty t^{u-1}e^{-t}dt, \tag{1.1.6}$$

and $\alpha(s)$ is the 'Regge trajectory', for which Veneziano postulated the linear form $\alpha(s) = \alpha(0) + \alpha's$; α' and $\alpha(0)$ are known in Regge-pole theory as the Regge slope and the intercept, respectively.

1.1.1 The Veneziano Amplitude and Duality

It is not evident at first sight that the Veneziano amplitude obeys duality, but we will now show that it does. First of all, we need to know something about the gamma function. This function obeys the identity

$$\Gamma(u + 1) = u\Gamma(u). \tag{1.1.7}$$

This is proved, starting from (1.1.6), by simple integration by parts:

$$\Gamma(u + 1) = -\int_0^\infty t^u \frac{d}{dt}e^{-t}dt = u\int_0^\infty t^{u-1}e^{-t}dt = u\Gamma(u). \tag{1.1.8}$$

It is evident from (1.1.6) that $\Gamma(1) = 1$. If u is a positive integer, then repeated use of (1.1.7) implies that

$$\Gamma(u) = (u - 1)!. \tag{1.1.9}$$

The integral representation of the Γ function in (1.1.6) is valid as long as the real part of u is positive, and shows that Γ has no singularities in this part of the complex u plane. The recursion relation (1.1.7) can be used to extend the domain of definition of Γ and determine its singularities. Writing (1.1.7) in the form

$$\Gamma(u) = \frac{\Gamma(u + 1)}{u} \tag{1.1.10}$$

gives a definition of the gamma function for $\mathrm{Re}\,u > -1$, since the right hand side of (1.1.10) has already been defined in that region. Equation

(1.1.10) also shows that Γ has a simple pole at $u = 0$ with residue 1. This process can be generalized; repeated use of (1.1.7) gives

$$\Gamma(u) = \frac{\Gamma(u + n)}{u(u+1)\ldots(u+n-1)} \qquad (1.1.11)$$

for any positive integer n. The right-hand side of (1.1.11) is uniquely defined by the integral representation (1.1.6) as long as Re $u > -n$, so we obtain a unique analytic continuation of the gamma function in this region. Since n is arbitrary, the gamma function actually has a unique analytic continuation throughout the whole complex u plane. From (1.1.11) we can see that the only singularities of Γ are simple poles at $u = 0, -1, -2, \ldots$. The behavior for u near $-n$ (n a non-negative integer) can be read off from (1.1.11) and is

$$\Gamma(u) \sim \frac{1}{u+n} \frac{(-1)^n}{n!}. \qquad (1.1.12)$$

Now we wish to discuss the analytic behavior of the function

$$B(u, v) = \frac{\Gamma(u)\Gamma(v)}{\Gamma(u + v)}, \qquad (1.1.13)$$

which is called the Euler beta function. It is related to the Veneziano amplitude by $A(s, t) = B(-\alpha(s), -\alpha(t))$. Evidently, (1.1.13) has a simple pole when u or v is a non-positive integer. There are no double poles in (1.1.13), since while $\Gamma(u)$ and $\Gamma(v)$ may simultaneously have poles, when this occurs the denominator in (1.1.13) has a pole at the same time. This is an important point, because simple poles are the only singularities allowed in tree amplitudes in relativistic quantum mechanics. The behavior of $B(u, v)$ for $v \sim -n$ (n being a non-negative integer) is evidently

$$B(u, v) \sim \frac{1}{v+n} \frac{(-1)^n}{n!}(u-1)(u-2)\ldots(u-n). \qquad (1.1.14)$$

Here we are using (1.1.7) to write the residue of the pole at $v = -n$ as a polynomial in u; this is again an important step since the residue of a pole in relativistic quantum mechanics must be a polynomial. As a function of v for fixed u, $B(u, v)$ has only the singularities indicated in (1.1.14). We claim now that (for Re $u > 0$ so that the following infinite sum converges) we can write

$$B(u, v) = \sum_{n=0}^{\infty} \frac{1}{v+n} \frac{(-1)^n}{n!}(u-1)(u-2)\ldots(u-n). \qquad (1.1.15)$$

The idea here is that the sum on the right of (1.1.15) reproduces all of the singularities of the beta function, so could differ from it only by an entire function of v, that is, a function without singularities in the complex v plane. Such a function could not vanish for large $|v|$. As the sum on the right-hand side of (1.1.15) vanishes for positive u and large $|v|$ (away from the real axis), and we will presently see that $B(u,v)$ has the same property, they must be equal.

We can immediately express (1.1.15) as a formula for the Veneziano amplitude:

$$A(s,t) = -\sum_{n=0}^{\infty} \frac{(\alpha(s)+1)(\alpha(s)+2)\ldots(\alpha(s)+n)}{n!} \frac{1}{\alpha(t)-n}. \quad (1.1.16)$$

While the Veneziano amplitude was defined originally to manifestly obey $A(s,t) = A(t,s)$, this symmetry is not at all apparent in (1.1.16). Because of the underlying symmetry, we can immediately write down the alternative expansion

$$A(s,t) = -\sum_{n=0}^{\infty} \frac{(\alpha(t)+1)(\alpha(t)+2)\ldots(\alpha(t)+n)}{n!} \frac{1}{\alpha(s)-n}. \quad (1.1.17)$$

Now, with the simple choice of 'Regge trajectory', $\alpha(t) = \alpha't + \alpha(0)$, the singularities of (1.1.16) are simple poles corresponding, as in (1.1.3), to t-channel exchange of particles of mass $M^2 = (n - \alpha(0))/\alpha'$, $n = 0, 1, 2, \ldots$. The residue of the pole at $\alpha(t) = n$ is (with the linear choice of Regge trajectory) an nth order polynomial in s, corresponding, in view of (1.1.3), to the fact that the particles of mass $(n - \alpha(0))/\alpha'$ have spin at most n. The smallest possible mass of a particle of spin J is thus $(J - \alpha(0))/\alpha'$, and this is why α' is called the 'Regge slope'; the particles of mass $M^2 = (J - \alpha(0))/\alpha'$ are said to lie on the 'leading Regge trajectory'. We are interested in the case $\alpha' > 0$, since these particles would otherwise be all or almost all tachyons.

The equality of (1.1.16) and (1.1.17) is the seemingly impossible property of 'duality': the *same* amplitude can be written as a sum of s-channel poles as in (1.1.17) *or* as a sum of t-channel poles as in (1.1.16).

One thing that is *not* obvious in either (1.1.17) or (1.1.16) is the sign of the residues of the s-channel and t-channel poles. In (1.1.2) for exchange of a spin J particle, the coefficient of $-(-1)^J/(t - M^2)$, which is called the residue of the pole, must be positive (since g^2 must be positive). More generally, the residues of poles must be positive in a relativistic quantum theory, for unitarity and absence of ghosts. We are thus led to the question of whether the residues in (1.1.16) and (1.1.17) are positive, something that is far from obvious. Much early work on dual models was concerned with this question, culminating eventually in the 'no-ghost theorem', which asserts that ghosts (or negative residues) are absent if certain rather surprising restrictions are placed on the value of $\alpha(0)$ and the dimension of space-time. In particular, it turned out that the dimension of space-time should be 26, and the constant $\alpha(0)$ in the Regge trajectory $\alpha(s) = \alpha's + \alpha(0)$ should be 1. (The no-ghost theorem suggests but by itself does not uniquely determine those values.) We shall return to these matters in the next chapter.

Next, we would like to work out an interesting integral representation for the Veneziano amplitude. Consider the function

$$C(u,v) = \int_0^1 dx\, x^{u-1}(1-x)^{v-1}. \qquad (1.1.18)$$

It obeys

$$C(u-1,v+1) = \int_0^1 dx\, x^{u-2}(1-x)^v = \frac{1}{u-1}\int_0^1 dx\, (\frac{d}{dx}x^{u-1})(1-x)^v$$

$$= \frac{v}{u-1}\int_0^1 dx\, x^{u-1}(1-x)^{v-1} = \frac{v}{u-1}C(u,v). \qquad (1.1.19)$$

where we have integrated by parts. The beta function obeys the same identity $B(u-1,v+1) = \frac{v}{u-1}B(u,v)$ by virtue of (1.1.7). C also obeys

$$C(u+1,v) = \int_0^1 dx\, x^u(1-x)^{v-1}$$

$$= \int_0^1 dx\, x^{u-1}(1-x)^{v-1} - \int_0^1 dx\, x^{u-1}(1-x)^v \qquad (1.1.20)$$

$$= C(u,v) - C(u,v+1).$$

The analogous beta function identity $B(u+1,v)+B(u,v+1) = B(u,v)$ is likewise a consequence of (1.1.7). These recursion relations together with the similar asymptotic behavior of the functions $B(u,v)$ and $C(u,v)$ and the fact that they are equal at $u = v = 1$ imply that in fact $B(u,v) = C(u,v)$. Therefore, we obtain an integral representation for the Veneziano amplitude:

$$A(s,t) = \int\limits_0^1 x^{-\alpha(s)-1}(1-x)^{-\alpha(t)-1}dx. \qquad (1.1.21)$$

This integral representation is quite important, since it is in this form that the Veneziano amplitude usually appears in most approaches to calculating string scattering amplitudes.

1.1.2 High-Energy Behavior of the Veneziano Model

Our next task is to understand the asymptotic behavior of the Veneziano amplitude for high energy. We consider first the Regge region of large s, fixed t. The physical region for elastic scattering is positive s, negative t or vice versa. Large s, fixed t corresponds to small angle scattering at high energy; it was the phenomenology of this region that gave birth to Regge-pole theory and eventually to dual models.

To explore the asymptotic behavior of the Veneziano amplitude, we first need to know the asymptotic behavior of the gamma function. The behavior of $\Gamma(u)$ for large u can be easily extracted from the integral representation

$$\Gamma(u) = \int\limits_0^\infty dt\, t^{u-1}e^{-t}. \qquad (1.1.22)$$

For large u, the integral is dominated by the region $t \approx u - 1$. A saddle point evaluation in this region gives Stirling's formula

$$\Gamma(u) \sim \sqrt{2\pi}u^{u-1/2}e^{-u}. \qquad (1.1.23)$$

Although our derivation assumed positive u, Stirling's formula is actually valid for large u throughout the u plane as long as one keeps away from the negative u axis, where $\Gamma(u)$ has poles. From (1.1.23), we see that the Veneziano amplitude

$$A(s,t) = \frac{\Gamma(-\alpha(s))\Gamma(-\alpha(t))}{\Gamma(-\alpha(s)-\alpha(t))} \qquad (1.1.24)$$

has in the region of large s, fixed t the asymptotic behavior

$$A(s,t) \sim \Gamma(-\alpha(t))(-\alpha(s))^{\alpha(t)}. \qquad (1.1.25)$$

For a linear Regge trajectory, $\alpha(s) \sim \alpha's$, the asymptotic behavior for large s and fixed t is

$$A(s,t) \sim s^{\alpha(t)}. \tag{1.1.26}$$

Equations (1.1.25) and (1.1.26) are valid throughout the complex s plane (for large $|s|$) as long as one does not get too close to the positive real s axis. Since the physical region is precisely the region that is excluded, the restriction merits discussion. The essence of the matter is that in the physical region $A(s,t)$ is, even for large s, a rapidly varying function with many zeros and poles. These reflect the existence of resonances. Equations (1.1.25) and (1.1.26) are valid in an average sense if one suitably averages over the zeros and poles. Giving s a moderate imaginary part is a simple and physically reasonable way to do this. It is reasonable because the resonances are actually unstable; this implies that quantum corrections will give an imaginary part to the positions of the poles in $A(s,t)$. We can simulate the effect of quantum corrections that move the poles off the real axis by leaving the poles on the real axis and giving s an imaginary part.

It is instructive to compare (1.1.26) with the general formula (1.1.2) for scattering by exchange of a single elementary particle of spin J. In this case, the large s, fixed t behavior of the amplitude is $A(s,t) \sim s^J$. Equation (1.1.26) thus corresponds to the asymptotic behavior that would arise from t-channel exchange of a particle of t-dependent angular momentum $J = \alpha(t)$. This is the magic of Regge-pole theory; an infinite sum of t-channel exchanges of particles of arbitrarily large angular momentum can be described effectively at high energies by exchange of a single fictitious particle of t-dependent effective angular momentum $J = \alpha(t)$. With $\alpha(t) = \alpha't + \alpha(0)$ and α' positive for reasons noted earlier, $\alpha(t)$ is negative in the physical region of elastic scattering (negative t, positive s), at least as long as t is sufficiently negative. For sufficiently negative t, $\alpha(t)$ is as negative as we wish, and the large s behavior is correspondingly as soft as is desired. In this region, therefore, the high-energy behavior of the Veneziano amplitude is extremely soft, as soft as we wish. This is an illustration of the fact that in certain respects the high-energy behavior of dual models is much softer than that of any field theory, even a super-renormalizable field theory.

Instead of considering the Regge limit, it is instructive to consider the behavior of the Veneziano amplitude at high energies for fixed scattering angle. It is easy to see that at high energies, s and t and the center-of-mass scattering angle θ_s are related by

$$2t = -s(1 - \cos\theta_s). \tag{1.1.27}$$

Stirling's formula can then be used to show that the behavior of the Veneziano amplitude for large s and fixed θ_s is

$$A(s,t) \sim [F(\theta_s)]^{-\alpha(s)}, \qquad (1.1.28)$$

where $F(\theta_s)$ is a certain (slightly messy) function of the scattering angle. Thus, the fixed angle, high-energy scattering amplitude falls off exponentially with s.

1.1.3 Ramifications of the Veneziano Model

At its inception, the prospects that the Veneziano amplitude would lead to something important were probably even worse than the prospects in 1900 that study of experimental data on black body radiation would be the first clue to a completely new theory. The duality hypothesis never had more than slender experimental support, and the Veneziano model was merely an *ad hoc* way of satisfying this not-so-well motivated hypothesis. Nevertheless, study of the Veneziano model revealed an extremely rich structure, whose partial elucidation absorbed the efforts of a large number of physicists and produced a whole chain of surprises. These surprises will occupy many chapters of this book. Among the surprises were the ease with which it was possible to write an n-body generalization of the Veneziano amplitude, the recognition that the Veneziano model was really a model of a relativistic string, the invention of what we would now call closed strings, the appearance of graded Lie algebras in the process of trying to incorporate fermions in the model, and the recognition that what we would now call bosonic string theory and supersymmetric string theory make sense only in 26 or 10 dimensions, respectively. At almost every turn dual models sprang a new set of surprises on those who studied them, but one thing did not happen; beyond the slender initial motivation, using dual models as a fundamental theory of the strong interactions never had any compelling successes.

In fact, crucial experimental developments in the post-1968 period showed that a different sort of theory of strong interactions was required. In the Regge region, the behavior of the Veneziano model may be considered to be in reasonable agreement with the actual high-energy behavior of strong interactions. There is another region, little explored experimentally as of 1968 but important from a modern perspective, in which the agreement is not so good. This is the region of high-energy scattering at fixed angles, or in other words $s \to +\infty$, $t \to -\infty$, with s/t fixed. In this region, the Veneziano amplitude falls off exponentially with s, as we have seen at the end of the last subsection. But modern experiments (and theory) indicate that the strong interaction scattering amplitudes fall off only according to a power law in high-energy fixed angle processes. This is understood in parton model terms. In one of history's curious coincidences, the SLAC experiments giving the first evidence of parton-like behavior of strong interactions (in deep inelastic scattering) were conducted at just about the same time that Veneziano published his famous amplitude.

The failure of dual models to incorporate the parton-like behavior of strong interactions in certain kinematic regimes was one of the chief reasons for their demise as theories of strong interactions. In a sense, we may say that the ultraviolet behavior of the Veneziano model is softer than that of the hadronic world, even though hadronic amplitudes in the high-energy fixed-angle region are themselves far softer (fall off with energy like a much larger power) than corresponding amplitudes for scattering of elementary quanta in quantum field theory.

1.2   Dual Models of Everything

After around 1973 or 1974, an alternative theory of the strong interactions emerged in the form of quantum chromodynamics, which among other things explained the parton behavior just cited. The original motivation for dual models disappeared. At about the same time, however, there emerged a new possible motivation for studying the remarkably rich structure embodied in dual models. The strong interactions are only one area of physics in which one is faced, at first sight, with elementary particles of embarrassingly high spin. Such a problem also arises in the quantum theory of gravity. In general relativity the gravitational field is described by a massless field of spin two, the graviton field. Its nonlinear interactions are determined by a nonabelian local symmetry group, the group of diffeomorphisms of space-time. General relativity was of course part of the inspiration for the invention of Yang–Mills theory, in which the nonlinearities are determined by an analogous local symmetry group. The Yang–Mills field is a massless field of spin one. Despite the many similarities between these two theories, there is a world of difference between spin one and spin two. According to (1.1.2), t-channel exchange of a massless spin one particle gives an amplitude proportional to s/t. In four dimensions, such a high-energy behavior is barely compatible with renormalizability. But massless spin two exchange in the t channel gives, according to (1.1.2), an amplitude proportional to s^2/t. In four dimensions, this is definitely an unacceptable high-energy behavior, corresponding to a hopelessly unrenormalizable theory.

Dual models, as we have sketched, give one way to incorporate particles of high spin without ultraviolet pathologies. In the case of the strong interactions, dual models are not the right way to do this; nature chose a different path. Could it be that dual models are instead the correct way to include particles of 'excessively high spin' in quantum gravity?

1.2.1 Duality and the Graviton

One of the difficulties of dual theories of strong interactions, apart from the failure to accommodate the parton properties, was that these models always predicted a variety of massless particles, none of which are present in the hadronic world. These massless particles show up, for instance, in the poles that appear at $s = 0$ and $t = 0$ in the Veneziano amplitude (1.1.5), once one sets $\alpha(0) = 1$ in order to eliminate ghosts. Dual models turn out to have massless particles of various spins. In particular, the closed-string sector of dual models turns out to have a massless spin two particle. Investigation reveals – as one might suspect on general grounds – that its couplings are similar to those of general relativity. Might we interpret this particle as the graviton?

Quantum gravity has always been a theorist's puzzle *par excellence*. Experiment offers little guidance except for the bare fact that both quantum mechanics and gravity do play a role in natural law. The characteristic mass scale in quantum gravity is the Planck mass $\sqrt{\hbar c/G} \simeq 10^{19}$ GeV. This is so far out of experimental reach that barring an unforeseeable stroke of good luck (like the discovery of a stable Planck mass particle left over from the big bang) we can hardly hope for direct experimental tests of a theory of quantum gravity. The real hope for testing quantum gravity has always been that in the course of learning how to make a consistent theory of quantum gravity one might learn how gravity must be unified with other forces. A consistent unified theory of gravity and other forces might someday confront experiment through its implications for already measured quantities like the mass of the electron or the Cabibbo angle or by predicting new phenomena at high but accessible energy scales, perhaps at an energy scale of order 1 TeV, where the secret of $SU(2) \times U(1)$ symmetry breaking may be hidden.

1.2.2 Unification in Higher Dimensions

The idea of viewing dual models as theories of quantum gravity gave an immediate bonus: a long time embarrassment of dual models was immediately turned into a virtue. The embarrassment was the fact that dual models do not seem to make sense in four space-time dimensions. Only 26 dimensions is possible for the Veneziano model of bosons, while the Ramond–Neveu–Schwarz model of bosons and fermions makes sense only in ten dimensions. Focusing on the more realistic case in which there are fermions as well as bosons, the six extra dimensions are an embarrassment if the goal is to present a theory of strong interactions.

If the goal is to make a unified theory of gravity and matter, things are very different. The earliest idea and one of the best ideas ever advanced about unifying general relativity with matter was Kaluza's suggestion in 1921 that gravity could be unified with electromagnetism by formulating general relativity not in four dimensions but in five dimensions. This suggestion was further developed by Klein beginning in 1926 (whence the name Kaluza–Klein theory) and was one of the main themes in Einstein's work on unified field theory. In a modern language, the idea is to assume that the ground state of five-dimensional general relativity is not five-dimensional Minkowski space M^5 but is the product $M^4 \times S^1$ of four-dimensional Minkowski space M^4 with a circle S^1. One assumes that the radius of the circle is so tiny that in everyday experience – and even in high-energy laboratories – observed phenomena always involve averages over the position on S^1, so that the world *appears* to be four-dimensional. A lucid discussion of how this might be possible was given by Einstein and Bergmann in 1938.

What is the virtue in the fifth dimension? In five dimensions the metric tensor is a 5×5 matrix g_{MN}, $M, N = 0, \ldots, 4$. We take the five-dimensional Einstein equations as the dynamical equations for all of the g_{MN}. The components $g_{\mu\nu}$, $\mu, \nu = 0, \ldots, 3$ have spin two from a four-dimensional viewpoint and are seen in four dimensions as the metric tensor of gravity. On the other hand the components $g_{\mu 4}$, $\mu = 0, \ldots, 3$ have spin one from the four-dimensional viewpoint and describe a massless photon – which is unified with the graviton under the rubric of five-dimensional general relativity. This unified theory makes a very interesting prediction: the component g_{44} is predicted to behave as a massless scalar.

Thus, as we might have hoped, this interesting attempt at unifying gravity with matter makes a very interesting prediction concerning the nature of the 'matter'. Actually, this prediction seems to have embarrassed the early writers; predicting a new particle (or actually a new field; the relation between particles and fields was not yet known when the fifth dimension was first proposed!) was not so accepted in those days. Most of the early writers mangled the Kaluza–Klein theory to eliminate the prediction of the massless scalar. Only with the work of Jordan and Brans and Dicke and others in the 1950s and later was the massless 'Brans–Dicke scalar' viewed as an interesting prediction that should be tested experimentally rather than a nuisance to be swept under the theoretical rug. Significant experimental bounds have been placed on a possible massless Brans–Dicke scalar, but its existence remains conceivable. In dual models something very similar to the Brans–Dicke scalar appears in the form of the dilaton field, though it is far from clear whether this field is really massless or receives mass from quantum corrections.

Nowadays, we know that electromagnetism and gravity are far from being the whole story. A satisfactory unified theory must accommodate a good deal more. In fact, five dimensions are not enough; with ten we might just manage. If dual theories are viewed as a way of making sense of quantum gravity, then the fact that the dual model of bosons and fermions makes sense only in ten dimensions has the earmarks of a blessing in disguise, a hint that dual models are not just consistent theories of quantum gravity but consistent unified theories of all interactions.

1.2.3 Supersymmetry

In 1974, such dreams were brought down to earth by the fact that even the so-called 'consistent' dual models of that day had at least one fault: they all predicted a tachyon. If not an actual physical inconsistency, this is certainly unappealing. It indicated at least that the calculations were being performed in an unstable vacuum state. Moreover, tachyon exchange contributed infrared divergences in loop diagrams, and these divergences made it hard to isolate the ultraviolet behavior of these 'unified quantum gravity theories' and determine whether it was really satisfactory.

The solution of the tachyon problem came about in a very surprising way. In 1974, inspired in part by the graded Lie algebras that had already entered in dual models, Wess and Zumino conceived the idea of space-time supersymmetry. This work was presented as a four-dimensional generalization of the two-dimensional world-sheet supersymmetry of the Ramond–Neveu–Schwarz string model.[*] With good reason the invention of supersymmetry motivated an enormous amount of subsequent work. For the first time one had a symmetry between bosons and fermions, a *sine qua non* for a truly unified theory of everything. Within a few years the global supersymmetry of the early work was extended to local supersymmetry or supergravity, a remarkable and remarkably rich extension of the symmetry principle of general relativity. At first it seemed that the

---

[*] A similar suggestion had been made by Gol'fand and Likhtman in 1971 in work that attracted wide attention only much later.

resemblance between string theory and space-time supersymmetry was merely an analogy. But in 1977, Gliozzi, Scherk and Olive showed that it was possible to modify the Ramond–Neveu–Schwarz model by a 'G parity projection' to get a model with no tachyon and with equal masses and multiplicities for bosons and fermions. It was very plausible to conjecture that this modified model might have space-time supersymmetry. In the early 1980s the subject was revived by two of us (MBG and JHS) and this conjecture was proved. In the completely consistent tachyon-free form of the spinning string theory it was then possible to show that the one-loop diagrams were completely finite and in particular free of any ultraviolet divergence.

One-loop finiteness in a generally covariant theory is not in itself a novelty. Ordinary general relativity is one-loop finite (on shell) in four space-time dimensions. But when generally covariant field theories are one-loop finite, this is so because of a kinematic accident, the absence of a possible operator with the correct dimension and quantum numbers to serve as a counterterm. On the other hand, the finiteness of super-symmetric string theory at the one-loop level holds for a good and easily understood reason. A loop diagram is merely a tree diagram with some legs sewn together, and the tree diagrams of string theory have an excellent high-energy behavior, softer in some respects than the high-energy behavior of the tree diagrams of any super-renormalizable field theory. Naturally, the loops made from such soft trees are finite, as long as the issue is not clouded by extraneous infrared divergences.

Differently put, string theories or dual models from their inception have always been manifestly free of ultraviolet problems. The question has always been whether these theories could exist in a form that is free of anomalies and other inconsistencies. Historically, many difficulties had to be overcome to eliminate the inconsistencies. The last link in this chain was space-time supersymmetry. With this link included one had for the first time completely consistent dual models in the form of supersymmetric string theories or superstring theories. With a completely consistent form of the theory, it was finally possible to check the ultraviolet behavior, at least at the one-loop level, without extraneous complications. While most workers on the subject believe that the finiteness will also hold – for the same reason – to all orders, complete and universally accepted proofs have not appeared as of this writing.

The mid 1980s have brought a further succession of dramatic develop-ments – some of the highlights being hexagon anomaly cancellation, the emergence of a new theory with the exceptional group $E_8 \times E_8$ as its gauge group, the beginnings of phenomenology, and perhaps the beginnings of an understanding of the symmetry structure of string theory, at a level deeper than hitherto possible. These developments will occupy much of our attention in this book. But here it is time to begin explaining what string theory is.

1.3   String Theory

String theory is a complex subject with many facets. Numerous approaches to the subject have been developed over the years. Many of them have yielded important results in one respect or another. Thorough development of any one approach to the subject can be lengthy, and it is important to learn several approaches since there is no telling what approach will be most useful in future. All of this may occasionally tend to give the subject a formidable reputation. The purpose of the next two sections is to give a mini-introduction to string theory in a way that will aim to be as simple as possible. If in the process we skip over some crucial points and make the subject seem simpler than it really is, we will have ample opportunity to remedy this in later chapters.

1.3.1 The Massless Point Particle

Let us consider a massless classical point particle moving in Minkowski space. A suitable action for describing such a particle is

$$S = \int d\tau\, e^{-1}(\tau)\eta_{\mu\nu}\frac{dx^\mu}{d\tau}\frac{dx^\nu}{d\tau}. \tag{1.3.1}$$

Here $\eta_{\mu\nu}$ is the Minkowski metric (the indices μ and ν take the values $0, 1, \ldots, (D-1)$ in D space-time dimensions), τ is an arbitrary parameter along the trajectory, $x^\mu(\tau)$ is the position of the particle, and $e(\tau)$ is a sort of 'metric' along the particle world line. The role of $e(\tau)$ is to ensure that the action S is invariant under reparametrizations of τ. Indeed, (1.3.1) is invariant under $\tau \to \tilde{\tau}(\tau)$, $e \to e(d\tau/d\tilde{\tau})$. The gauge invariance can be used to pick a gauge with $e = 1$. In this gauge (1.3.1) reduces simply to the quadratic action

$$\tilde{S} = \int d\tau\, \eta_{\mu\nu}\frac{dx^\mu}{d\tau}\frac{dx^\nu}{d\tau}. \tag{1.3.2}$$

The variational equation derived from $\tilde{S}$ is simply $d^2x^\mu/d\tau^2 = 0$; the solutions are, of course, the straight lines in Minkowski space. It is not true, however, that an arbitrary straight line in Minkowski space is a solution of the equations of motion derived from the original action (1.3.1).

We are not entitled to make the gauge choice $e = 1$ and forget about e. We must also impose the equation of motion $\delta S/\delta e = 0$. One way to look at it is that one might first impose the gauge invariant equation $\delta S/\delta e = 0$ and then fix the gauge $e = 1$. An analogous phenomenon arises in any system with a local gauge invariance. In electrodynamics, one can if one wishes set the time component of the vector potential to zero with the gauge condition $A_0 = 0$, but one must not forget to impose Gauss's law, $\delta S/\delta A_0 = 0$. At any rate, the equation $\delta S/\delta e = 0$ tells us in this case that the gauge invariant quantity

$$T = \eta_{\mu\nu} \frac{dx^\mu}{d\tau} \frac{dx^\nu}{d\tau} \tag{1.3.3}$$

must vanish. So the solutions of our equations are precisely the *lightlike* geodesics in Minkowski space, showing finally that (1.3.1) is indeed the action for a *massless* classical point particle.

The variational equation $d^2 x^\mu/d\tau^2 = 0$ derived from the gauge fixed action (1.3.2) does not imply $T = 0$, but it does imply that T is a conserved quantity, $dT/d\tau = 0$. Thus, from the point of view of the gauge fixed action, $T = 0$ is a constraint on the initial data. If we impose this constraint at $\tau = 0$ it will automatically remain true for all τ. This again is typical of systems with local gauge symmetries. In Maxwell theory, for instance, Gauss's law $\delta S/\delta A_0 = 0$ does not follow from the other Maxwell equations, but the other equations do imply that $\delta S/\delta A_0$ is time independent so that it vanishes at all times if it vanishes at time zero.

What happens if we quantize this system? The canonical momenta are $p_\mu = \delta S/\delta(dx^\mu/d\tau) = \eta_{\mu\nu} dx^\nu/d\tau$. Upon quantization these become $p_\mu = -i\partial/\partial x^\mu$. This means that T becomes the Lorentz invariant wave operator $\eta^{\mu\nu} \partial^2/\partial x^\mu \partial x^\nu$, which we will call $\Box$. A quantum state is simply a function $\phi(x^\mu)$ of the spacetime coordinates x^μ. However, an arbitrary function ϕ should not be considered physically acceptable. Classically, we allowed only orbits of $T = 0$; quantum mechanically we should require that a physical state is a state annihilated by T. Since T is the wave operator, the constraint $T\phi = 0$ is just the massless Klein–Gordon equation $\Box\phi = 0$.

Thus, the massless Klein–Gordon equation is the Schrödinger equation for the quantum system derived from (1.3.1). That this equation is linear is no surprise; the Schrödinger equation of any quantum system is linear. The Poincaré invariance of the equation is also an obvious consequence of the fact that (1.3.1) is covariant under Poincaré transformations $x^\mu \to a^\mu{}_\nu x^\nu + b^\mu$, a being a Lorentz transformation and b a constant vector. The fact that the 'time' derivative $d/d\tau$ does not appear in the Schrödinger equation may be surprising at first sight, but this is the typical (though perhaps not so widely known) behavior of quantum systems that have symmetries under reparametrization of the time coordinate; similar behavior is encountered in the quantum theory of the free string (as we will see) and in formal attempts to quantize general relativity.

Before moving on to string theory we might pause to discuss briefly the supersymmetric generalization of the point particle theory. In fact, there are several possible supersymmetric generalizations, with supersymmetry either on the particle world line or in space-time. Of more interest to us at the moment is the Lagrangian with space-time supersymmetry:

$$S = \int d\tau \, \eta_{\mu\nu} (\frac{dx^\mu}{d\tau} - i\bar{\theta}\Gamma^\mu \frac{d\theta}{d\tau})(\frac{dx^\nu}{d\tau} - i\bar{\theta}\Gamma^\nu \frac{d\theta}{d\tau}). \qquad (1.3.4)$$

This action describes a point particle propagating not in Minkowski space but in a superspace with coordinates (x^μ, θ^a). Here the θ^a are anticommuting coordinates that transform as spinors under Lorentz transformations of the x^μ; the Γ^μ are Dirac gamma matrices. Quantization of (1.3.4) is slightly subtle; we will discuss this in chapter 5. For the moment let us simply note that the quantum theory will inevitably involve some supersymmetric extension of the massless Klein–Gordon equation that we found for the 'bosonic' point particle. In the interesting case of ten dimensions, with θ^a taken to be a single Majorana–Weyl spinor, one finds that the quantum system derived from (1.3.4) is the massless multiplet of fields of spin[*] $(1, 1/2)$; this multiplet consists of a 'photon' A_μ and a positive chirality spinor field ψ obeying the massless Maxwell and Dirac equations

$$0 = \partial_\mu F^{\mu\nu} = \Gamma^\mu \partial_\mu \psi. \qquad (1.3.5)$$

If instead we take for θ a pair of Majorana–Weyl spinors in ten dimensions, the quantum theory derived from (1.3.4) gives rise instead to the linearized approximation of the super-Einstein equations – the linearized approximation, that is, of ten-dimensional supergravity.

Now, what is somewhat unusual about this is that in quantizing the point particle we encounter equations – the Klein–Gordon equation in the bosonic case, the super-Maxwell and super-Einstein equations in the supersymmetric case – that are usually viewed as *classical* equations. What is more, while these equations are linear equations, they have natural nonlinear generalizations, which usually are our real interest. The massless Klein–Gordon equation has the more or less natural generalization

---

[*] We refer to the spin as if we were describing the representations of the four-dimensional Lorentz group where the terminology is standard. For example, in any dimension 'spin 1' refers to a Lorentz vector, 'spin 1/2' to a spinor, etc.

$\Box\phi + \lambda\phi^3 = 0$ of ϕ^4 theory. The Maxwell equations and the linearized Einstein equations have the extremely natural generalizations $D_\mu F^{\mu\nu} = 0$ and $R_{\mu\nu} = 0$ of Yang–Mills theory and general relativity. Usually, when we quantize a classical theory we do not go on to add nonlinear terms to the linear Schrödinger equation, but in the case of the massless point particle or superparticle, this is what we should do.

The innocent idea of a light ray in Minkowski space should be viewed as a hint of Yang–Mills theory and general relativity. The 'proper' theory of the massless point particle – namely the supersymmetric theory – gives rise upon quantization to the Maxwell and linearized Einstein equations, and these should then be generalized to the nonlinear Yang–Mills and Einstein equations.

1.3.2 Generalization to Strings

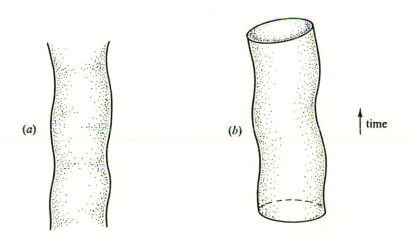

Figure 1.3. An open string (a) or closed string (b) propagating in Minkowski space sweeps out a two-dimensional surface known as the world sheet of the string. The classical equations of the free string theory require that this should be a minimal surface or surface of least area.

Now we are ready to discuss strings. The string is a one-dimensional object, a mathematical curve. We consider both open strings, which have endpoints, and closed strings, which from a topological viewpoint are circles. The open string is conventionally described by a coordinate σ that runs from 0 to π; for closed strings we also take $0 \leq \sigma \leq \pi$. To describe the motion of the string it is useful to introduce in addition a timelike evolution parameter τ, which is a sort of time coordinate for an observer sitting at the position σ along the string. As the string propagates in space-time (fig. 1.3), it sweeps out a world sheet that is the generalization of the world line of a point particle. The world sheet is described mathematically by specifying $X^\mu(\sigma, \tau)$, the position of the string at given values of σ and τ. We sometimes combine σ and τ as a two-vector $\sigma^\alpha = (\tau, \sigma)$, and use $d^2\sigma$ as a synonym for $d\tau d\sigma$.

In the form originally advocated by Nambu and Goto, the action for the string is simply proportional to the area of its world sheet. Mathematically, one formula for the area of a sheet embedded in Minkowski space is[*]

$$S = T \int d\sigma d\tau \sqrt{\dot{X}^2 X'^2 - (\dot{X} \cdot X')^2}, \qquad (1.3.6)$$

where

$$\dot{X}^\mu = \frac{\partial X^\mu(\sigma, \tau)}{\partial \tau}, \qquad X'^\mu = \frac{\partial X^\mu(\sigma, \tau)}{\partial \sigma}. \qquad (1.3.7)$$

With the action of the string taken to be the area of its world sheet, the solutions of the classical equations of the free string are the world sheets of minimal (or at least extremal) area. This generalizes the fact that the orbits of the point particle are geodesics or curves of minimal length. It is awkward to work with (1.3.6) because it is so highly nonlinear and especially because of the square root. An equivalent but far more convenient form of the action can be written if (as in the point particle case) we introduce in addition to $X^\mu(\sigma, \tau)$ a new variable $h_{\alpha\beta}$, which will be a metric tensor of the string world sheet. The more convenient form is

$$S = -\frac{T}{2} \int d^2\sigma \sqrt{h}\, h^{\alpha\beta} \eta_{\mu\nu} \partial_\alpha X^\mu \partial_\beta X^\nu. \qquad (1.3.8)$$

Here $\sqrt{h}$ is the square root of the absolute value of the determinant of $h_{\alpha\beta}$ and $h^{\alpha\beta}$ is the inverse of $h_{\alpha\beta}$. Since the derivatives of $h_{\alpha\beta}$ do not appear in (1.3.8) its equation of motion is a constraint equation and h can be eliminated or integrated out if one desires, giving back (1.3.6), as we will discuss further in chapter 2. Classically, (1.3.8) describes the propagation

---

[*] T is the constant of proportionality, required to make the expression dimensionless. Setting Planck's constant and the speed of light equal to one ($\hbar = c = 1$) gives T the dimension of (length)$^{-2}$. In chapter 2 we show that T is actually the tension in the string, which is related to the Regge slope parameter of open strings by $2\pi\alpha'T = 1$. It will often be convenient to choose special units in which $T = 1/\pi$ in order to simplify the notation. It is always easy to reinstate the general value of T by simple dimensional analysis.

of a string in Minkowski space of any number of dimensions, but quantum mechanically it turns out that 26 dimensions is the case of most interest.

Either (1.3.6) or (1.3.8) is invariant under general coordinate transformations of the string world sheet, $\tau, \sigma \to \tau'(\tau, \sigma), \sigma'(\tau, \sigma)$. (In the case of (1.3.8) the world-sheet metric $h_{\alpha\beta}$ should be taken to transform under such transformations according to the standard transformation law of a metric tensor.) From the point of view of a one-dimensional ant who lives on the string, (1.3.6) and (1.3.8) describe a field theory in $1+1$ dimensions that is generally covariant. In this $(1 + 1)$-dimensional field theory, the X^μ enter as scalar fields; they transform as vectors under 26-dimensional Poincaré transformations, but as scalars under reparametrizations of the world sheet. In fact, (1.3.8) is the standard form for coupling of massless scalar fields X^μ to $(1 + 1)$-dimensional gravity. The two-dimensional actions in (1.3.6) and (1.3.8) and their supersymmetric generalizations describe the only generally covariant field theories that are known to make sense in any number of space-time dimensions. One might be tempted to add to (1.3.6) a $(1+1)$-dimensional Einstein term $\sqrt{h}R$, with R being the $(1+1)$-dimensional curvature scalar; for our present purposes this term is irrelevant, since $\sqrt{h}R$ is a total divergence (more precisely, a topological invariant) in $1 + 1$ dimensions. In later developments it is useful to add this term.

1.3.3 Constraint Equations

Invariance under reparametrizations of the world sheet is essential for solving the minimal surface equations derived from (1.3.8). The 2×2 symmetric tensor $h_{\alpha\beta}$ has three independent components. In two dimensions a general coordinate transformation $\sigma, \tau \to \sigma', \tau'$ depends on two free functions, namely the new coordinates σ' and τ'. By means of such a transformation any two of the three independent components of h can be eliminated. A standard and convenient choice is a parametrization of the world sheet such that $h_{\alpha\beta} = \eta_{\alpha\beta}e^\phi$, where $\eta_{\alpha\beta}$ is the metric of a flat world sheet, and e^ϕ is an unknown conformal factor. It is always possible to make such a choice at least locally. We will refer to this as the choice of conformal gauge.

Now we meet a happy surprise. In conformal gauge, the conformal factor e^ϕ drops out of (1.3.8), the reason being that in two dimensions $\sqrt{h}$ is proportional to e^ϕ, while $h^{\alpha\beta}$ is proportional to $e^{-\phi}$. Hence, in conformal gauge (1.3.8) reduces to a simple free field action

$$S = -\frac{T}{2} \int d^2\sigma \, \eta_{\mu\nu} \, \eta^{\alpha\beta} \, \partial_\alpha X^\mu \partial_\beta X^\nu. \qquad (1.3.9)$$

The equation of motion derived from (1.3.9) is a simple linear wave equation

$$\left(\frac{\partial^2}{\partial\tau^2} - \frac{\partial^2}{\partial\sigma^2}\right)X^\mu = 0. \tag{1.3.10}$$

However, just as in the point-particle case, the wave equation derived from the gauge-fixed action of (1.3.9) must be supplemented with certain constraint equations. The constraint equations that arise are the equations $\delta S/\delta h_{\alpha\beta} = 0$. In $(1+1)$-dimensional quantum field theory, one usually defines the energy–momentum tensor as[*]

$$T_{\alpha\beta} = -\frac{2\pi}{\sqrt{h}}\frac{\delta S}{\delta h^{\alpha\beta}}, \tag{1.3.11}$$

so the constraint equations are simply $T_{\alpha\beta} = 0$. The energy–momentum tensor takes its simplest form in terms of light-cone world-sheet coordinates $\sigma^\pm = (\tau \pm \sigma)$. In these coordinates, we have $T_{++} = \partial_+ X^\mu \partial_+ X_\mu$, with T_{--} differing from T_{++} by $+ \leftrightarrow -$, and $T_{+-} = 0$. Here T_{+-} is the trace of the two-dimensional energy–momentum tensor, and the fact that this vanishes, or in other words that the massless scalar field theory of (1.3.9) is conformally invariant in $1+1$ dimensions, turns out to have far-reaching consequences.

Classically, the constraint equations are quite tractable, and the general solution of the wave equation plus the constraint equations can be found without undue difficulty, as we will see in chapter 2. Quantum mechanically, the story is more complicated. A proper treatment of the conformal gauge quantum mechanically requires that one introduce ghosts, a complication that we will try to suppress in this introductory chapter. Ignoring the constraints, it is of course quite straightforward to quantize the free field theory with Lagrangian (1.3.9). Imposing the constraints is not so straightforward. The constraint equations mean that a physical state $|\phi\rangle$ must obey $T_{\alpha\beta}|\phi\rangle = 0$. To understand the significance of this equation we must compute the equal τ commutation relations of the $T_{\alpha\beta}$. These turn out to be

$$\begin{aligned}[T_{++}(\sigma), T_{++}(\sigma')] &= i(T_{++}(\sigma) + T_{++}(\sigma'))\delta'(\sigma - \sigma') \\ &\quad + \frac{i}{24}(26 - D)\delta'''(\sigma - \sigma'),\end{aligned} \tag{1.3.12}$$

with a similar equation for T_{--} and with $[T_{++}, T_{--}] = 0$. Here D is the number of space-time dimensions (the number of X^μ).

---

[*] Here $h = -\det h_{\alpha\beta}$. The factor of 2π in (1.3.11) is not conventional in field theory, but has become standard in string theory.

The first term on the right-hand side of (1.3.12) would arise in the classical theory in taking Poisson brackets; the second term, proportional to $(26 - D)\delta'''(\sigma - \sigma')$, arises quantum mechanically via an anomaly. As we will see in chapter 3, it is necessary to include properly the ghosts in order to compute the coefficient $(26 - D)$ that appears in (1.3.12).[†] It follows directly from (1.3.12) that the constraint equations can only make sense in 26 dimensions. For a physical state $|\phi\rangle$, obeying $T_{++}|\phi\rangle = 0$, is automatically annihilated by the left-hand side of (1.3.12) and by the first term on the right-hand side. It must therefore be annihilated also by the second term on the right of (1.3.12). But the second term on the right-hand side of (1.3.12), the c-number $(26 - D)\delta'''(\sigma - \sigma')$, does not annihilate anything unless $D = 26$; this is the only case in which there are any physical states at all. It is for this reason that the Veneziano model of open strings, and the corresponding Shapiro–Virasoro model of closed strings, only make sense in 26 dimensions.

Apart from local reparametrization invariance on the string world sheet, this theory also is invariant under global $(\sigma, \tau$ independent) transformations $X^\mu \to \Lambda^\mu{}_\nu X^\nu + b^\mu$ (where Λ is a constant orthogonal matrix and b is a constant vector). To an ant living on the string, these are merely internal symmetries of the 26 free, massless quantum fields propagating on the string, but for an observer in space–time they are Poincaré transformations. Since the theory is Poincaré invariant, the Hilbert space will furnish a unitary representation of the Poincaré group. This means that the particle states will be labeled by their mass and spin. Since we are dealing with a string that is free to oscillate and has an infinite number of harmonics like any other string, an infinite number of particle states will appear.

For the bosonic string that we have been discussing, the ground state is unfortunately a tachyon whose mass (in the case of closed strings) is $m^2 = -8$ (in units where $T = 1/\pi$). This result emerges from careful study of the constraint equations, as we will see in a systematic treatment in chapter 2. The first excited level is more appealing; it consists of massless particles. Specifically, the massless particles that one finds are a spin two state, which we might try to identify as the graviton $g_{\mu\nu}$, a scalar known as the 'dilaton', and an additional state that transforms as the second rank antisymmetric tensor of $SO(24)$ (which enters since it is the little group of a massless particle in 26 dimensions). These are all of utmost importance; the graviton was the motivation for the idea of 'dual models of everything',

---

[†] In formalisms that do not contain the ghosts, the restriction to 26 dimensions arises not in the fashion described here, but in other ways explained in chapter 2.

the dilaton expectation value turns out to determine the fine structure constant, and the antisymmetric tensor (or rather its superstring cousin) plays a crucial role in anomaly cancellation. For higher excitations, one finds an infinite tower of massive states, extending to higher and higher masses and spins; for instance, the second excited level contains particles of $m^2 = +8$ and spin ranging up to four. This spectrum of excited states goes on to infinity, as is inevitable for the spectrum of an extended body with an infinite number of harmonics.

Like the bosonic point particle, the bosonic string has supersymmetric generalizations. These are the supersymmetric string theories or superstring theories, which will be our main interest in this book. Here we note only that in the supersymmetric case the tachyon is necessarily absent, since the Hamiltonian of a supersymmetric theory is positive; the lowest mass levels of superstring theories consist of the same massless Maxwell and Einstein multiplets that enter in quantizing suitable point particle theories. For instance, the open superstring gives rise at the massless level precisely to a massless Maxwell supermultiplet; there are in addition an infinite tower of massive states.

The constraint equations $T_{\alpha\beta} |\phi\rangle = 0$ are linear, like the Schrödinger equation of any quantum system; indeed, these equations play the role of the Schrödinger equation for the string. Usually, one studies the Schrödinger equation as a linear equation. But earlier, we saw that in the case of the point particle the 'right' thing to do is to search for a non-linear generalization of the Schrödinger equation; this leads to the most beautiful equations of physics, the supersymmetric Yang–Mills and Einstein equations. Should we do the same thing for the string? Should we search for a nonlinear theory governing the whole infinite tower of string harmonics? Is there a generalization of Yang–Mills theory and general relativity that is related to the string in the way that those theories are related to the point particle?

That there is such a generalization is abundantly clear; it has been constructed, by trial and error, in nearly 20 years of work on string theory. The mystery to this day is 'why' Yang–Mills theory and general relativity have such generalizations. The local gauge invariances of Yang–Mills theory and general relativity have a satisfying basis in fundamental physical and mathematical concepts: local symmetry, connections and curvature, vector bundles, Riemannian geometry. In fact, historically, in the case of general relativity, it was the concepts that came first; Einstein first identified the concepts on which a relativistic theory of gravity should be based, and then found the theory. String theory has been the other way around. As we have sketched in the first section, string theory was origi-

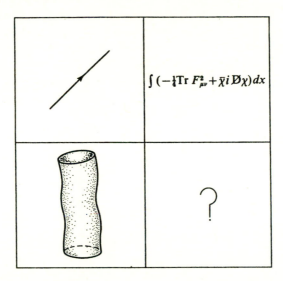

Figure 1.4. This figure illustrates what might be called the 'magic square' of string theory. In the upper left-hand corner we sketch a light ray in Minkowski space, corresponding to the massless superparticle. It is to be viewed as a hint of supersymmetric Yang–Mills theory (whose action is written in the upper right-hand corner) and supergravity. In the lower left-hand corner we sketch a minimal surface in Minkowski space – the classical orbit of a string. This is to be viewed as a hint of a new class of theories, the string generalizations of the Yang–Mills and Einstein equations. Though much is known about them, they are indicated in the lower right-hand corner of the square with a question mark to underscore the fact that their conceptual framework is still largely mysterious.

nally invented for other reasons entirely, in an unsuccessful assault on the strong interactions. It eventually became clear that string theory should be used, instead, to give a fundamental generalization of general relativity and Yang–Mills theory. But the concepts behind this generalization remain largely mysterious. At best we have perhaps just begun to scratch the surface of this question. For this reason the stringy generalization of Yang–Mills theory and gravity, though we know much about it, is indicated by a question mark in fig. 1.4. The missing circle of ideas is bound to one day have a far reaching impact on physics, and perhaps on certain branches of mathematics.

1.4   String Interactions

So far we have been discussing the free string. In this section we will undertake a lightning introduction to the interacting or nonlinear theory. First let us review the Feynman diagrams of field theory. For, say, a massless scalar field in n space–time dimensions, the standard propagator between space–time points x and y is $\langle y|\square^{-1}|x\rangle$. Here $\square$ is the d'Alembertian or wave operator $\square = \eta^{\mu\nu}(\partial^2/\partial x^\mu \partial x^\nu)$. The propagator, which is the inverse of the wave operator, has the representation

$$\langle y|\square^{-1}|x\rangle = \int_0^\infty d\tau \, \langle y| \, e^{-\tau\square}|x\rangle. \tag{1.4.1}$$

Now, the Hamiltonian of a nonrelativistic particle of mass m in $n+1$ dimensions is simply $p^2/2m = \square/2m$; for $m = 1/2$ this is simply $\square$. The operator $e^{-\tau\square}$ on the right-hand side of (1.4.1) is thus merely the operator that propagates the nonrelativistic particle through imaginary proper time τ; for this there is a well-known path integral formula

$$\langle y|\square^{-1}|x\rangle = \int_0^\infty d\tau \int_x^y Dx(t) \exp\{-\frac{1}{4}\int_0^\tau dt \, \dot{x}^2\}. \tag{1.4.2}$$

Here the exponent is just the action of the classical point particle; the symbol $\int_x^y Dx(t)$ represents an integral over all paths $x(t)$ that start at x at $t = 0$ and end at y at $t = \tau$. The right-hand side of (1.4.2) has a simple intuitive meaning. We integrate over the arbitrary positive proper time τ of a particle that propagated from x to y, and over all orbits $x(t)$ that propagate from x to y in proper time τ.

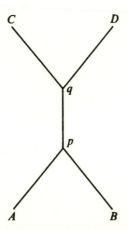

Figure 1.5. A typical Feynman diagram in quantum field theory. Four external particles, originating at space–time points A, B, C and D, undergo a tree level scattering process, the interactions occurring at p and q.

Now let us look at a typical Feynman diagram, as in fig. 1.5. Instead of evaluating this diagram in momentum space, as is customary, let us think of fig. 1.5 in coordinate space, so that the external particles originate at space–time points A, B, C and D, and the interactions occur at points p and q. According to the usual rules for computing an amplitude from the Feynman diagram, each line in fig. 1.5 corresponds to a propagator. With the representation (1.4.2) for the propagator, each line in the figure represents an integration over the trajectory of a particle that propagated in space–time between the indicated points. In evaluating the diagram one is also instructed to integrate over the interaction points p and q, and to include at the vertices certain factors that depend on the precise theory considered. The point is that a Feynman diagram can be viewed as an actual history of particles propagating in space–time and joining and splitting at interaction vertices.

1.4.1 Splitting of Strings

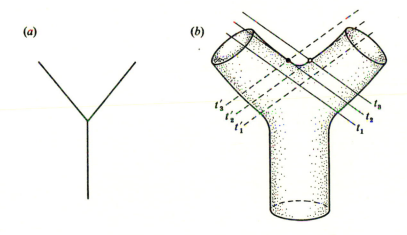

Figure 1.6. Interaction vertices in field theory and in string theory: in (*a*) a point particle splits into two; in (*b*) a closed string splits into two. In (*b*) the surfaces of constant time in two different Lorentz frames 1 and 2 are indicated with solid and dashed lines respectively.

Now let us try to formulate an analog of this in string theory. Just as a point particle can split into two, as in fig. 1.6*a*, so can a string; this is indicated in fig. 1.6*b*, in the case of closed strings. The crucial difference is that when a point particle splits into two, there is a well-defined Lorentz

invariant notion of the space-time point at which the splitting occurred. It
is simply the interaction vertex in the Feynman diagram. However, when
a string splits into two, there is no well-defined notion of when and where
this happened. In fig. 1.6b we have sketched the surfaces of constant time
in two different Lorentz frames 1 and 2. In frame 1, the splitting occurred
at the point indicated with the solid dot. To the past of this there was
only one string, and in the future there are two. In frame 2, there is
nothing special about the point with the solid dot. In this frame, the
point with the open dot is the point where the interaction occurred.

This difference has many consequences. First of all, it is part of the
reason that there are many quantum field theories of point particles but
not many string theories. Since there is a Lorentz invariant interaction
vertex in fig. 1.6a, we can choose some special factors to be included at
this vertex in defining the Feynman amplitude. The choice of these factors
corresponds to the choice of which quantum field theory one wishes to
study. On the other hand, in fig. 1.6b, any part of the diagram looks
locally like the propagation of a free string; once we have decided upon
the rules for the propagation of the free string, there are no additional
choices to be made. It is still necessary to understand why there are not
many possibilities for the free string; this follows from the subtleties that
arise in trying to quantize (1.3.8) and its various generalizations. But
once we settle on a free theory, the form of the interactions is uniquely
determined, simply because there is no Lorentz invariant interaction point
in fig. 1.6b.

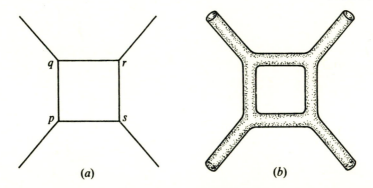

Figure 1.7. Here we sketch in (a) a one-loop Feynman diagram (with interaction points
marked p, q, r and s), and in (b) the corresponding string diagram for closed strings.

Closely related to this is a heuristic explanation of why string theories are free of the ultraviolet divergences that plague the quantum field theories of point particles. In fig. 1.7, we have sketched a one-loop Feynman diagram, which in quantum gravity would diverge in the ultraviolet, and the corresponding closed-string diagram. The string diagram of fig. 1.7*b* differs from the field-theory diagram of fig. 1.7*a* in that every world line or propagator of a point particle has been replaced by the world tube of a propagating closed string. The two diagrams are evaluated by integrating over the trajectories in space–time of the propagating points or strings. Insofar as the strings in fig. 1.7*b* have a very small radius, this figure reduces approximately to fig. 1.7*a*; this is how field theory emerges as the long wavelength limit of string theory.

Why are there ultraviolet divergences in fig. 1.7*a* but not in fig. 1.7*b*? The crucial difference is that in fig. 1.7*a* there are well-defined interaction vertices, labeled p, q, r and s. The ultraviolet divergence occurs because when $p = q = r = s$ the propagators connecting the vertices simultaneously blow up. In the string diagram of fig. 1.7*b*, there is no well-defined analog of the interaction vertices p, q, r and s of the field-theory diagram. There is consequently no analog of the dangerous region $p = q = r = s$. While this certainly does not prove the finiteness of fig. 1.7*b*, which can only be established by a computation once the necessary machinery is developed, it is a strong hint.

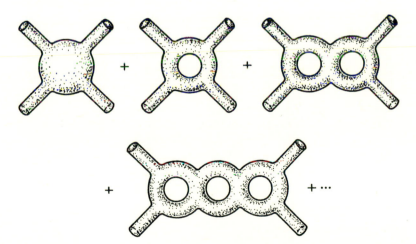

Figure 1.8. Here we sketch *all* of the closed-string diagrams that contribute to the four-particle amplitude; there is one and only one diagram in each order of perturbation theory.

Another basic difference between field-theory diagrams and string diagrams is that there are far fewer string diagrams. For every field-theory

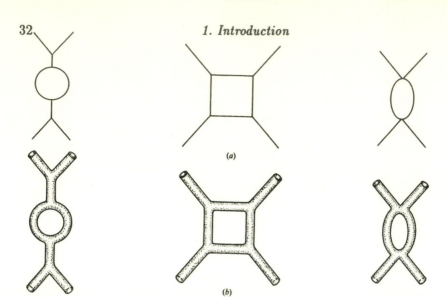

Figure 1.9. Several distinct field-theory diagrams may become isomorphic as string diagrams. In (*a*) we show several Feynman diagrams that in field theory represent one-loop corrections to a four-particle amplitude; in (*b*) we show the corresponding closed-string diagrams. The closed-string diagrams of (*b*) all have the same topology, so they represent different integration regions in the same string diagram.

diagram there is a corresponding string theory diagram made by blowing up world lines into world tubes. For closed strings this is indicated in fig. 1.9. But several different Feynman diagrams may give rise in this way to the same string diagram, as illustrated in that figure. In fact, in the theory of oriented closed strings, there is one and only one Feynman diagram in any given order of perturbation theory, shown in fig. 1.8, corresponding to the fact that oriented two-dimensional manifolds are completely specified by giving the number of handles (or loops) and the number of holes (or external particles). The classification of string diagrams is more complicated in theories with open strings or unoriented closed strings, but still there are far fewer string diagrams than Feynman diagrams. The reason that it is possible for a string theory with just a few diagrams to reproduce at low energies a field theory with many Feynman diagrams is that, in different limits of the integration over the size and shape parameters of the string world sheet, a single string diagram can approximately reproduce many Feynman diagrams. For instance, the one-loop diagram in fig. 1.8 reduces in different limits to any of the diagrams in fig. 1.9.

1.4.2 Vertex Operators

At first sight it would appear rather formidable to evaluate an integral over string world sheets corresponding to one of the diagrams of fig. 1.9.

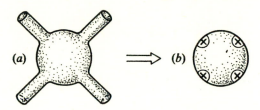

Figure 1.10. Conformal invariance makes it feasible to evaluate string diagrams. Among other things this makes it possible to compactify the world sheet, closing off the holes corresponding to incoming and outgoing particles. The external string states in, say, part (a) of this figure are thereby projected to points, indicated as $\otimes$ in (b); at these points insertions of suitable local operators must be understood.

What makes this feasible is the invariance of (1.3.8) under a conformal rescaling of the world-sheet metric $h_{\alpha\beta} \to e^{\phi}h_{\alpha\beta}$. By a suitable choice of ϕ, fig. 1.10a in which the world sheet has tubes extending into the far past and the far future corresponding to incoming and outgoing strings can be converted into fig. 1.10b in which the world sheet is compact. At the same time, the holes in the string world sheet corresponding to external states are closed up, and the external string states appear as points, as in fig. 1.10b.

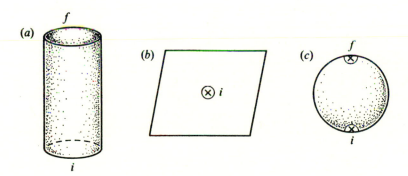

Figure 1.11. A world sheet with one incoming and one outgoing string (as in (a)) can as described in the text be conformally mapped to a variety of other figures. It can be mapped to the plane of (b) with the incoming string appearing at the origin and the outgoing string at infinity (not shown) or to the sphere of (c) with the incoming and outgoing strings appearing at the south and north poles.

What sort of conformal change of metric can bring about this magic? Consider the simplest case of a world sheet with only one incoming string and one outgoing string, described by the cylinder of fig. 1.11a, with metric $ds^2 = dz^2 + d\varphi^2$, $-\infty < z < \infty$, $0 \leq \varphi \leq 2\pi$. If we let $z = \ln r$, this becomes $ds^2 = r^{-2}(dr^2 + r^2 d\varphi^2)$. By a conformal change of metric $ds^2 \to \tilde{ds}^2 = r^2 ds^2$ we get a new metric $\tilde{ds}^2 = dr^2 + r^2 d\varphi^2$, which we can recognize as the metric of the plane. In effect the incoming string, which was a circle in the far past ($z = -\infty$, $0 \leq \varphi \leq 2\pi$), has been projected to a point at a finite distance ($r = 0$) as shown in fig. 1.11b, and the outgoing string has been projected to the point at infinity. If we wish to project both the incoming and outgoing strings to a finite distance, we must pick the conformal factor to be not r^2 but something that behaves like r^2 for small r and like r^{-2} for large r. If for instance we rescale the metric ds^2 by a conformal factor $r^2(1 + r^2/a^2)^{-2}$, then the new metric $\tilde{ds}^2 = (dr^2 + r^2 d\varphi^2)/(1 + r^2/a^2)^2$ is the standard round metric on the sphere. The incoming and outgoing strings are now finite points, namely the south and north poles in fig. 1.11c. For a more complicated string diagram with many external lines, the conformal factor e^ϕ can be chosen to map each of them to finite points. The essence of the matter is that to map a given incoming or outgoing string L to a finite point, only the asymptotic behavior of e^ϕ far out on L is relevant; the asymptotic behavior of e^ϕ can be chosen independently for each L. We will discuss the conformal mappings of string diagrams in more detail in later chapters, especially in chapter 11.

If we do conformally map the external string states to finite points, their quantum numbers will not be simply lost. At each point, marked $\otimes$ in fig. 1.10, to which an external string has been mapped, there must appear some local operator with the quantum numbers of the string state that was mapped to that point. We thus are led to the idea that for each string state we must find some local operator in the (1+1)-dimensional quantum field theory that describes string propagation. The local operator that corresponds in this way to a given string state $|\Lambda\rangle$ is called the 'vertex operator' for emission or absorption of $|\Lambda\rangle$. These vertex operators turn out to be remarkably useful tools.

Without pretense of rigor, let us simply try to guess suitable vertex operators in the case of closed strings. First of all, for each particle type Λ in the closed-string theory we find a local operator $W_\Lambda(\sigma, \tau)$ that is a scalar under reparametrizations of σ and τ (since there is no preferred parametrization in fig. 1.10) and has the same Lorentz quantum numbers as Λ. W_Λ will be a suitable polynomial in X^μ and its derivatives. For instance, if Λ is the tachyon, which has spin zero under 26-dimensional Lorentz transformations, we can simply take $W = 1$. If Λ is the graviton G, we must pick W to have spin two; the minimal spin two operator would be $W_G^{\mu\nu} = \partial_\alpha X^\mu \partial^\alpha X^\nu$ for a graviton of polarization $\mu\nu$. Normal ordering for this operator and others considered later (such as $e^{ik\cdot X}$) will be assumed and not indicated explicitly. If Λ is the massless dilaton D, which has spin zero, we must again pick a spin zero operator; the minimal choice that is orthogonal to the tachyon operator is $W_D \sim \partial_\alpha X_\mu \partial^\alpha X^\mu$.

The operators W_Λ that we have just defined transform correctly under Lorentz transformations, but we must also take account of space-time translations. Under the global symmetry $X^\mu \to X^\mu + a^\mu$ in which the position of each string is shifted by a constant a^μ, the wave function of an external state of momentum k^μ is multiplied by $e^{ik\cdot a}$. The simplest quantum field operator that transforms in this way under $X^\mu \to X^\mu + a^\mu$ is the operator $e^{ik\cdot X}$, so we postulate that this factor is present for emission or absorption of a string of momentum k^μ. In addition, we must note that in fig. 1.10 the point marked $\otimes$ at which a given vertex operator is inserted may appear anywhere on the surface. Putting these facts together, we are led to define the operator

$$V_\Lambda(k) = \int d^2\sigma \sqrt{h}\, W_\Lambda(\sigma,\tau) e^{ik\cdot X} \qquad (1.4.3)$$

for emission or absorption of a string state of type Λ and momentum k^μ.

1.4.3 Use of Vertex Operators

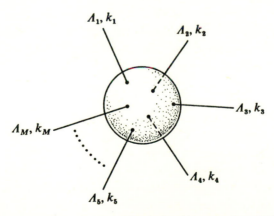

Figure 1.12. Representation of the amplitude for the scattering of M external particles of types $\Lambda_1, \Lambda_2, \ldots, \Lambda_M$ and momenta $k_1, k_2, \ldots, k_M$.

How do we use these vertex operators, in practice? Figure 1.12 suggests that the scattering amplitude for scattering of particles of types $\Lambda_1, \Lambda_2, \ldots \Lambda_M$ and momenta $k_1, k_2, \ldots k_M$ should be a path integral in the $(1+1)$-dimensional quantum field theory that governs string propagation with insertions of the operators V_Λ. This would be (setting $T = 1/\pi$)

$$A(\Lambda_1, k_1; \Lambda_2, k_2; \ldots; \Lambda_M, k_M) = \kappa^{M-2} \int DX(\sigma, \tau) Dh_{\alpha\beta}(\sigma, \tau)$$

$$\times \exp -\{\frac{1}{2\pi} \int d^2\sigma \sqrt{h} h^{\alpha\beta} \partial_\alpha X^\mu \partial_\beta X_\mu\} \cdot \prod_{i=1}^{M} V_{\Lambda_i}(k_i). \qquad (1.4.4)$$

Here κ is a coupling constant, and the symbols DX^μ and $Dh_{\alpha\beta}$ denote path integrals on the compact string world sheet of fig. 1.11. We require a surface that is topologically a sphere to evaluate tree diagrams, a torus to evaluate one-loop diagrams or a surface with n handles (often called a Riemann surface of genus n) to evaluate n-loop diagrams.

While it is possible to relate rigorously (1.4.4) to string diagrams such as fig. 1.7b, we will not attempt this here. (It will be one of our tasks in later chapters, especially chapter 11.) Actually, in a first approach to string theory the reader may simply think of (1.4.4) as the *definition* of what a string scattering amplitude is supposed to be. We certainly do not know at present which formalism of string theory is more fundamental, and the vertex operator formulation in (1.4.4) is as good a starting point as any other.

Equation (1.4.4) should seem like a rather startling formula. In (1.4.4), a scattering amplitude *in 26-dimensional space–time* is expressed in terms of a correlation function *in an auxiliary $(1+1)$-dimensional quantum field theory*. According to the standard *LSZ* formalism of quantum field theory, correlation functions in a $(1+1)$-dimensional field theory may very well be related to scattering processes in a $(1+1)$-dimensional world. That they can be interpreted instead to give scattering amplitudes in 26 dimensions is one of the wonders of string theory. It is one of many surprising and still largely mysterious relations and analogies between phenomena that occur on a string world sheet and phenomena that occur in space–time.

Actually evaluating (1.4.4) would be a hopeless task were it not for the conformal invariance that we discussed earlier. The 2×2 symmetric tensor h has three independent components. By a reparametrization of the world sheet we can eliminate two of these three components. Locally, as we discussed before, one can by reparametrization of σ, τ put h in the form $h = e^\phi h_0$, where h_0 is any desired metric on the string world sheet. According to a classical theorem of Riemann, the same is also true globally in the case that the world sheet is a sphere or, in other words, for tree diagrams. We will now concentrate on this case, with the goal of explicitly evaluating the tree diagrams.

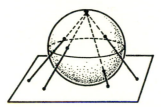

Figure 1.13. A stereographic projection from the two sphere S^2 onto the plane.

The theorem just cited ensures that by a choice of parametrization we can write $h = e^\phi h_0$, where h_0 is, for instance, the standard round metric on the sphere S^2. Actually, it is even more convenient to use the conformal invariance to make a stereographic projection of the sphere onto the $x - y$ plane R^2, as shown in fig. 1.13. Thus, in effect, we can take $h^{\alpha\beta} = e^\phi \delta^{\alpha\beta}$ (*i.e.*, $ds^2 = e^\phi(dx^2 + dy^2)$), where now x and y are coordinates on the plane. Moreover, by virtue of conformal invariance the e^ϕ factor drops out of (1.4.4), which simplifies to

$$A = \kappa^{M-2} \int DX(x,y) \exp -\left\{ \frac{1}{2\pi} \int d^2x \, \partial_\alpha X_\mu \partial^\alpha X^\mu \right\} \cdot \prod_{i=1}^{M} V_{\Lambda_i}(k_i)$$

$$= \kappa^{M-2} \langle \prod_{i=1}^{M} V_{\Lambda_i}(k_i) \rangle.$$

$$(1.4.5)$$

The important point here is that now we are doing free field theory on a *flat* two-dimensional world, so we can expect to be able to evaluate (1.4.5).

In going from (1.4.4) to (1.4.5), the h integral has disappeared rather effortlessly. Actually, a few subtleties in this procedure should be noted. First of all, we must ask whether the conformal invariance used to discard ϕ in $h = e^\phi h_0$ is really valid or suffers from a quantum anomaly. In checking this it is important to take account of Faddeev–Popov ghosts that arise when one imposes the gauge choice $h = e^\phi h_0$. We postpone until chapter 3 the analysis, which is slightly technical. Here we note merely the result: the conformal invariance and with it the derivation of (1.4.5) hold in, and only in, 26 dimensions.

Of more immediate concern, requiring $ds^2 = e^\phi(dx^2 + dy^2)$ does not quite uniquely fix the reparametrization invariance. This form of h is preserved by global conformal transformations of the world sheet S^2. The relevant conformal transformations are most easily written in terms of $z = x + iy$. In terms of this complex coordinate, the world-sheet metric is $ds^2 = e^\phi dz\,d\bar{z}$. If we change coordinates from z to some analytic function $w(z)$, so that $dz = (\partial z/\partial w)dw$, the metric becomes $ds^2 = e^\phi|\partial z/\partial w|^2 dw\,d\bar{w}$, which is still in the conformal gauge. These are the coordinate transformations that are permitted by the gauge choice $ds^2 = e^\phi(dx^2 + dy^2)$.

Infinitesimally, the residual gauge invariances are thus transformations $\delta z = \epsilon(z)$, where ϵ is an *analytic* function of z. Actually, ϵ is not an arbitrary analytic function of z, but is subject to a strong restriction, for reasons that we will now describe. Though we have made a stereographic projection to the complex z plane, the string world sheet was originally the sphere S^2 – which we may think of as the Riemann sphere consisting of the z plane plus a 'point at infinity.' We must require that the infinitesimal coordinate transformation $\delta z = \epsilon(z)$ does not have a pole at the point at infinity. The analysis is conveniently carried out in terms of a new coordinate $\tilde{z} = 1/z$, in terms of which the point at infinity is an ordinary point, namely the origin, $\tilde{z} = 0$. The coordinate transformation $\delta z = \epsilon(z)$ becomes in the new coordinate system $\delta\tilde{z} = -\delta z/z^2 = -\epsilon(z)/z^2$. It is nonsingular at $\tilde{z} = 0$ if and only if $\epsilon(z)/z^2$ is finite for $z \to \infty$. Consequently, ϵ must be a quadratic polynomial; the residual symmetries not removed by the choice of the conformal gauge are infinitesimally of the form $\delta z = a + bz + cz^2$, with a, b, c being three arbitrary complex parameters. These transformations generate a group isomorphic to the group $SL(2, C)$ of 2×2 complex matrices of determinant one. This subgroup of the original reparametrization group has not been removed by the gauge fixing that led to (1.4.5), and therefore we will have to complete the gauge fixing shortly.

1.4.4 Evaluation of the Scattering Amplitude

It is actually not difficult to evaluate (1.4.5) in the case in which the external particles are tachyons, with $V_0 = \int d^2z e^{ik\cdot X}$. In this case, (1.4.5) reduces to

$$A = \kappa^{M-2} \int \prod_{i=1}^{M} d^2z_i \cdot \langle \prod_{i=1}^{M} \exp\{ik_i \cdot X(z_i)\}\rangle, \qquad (1.4.6)$$

where $\langle\ \rangle$ represents an expectation value with respect to the gaussian measure defined by the free field path integral as shown in (1.4.5). To evaluate (1.4.6), recall the standard formula for gaussian integrals (obtained by completing a square)

$$\langle\exp\{\int id^2z\, J_\mu(z)X^\mu(z)\}\rangle = \exp\{\frac{1}{4}\int d^2z d^2z'\, J_\mu(z)G(z,z')J^\mu(z')\},$$

$$(1.4.7)$$

where $J_\mu(z)$ is an arbitrary source and $G(z,z')$ is the propagator of the free field X^μ. We are dealing in (1.4.6) with the special case $J^\mu(z) = \sum_{i=1}^{M} k_i^\mu \delta^2(z - z_i)$. Hence (1.4.6) reduces to

$$A = \kappa^{M-2}\int\prod_{i=1}^{M}d^2z_i\prod_{i<j}\exp\{\tfrac{1}{2}k_i\cdot k_j G(z_i, z_j)\}.\qquad(1.4.8)$$

Because of normal ordering, we do not include terms with $i = j$ in the product in (1.4.8). As for the propagator in (1.4.8), it is the Green function for the two-dimensional Laplace equation, satisfying

$$\Delta_z G(z,z') = 2\pi\delta^2(z - z').\qquad(1.4.9)$$

(Δ_z is the Laplacian with respect to the variable z.) This equation has the solution

$$G(z,z') = -2\pi\int\frac{d^2q}{4\pi^2}\frac{e^{iq\cdot(z-z')}}{q^2} = \ln(\mu|z - z'|),\qquad(1.4.10)$$

where μ is an arbitrary infrared cutoff needed to handle the divergence at $q = 0$ in (1.4.10). From (1.4.8) we therefore get

$$A = \kappa^{M-2}\int\prod d^2z_i\prod_{i<j}|z_i - z_j|^{k_i\cdot k_j/2},\qquad(1.4.11)$$

where a μ-dependent factor has been absorbed in the unknown coupling constant. (Properly, the μ dependence cancels a similar dependence coming from the normal ordering or in other words from the terms with $i = j$ in (1.4.8).)

Equation (1.4.11) is very nearly our final form for the scattering amplitude. It expresses the M-point amplitude not as a path integral but as an integral over a finite number of variables, the points z_i, $i = 1, \ldots, M$ at which the external strings were attached to the world sheet. However, the integral in (1.4.11) is infinite. The reason for the infinity is that, as mentioned earlier, the gauge fixing that we used in deriving (1.4.11) did not completely remove the reparametrization invariance, but left a residual symmetry $\delta z = a + bz + cz^2$, where $z = x + iy$. Because of the failure to remove this residual invariance, the integral in (1.4.11) contains an integral over the infinite volume of the group $SL(2, C)$. The remaining gauge fixing is easily carried out. The three complex parameters of $SL(2, C)$ can be used to set any three of the z_i to any desired values. It is convenient (and conventional) to choose $z_1 = 0, z_2 = 1, z_3 = \infty$. In the limit as $z_3 \to \infty$, terms $|z_3 - z_j|^{k_3 \cdot k_j/2}$ can be dropped in (1.4.11). The reason for this is that for $|z_3| \to \infty$ these terms become independent of the z_j for $j \neq 3$, and they are also momentum independent since (using momentum conservation)

$$\prod_{j \neq 3} |z_3|^{k_3 \cdot k_j/2} = |z_3|^{-k_3^2/2} = |z_3|^{m^2/2}, \qquad (1.4.12)$$

where m^2 is the mass squared of the ground state. (As we shall see shortly, the requirement of $SL(2, C)$ invariance demands that $m^2 = -8$.) We will simply discard this factor since it is independent of external momenta; properly it cancels the mini-Faddeev–Popov determinant that enters in $SL(2, C)$ gauge fixing. The scattering amplitude thus reduces to

$$A = \kappa^{M-2} \int \prod_{l=4}^{M} d^2 z_l \prod_{j=4}^{M} |z_j|^{k_1 \cdot k_j/2} |1 - z_j|^{k_2 \cdot k_j/2} \prod_{4 \leq i < j \leq M} |z_i - z_j|^{k_i \cdot k_j/2}.$$
$$(1.4.13)$$

For the four-point function, this reduces to

$$A = \kappa^2 \int d^2 z_4 |z_4|^{k_1 \cdot k_4/2} |1 - z_4|^{k_2 \cdot k_4/2}. \qquad (1.4.14)$$

This four-point function was first introduced by Virasoro; it was generalized to an M-point amplitude by Shapiro.

1.4.5 The Mass of the Graviton

The reader will immediately note that (1.4.14) is not manifestly crossing symmetric; that is, it does not possess any obvious symmetry under permutation of the momenta $k_1, \ldots, k_4$. It is possible to check by hand that (1.4.14) is crossing symmetric if and only if the external tachyons are all on mass shell, $k_j^2 = 8$, $j = 1, \ldots, 4$. This is one way to determine the tachyon mass, and it gives the same answer that comes from quantization of the string action as sketched earlier. A more satisfying way to understand this important point is to note that (1.4.8), while unacceptable because it contains the unwanted integration over the $SL(2, C)$ manifold, is manifestly crossing symmetric. The crucial step in deriving (1.4.14) from (1.4.8) was $SL(2, C)$ gauge fixing, and (1.4.8) will be crossing symmetric if this $SL(2, C)$ symmetry is really valid. To make sure that (1.4.14) is crossing symmetric, we must make sure that (at least at this level) there is no anomaly in $SL(2, C)$ invariance.

One aspect of $SL(2, C)$ symmetry is that the integrated vertex operator $V = \int d^2\sigma e^{ik\cdot X(\sigma)}$ should be $SL(2, C)$ invariant. After all, an $SL(2, C)$ transformation is just a special case of a reparametrization of the world sheet, and V, which describes the amplitude for emission or absorption of a string from *anywhere* on the world sheet, should be invariant under reparametrizations. A special case of an $SL(2, C)$ transformation is the global rescaling of the world sheet $z \to tz$ (or infinitesimally $\delta z = bz$). As the integration measure d^2z picks up a factor of t^2 under such scaling, V can be invariant only if $e^{ik\cdot X}$ transforms with a factor of t^{-2}. This amounts to saying that the quantum field operator $e^{ik\cdot X}$ should be an operator of dimension two. At first sight, this looks impossible, since global scale invariance of the gauge-fixed string action (1.3.9) requires that X^μ be dimensionless; classically if X is dimensionless then $e^{ik\cdot X}$ is dimensionless as well. The only hope is to find a suitable quantum anomalous dimension of the operator $e^{ik\cdot X}$. It is unusual to find anomalous dimensions in free field theory, but it turns out that for free spinless fields in $1+1$ dimensions, this does occur.

The quickest way to determine the anomalous dimension of the operator $e^{ik\cdot X}$ is to study the two-point function of this operator. In general, in a scale invariant theory the two-point function of an operator Y of dimension p is $\langle Y(z)Y(0)\rangle = C|z|^{-2p}$ for some constant C. From our above discussion, the two-point function of the operator $e^{ik\cdot X}$ is $\langle e^{ik\cdot X(z)}e^{-ik\cdot X(0)}\rangle = |z|^{-k^2/2}$. From this we can read off the fact that $e^{ik\cdot X}$ is an operator of (anomalous) dimension $k^2/4$. Requiring it to have dimension two determines the tachyon mass squared to be $m^2 = -k^2 = -8$. In fact, the reader can easily check that the M-point amplitude (1.4.8) is invariant under global scale transformations, provided $m^2 = -8$ and momentum is conserved, $\sum k_i = 0$. It is also possible to verify that full $SL(2, C)$ invariance holds in (1.4.8) under the same conditions.

What we have determined in (1.4.13) is the amplitude for scattering of n tachyons. This amplitude has the excellent ultraviolet behavior for which string theory is famous. Of more theoretical interest, perhaps, is to construct a *graviton* scattering amplitude with the same excellent ultraviolet behavior. This can be done in a similar way. One simply

replaces the tachyon operator $V = \int d^2\sigma e^{ik \cdot X}$ with a graviton opera-
tor $V^{\mu\nu} = \int d^2\sigma \partial_\alpha X^\mu \partial^\alpha X^\nu e^{ik \cdot X}$, and otherwise one repeats the above
calculation. The algebra involved is much more complicated than in
the tachyon case (but far simpler than computation of graviton–graviton
scattering in field theory!). We will postpone actual computation of
graviton–graviton scattering until chapter 7, where we will carry out such
computations in the even more interesting case of superstring theories,
which are completely consistent and tachyon free. Here we may simply pause to determine the graviton mass. Like the tachyon mass, it
can be determined by requiring global scale invariance of the integrated
vertex operator. This amounts to saying that $W^{\mu\nu} = \partial_\alpha X^\mu \partial^\alpha X^\nu e^{ik \cdot X}$
should have dimension two. The difference from the tachyon case is
that because of the two derivatives that are present, $W^{\mu\nu}$ has dimension two already at the *classical* level, and therefore in this case we wish
the anomalous dimension of $e^{ik \cdot X}$ to be *zero*. Since this operator actually has anomalous dimension $k^2/4$, requiring its anomalous dimension
should be zero means that $k^2 = 0$ or in other words that the graviton is massless. This is certainly one of the most efficient ways to see
that string theory gives rise to a massless spin two particle. In the same
way we can see that the dilaton with $V_D = \int d^2\sigma \partial_\alpha X_\mu \partial^\alpha X^\mu e^{ik \cdot X}$ or
the antisymmetric tensor with $V_A^{\mu\nu} = \int d^2\sigma \epsilon^{\alpha\beta} \partial_\alpha X^\mu \partial_\beta X^\nu e^{ik \cdot X}$ must be
massless. This completes the list of massless particles (in the closed-
string theory) since there are no other suitable operators of dimension
two. Other possible vertex operators correspond to particles of positive
mass squared. For instance, a spin four particle Y with vertex operator
$V_Y^{\mu\nu\lambda\rho} = \int d^2\sigma \partial_\alpha X^\mu \partial^\alpha X^\nu \partial_\beta X^\lambda \partial^\beta X^\rho e^{ik \cdot X}$ would have to be a massive
particle of $m^2 = +8$.

1.5 Other Aspects of String Theory

1.5.1 Gravitational Ward Identities

The masslessness of the graviton appeared by surprise in the above discussion, just as it did historically. Nothing in the discussion justifies any
expectation that this 'graviton' will couple in a gauge invariant way, yet
one might conjecture that this will turn out to be true simply because
general covariance is one of the few possible ways to make a consistent
theory of a massless spin two particle.[*] It is, in fact, not hard to show the

---

[*] The only other possibility seems to be to have a massless spin two particle with
a *linear* gauge invariance that couples via derivative interactions only. This is
analogous to a massless vector meson that couples to neutral particles only via

appearance of Ward identities analogous to those that are deduced from general covariance in generally covariant field theories.

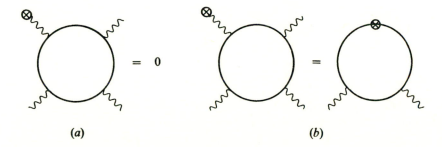

Figure 1.14. Depicted here is the scattering amplitude for M gauge bosons for one of which (shown with $\otimes$) we pick a longitudinal polarization state. In QED this amplitude (a) would simply vanish, while in a nonabelian theory the structure of the Ward identity is more complicated, as in (b).

Let us first recall the nature of Ward identities in field theory. Consider in QED the M-point function $A_{\mu_1\mu_2...\mu_M}(k_1, k_2, ..., k_M)$ for M external photons of polarizations $\mu_1, ..., \mu_M$ and momenta $k_1, ..., k_M$. In QED this amplitude vanishes (even off shell) if one of the photons is taken to have a longitudinal polarization depicted by $\otimes$ in fig. 1.14. Thus $k^{\mu_1} A_{\mu_1\mu_2...\mu_M}(k_1, k_2, ..., k_M) = 0$. It is instructive to recall how this is proved. The M-point function is written in terms of a current correlation function,

$$A_{\mu_1\mu_2...\mu_M}(k_1, k_2, ..., k_M) = \int d^4x_1 d^4x_2 ... d^4x_M e^{i \sum k_i \cdot x_i}$$
$$\times \langle T(J_{\mu_1}(x_1) J_{\mu_2}(x_2) ... J_{\mu_M}(x_M)) \rangle,$$
$$(1.5.1)$$

with J_μ being the electromagnetic current, T denoting a time ordered product, and $\langle \ \rangle$ denoting a vacuum expectation value. Because the elec-

derivative interactions such as magnetic moment couplings. Theories of massless spin two particles with derivative couplings only seem to be unrenormalizable and to suffer from certain other pathologies, but there is no general theorem guaranteeing that there cannot be a consistent theory of this kind.

tromagnetic current is conserved, we have

$$\langle T(\partial_{\mu_1} J^{\mu_1}(x_1) J_{\mu_2}(x_2) \ldots J_{\mu_n}(x_n)) \rangle = 0. \tag{1.5.2}$$

Because the electromagnetic current commutes with itself, we can bring the derivative out of the T product and write

$$0 = \partial_{\mu_1} \langle T(J^{\mu_1}(x_1) J_{\mu_2}(x_2) \ldots J_{\mu_n}(x_n)) \rangle. \tag{1.5.3}$$

Inserting this latter equation in (1.5.1) and integrating ∂_{μ_1} by parts, one arrives at the Ward identity

$$k^{\mu_1} A_{\mu_1 \mu_2 \ldots \mu_n}(k_1, k_2, \ldots, k_n) = 0. \tag{1.5.4}$$

In the nonabelian case the structure of Ward identities is more complicated. Roughly speaking, the electromagnetic current is replaced by Yang–Mills currents J_μ^a, which, while still conserved, do not commute with each other. Hence, in trying to remove the derivative ∂_{μ_1} from inside the T product, one picks up extra terms involving equal-time commutators. Therefore, the M-point function with one longitudinal polarization vector is not zero off shell but is expressed in terms of certain unphysical $(M-1)$-point functions in which the external gluon with longitudinal polarization has been contracted with one of the others, as in fig. 1.14. A further analysis is required to show that these $(M-1)$-point functions vanish on shell.

Let us now sketch the basic idea of how Ward-like identities arise in string theory. An external graviton of momentum k^μ and polarization $\zeta_{\mu\nu}$ would be represented in (1.4.5) by insertion of

$$V = \int dz \, d\bar{z} \, \zeta_{\mu\nu} \frac{\partial X^\mu}{\partial z} \frac{\partial X^\nu}{\partial \bar{z}} e^{ik \cdot X}. \tag{1.5.5}$$

(Here, as in our evaluation of the string scattering amplitude, we have made a stereographic projection of the string world sheet onto the complex z plane, the integration measure being then $dz \, d\bar{z}$.) We wish to show that the amplitude vanishes if $\zeta_{\mu\nu}$ is a longitudinal polarization tensor, $\zeta_{\mu\nu} = k_\mu \zeta_\nu + k_\nu \zeta_\mu$. It is enough to consider the case in which $\zeta_{\mu\nu}$ is replaced by $k_\mu \zeta_\nu$ since the other term may be treated similarly. With $\zeta_{\mu\nu} = k_\mu \zeta_\nu$ we get

$$V = \int dz \, d\bar{z} \, k_\mu \zeta_\nu \frac{\partial X^\mu}{\partial z} \frac{\partial X^\nu}{\partial \bar{z}} e^{ik \cdot X} = -i \int dz \, d\bar{z} \, \zeta_\nu \frac{\partial X^\nu}{\partial \bar{z}} \frac{\partial e^{ik \cdot X}}{\partial z}. \tag{1.5.6}$$

Integrating by parts and discarding a total divergence (which vanishes because we are really integrating over a compact string world sheet), this reduces to

$$V = +i \int dz d\overline{z} \frac{\partial^2 X^\nu}{\partial z \partial \overline{z}} \zeta_\nu e^{ik \cdot X}. \qquad (1.5.7)$$

Now, the field equation of the $(1 + 1)$-dimensional quantum field theory that governs string propagation is – in the conformal gauge we are using – simply $\partial^2 X^\nu / \partial z \partial \overline{z} = 0$. If therefore we were entitled to use the field equations inside the path integral of (1.4.5), we would conclude that $V = 0$ for gravitons with longitudinal polarization. Trying to use the equations of motion inside the path integral is a step analogous to claiming in QED that $\langle \partial_{\mu_1} T(J^{\mu_1} J_{\mu_2} \dots J_{\mu_n}) \rangle = 0$ because $\partial_\mu J^\mu = 0$. While that step is valid in QED, in Yang–Mills theory and in string theory things are not so simple. In particular, in string theory a path integral with an insertion of $\partial^2 X^\mu / \partial z \partial \overline{z}$ is not zero but gives rise to equal time commutator terms similar to those in the Ward identities of Yang–Mills theory. In Yang–Mills theory one proves with a somewhat technical analysis that these equal time commutator terms vanish if all other external lines are on mass shell. In string theory the same conclusion can be reached by a short cut. The equal-time commutator terms do not depend on the full set of kinematic variables but only on a subset (an equal-time commutator term made by contracting two external lines of, say, momentum k_i and k_j, respectively, would depend on $k_i + k_j$ but not on k_i and k_j separately). An amplitude that depends only on a subset of the kinematic variables cannot have the Regge-like asymptotic behavior that the on shell scattering amplitudes are known to have in string theory. Consequently, the equal-time commutator terms must vanish when all external lines are on shell. The argument we have just made is usually called the 'canceled propagator' argument, for reasons that will be more clear when we return to the subject in chapter 7.

The decoupling of longitudinal gravitons is only one of the conditions in general relativity that follows from the underlying general covariance. General relativity implies that a graviton of polarization $\zeta_{\mu\nu}$ and momentum k^σ must obey not only $k^2 = 0$ but also $k^\mu \zeta_{\mu\nu} = \zeta_\lambda^\lambda = 0$. How do these additional conditions emerge in string theory? The essence of the matter is that in determining the dimension of the operator $V = \int d^2\sigma \, \zeta_{\mu\nu} (\partial X^\mu / \partial \sigma^\alpha)(\partial X^\nu / \partial \sigma_\alpha) e^{ik \cdot X}$ we merely added the dimension (two) of $\zeta_{\mu\nu}(\partial X^\mu / \partial \sigma^\alpha)(\partial X^\nu / \partial \sigma_\alpha)$ and the dimension $(k^2/4)$ of $e^{ik \cdot X}$. In general, this would not be valid. The dimension of a product $A(\sigma)B(\sigma)$ of local operators A and B at the same point σ is not always the sum of the dimensions of A and B, because of short distance singularities in the operator product $A(\sigma)B(\sigma)$, which must be removed by normal ordering in order to define what one means by AB. In the case of the graviton vertex operator, with $A = \zeta_{\mu\nu}(\partial X^\mu / \partial \sigma^\alpha)(\partial X^\nu / \partial \sigma_\alpha)$, and $B = e^{ik \cdot X}$, the

operator A is free of normal ordering ambiguities (which would prevent it
from having definite dimension two) if $\zeta^\mu_\mu = 0$, and the operator product
AB is free of singularities if $k^\mu \zeta_{\mu\nu} = 0$. If these conditions are obeyed,
$\int d^2\sigma AB$ is an acceptable vertex operator provided $k^2 = 0$. If they are not
obeyed, it is not possible to make a conformally invariant vertex operator
for any k^2. This, together with our previous discussion of Ward identities,
shows how (in this formalism) string theory recovers the conditions that
follow from general covariance in general relativity.

1.5.2 Open Strings

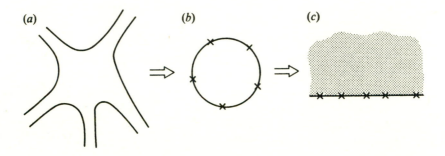

Figure 1.15. A 'planar' world sheet for scattering of open strings is sketched in (a). It
can be conformally mapped onto the disk of (b) or onto the upper half of the complex
plane as in (c). In either case external open-string states appear as insertions on the
boundary; these are denoted by $\otimes$ in (b) and (c).

Let us now leave the subject of closed strings, and discuss how similar
ideas can be applied to open strings. In fig. 1.15 we have sketched three
equivalent representations of a tree level diagram for scattering of open
strings. Particularly convenient are the pictures of fig. 1.15b and c in
which the world sheet is mapped onto the disk or the upper half plane
with the external strings appearing as finite points on the boundary. In
such formalisms, since the external open string is inserted only on the
boundary of the world sheet, its insertion is described by an operator of
the form $V = \int d\tau \sqrt{h_{\tau\tau}} U(\tau)$, where τ is a parameter on the boundary of
the world sheet. Invariance of V under conformal rescaling of the metric
$h_{\tau\tau} \to e^\phi h_{\tau\tau}$ requires now that U should have dimension *one* (and not two
as for closed strings). Like we did for closed strings, we write $U = W \cdot e^{ik \cdot X}$
and try to choose for W a polynomial in X^μ and its derivatives. For a

spin zero particle we can take $W = 1$; this necessitates $k^2 = 2,$[*] so we are dealing with a tachyon of mass squared -2. For spin one we try $W = dX^\mu/d\tau$. This requires $k^2 = 0$, so we are dealing with a spin one massless 'photon.' Other choices of W correspond to particles of positive mass squared.

The evaluation of open-string scattering amplitudes is similar to the previous analysis for closed strings. For instance, to evaluate the M-point function for scattering of tachyons, one conformally maps the world sheet onto the upper half plane as in fig. 1.15c. Representing a tachyon of momentum k by an insertion of $V(k) = \int_{-\infty}^{+\infty} dx\, e^{ik \cdot X}$ (with x running over the real axis) the analog of (1.4.6) and (1.4.11) is

$$A(k_1, k_2 \ldots k_M) = g^{M-2} \int dx_1 dx_2 \ldots dx_M \langle \prod_{i=1}^{M} e^{ik_i \cdot X(x_i)} \rangle, \qquad (1.5.8)$$

where g is a coupling constant. As in the closed-string case, (1.5.8) needs some modification because the conformal mapping of the world sheet onto the upper half plane is not quite unique. The residual symmetry group consists of conformal mappings of the upper half plane into itself. With $z = x + iy$ these are generated by $\delta z = a + bz + cz^2$, where now a, b and c must be real so that the boundary of the upper half plane (the real axis) is mapped into itself. These transformations generate the group $SL(2, R)$ of 2×2 real-valued matrices of determinant one. The three real parameters of this group can be used to set any three integration variables to prescribed values (rather as in the closed-string case where the group parameters and the integration variables in the integral representation of the scattering amplitude were both complex).

1.5.3 Internal Symmetries of Open Strings

Before trying to fix this residual gauge invariance, let us discuss what is the appropriate range for the integral over the x_i in (1.5.8). For closed

---

[*] The anomalous dimension of the operator $e^{ik \cdot X}$ inserted on the *boundary* of an open string world sheet can be determined from the propagator, which we compute in §1.5.4 below, and is $k^2/2$. This is twice as large as the anomalous dimension of the operator $e^{ik \cdot X}$ regarded as a closed string vertex operator and inserted on interior points. The factor of two arises because the short distance singularity in the propagator $\langle X^\mu(\sigma)X^\nu(\sigma') \rangle$ is twice as large when the points σ and σ' are boundary points, due to the contribution of 'image charges.' This will become evident when we compute the propagator on the upper half plane in §1.5.4.

strings, the underlying reparametrization invariance did not permit us to place any natural restriction on where on the string world sheet an external closed string is to be inserted. For open strings, matters are different. The M external open strings appear in fig. 1.15b in some order, say 1 2 3 ... M. A rotation of the world sheet can turn 1 2 3 ... M into, say, 2 3 4 ... M 1, but no reparametrization can turn 1 2 3 ... M into, say, 2 1 3 ... M. So the cyclic order in which the external strings appear is invariant under reparametrization, and it makes sense to try to define an amplitude in which the integral in (1.5.8) is carried out only over values of the x_i corresponding to a given cyclic order.

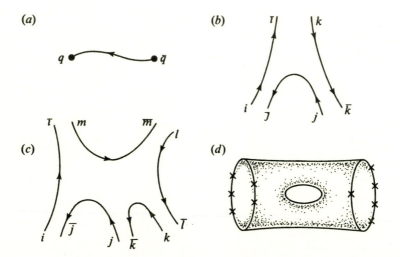

Figure 1.16. We can suppose that an oriented open string has a 'quark' at one end and an 'antiquark' at the other end, as sketched in (a). They can be assumed to transform, respectively, in the n and $\bar{n}$ representations of a $U(n)$ symmetry group. When open strings join, the quark and antiquark charges are required to match as in (b). The group-theory factor associated with a general planar open-string amplitude, sketched in (c), is then $\text{tr}\,(\lambda_1\lambda_2 \ldots \lambda_M)$. In a more general string diagram, as in (d), a similar group theory factor is assigned to each boundary component.

This possibility is related to some very important physics. Unlike the closed string, the open string has two special points, the endpoints. It is possible to assume that the open string carries 'charges' at its endpoints. For instance, for oriented open strings we can suppose that there is a 'quark' at one end of the string and an 'antiquark' at the other end. We introduce a $U(n)$ symmetry group that acts only on these 'quarks' and 'antiquarks', not on any of the other degrees of freedom, and we postulate that the 'quark' and 'antiquark' transform respectively in the n and $\overline{n}$ representations of $U(n)$. When strings join, we require the charges to match in the fashion indicated in fig. 1.16b. The tensor product $n \otimes \overline{n}$ is the adjoint representation of $U(n)$, so an open string with a 'quark' at one end and an 'antiquark' at the other end transforms in this representation; its $U(n)$ quantum numbers can therefore be specified by giving a $U(n)$ generator λ. Such a generator is concretely an $n \times n$ matrix $\lambda^i{}_j$, whose i and j indices correspond to the $U(n)$ states of the quark and antiquark, respectively. If M external strings are attached to a disk in the cyclic order 1 2 3...M, then the rule of fig. 1.16b says that the 'antiquark' index of each external string is to be contracted with the 'quark' index of the next one; this produces in fig. 1.16c the group-theory factor

$$(\lambda_1)^{i_1}{}_{i_2}(\lambda_2)^{i_2}{}_{i_3} \ldots (\lambda_M)^{i_M}{}_{i_1} = \text{tr}(\lambda_1 \lambda_2 \ldots \lambda_M). \qquad (1.5.9)$$

More generally, we can consider an arbitrary string world sheet whose boundary may have more than one component, as in fig. 1.16d. For each boundary component we include a group-theory factor $\text{tr}(\lambda_1 \lambda_2 \ldots \lambda_M)$, the product running over all open strings inserted on that boundary component. This factor is known as the Chan–Paton factor. Chan–Paton factors have no analog for closed strings, although internal symmetry groups can arise for closed strings in a more subtle way that we will discuss in chapter 6.

1.5.4 Recovery of the Veneziano Amplitude

Returning now to (1.5.8), let us calculate the coefficient of the group theory factor $\text{tr}(\lambda_1 \lambda_2 \ldots \lambda_M)$ or in other words the amplitude corresponding to the cyclic ordering 1 2 3...M of the external lines. We must eliminate the residual $SL(2, R)$ reparametrization symmetry. A convenient way to do this is to use the $SL(2, R)$ invariance to fix, say, $x_1 = 0, x_{M-1} = 1, x_M = \infty$. Evaluation of (1.5.8) then requires only a knowledge of the free-field propagator on the upper half plane. This can be easily determined by the method of images. On the whole plane the propagator was determined earlier to be $G(z, z') = \ln |z - z'|$; on the upper half plane we must include an image charge so that G satisfies the boundary condition that its derivative normal to the real axis vanish when $z(= x + iy)$ is real, $i.e.$,

$$\frac{\partial G(z, z')}{\partial y} = 0, \qquad \text{when } y = 0. \qquad (1.5.10)$$

This has the solution $\tilde{G}(z,z') = \ln|z - z'| + \ln|z - \bar{z}'|$, where $\bar{z}'$ is the image point. (If $z' = (x, y)$, then $\bar{z}' = (x, -y)$.) What we actually need in (1.5.8) is the propagator between two points that are both on the real axis, and this is $\tilde{G}(x, x') = 2\ln|x - x'|$. Allowing for this factor of two, the evaluation of (1.5.8) gives

$$A = g^{M-2} \int\limits_{0 < x_2 < x_3 < ... < x_{M-2} < 1} dx_2 dx_3 \ldots dx_{M-2}$$

$$\prod_{j=2}^{M-2} |x_j|^{k_1 \cdot k_j} |1 - x_j|^{k_j \cdot k_{M-1}} \prod_{2 \le l < m \le M-2} |x_l - x_m|^{k_l \cdot k_m}. \tag{1.5.11}$$

This is the Koba–Nielsen M-particle generalization of the Veneziano amplitude. For the four-point function it simplifies using (1.1.1) to

$$A = g^2 \int_0^1 dx\, x^{k_1 \cdot k_2} (1 - x)^{k_2 \cdot k_3} = g^2 B(-\frac{s}{2} - 2, -\frac{t}{2} - 2). \tag{1.5.12}$$

This is the original Veneziano amplitude.

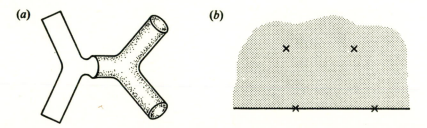

(a)   (b)

Figure 1.17. In (a) a string diagram with both closed and open strings as external particles is sketched. After a conformal mapping that projects the external strings to finite points, the open strings are represented by vertex operators inserted on the boundary of the world sheet, and the closed strings are represented by vertex operators inserted in the interior, as in (b).

More generally, we can calculate a scattering amplitude with both open and closed strings in the initial and final states. For example, in fig. 1.17a we sketch a string diagram with external open and closed strings, which can be conformally mapped to the upper half plane. After this mapping, the external open strings are represented by vertex operators that are integrated over the real axis (with some cyclic ordering), while the closed strings are represented by vertex operators integrated over the whole upper half plane. This is depicted in fig. 1.17b.

1.5.5 Comparison With QCD

In this discussion, we have simply pulled the Chan–Paton factors 'out of a hat', noting that they did not in any obvious way ruin the consistency of the theory. Historically, however, the Chan–Paton factors played an important role in the attempt to interpret string theory as a theory of hadrons. In fact, the unitary Chan–Paton symmetry group was interpreted as what we would now call the 'flavor group' of strong interactions. For instance, with three 'flavors', that is if the quarks at the end of the string come in three types (say u, d, s), the Chan–Paton group is $U(3) \simeq SU(3) \times U(1)$. Here $SU(3)$ corresponds to the symmetry of the eight-fold way.

In this historical interpretation of the model, the open string corresponds to a meson, with a quark at one end and an antiquark at the other end. The 'string' is some sort of a force that holds the quark and antiquark together. The closed string is what would now be called a glueball, that is a neutral state made out of the force carriers that hold together quarks and antiquarks. Of course, the model has a few problems. First, there are tachyons; second, it seems to be impossible to introduce explicit $U(3)$ symmetry breaking without pathologies; third, there are embarrassing massless particles that do not have cousins in the strongly interacting world. What is more, while we might want to interpret the $U(1)$ factor in the Chan–Paton group $U(3) \simeq SU(3) \times U(1)$ as baryon number, the model is regrettably lacking in baryons or fermions of any kind. In 1971, Ramond made an interesting attempt to incorporate fermions in the model. This led, however, not to a theory of strong interactions but (after many twists and turns) to supersymmetry, supergravity and superstrings.

Despite the many failures of string theory as a theory of strong interactions, there is a striking analogy between the open string and the modern view of the meson as a $q\bar{q}$ pair held together by a QCD flux tube. Insofar as the invention of string theory was not merely a lucky accident, this analogy is the reason that hadronic physics inspired the invention of string theory. The classification of string diagrams in terms of the topology of the world sheet actually has a close counterpart in QCD. If one generalizes the *color* group of QCD from $SU(3)$ to $SU(n)$, it is possible to argue that the $SU(n)$ theory should have an expansion in powers of $1/n$ quite analogous to the topological expansion in string theory. The large-n limit of QCD involves planar diagrams like those in fig. 1.18. So far the mathematics of the $1/n$ expansion in QCD has been intractable – like all other approaches to a quantitative understanding of the dynamical aspects of QCD. The hope remains nonetheless that progress in string theory will one day shed light on the $1/n$ expansion in QCD.

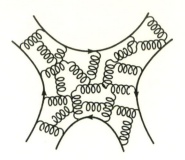

Figure 1.18. Meson scattering in the large-n limit of QCD is described by 'planar' Feynman diagrams with quarks on the boundary. Describing the flavor state of a meson by the flavor matrix λ_i, the planar amplitude for scattering of M mesons in the cyclic order 1 2 3 ... M involves a contraction of each quark with the antiquark of the adjoining meson, and so involves the Chan–Paton factor $\mathrm{tr}(\lambda_1\lambda_2 \ldots \lambda_M)$.

(a) (b)

Figure 1.19. If a closed-string world sheet with the topology of the two sphere S^2 is 'cut', as in (a), one finds an intermediate state containing a single closed string. The planar open-string diagram of (b) can be cut to reveal a single open string.

1.5.6 Unitarity and Gravity

String theory is certainly an unusual approach to formulating a relativistic quantum theory, and one consequence of this is that, while the rules we have sketched for calculating scattering amplitudes are manifestly Lorentz invariant, it is not too obvious that unitarity will be obeyed. In fact, in studying unitarity one finds a number of surprises. Among them is the restriction to ten or 26 dimensions, of which we will give a proper account in the following chapters. But one surprise can easily be described here.

Consider a complex string scattering process with strings 1 2 3 ... M in the initial state and strings $1'$ $2'$ $3'$ ... M' in the final state. A preliminary step before trying to give a proper proof of unitarity is to ask, for each given string diagram, what intermediate states occur in the reaction 1 2 3 ... $M \to 1'\ 2'\ 3' \ldots M'$. The way to answer this is to 'cut' the diagram in some way to separate the initial states from the final states. The cut reveals the intermediate states. For instance, if one 'cuts' the closed-string tree diagram of fig. 1.19 a to separate the initial states from the final states, the cut is along a single circle, so the intermediate state consists of a single closed string. Thus, the only singularities of the amplitude computed from this diagram will be single particle (closed string) poles. That is why this diagram is known as a tree diagram. Likewise, a 'cut' of the open-string diagram of fig. 1.19 b reveals a single open string propagating in the intermediate state. Of course, these heuristic pictures are only a starting point in discussing unitarity, a subject to which we will return in chapter 7.

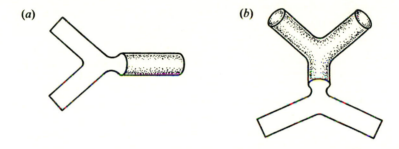

Figure 1.20. The basic couplings of closed and open strings are sketched in (a). Equality of the coupling κ of three closed strings and the coupling κ' of a closed string to two open strings follows from consideration of the diagram in (b).

Before moving on to a case that will give a more surprising answer, let us discuss the coupling constants in string diagrams. We have so far introduced an arbitrary coupling κ for the interaction of three closed strings and an arbitrary coupling g for the interaction of three open strings. From the discussion, they appear to be generalizations of the gravitational and Yang–Mills couplings, respectively, and they appear to be independent of each other. It is also possible to consider as in fig. 1.20 a a process in which a closed string couples to an open-string pair. Let us call this coupling constant κ'. In general relativity $\kappa = \kappa'$, since general covariance requires that the graviton (closed string) couples universally to energy–momentum whether this is in the form of gravitons (closed strings) or matter (open strings). We do not understand the analog of general covariance for string theory well enough to make the analogous argument in string theory, but we can prove that $\kappa = \kappa'$ from unitarity. Consider the planar diagram with two external open strings and two external closed strings shown in fig. 1.20b. This diagram has a closed-string pole in one channel and an open-string pole in the crossed channel. The residue of the closed-string pole should be $\kappa\kappa'$, while the residue of the open-string pole should be κ'^2; thus we are forced to take $\kappa = \kappa'$, as suggested by the analogy with general relativity.

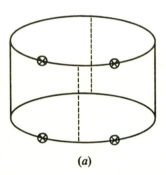
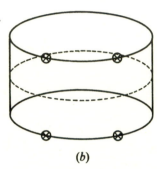

(a)   (b)

Figure 1.21. A one-loop open-string diagram can be cut as in (a) to reveal two open strings in the intermediate state or as in (b) to reveal a single closed string.

Now, in field theory of point particles the gravitational and Yang–Mills couplings are entirely independent of each other, so field theory would not suggest any relation between the closed-string coupling κ and the open-string coupling g. It is here that considerations of unitarity lead to a surprise. The one-loop diagram with external open strings, shown in fig. 1.21, can be cut in quite inequivalent ways to reveal either a pair of open strings or a single closed string in the intermediate state. Thus, this one amplitude has both a two-particle cut shown in fig. 1.21a and a one-particle pole shown in fig. 1.21b. The coefficient of the cut should be g^4, while the residue of the pole should be κ'^2 (or κ^2, since we know that these are equal), so unitarity requires a relation $\kappa \sim g^2$. This relation between gravitational and Yang–Mills couplings has no real counterpart in field theory of point particles.

Quantum field theory of point particles seems to be inconsistent in the presence of gravity; at any rate (with the sole exception of $(1 + 1)$-dimensional theories of propagating strings!) there has been little progress towards constructing generally covariant quantum field theories that are free of infinities. We have now learned that in string theory just the reverse holds. Not only is it possible to incorporate gravity in string theory; it is necessary to do so. A string theory that does *not* describe gravity would have to be a theory of open strings only. At the classical level (tree diagrams), there is nothing wrong with such a theory, but at the one-loop level one finds a graviton pole. The graviton pole in fig. 1.21*b* means that at the one-loop level there simply is not a unitary string theory without gravity.

1.6 Conclusion

String theory was invented as an approach to the strong interactions, but as such it was not the right theory. String theory has given rise instead to remarkable generalizations of Yang–Mills theory and gravity. In particular, supersymmetric string theories or superstring theories seem to be entirely free of the inconsistencies that plague quantum field theories of gravity.

We have tried in this introductory chapter to convey some idea of what string theory is, concentrating on the bosonic case for simplicity. In the rest of this book we will attempt a systematic exposition of some of the major approaches to this vast subject.

The logical foundations of the string theoretic generalization of Yang–Mills theory and general relativity remain clouded in mystery. For this reason among others, the coming decades are likely to be an exceptional period of intellectual adventure.

2. Free Bosonic Strings

In string theory, just as in other theories, it is necessary to understand the free theory well before trying to describe interactions. Our first task in a systematic exposition of string theory is to understand thoroughly the propagation of a single free string in space-time at both the classical and quantum levels. We begin in this chapter with a study of bosonic strings. In the course of this discussion, we will approach the bosonic string from many different points of view, corresponding to many different formalisms that have been developed over the years. These include various approaches to covariant and to light-cone quantization. Each adds important ingredients to an overall understanding of string theory, so it really is useful to become familiar with all of them.

2.1   The Classical Bosonic String

It may be helpful, as in the introduction, to begin with a discussion of point particles. Thus, let us consider the motion of a point particle of mass m in a background gravitational field, $i.e.$, in a curved Riemannian geometry described by a metric tensor $g_{\mu\nu}(x)$. The metric is assumed to have $D-1$ positive eigenvalues and one negative eigenvalue corresponding to the Minkowski signature of D-dimensional space-time. We always use units in which $\hbar = c = 1$.

The action principle that describes the motion of a massive point particle is well-known, and already entered in the introduction. It is simply proportional to the invariant length of the world line, $i.e.$,

$$S = -m \int ds \qquad (2.1.1)$$

where the invariant interval is given, as usual, by

$$ds^2 = -g_{\mu\nu}(x)dx^\mu dx^\nu. \qquad (2.1.2)$$

Suppose that a classical trajectory is written as $x^\mu(\tau)$, where τ is an arbitrary parameter that labels the points along the world line. Then

(2.1.1) may be written in the form

$$S = -m \int d\tau \sqrt{-\dot{x}^2}, \tag{2.1.3}$$

where

$$\dot{x}^2 \equiv g_{\mu\nu}(x) \frac{dx^\mu}{d\tau} \frac{dx^\nu}{d\tau}. \tag{2.1.4}$$

The action principle (2.1.3) has a very important property. It is invariant under reparametrizations $\tau \to \tilde{\tau}(\tau)$ of the particle trajectory. Equation (2.1.3) therefore really characterizes the world line of the particle and not a particular choice of coordinates. However, the square-root in the formula is somewhat awkward. Also, the formula does not apply to massless particles. To overcome these difficulties we introduce an auxiliary coordinate $e(\tau)$, which can be interpreted as an 'einbein' for the one-dimensional geometry of the world line. In terms of the einbein (2.1.1) can be re-expressed in the classically equivalent form

$$S = \frac{1}{2} \int (e^{-1}\dot{x}^2 - em^2) d\tau. \tag{2.1.5}$$

If one solves the e equation of motion

$$\dot{x}^2 + e^2 m^2 = 0 \tag{2.1.6}$$

and substitutes back in (2.1.5), then the action (2.1.3) is recovered. In the present form the τ reparameterization symmetry can be described infinitesimally as invariance of (2.1.5) under the transformations

$$\delta x = \xi \dot{x} \tag{2.1.7}$$

$$\delta e = \frac{d}{d\tau}(\xi e), \tag{2.1.8}$$

where $\xi(\tau)$ is an infinitesimal parameter with arbitrary τ dependence. This formulation is applicable to the massless case.

The reparametrization invariance can be used to make the gauge choice $e = 1/m$. With this choice the conjugate momentum is simply

$$p_\mu = mg_{\mu\nu}\dot{x}^\nu \tag{2.1.9}$$

and the equation of motion is obtained in the usual manner. However, (2.1.6) remains as a constraint. It can be interpreted as the mass-shell condition, generalized to propagation in a curved background.

Quantum-mechanical propagation of point particles could then be described by path integrals of the form

$$\int Dx De\, e^{iS(x,e)},\qquad(2.1.10)$$

where the 'gauge symmetry' of (2.1.5) would still have to be dealt with. Interactions among particles could be described by rules for the branchings and joinings of world lines so as to build up Feynman diagrams. There are many issues that could be pursued here, but perhaps this will suffice to prepare the ground for a study of strings.

2.1.1 String Action and Its Symmetries

Figure 2.1. One might generalize a point particle (a) to a string (b) or a membrane (c).

The point-particle actions can be generalized to objects of higher dimensionality. Our interest in this book is really in strings, but before committing ourselves to this case let us briefly contemplate replacing point particles by more general extended objects. If the objects are to be n-dimensional (illustrated for the cases $n = 0, 1, 2$ in fig. 2.1), then the obvious generalization of (2.1.1) is the invariant $n+1$-dimensional spacetime volume that it sweeps out. The coefficient must have dimensionality (mass)$^{n+1}$ to leave a dimensionless action. Alternatively, a formulation free from square roots analogous to (2.1.5) can be used, and that is what we discuss here. Since we are only describing bosonic degrees of freedom, the intrinsic geometry of the ($n + 1$)-dimensional manifold can be described by a metric $h_{\alpha\beta}(\sigma)$. Specifically, the generalization of the first term in (2.1.5) is

$$S = -\frac{T}{2}\int d^{n+1}\sigma\sqrt{h}h^{\alpha\beta}(\sigma)g_{\mu\nu}(X)\partial_\alpha X^\mu\partial_\beta X^\nu,\qquad(2.1.11)$$

where $\sigma^0 = \tau$ and the spatial coordinates $\sigma^i\,(i = 1, 2, \ldots, n)$ describe an n-dimensional object. $h^{\alpha\beta}$ is the inverse of $h_{\alpha\beta}$ and h is the absolute value of the determinant of $h_{\alpha\beta}$. The metric $h_{\alpha\beta}$ has Minkowski signature so that one of its eigenvalues is negative (timelike) and n are positive (spacelike). The functions $X^\mu(\sigma)$ give a map of the 'world-manifold' (line, sheet, tube, ...) into the physical space-time. $h_{\alpha\beta}(\sigma)$ describes the geometry of the $(n+1)$-dimensional manifold and $g_{\mu\nu}(x)$ describes the geometry of the D-dimensional space-time. It is necessary, of course, that $D \geq n+1$.

Equation (2.1.11) has a geometric character; it is independent of any particular choice of coordinates σ^α. This is manifest from the usual calculus of general relativity: $\sqrt{h}d^{n+1}\sigma$ is the invariant volume element, and $h^{\alpha\beta}\partial_\alpha X^\mu \partial_\beta X^\nu$ is also invariant since tensor indices are properly contracted. One feature that allowed for a sensible physical interpretation in the point-particle case was the possibility of choosing a gauge in which the metric $h_{\alpha\beta}$ (corresponding to e) was eliminated. Now in the general case $h_{\alpha\beta}$ has $\frac{1}{2}(n+1)(n+2)$ components and there are $n+1$ independent reparametrization gauge invariances. Thus after they are used up $\frac{1}{2}n(n+1)$ components of h still remain. Hence for $n > 0$, h cannot be eliminated simply by reparametrizations of the world surface. However, there is one more local symmetry that occurs only in the particular case of a string ($n = 1$) and must still be taken into account. There is a local Weyl scaling of the metric

$$h_{\alpha\beta} \rightarrow \Lambda(\sigma)h_{\alpha\beta} \qquad (2.1.12)$$

under which

$$\sqrt{h}h^{\alpha\beta} \rightarrow \Lambda^{\frac{1}{2}(n+1)-1}\sqrt{h}h^{\alpha\beta}. \qquad (2.1.13)$$

This leaves S invariant when $n = 1$. Therefore in the case of strings the counting suggests that it should still be possible to gauge away all the $h_{\alpha\beta}$ dependence with the help of this extra symmetry.

Weyl invariance, or at least the ability to locally gauge away the $h_{\alpha\beta}$ dependence, is central in the physics of strings. This is one of the things that singles out strings as opposed to, say, membranes. Membranes and objects of still higher dimensionality have another glaring problem, as follows. Equation (2.1.11) defines an $(n+1)$-dimensional quantum field theory, which by power counting is renormalizable for $n = 1$ and unrenormalizable for $n > 1$. Making sense of (2.1.11) as a quantum theory for $n > 1$ is as difficult a problem as making sense of general relativity as a quantum theory. Thus, membranes or higher-dimensional objects would hardly be a promising start toward quantum gravity.

Henceforth only strings ($n = 1$) are considered. The parameter T has dimensions of (length)$^{-2}$ or (mass)^2 and can be identified as the string tension. It turns out to be related to the universal Regge slope parameter (for open strings) by

$$T = (2\pi\alpha')^{-1}. \tag{2.1.14}$$

This will be established by quantizing (2.1.11) and determining the resulting spectrum.

2.1.2 The Free String in Minkowski Space

While the action (2.1.11) is formulated for string propagation on a general space-time manifold, this chapter concentrates on the case of flat Minkowski space. A thorough understanding of this situation is a prerequisite for generalizations. Equation (2.1.11) reduces for flat Minkowski space to

$$S = -\frac{T}{2} \int_M d^2\sigma \sqrt{h} h^{\alpha\beta}(\sigma) \partial_\alpha X^\mu \partial_\beta X_\mu. \tag{2.1.15}$$

The coordinate σ has a range that can be chosen to be $0 \leq \sigma \leq \pi$ as a matter of convenience. Even in a flat background, one could add additional terms to the string action (2.1.15). The possibilities that are compatible with D-dimensional Poincaré invariance and with power-counting renormalizability of the two-dimensional theory are

$$S_1 = \lambda \int d^2\sigma \sqrt{h}, \tag{2.1.16}$$

and

$$S_2 = \frac{1}{2\pi} \int d^2\sigma \sqrt{h} R^{(2)}(h). \tag{2.1.17}$$

The first one, S_1, is a two-dimensional 'cosmological constant' term. It does not have the Weyl symmetry of S and as a result leads to inconsistent classical field equations. Specifically, the trace of the $h_{\alpha\beta}$ equation of motion for $S + S_1$ implies that $h_{\alpha\beta} = 0$, which is hardly acceptable, unless $\lambda = 0$. In S_2, the expression $R^{(2)}(h)$ denotes the intrinsic two-dimensional scalar curvature of the world sheet formed from the metric $h_{\alpha\beta}$. Though it plays a significant role in relation to string interactions, S_2 is not important for our purposes here because in two dimensions the combination $\sqrt{h} R^{(2)}(h)$ is a total derivative. As a result, S_2 does not contribute to the classical field equations and does not enter in our current task of quantizing the free string.

Let us turn now to the symmetries of (2.1.15). It has the local symmetries mentioned earlier regardless of the choice of background. These are the reparametrization invariances

$$\delta X^\mu = \xi^\alpha \partial_\alpha X^\mu \tag{2.1.18}$$

$$\delta h^{\alpha\beta} = \xi^\gamma \partial_\gamma h^{\alpha\beta} - \partial_\gamma \xi^\alpha h^{\gamma\beta} - \partial_\gamma \xi^\beta h^{\alpha\gamma} \tag{2.1.19}$$

$$\delta(\sqrt{h}) = \partial_\alpha(\xi^\alpha \sqrt{h}) \tag{2.1.20}$$

and the Weyl scaling

$$\delta h^{\alpha\beta} = \Lambda h^{\alpha\beta}. \tag{2.1.21}$$

In addition there are global symmetries that reflect the symmetry of the background in which the string is propagating. For flat Minkowski space this is just Poincaré invariance, described by

$$\delta X^\mu = a^\mu{}_\nu X^\nu + b^\mu \tag{2.1.22}$$

and

$$\delta h^{\alpha\beta} = 0, \tag{2.1.23}$$

where $a_{\mu\nu} = \eta_{\mu\rho} a^\rho{}_\nu$ is antisymmetric. ($\eta_{\mu\rho}$ is the Minkowski metric.) We emphasize that ξ^α and Λ are arbitrary (infinitesimal) functions of σ^α, whereas $a_{\mu\nu}$ and b^μ are constants.

2.1.3 Classical Covariant Gauge Fixing and Field Equations

The two-dimensional energy–momentum tensor is given by the variational derivative of S with respect to the two-dimensional metric $h^{\alpha\beta}$, so that

$$T_{\alpha\beta} = -\frac{2}{T} \frac{1}{\sqrt{h}} \frac{\delta S}{\delta h^{\alpha\beta}}. \tag{2.1.24}$$

One finds

$$T_{\alpha\beta} = \partial_\alpha X^\mu \partial_\beta X_\mu - \frac{1}{2} h_{\alpha\beta} h^{\alpha'\beta'} \partial_{\alpha'} X^\mu \partial_{\beta'} X_\mu. \tag{2.1.25}$$

This is automatically traceless, $h^{\alpha\beta} T_{\alpha\beta} = 0$, as a consequence of the Weyl symmetry. The field equation $\delta S/\delta h^{\alpha\beta} = 0$ is the requirement $T_{\alpha\beta} = 0$.

If we define $G_{\alpha\beta} = \partial_\alpha X^\mu \partial_\beta X_\mu$, and $G = |\det G_{\alpha\beta}|$, then vanishing of $T_{\alpha\beta}$ gives

$$G_{\alpha\beta} = \frac{1}{2} h_{\alpha\beta} h^{\alpha'\beta'} G_{\alpha'\beta'} \tag{2.1.26}$$

$$G = \frac{1}{4} h (h^{\alpha\beta} G_{\alpha\beta})^2 \tag{2.1.27}$$

and thus

$$\frac{1}{2} \int_\Sigma d^2\sigma \sqrt{h} h^{\alpha\beta} G_{\alpha\beta} = \int_\Sigma d^2\sigma \sqrt{G}, \tag{2.1.28}$$

which is precisely the formula for the area of the world sheet Σ, as first proposed by Nambu.

The subsequent analysis of string dynamics and quantization is expedited by making a convenient choice of gauge. The three local symmetries (two reparametrizations and one Weyl scaling) are used to choose the three independent elements of $h_{\alpha\beta}$ so that $h_{\alpha\beta} = \eta_{\alpha\beta} = \left(\begin{smallmatrix} -1 & 0 \\ 0 & 1 \end{smallmatrix}\right)$, the two-dimensional Minkowski metric. (This has to be considered more carefully in the quantum theory.) Making this choice, the action simplifies to

$$S = -\frac{T}{2} \int d^2\sigma \eta^{\alpha\beta} \partial_\alpha X \cdot \partial_\beta X. \tag{2.1.29}$$

The Euler–Lagrange equation derived from (2.1.29) is simply the free two-dimensional wave equation

$$\Box X^\mu \equiv \left(\frac{\partial^2}{\partial\sigma^2} - \frac{\partial^2}{\partial\tau^2} \right) X^\mu = 0. \tag{2.1.30}$$

In the case of open strings, (2.1.30) is necessary but not sufficient to ensure that (2.1.29) is invariant under a general variation

$$X^\mu \to X^\mu + \delta X^\mu \tag{2.1.31}$$

The variation of (2.1.29) under (2.1.31) contains a volume term proportional to (2.1.30), but also a 'surface term':

$$-T \int d\tau \left[X'_\mu \delta X^\mu |_{\sigma=\pi} - X'_\mu \delta X^\mu |_{\sigma=0} \right] = 0, \tag{2.1.32}$$

The vanishing of this surface term gives the open-string boundary condition. For closed strings, (2.1.30) and periodicity of X is necessary and sufficient to ensure that (2.1.29) is stationary.

As usual in two dimensions, the general solution to the massless wave equation can be written as a sum of two arbitrary functions

$$X^\mu(\sigma) = X_R^\mu(\sigma^-) + X_L^\mu(\sigma^+), \qquad (2.1.33)$$

where

$$\sigma^- = \tau - \sigma \qquad (2.1.34)$$

$$\sigma^+ = \tau + \sigma. \qquad (2.1.35)$$

X_R^μ describes 'right-moving' modes of the string and X_L^μ describes 'left-moving' modes. It is convenient to introduce the world-sheet 'light-cone coordinates' σ^+ and σ^-, since the X_R and X_L are functions only of σ^- and σ^+, respectively. The derivatives conjugate to $\sigma^\pm$ are defined by

$$\partial_\pm = \tfrac{1}{2}(\partial_\tau \pm \partial_\sigma) \qquad (2.1.36)$$

so that in light-cone coordinates, the Minkowski world-sheet metric tensor becomes

$$\eta_{+-} = \eta_{-+} = -\tfrac{1}{2}, \qquad \eta_{++} = \eta_{--} = 0, \qquad (2.1.37)$$

so that its inverse is $\eta^{+-} = \eta^{-+} = -2$. World-sheet indices are accordingly raised and lowered by the rule $U^+ = -2U_-$, $U^- = -2U_+$.

The wave equation (2.1.30) must still be supplemented by the constraint equations $T_{\alpha\beta} = 0$. Using dots for τ derivatives and primes for σ derivatives, the latter take the form

$$T_{10} = T_{01} = \dot{X} \cdot X' = 0 \qquad (2.1.38)$$

$$T_{00} = T_{11} = \frac{1}{2}(\dot{X}^2 + X'^2) = 0. \qquad (2.1.39)$$

Here, for instance, $\dot{X} \cdot X'$ is short for $\dot{X}^\mu X'_\mu$. If one writes the world-sheet energy–momentum tensor $T_{\alpha\beta}$ in the $\sigma^\pm$ coordinate system according to the standard rules of tensor analysis, one finds using (2.1.38) and (2.1.39) that

$$T_{++} = \frac{1}{2}(T_{00} + T_{01}) = \partial_+ X \cdot \partial_+ X \qquad (2.1.40)$$

and

$$T_{--} = \frac{1}{2}(T_{00} - T_{01}) = \partial_- X \cdot \partial_- X. \qquad (2.1.41)$$

Tracelessness of the energy–momentum tensor, $h^{\alpha\beta} T_{\alpha\beta} = 0$, therefore becomes the statement that $T_{+-} = T_{-+} = 0$. This is equivalent to the assertion, already made in (2.1.39), that $T_{00} = T_{11}$. Using the above-stated facts, the constraint equations $T_{++} = T_{--} = 0$ become the statements

$$\dot{X}_R^2 = \dot{X}_L^2 = 0. \tag{2.1.42}$$

In two-dimensional quantum field theory, the law of energy-momentum conservation takes the general form $\partial_- T_{++} + \partial_+ T_{-+} = 0$, with a similar equation for $- \leftrightarrow +$. In the conformally invariant case, $T_{+-} = 0$, energy-momentum conservation reduces to

$$\partial_- T_{++} = 0. \tag{2.1.43}$$

This is a very powerful statement; it corresponds to the existence of an infinite set of conserved quantities. Let $f(x^+)$ be any function of x^+ (so $\partial_- f = 0$). Then (2.1.43) implies that the current fT_{++} is conserved, $\partial_-(fT_{++}) = 0$. The charge $Q_f = \int d\sigma \, f(x^+) T_{++}$ is then likewise conserved. As f is arbitrary, this is an infinite set of conserved quantities. It is only in two space-time dimensions that conformal invariance leads in this way to an infinite set of conserved quantities. The conserved quantities that we have just found are the constraints in (2.1.42). It is because they are conserved that it makes sense to set them to zero; if they vanish at one time, they will vanish at all later times.

The comments in the last paragraph hold for any conformally invariant theory in two dimensions. In the case of string theory the conserved quantities in question correspond to residual symmetries left over by the gauge fixing. As is typically the case for covariant gauge choices, setting $h^{\alpha\beta} = \eta^{\alpha\beta}$ does not completely use up the gauge freedom. To see this we note (using (2.1.19) and (2.1.21)) that any combined reparametrization and Weyl scaling for which

$$\partial^\alpha \xi^\beta + \partial^\beta \xi^\alpha = \Lambda \eta^{\alpha\beta} \tag{2.1.44}$$

preserves the gauge choice. In terms of the combinations $\xi^\pm = (\xi^0 \pm \xi^1)$ this implies that ξ^+ may be an arbitrary function of $\sigma^+ = (\tau + \sigma)$ and ξ^- an arbitrary function of $\sigma^- = (\tau - \sigma)$. If we think of the world-sheet reparametrization $\delta\sigma^\alpha = \xi^\alpha$ as being generated by the operator $V = \xi^\alpha \partial/\partial\sigma^\alpha$, then the generators of the residual symmetries are

$$V^+ = \xi^+(\sigma^+)\partial/\partial\sigma^+, \qquad V^- = \xi^-(\sigma^-)\partial/\partial\sigma^-. \tag{2.1.45}$$

This residual symmetry will be exploited to introduce the light-cone gauge in a later section. With $f^+ \sim \xi^+$, the conserved charges found in the last paragraph are the ones that generate (2.1.45). The operators written in

(a)   (b)

Figure 2.2. For a one-dimensional compact manifold, two topologies are possible – corresponding to the closed and open strings of (*a*) and (*b*).

(2.1.45) are the generators of the group of conformal transformations of two dimensional Minkowski space; only in two dimensions is the conformal group infinite dimensional.

There are two types of boundary conditions that we will have to consider, corresponding to closed strings and open strings, respectively, depicted in fig. 2.2. Closed strings are loops with no free ends, topologically equivalent to circles shown in fig. 2.2*a*. For them the appropriate boundary condition is just periodicity of the coordinates

$$X^\mu(\tau, \sigma) = X^\mu(\tau, \sigma + \pi). \tag{2.1.46}$$

The general solution of (2.1.33) compatible with the periodicity requirement is

$$X_R^\mu = \frac{1}{2}x^\mu + \frac{1}{2}l^2 p^\mu(\tau - \sigma) + \frac{i}{2}l \sum_{n \neq 0} \frac{1}{n} \alpha_n^\mu e^{-2in(\tau-\sigma)} \tag{2.1.47}$$

$$X_L^\mu = \frac{1}{2}x^\mu + \frac{1}{2}l^2 p^\mu(\tau + \sigma) + \frac{i}{2}l \sum_{n \neq 0} \frac{1}{n} \tilde{\alpha}_n^\mu e^{-2in(\tau+\sigma)}, \tag{2.1.48}$$

where the α_n^μ are Fourier components, which will be interpreted as oscillator coordinates. A fundamental length, called l, has been introduced in these equations. It is related to α' and the string tension T (in units $\hbar = c = 1$) by

$$l = \sqrt{2\alpha'} = 1/\sqrt{\pi T}. \tag{2.1.49}$$

Later it will be set equal to one. As for x^μ and p^μ, they may be interpreted as the center of mass position and momentum of the string. Normalization constants in (2.1.47) and (2.1.48) have been chosen for later convenience. Note that the terms linear in σ cancel in the sum $X^\mu = X_L^\mu + X_R^\mu$, so that the closed-string boundary condition is indeed obeyed. The requirement that X_R^μ and X_L^μ are real functions implies that x^μ and p^μ are real and that α^μ_{-n} is the adjoint of α^μ_n, i.e.,

$$\alpha^\mu_{-n} = (\alpha^\mu_n)^\dagger, \qquad \tilde{\alpha}^\mu_{-n} = (\tilde{\alpha}^\mu_n)^\dagger. \tag{2.1.50}$$

It is also essential to determine the Poisson brackets of the α^μ_n. To do so, we note from (2.1.29) that the Poisson brackets of the X^μ and $\dot{X}^\mu$ at equal τ are

$$[X^\mu(\sigma), X^\nu(\sigma')]_{\text{P.B.}} = [\dot{X}^\mu(\sigma), \dot{X}^\nu(\sigma')]_{\text{P.B.}} = 0 \tag{2.1.51}$$

$$[\dot{X}^\mu(\sigma), X^\nu(\sigma')]_{\text{P.B.}} = T^{-1}\delta(\sigma - \sigma')\eta^{\mu\nu}. \tag{2.1.52}$$

Insertion of (2.1.47) and (2.1.48) shows that the Poisson brackets of the α^μ_n are

$$\begin{aligned}
[\alpha^\mu_m, \alpha^\nu_n]_{\text{P.B.}} &= [\tilde{\alpha}^\mu_m, \tilde{\alpha}^\nu_n]_{\text{P.B.}} = im\delta_{m+n}\eta^{\mu\nu} \\
[\alpha^\mu_m, \tilde{\alpha}^\nu_n]_{\text{P.B.}} &= 0.
\end{aligned} \tag{2.1.53}$$

(The i will disappear soon when we replace the Poisson brackets by commutators.) Thus, the Fourier modes α^μ_n for $n \neq 0$ are harmonic-oscillator coordinates, as we might have anticipated from experience with other free field theories or for that matter from experience with violin strings. Equation (2.1.53) remains valid when $n = 0$ or $m = 0$ if we adopt the useful convention $\tilde{\alpha}^\mu_0 = \alpha^\mu_0 = \frac{1}{2}lp^\mu$. Comparing (2.1.47) and (2.1.48) with (2.1.52) also gives the Poisson bracket

$$[p^\mu, x^\nu]_{\text{P.B.}} = \eta^{\mu\nu}, \tag{2.1.54}$$

so that the center-of-mass position and momentum of the string are canonically conjugate variables, as we might have suspected.

The analysis for open strings shown in fig. 2.2b is similar, except that we must determine the proper boundary condition to be used at the endpoints of the string, $\sigma = 0, \pi$. Requiring that the boundary terms (2.1.32) in the variation of the action must vanish gives us the open string boundary conditions, which say that

$$X'^\mu = 0, \qquad \text{for } \sigma = 0 \text{ and } \sigma = \pi \tag{2.1.55}$$

or in other words that the normal derivative of X^μ must vanish at the string boundary. These are 'free boundary conditions', which prevent momentum from flowing off the ends of the string. The general solution of the wave equation with these boundary conditions is given by

$$X^\mu(\sigma,\tau) = x^\mu + l^2 p^\mu \tau + il \sum_{n\neq 0} \frac{1}{n}\alpha_n^\mu e^{-in\tau}\cos n\sigma. \qquad (2.1.56)$$

The open-string boundary conditions cause the left- and right-moving components to combine into standing waves. In particular,

$$2\partial_\pm X^\mu = \dot{X}^\mu \pm X'^\mu = l\sum_{-\infty}^{\infty}\alpha_n^\mu e^{-in(\tau\pm\sigma)}, \qquad (2.1.57)$$

where we have set $\alpha_0^\mu = lp^\mu$.

Analogous formulas for closed strings are

$$\partial_- X_R^\mu = \dot{X}_R^\mu = l\sum_{-\infty}^{\infty}\alpha_n^\mu e^{-2in(\tau-\sigma)} \qquad (2.1.58)$$

$$\partial_+ X_L^\mu = \dot{X}_L^\mu = l\sum_{-\infty}^{\infty}\tilde{\alpha}_n^\mu e^{-2in(\tau+\sigma)}. \qquad (2.1.59)$$

The important difference is that the left- and right-moving modes are independent in this case. Also $\alpha_0^\mu = \tilde{\alpha}_0^\mu = \frac{1}{2}lp^\mu$ in the closed-string case, and there is an extra factor of two in the exponents.

We turn next to a description of the D-dimensional Poincaré invariance. Since the Poincaré transformations $\delta X^\mu = a^\mu{}_\nu X^\nu + b^\mu$ are simply global symmetries from the point of view of the two-dimensional theory, they are associated with conserved 'Noether currents'. There is a standard procedure in field theory, known as the 'Noether method', for constructing the conserved current J_α associated with the global symmetry transformation $\phi(\sigma) \to \phi(\sigma) + \epsilon\delta\phi(\sigma)$, where $\phi(\sigma)$ is any field in the theory and ϵ is a constant infinitesimal parameter. One considers the transformation

$$\phi(\sigma) \to \phi(\sigma) + \epsilon(\sigma)\delta\phi(\sigma) \qquad (2.1.60)$$

where ϵ is an infinitesimal parameter that is *not* constant on the world sheet. The action is not invariant under such transformations for general ϵ, since the symmetry we are considering is only a global symmetry. Since the action would be invariant for constant ϵ, its variation is proportional to the derivative of ϵ and so is of the general form

$$\delta S = \int d^2\sigma J_\alpha \partial^\alpha \epsilon, \qquad (2.1.61)$$

for some current J_α. The current defined in this way is always conserved if the equations of motion are obeyed. Indeed, when the equations are obeyed, the action is stationary under any variation and in particular under a variation of the form (2.1.60). Thus, when the equations of motion are obeyed, δS in (2.1.61) is zero for any ϵ. This is possible only if $\partial_\alpha J^\alpha = 0$.

One can readily apply this method to derive the conserved currents associated with the Poincaré transformations of X^μ:

$$P_\alpha^\mu = T\partial_\alpha X^\mu \qquad (2.1.62)$$

$$J_\alpha^{\mu\nu} = T(X^\mu \partial_\alpha X^\nu - X^\nu \partial_\alpha X^\mu). \qquad (2.1.63)$$

Here P_α is the current associated with translation invariance while $J_\alpha^{\mu\nu}$ is the current associated with Lorentz invariance. Current conservation is the statement that

$$\partial_\alpha P^{\alpha\mu} = \partial_\alpha J^{\alpha\mu\nu} = 0. \qquad (2.1.64)$$

These currents describe the density of D-dimensional momentum and angular momentum on the two-dimensional world sheet. The amount of momentum flowing across an arbitrary line segment in the world sheet, $(d\sigma, d\tau)$ is given by

$$dP^\mu = P_\tau^\mu d\sigma + P_\sigma^\mu d\tau, \qquad (2.1.65)$$

so that the boundary conditions at the ends of an open string, (2.1.32), indeed imply that there is no momentum flowing out of the ends of the string. Similar statements apply to the current of angular momentum, $J_\alpha^{\mu\nu}$.

The total conserved momentum and angular momentum of a string are found by integrating the currents of (2.1.62) and (2.1.63) over σ at $\tau = 0$. For example, the total momentum of a closed string is given by

$$P^\mu = T\int_0^\pi d\sigma \frac{dX^\mu(\sigma)}{d\tau} = \pi T(l\alpha_0^\mu + l\tilde{\alpha}_0^\mu) = p^\mu, \qquad (2.1.66)$$

so that the total momentum of the string is the same as the 'momentum' p^μ of the zero mode. This also holds for open strings. The total angular momentum is given by

$$J^{\mu\nu} = T \int_0^\pi d\sigma \left(X^\mu \frac{dX^\nu}{d\tau} - X^\nu \frac{dX^\mu}{d\tau} \right). \tag{2.1.67}$$

Inserting the mode expansions one obtains

$$J^{\mu\nu} = l^{\mu\nu} + E^{\mu\nu} \tag{2.1.68}$$

for open strings and

$$J^{\mu\nu} = l^{\mu\nu} + E^{\mu\nu} + \tilde{E}^{\mu\nu} \tag{2.1.69}$$

for closed strings, where

$$l^{\mu\nu} = x^\mu p^\nu - x^\nu p^\mu \tag{2.1.70}$$

and

$$E^{\mu\nu} = -i \sum_{n=1}^\infty \frac{1}{n} (\alpha_{-n}^\mu \alpha_n^\nu - \alpha_{-n}^\nu \alpha_n^\mu) \tag{2.1.71}$$

with an identical expression for $\tilde{E}^{\mu\nu}$ in terms of $\tilde{\alpha}_n^\mu$. Though we obtained the total momentum and angular momentum of the string by integrating the conserved currents over σ at $\tau = 0$, current conservation means that we would get the same result by integrating over any space-like curve that cuts once across the string world sheet.

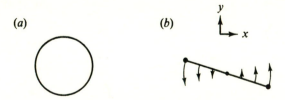

Figure 2.3. In (a) we sketch a closed string initially at rest with radius R. (b) depicts an open string that is spinning in the $x - y$ plane.

As an example of the use of these formulas, let us verify that T is really the string tension. Consider, as in fig. 2.3a, a closed string that at time

$t = 0$ is at rest and is a circle in the x–y plane of radius R. Let σ, at $\tau = t = 0$, be proportional to the arc length of the string:

$$x = R\cos 2\sigma, \qquad y = R\sin 2\sigma. \tag{2.1.72}$$

It is then easy to see that the equations of motion, (2.1.30), and the constraint equations, (2.1.38) and (2.1.39), can be obeyed by assuming that $t = 2R\tau$ near $\tau = t = 0$. It is easy to see from (2.1.62) that such a string has $p^0 = 2\pi RT$, confirming that T is the energy per unit length of the string.

As another example, let us verify at the classical level that $\alpha' = 1/(2\pi T)$ is the Regge slope of open strings. The Regge slope is defined to be the maximum possible angular momentum per unit energy squared. For open strings, angular momentum per unit energy is maximized by a string that is spinning in, say, the $x - y$ plane, as shown in fig. 2.3b, according to the formula

$$x = A\cos\tau\cos\sigma, \quad y = A\sin\tau\cos\sigma, \qquad t = A\tau. \tag{2.1.73}$$

It is instructive to verify that this formula obeys the equations of motion, (2.1.30) and the constraint equations. (2.1.38) and (2.1.39). Computing from (2.1.62) and (2.1.63) that the energy of this configuration is πAT and the angular momentum is $\pi A^2 T/2$, we see that the maximum angular momentum per unit energy squared is $1/(2\pi T)$, confirming the interpretation of α' as the Regge slope. Furthermore, we see that at the string endpoints $|\partial x/\partial t|^2 + |\partial y/\partial t|^2 = 1$, which means that they move at the speed of light (recalling that $c = 1$ in our units). It is a general consequence of the boundary condition $X^{\mu\prime} = 0$ together with the constraint equations that the endpoints move at the speed of light.

Returning from space-time to world-sheet concepts, the Hamiltonian of the two-dimensional theory is given by

$$H = \int_0^\pi d\sigma(\dot{X}\cdot P_\tau - L) = \frac{T}{2}\int_0^\pi (\dot{X}^2 + X^{\prime 2})d\sigma. \tag{2.1.74}$$

This gives

$$H = \frac{1}{2}\sum_{-\infty}^{\infty}\alpha_{-n}\cdot\alpha_n \qquad \text{(open strings)} \tag{2.1.75}$$

$$H = \frac{1}{2}\sum_{-\infty}^{\infty}(\alpha_{-n}\cdot\alpha_n + \tilde{\alpha}_{-n}\cdot\tilde{\alpha}_n) \qquad \text{(closed strings).} \tag{2.1.76}$$

The Hamiltonian generates (by Poisson brackets) the τ evolution of the string. It is dimensionless, because τ has been chosen to be dimensionless.

Let us consider now the mode expansions of the constraints $T_{\alpha\beta} = 0$. In the case of closed strings we saw that the full content of these equations is given by $\dot{X}_R^2 = \dot{X}_L^2 = 0$. Using the mode expansions of (2.1.58) and (2.1.59) these have Fourier components (evaluated at $\tau = 0$)

$$
\begin{aligned}
L_m &= \frac{T}{2} \int_0^\pi e^{-2im\sigma} T_{--} d\sigma \\
&= \frac{T}{2} \int_0^\pi e^{-2im\sigma} \dot{X}_R^2 d\sigma = \frac{1}{2} \sum_{-\infty}^{\infty} \alpha_{m-n} \cdot \alpha_n
\end{aligned}
\tag{2.1.77}
$$

$$
\begin{aligned}
\tilde{L}_m &= \frac{T}{2} \int_0^\pi e^{2im\sigma} T_{++} d\sigma \\
&= \frac{T}{2} \int_0^\pi e^{2im\sigma} \dot{X}_L^2 d\sigma = \frac{1}{2} \sum_{-\infty}^{\infty} \tilde{\alpha}_{m-n} \cdot \tilde{\alpha}_n.
\end{aligned}
\tag{2.1.78}
$$

In the case of open strings this requires some modification since the $e^{in\sigma}$ are not orthogonal functions for the interval $0 \leq \sigma \leq \pi$. The open-string constraint equations can be described conveniently if we formally extend the definition of X_L and X_R (introduced in (2.1.33)) beyond the interval $0 \leq \sigma \leq \pi$ by saying that $X_R(\sigma + \pi) = X_L(\sigma)$, $X_L(\sigma + \pi) = X_R(\sigma)$. Open-string boundary conditions then imply that X_R (or X_L) is a periodic function of σ with period 2π. The constraint equations then amount to the vanishing of T_{++} for $-\pi \leq \sigma \leq \pi$, or equivalently to the vanishing of its Fourier components

$$
\begin{aligned}
L_m &= T \int_0^\pi (e^{im\sigma} T_{++} + e^{-im\sigma} T_{--}) d\sigma \\
&= \frac{T}{4} \int_{-\pi}^\pi e^{im\sigma} (\dot{X} + X')^2 d\sigma = \frac{1}{2} \sum_{-\infty}^{\infty} \alpha_{m-n} \cdot \alpha_n.
\end{aligned}
\tag{2.1.79}
$$

We note in particular that for open strings $H = L_0$ and for closed strings $H = L_0 + \tilde{L}_0$. In the case of closed strings the combination $L_0 - \tilde{L}_0$, which must vanish according to the constraint equations, does not contain the momentum p^μ. This combination, which generates rigid rotations of the closed string, $\sigma \to \sigma +$ constant, will play an important role later.

A string in a given state of oscillation has a mass squared $M^2 = -p_\mu p^\mu$. The constraint equation $L_0 = 0$ translates into a very important equation that determines M^2 in terms of the internal modes of oscillation of the string. This is

$$M^2 = \frac{1}{\alpha'} \sum_{n=1}^{\infty} \alpha_{-n} \cdot \alpha_n \tag{2.1.80}$$

for open strings, and

$$M^2 = \frac{2}{\alpha'} \sum_{n=1}^{\infty} (\alpha_{-n} \cdot \alpha_n + \tilde{\alpha}_{-n} \cdot \tilde{\alpha}_n) \tag{2.1.81}$$

for closed strings. Equations (2.1.80) and (2.1.81) are known as the mass-shell conditions for open and closed strings, respectively. The mass-shell condition is the relativistic analog of the equation that expresses the energy of, say, a nonrelativistic violin string in terms of its oscillator coordinates. In the quantum theory these equations become slightly modified due to normal-ordering effects. The fact that $L_0 = \tilde{L}_0$ for closed strings implies that the two terms in (2.1.76) or (2.1.81) give equal contributions.

The Fourier modes of the energy-momentum tensor, L_m and $\tilde{L}_m$, are called the Virasoro operators. The Poisson brackets of the Virasoro operators can be straightforwardly (though somewhat tediously) calculated from the Poisson brackets of the individual oscillators. From the definition of the L_n, we have

$$[L_m, L_n]_{\text{P.B.}} = \frac{1}{4} \sum_{k,l} [\alpha_{m-k} \cdot \alpha_k, \alpha_{n-l} \cdot \alpha_l]_{\text{P.B.}} \tag{2.1.82}$$

Using the well-known identity $[AB, CD] = A[B, C]D + AC[B, D] + [A, C]DB + C[A, D]B$, and the Poisson brackets of the oscillators, this becomes

$$[L_m, L_n]_{\text{P.B.}} = \frac{i}{4} \sum_{k,l} (k\alpha_{m-k} \cdot \alpha_l \, \delta_{k+n-l} + k\alpha_{m-k} \cdot \alpha_{n-l} \, \delta_{k+l}$$
$$+ (m-k)\alpha_l \cdot \alpha_k \, \delta_{m-k+n-l} + (m-k)\alpha_{n-l} \cdot \alpha_k \, \delta_{m-k+l}) \tag{2.1.83}$$

where, as before, δ_n is 1 if $n = 0$, and 0 otherwise. Equation (2.1.83) reduces to

$$[L_m, L_n]_{\text{P.B.}} = \frac{i}{2} \sum_k k\alpha_{m-k} \cdot \alpha_{k+n} + \frac{i}{2} \sum_k (m-k)\alpha_{m-k+n} \cdot \alpha_k. \tag{2.1.84}$$

Changing variables $k \to k' = k + n$ in the first sum, (2.1.84) reduces to the Virasoro algebra

$$[L_m, L_n]_{\text{P.B.}} = i(m - n)L_{m+n}. \tag{2.1.85}$$

This is a crucial formula whose modification by quantum anomalies will concern us later. The $\tilde{L}$, of course, obey the same algebra. The Virasoro algebra (2.1.85) has a very simple interpretation. Let θ be an ordinary angular variable, $0 \le \theta \le 2\pi$, which we view as parametrizing a circle S^1. An infinitesimal general coordinate transformation of the circle, $\theta \to \theta + a(\theta)$, would be generated by $D_a = ia(\theta)d/d\theta$. A complete basis for such 'diffeomorphisms' of the circle is provided by the operators

$$D_n = i\,e^{in\theta}\frac{d}{d\theta}. \tag{2.1.86}$$

These are readily seen to obey the Virasoro algebra (2.1.85). Thus, the Virasoro algebra is the same as the algebra of infinitesimal diffeomorphisms of S^1. The *raison d'etre* for the appearance of a Virasoro algebra is easily understood by noting that, with $e^{in\theta}$ replaced by $\xi^\pm$, and $id/d\theta$ replaced by $\partial/\partial\sigma^\pm$, (2.1.86) coincides with the formula (2.1.45) for the generators of the residual symmetries that preserve the conformal gauge condition. Although the $\sigma^\pm$ in (2.1.45) are not angular variables *a priori*, they become angular variables upon imposing the equations of motion of the theory, since our mode expansions of open and closed strings contain $\exp(in\sigma^\pm)$ with integer n only.

2.2   Quantization – Old Covariant Approach

We now turn to the quantization of the bosonic string theory. Many different procedures have been used to describe its quantization. When used correctly, they are all equivalent. The relations are quite nontrivial, however, and each has certain advantages, so it is desirable to become familiar with all of them. There are two basic types of covariant approaches that are used. The first (and oldest) is based on a description in terms of the X^μ coordinates only with restrictions on the physical Fock space corresponding to the Virasoro constraint conditions. These restrictions are analogous to the Gupta–Bleuler condition in electrodynamics, in which the classical constraint $\partial_\mu A^\mu = 0$ is replaced by the requirement that the positive-frequency components of the corresponding quantum operator should annihilate physical photon states. The modern covariant quantization approach has a deeper geometric basis. It involves the introduction of Faddeev–Popov ghosts and the identification of BRST symmetries and currents. Those methods will be described in the next chapter.

2.2.1 Commutation Relations and Mode Expansions

In the classical gauge fixing of §2.1.3 we used the reparametrization and Weyl scaling symmetries to set the world-sheet metric $h_{\alpha\beta}$ equal to the flat two-dimensional Minkowski metric $\eta_{\alpha\beta}$. In the quantum theory one has to consider the validity of this procedure more carefully. The Weyl scaling symmetry is responsible for the tracelessness of the energy–momentum tensor obtained by varying the action with respect to $h^{\alpha\beta}$. In general, the quantum theory gives rise to an anomaly in the trace of $T_{\alpha\beta}$. Only under very special circumstances does this anomaly cancel in the quantum theory. Historically, the first approaches to the problem involved setting $h_{\alpha\beta} = \eta_{\alpha\beta}$ without worrying about anomalies. The subsequent analysis then revealed (after much toil) that a satisfactory theory only emerged after imposing certain consistency requirements on the dimension of space-time and the mass of the ground state. The light-cone quantization, described in §2.3, is a very 'physical' approach in which one starts with $h_{\alpha\beta} = \eta_{\alpha\beta}$ and then imposes additional gauge restrictions. It leads to the same restrictions on masses and the space-time dimensions, as we will see. In an alternative covariant approach, which we will develop in chapter 3, the conditions on D and a are interpreted as the conditions for the cancellation of the trace anomaly.

Beginning with the most traditional approach, let us try to set $h_{\alpha\beta} = \eta_{\alpha\beta}$ in the quantum theory, and explore where it leads. It was shown earlier that in this gauge the classical string dynamics is described by the action

$$S = -\frac{T}{2} \int d^2\sigma \partial^\alpha X \cdot \partial_\alpha X \qquad (2.2.1)$$

supplemented by the subsidiary conditions

$$(\dot{X} \pm X')^2 = 0, \qquad (2.2.2)$$

corresponding to $T_{++} = T_{--} = 0$, and suitable open-string or closed-string boundary conditions. The momentum conjugate to X^μ is

$$P_\tau^\mu = T\dot{X}^\mu, \qquad (2.2.3)$$

the τ component of the momentum current introduced in (2.1.62). A standard method of passing from classical physics to quantum physics is to replace the Poisson brackets by commutator brackets *via* the substitution

$$[\ldots]_{\text{P.B.}} \rightarrow -i[\ldots]. \qquad (2.2.4)$$

We can now interpret X^μ as a quantum operator with (2.1.52) replaced by canonical commutation relations at equal τ.

$$[P_\tau^\mu(\sigma,\tau), X^\nu(\sigma',\tau)] = -i\delta(\sigma - \sigma')\eta^{\mu\nu} \tag{2.2.5}$$

$$[X^\mu(\sigma,\tau), X^\nu(\sigma',\tau)] = [P_\tau^\mu(\sigma,\tau), P_\tau^\mu(\sigma',\tau)] = 0. \tag{2.2.6}$$

Equations (2.1.53) and (2.1.54) are likewise replaced by equal-time commutators:

$$[x^\mu, p^\nu] = i\eta^{\mu\nu} \tag{2.2.7}$$

and the α_m^μ and $\tilde{\alpha}_m^\mu$ have commutation relations

$$[\alpha_m^\mu, \alpha_n^\nu] = m\delta_{m+n}\eta^{\mu\nu} \tag{2.2.8}$$

$$[\alpha_m^\mu, \tilde{\alpha}_n^\nu] = 0 \tag{2.2.9}$$

$$[\tilde{\alpha}_m^\mu, \tilde{\alpha}_n^\nu] = m\delta_{m+n}\eta^{\mu\nu}. \tag{2.2.10}$$

The α_m are therefore naturally interpreted as harmonic oscillator raising and lowering operators for negative or positive m, respectively. The oscillator ground state $|0\rangle$ is defined to be annihilated by the α_m for $m > 0$. Actually, specifying that the oscillators are in their ground states does not completely determine a state of the string. The other degree of freedom is the center-of-mass momentum p^μ. When we wish to refer specifically to a state annihilated by the α_m with $m > 0$, and which has center-of-mass momentum p^μ, we will call it $|0; p^\mu\rangle$.

The α_m are related to conventionally normalized harmonic oscillator operators by

$$\alpha_m^\mu = \sqrt{m}\, a_m^\mu, \qquad m > 0 \tag{2.2.11}$$

$$\alpha_{-m}^\mu = \sqrt{m}\, a_m^{\mu\dagger}, \qquad m > 0. \tag{2.2.12}$$

An elementary point, of fundamental importance, is that the Fock space built up by applying the raising operators $a_m^{\mu\dagger}$ to the ground state $|0\rangle$ is not positive definite. The time components have an unusual minus sign in their commutation relations, $[a_m^0, a_m^{0\dagger}] = -1$, and therefore a state of the form $a_m^{0\dagger}|0\rangle$ has negative norm since $\langle 0| a_m^0 a_m^{0\dagger}|0\rangle = -1$. The physical space of allowed string states is a subspace of the complete Fock space; this subspace is specified by certain subsidiary conditions. In order to have a sensible causal theory it is necessary that the physical subspace be free from negative-norm states, which are usually called 'ghosts'. [*]

---

[*] These ghosts should not be confused with BRST ghosts, which we will encounter later.

The subsidiary conditions of the classical theory were shown to correspond to the vanishing of the energy–momentum components T_{++} and T_{--}, whose Fourier modes give the Virasoro generators

$$L_m = \tfrac{1}{2} \sum_{-\infty}^{\infty} \alpha_{m-n} \cdot \alpha_n, \qquad (2.2.13)$$

as well as a similar expression $\tilde{L}_m$ in the case of closed strings. In the quantum theory the α_m are operators, so one must resolve ordering ambiguities. Since α_{m-n} commutes with α_n unless $m = 0$, the only such ambiguity arises in the expression for L_0. As we have at the moment no natural way to resolve this ambiguity, we simply *define* L_0 to be given by the normal-ordered expression

$$L_0 = \frac{1}{2} \alpha_0^2 + \sum_{n=1}^{\infty} \alpha_{-n} \cdot \alpha_n. \qquad (2.2.14)$$

Since an arbitrary constant could have been present here, we must add a to-be-determined constant to all formulas containing L_0. In the classical theory an important example of the imposition of the constraints is the condition that L_0 must vanish for the allowed motions of the string. It is this condition that gives a formula for the mass. The most naive quantum mechanical analog of that requirement would be the statement that L_0 should annihilate physical states. Because of the normal-ordering ambiguity, we include an undetermined constant and say that a physical state $|\phi\rangle$ must satisfy

$$(L_0 - a)|\phi\rangle = 0, \qquad (2.2.15)$$

where a will be determined later. As we learned in the classical theory, this equation determines the mass of a string state in terms of its internal state of oscillation. In fact, (2.2.15) states more explicitly in the case of open strings that (with $\alpha' = 1/2$)

$$M^2 = -2a + 2 \sum_{n=1}^{\infty} \alpha_{-n} \cdot \alpha_n \qquad (2.2.16)$$

showing that the oscillator ground state has mass squared $-2a$, and excitations have mass squared larger than this by any multiple of 2. For closed strings the conditions $(L_0 - a) |\phi\rangle = (\tilde{L}_0 - a) |\phi\rangle = 0$ give

$$M^2 = -8a + 8 \sum_{n=1}^{\infty} \alpha_{-n} \cdot \alpha_n = -8a + 8 \sum_{n=1}^{\infty} \tilde{\alpha}_{-n} \cdot \tilde{\alpha}_n \qquad (2.2.17)$$

The fact that the mass squared of the ground state for closed strings is four times that for open strings was explained in another way in chapter 1. Subtracting the two formulas in (2.2.17), or equivalently imposing the condition $(L_0 - \tilde{L}_0) |\phi\rangle = 0$, we learn that

$$\sum_{n=1}^{\infty} \alpha_{-n} \cdot \alpha_n = \sum_{n=1}^{\infty} \tilde{\alpha}_{-n} \cdot \tilde{\alpha}_n \qquad (2.2.18)$$

Of all of the constraint equations, this is the only one that couples left-moving modes to right-moving modes. Physical states are found by choosing independently the left-moving and right-moving states of oscillation, subject to this one constraint.

The other L_m and $\tilde{L}_m$ correspond to terms of definite nonzero frequency in T_{++} and T_{--}. Just as in the Gupta–Bleuler treatment of electrodynamics, the vanishing of these terms in the classical theory is replaced in the quantum theory by the weaker requirement that the positive frequency components annihilate a physical state, *i.e.*,

$$L_m |\phi\rangle = 0 \quad m = 1, 2, \ldots . \qquad (2.2.19)$$

This is sufficient to ensure that with a suitable ordering convention, the operators $L_m - a\delta_m$ for both positive and negative m have vanishing matrix elements between pairs of physical states. To see this, let $|\phi\rangle$ and $|\chi\rangle$ be two physical states, obeying (2.2.15) and (2.2.19). Consider an expression

$$\langle \chi | L_{n_1} L_{n_2} \ldots L_{n_p} | \phi \rangle \qquad (2.2.20)$$

(If one of the terms in (2.2.20) has $n_k = 0$, the corresponding L_0 should be replaced by $L_0 - a$.) Since the L_n do not commute (and we will shortly find a quantum anomaly that will bring a c-number term in their commutators), the value of (2.2.20) depends on the operator ordering. If, however, the L_{n_k} are to the right for n_k positive and to the left for n_k negative, then (2.2.20) vanishes by virtue of the physical state conditions and the hermiticity property $L_{-m} = L_m^\dagger$. This is the closest we can come at the quantum level to the classical statement that the L_n are all zero for allowed classical motions of the string. The anomalous commutation relations of the L_n make it impossible to find states annihilated by all of them.

There is no ordering ambiguity in the angular momentum operators

$$J^{\mu\nu} = x^\mu p^\nu - x^\nu p^\mu - i \sum_{n=1}^{\infty} \frac{1}{n}(\alpha^\mu_{-n}\alpha^\nu_n - \alpha^\nu_{-n}\alpha^\mu_n) \qquad (2.2.21)$$

These were introduced in (2.1.67) – (2.1.71). Since there is no ordering problem, they can be interpreted unambiguously as quantum operators. It is not too difficult to use the canonical commutation relations to verify that the Poincaré algebra

$$[p^\mu, p^\nu] = 0 \qquad (2.2.22)$$

$$[p^\mu, J^{\nu\rho}] = -i\eta^{\mu\nu}p^\rho + i\eta^{\mu\rho}p^\nu \qquad (2.2.23)$$

$$[J^{\mu\nu}, J^{\rho\lambda}] = -i\eta^{\nu\rho}J^{\mu\lambda} + i\eta^{\mu\rho}J^{\nu\lambda} + i\eta^{\nu\lambda}J^{\mu\rho} - i\eta^{\mu\lambda}J^{\nu\rho} \qquad (2.2.24)$$

is satisfied. Indeed, the construction of the Poincaré generators from Noether currents guarantees that the Poincaré algebra is obeyed classically (with Poisson brackets instead of commutators, of course), and no real possibility of an anomaly arises in checking the algebra quantum mechanically. Since $[L_n, J^{\mu\nu}] = 0$, the physical state conditions are invariant under Lorentz transformations, and the physical states are guaranteed to form Lorentz multiplets.

2.2.2 Virasoro Algebra and Physical States

As already explained, the Fock space built up with the oscillators α^μ_m (and $\tilde{\alpha}^\mu_m$) is not positive definite because of the negative metric in the commutation relations of the time components. However, the physical states correspond to the subspace satisfying the Virasoro conditions $(L_m - a\delta_m)|\phi\rangle = 0$ for $m \geq 0$. Since these conditions are in one-to-one correspondence with timelike oscillators their number is just sufficient to have a chance of leaving a positive-definite Fock space. To see this note that $L_m \sim p \cdot \alpha_m+$ *terms quadratic in oscillators*. If the quadratic terms were absent, the L_n conditions would (as p^μ is timelike, except for a few low-lying states) decouple the timelike modes. In the rest frame, the physical states would then be generated by the space components of the oscillators. This demonstrates that the counting of conditions is sufficient to have a chance of decoupling of ghosts. The quadratic terms do play an important role, however, and the truth is a good deal more subtle and interesting. A ghost-free spectrum is only possible for certain values of the constant a and the space-time dimension D. To investigate the matter more closely it is necessary to study the algebra of the Virasoro operators.

We have already worked out the classical form of the Virasoro algebra, namely

$$[L_m, L_n] = (m - n)L_{m+n}. \qquad (2.2.25)$$

Now we would like to compute whatever quantum mechanical correction may appear in (2.2.25). In our previous discussion, we obtained (2.1.84) by steps that are valid even at the quantum mechanical level. Moreover, as long as $m+n \neq 0$, the reasoning leading from (2.1.84) to (2.1.85) is valid quantum mechanically, so there is no quantum modification of (2.1.85) for $m + n \neq 0$. For $m + n = 0$, the two infinite sums in (2.1.84) each suffer from infinite normal-ordering ambiguities at the quantum level. Since each of these two infinite sums is ill-defined, the shift in the summation variable in one of these two sums to give (2.1.85) is a dangerous step. On the other hand, since any normal-ordering ambiguity that might arise in (2.1.85) for $m + n = 0$ would only involve a c-number, we are guaranteed to have

$$[L_m, L_n] = (m - n)L_{m+n} + A(m)\delta_{m+n}. \qquad (2.2.26)$$

where A_m is an m-dependent c-number. This generalized algebra is known as a central extension of the Virasoro algebra; the c-number term is called the anomaly term in that algebra. From (2.2.26), evidently $A(m) = -A(-m)$, and $A(0) = 0$, so it suffices to determine $A(m)$ for positive m.

To directly calculate $A(m)$ by studying the normal ordering of the two infinite sums in (2.1.84) is surprisingly tricky. The following approach is simpler and more useful. By studying the Jacobi identity $0 = [L_k, [L_n, L_m]] + [L_n, [L_m, L_k]] + [L_m, [L_k, L_n]]$ one learns for $k + n + m = 0$ that

$$(n - m)A(k) + (m - k)A(n) + (k - n)A(m) = 0. \qquad (2.2.27)$$

Setting $k = 1$ and $m = -n - 1$ in (2.2.27) gives

$$A(n + 1) = \frac{(n + 2)A(n) - (2n + 1)A(1)}{(n - 1)}. \qquad (2.2.28)$$

This recursion relation is enough to determine all of the $A(n)$ in terms of $A(1)$ and $A(2)$, so the general form of the $A(n)$ is determined in terms of two unknown coefficients. The general solution is in fact

$$A(m) = c_3 m^3 + c_1 m, \qquad (2.2.29)$$

where c_1 and c_3 are constants. This can indeed be verified to obey (2.2.27). The constant c_1 can be changed by shifting the definition of L_0 by a constant (an operation that otherwise does not disturb the Virasoro algebra).

Shifting L_0 by a constant would also shift the constant a in (2.2.15), so it is only the relation between a and c_1 that has an invariant meaning.

The commutator of L_m and L_{-m} has to be evaluated very carefully to obtain the anomaly contribution correctly. The simplest and safest method to determine c_1 and c_3 is to calculate the expectation value of $[L_m, L_{-m}]$ in a suitable state. The most convenient choice of this state is an oscillator ground state $|0; 0\rangle$ with $p^\mu = 0$. For $m = 1$ we find

$$\langle 0; 0| [L_1, L_{-1}] |0; 0\rangle = 0, \qquad (2.2.30)$$

since every term in L_1 or L_{-1} annihilates a zero momentum ground state. But for $m = 2$ we obtain

$$
\begin{aligned}
\langle 0; 0|[L_2, L_{-2}]|0; 0\rangle &= \langle 0; 0|L_2 L_{-2}|0; 0\rangle \\
&= \frac{1}{4}\langle 0; 0|\alpha_1 \cdot \alpha_1 \alpha_{-1} \cdot \alpha_{-1}|0; 0\rangle \\
&= \tfrac{1}{2}\eta_{\mu\nu}\langle 0; 0|\alpha_1^\mu \alpha_{-1}^\nu|0; 0\rangle = \tfrac{1}{2}\eta_{\mu\nu}\eta^{\mu\nu} = \tfrac{1}{2}D.
\end{aligned}
\qquad (2.2.31)
$$

This is sufficient information to determine c_1 and c_3, obtaining

$$A(m) = \frac{1}{12}D(m^3 - m). \qquad (2.2.32)$$

Notice that the structure of the Virasoro algebra and of the anomaly is such that L_1, L_0 and L_{-1} generate a closed subalgebra, without anomaly, isomorphic to $SU(1,1)$ or $SL(2,R)$.

Let us next make a preliminary investigation of the conditions that ensure that there are no negative-norm physical states. The result will be that there are negative-norm states for certain regions of the parameter a and the space-time dimension D and not for others. To delineate the regions of the parameters a and D in which there are no negative-norm states in the physical Hilbert space, it is very useful to look for physical states of zero norm. If one varies a and D to cross from a region where the physical Hilbert space is positive semi-definite to a region where it has negative-norm states, extra physical states of zero norm are always present at the boundaries between the two regions. It turns out, for reasons that we will discuss, that these extra states of zero norm are related to important physical principles and that the most interesting case is the 'critical' case in which the physical Hilbert space is on the verge of developing ghosts.

The discussion will be presented for open strings, but the closed-string story is almost identical, with a doubling of the α oscillators and the

Virasoro conditions. In fact, the physical states at the nth closed-string mass level can be expressed as a tensor product of a Hilbert space of physical states made from left-moving oscillators with a Hilbert space of physical states made from right-moving oscillators, except for the one constraint (2.2.18). The left-moving and right-moving physical spaces are equivalent to the physical space of an open string, so we may as well search for states of negative norm in this case.

Let us denote the open-string ground state of momentum k^μ as $|0; k\rangle$. The mass-shell condition $L_0 = a$ implies that $\alpha' k^2 = a$. Now consider states at the first excited level. These are given by $\zeta \cdot \alpha_{-1} |0; k\rangle$ where $\zeta^\mu(k)$ is a polarization vector with D independent components before the gauge constraints are taken into account. The mass-shell condition now gives $\alpha' k^2 = a - 1$, and the L_1 subsidiary condition implies that $\zeta \cdot k = 0$. This condition leaves $D - 1$ allowed polarizations. The norm of these states is given by $\zeta \cdot \zeta$. If we choose the vector k to lie in the $(0, 1)$ plane then the $(D - 2)$ states with (spacelike) polarizations normal to that plane obviously have positive norms. If, on the other hand, a is such that the first excited state is a tachyon, with $k^2 > 0$, then k can be chosen to have no time component. Then the last ζ is timelike and has negative norm. If $k^2 < 0$, k can be chosen to have only a time component and the last ζ is spacelike with positive norm. Finally, if $k^2 = 0$ the last ζ is proportional to k and has zero norm. Thus we obtain the first condition for the absence of ghosts

$$a \leq 1. \tag{2.2.33}$$

In the boundary case ($a = 1$) the vector particle is massless and the scalar ground state is a tachyon. In this case, the L_1 subsidiary condition corresponds to the covariant gauge condition $\partial_\mu A^\mu = 0$ of electrodynamics. Just as in the covariant Gupta–Bleuler quantization of electrodynamics, this condition leaves $D - 2$ positive-norm states with transverse polarization, and one longitudinal state $\zeta^\mu = k^\mu$ of zero norm; and one must aim to show that this zero norm state decouples from the S matrix. In field theory this follows from gauge invariance and current conservation; in string theory it can be proved (and we indeed sketched one approach in our discussion of gravitational Ward identities in the introduction), but the deeper structure of which it is a manifestation is as yet imperfectly understood.

The 'null' state that appears at the first excited level when $a = 1$ is just the first of an infinite number of such states. The result found at the first excited level can be generalized by the following considerations. An arbitrary state $|\phi\rangle$ is called a *physical state* if it satisfies the constraints

$L_m|\phi\rangle = 0$ for $m > 0$ and $(L_0 - a)|\phi\rangle = 0$. A state $|\psi\rangle$ which obeys $(L_0 - a)|\psi\rangle = 0$ is called a *spurious* state if it is orthogonal to *all* physical states, *i.e.*,

$$\langle\phi|\psi\rangle = 0, \qquad (2.2.34)$$

for all physical states $|\phi\rangle$. Spurious states can always be written in the form

$$|\psi\rangle = \sum_{n>0} L_{-n}|\chi_n\rangle, \qquad (2.2.35)$$

where $|\chi_n\rangle$ is some state that obeys

$$(L_0 - a + n)|\chi_n\rangle = 0. \qquad (2.2.36)$$

Actually, the infinite series in (2.2.35) can be truncated, since the L_{-n} for $n \geq 3$ can be represented as iterated commutators of L_{-1} and L_{-2}, e.g. $L_{-3} \sim [L_{-1}, L_{-2}]$. So we can simply write a spurious state as

$$|\psi\rangle = L_{-1}|\chi_1\rangle + L_{-2}|\chi_2\rangle, \qquad (2.2.37)$$

where $|\chi_1\rangle$ and $|\chi_2\rangle$ obey (2.2.36). States of the form (2.2.37) are orthogonal to physical states, since

$$\langle\phi|\psi\rangle = \sum_{m=1}^{2}\langle\phi|L_{-m}|\chi_m\rangle = \sum_{m=1}^{2}\langle\chi_m|L_m|\phi\rangle^* = 0. \qquad (2.2.38)$$

To see that a spurious state must be expressible in the form (2.2.35) or (2.2.37), note that if $|\psi\rangle$ is a spurious state, then the operator $O = |\psi\rangle\langle\psi|$ annihilates all physical states. Since the only restriction on general physical states is that they are annihilated by L_m for $m > 0$, this implies that O can be written

$$O = \sum_{n>0} X_{-n}L_n \qquad (2.2.39)$$

for some operators X_{-n}. With $O = |\psi\rangle\langle\psi|$, (2.2.39) implies that $|\psi\rangle$ has a representation of the form (2.2.35).

Something special arises when a state $|\psi\rangle$ is both spurious *and* physical, *i.e.*,

$$\langle\phi|\psi\rangle = 0, \qquad L_m|\psi\rangle = 0, \quad m > 0, \qquad (L_0 - a)|\psi\rangle = 0. \qquad (2.2.40)$$

By (2.2.35) such states have zero norm since

$$\langle\psi|\psi\rangle = \sum_{m>0}\langle\chi_m|L_m|\psi\rangle = 0. \qquad (2.2.41)$$

These states are orthogonal to all physical states, including themselves (and are sometimes said to be 'null' physical states).

We can construct states of this type by considering spurious states of the form

$$|\psi\rangle = L_{-1}|\tilde{\chi}\rangle, \qquad (2.2.42)$$

where $|\tilde{\chi}\rangle$ is an arbitrary state satisfying $L_m|\tilde{\chi}\rangle = 0$ for $m > 0$ and $(L_0 - a + 1)|\tilde{\chi}\rangle = 0$. Here $|\tilde{\chi}\rangle$ could be the zero momentum state $|0;0\rangle$ or for that matter any physical state with suitably shifted p^μ. In addition to being spurious, the state $|\psi\rangle$ also satisfies all the conditions for being physical apart from the L_1 condition. The action of L_1 is given by

$$L_1|\psi\rangle = L_1 L_{-1}|\tilde{\chi}\rangle = 2L_0|\tilde{\chi}\rangle, \qquad (2.2.43)$$

which vanishes for $a = 1$. In this case the $|\psi\rangle$ states are both spurious and physical and hence have zero norm by our general argument. Evidently, by applying L_{-1} to an arbitrary state $|\tilde{\chi}\rangle$ an infinite number of zero-norm states can be made this way. The massless vector state considered earlier is just the simplest example, in which $|\tilde{\chi}\rangle = |0;k\rangle$.

The number of zero-norm states increases dramatically, however, in $D = 26$ dimensions. This is discovered by considering spurious states having the structure

$$|\psi\rangle = (L_{-2} + \gamma L_{-1}^2)|\tilde{\chi}\rangle, \qquad (2.2.44)$$

where now we take $a = 1$ and require that $L_m|\tilde{\chi}\rangle = 0$ for $m > 0$ and $(L_0 + 1)|\tilde{\chi}\rangle = 0$, so that $(L_0 - 1)|\psi\rangle = 0$. For $|\psi\rangle$ to have zero norm it must be physical and, in particular, it should be annihilated by L_m for $m > 0$. Since it is trivially annihilated by L_m with $m \geq 3$ we need only consider whether it is possible to impose the conditions $L_1|\psi\rangle = L_2|\psi\rangle = 0$ making use of the Virasoro algebra with the anomaly (2.2.32). This gives the equations $3 - 2\gamma = 0$ and $\frac{1}{2}D - (4 + 6\gamma) = 0$, which imply that $\gamma = 3/2$ and $D = 26$. Thus for $D = 26$ (with $a = 1$) there are many more zero-norm states of the form

$$\left(L_{-2} + \frac{3}{2}L_{-1}^2\right)|\tilde{\chi}\rangle. \qquad (2.2.45)$$

Unlike the first infinite class of zero-norm states, the norm of states of the type (2.2.45) is zero when and only when $D = 26$.

The first example of a state of the form (2.2.44) is

$$(L_{-2} + \frac{3}{2}L_{-1}^2) \, |0; p\rangle$$

$$= [\frac{1}{2}\alpha_{-1} \cdot \alpha_{-1} + \frac{5}{2}p \cdot \alpha_{-2} + \frac{3}{2}(p \cdot \alpha_{-1})^2] \, |0; p\rangle \,, \tag{2.2.46}$$

where $p^2 = -2$. This state has norm $(D - 26)/2$, which vanishes for $D = 26$, as expected. The fact that it has negative norm for $D < 26$ does not matter, since it then does not obey the physical state conditions.

Physical states of negative norm can be constructed for $D > 26$. As an example, consider states of the form

$$|\phi\rangle = [c_1\alpha_{-1} \cdot \alpha_{-1} + c_2 p \cdot \alpha_{-2} + c_3(p \cdot \alpha_{-1})^2] \, |0; p\rangle \,, \tag{2.2.47}$$

with $p^2 = -2$ so that $(L_0 - 1)|\phi\rangle = 0$. Such states obey $L_1 |\phi\rangle = L_2 |\phi\rangle = 0$ for

$$c_2 = c_1 \frac{D - 1}{5} \quad \text{and} \quad c_3 = c_1 \frac{D + 4}{10}. \tag{2.2.48}$$

In this case the norm is

$$\langle \phi | \phi \rangle = \frac{2c_1^2}{25}(D - 1)(26 - D), \tag{2.2.49}$$

so that we find ghosts in the physical spectrum for $D > 26$.

The general rule, which will be proved later, is that the spectrum is ghost-free provided that $a = 1$ and $D = 26$ or $a \leq 1$ and $D \leq 25$. In the former case there are many zero-norm states and the physical spectrum has as many propagating modes as would be generated by 24 sets of α oscillators, whereas in the latter case there are far fewer zero-norm states, and the physical spectrum corresponds to $D - 1$ sets of oscillators. One may say that in the $a = 1$, $D = 26$ case the string has only transverse excitations, whereas in the $a \leq 1$, $D \leq 25$ case it possesses longitudinal modes as well. These facts will emerge with the proof of the no-ghost theorem.

It is impossible to prove just from study of the free theory in the formalism under discussion here that D must equal 26. The reason for this is that if the theory has no ghosts for $D = 26$, then, by studying the subspace of states in which the α_m^μ oscillators are in their ground state for $\mu = 25,^*$ we recover the physical states of the $D = 25$ theory, which

---

* Recall that μ is a space-time index running from 0 to 25.

must also be free of ghosts.[†] The most that we can expect to show at
tree level in this formalism is that $D = 26$ is the most natural case, with
$D < 26$ being an arbitrary truncation of a theory that really belongs in
26 dimensions. The discovery above of extra zero-norm states that only
arise for $D = 26$ is the first sign of this. For these extra zero-norm states
have a very particular significance. Like the longitudinal mode of the
massless vector meson discussed earlier, zero-norm states in a physically
sensible theory must be decoupled from the S matrix by some underlying
principle analogous to gauge invariance in field theory. So the occurrence
of extra zero-norm states for $D = 26$ suggests that this theory has an
enlarged gauge invariance and may be the most interesting one to study.
Likewise, the occurrence of an infinite series of zero-norm states precisely
for $a = 1$ suggests that this is the interesting case. So we will tentatively
think of the open-string ground state as a tachyon of mass squared -2,
corresponding to $a = 1$, and the first excited level as a massless vector
meson. The existence of this massless gauge particle is one aspect of the
very special properties of string theory in the critical dimension.

2.2.3 Vertex Operators

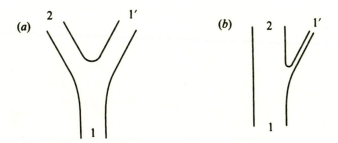

Figure 2.4. When one open string splits into two, as in (a), all of them may in general
be off mass shell. If one of the final states is on mass shell, as in (b), it has a width of
order $\hbar$ and is in some ways pointlike.

As we described in the introduction, the basic open-string interaction
can be viewed as a process in which a single string breaks into two or else

---

[†] While the arbitrary truncation to $D < 26$ does not introduce ghosts at tree level,
it gives trouble with unitarity at one loop level, and this is how the restriction to
$D = 26$ was first discovered.

two strings join to give a single one. In general, as in fig. 2.4a, we can contemplate a string interaction in which all three participating strings are off mass shell. However, it is interesting to consider the case in which, as in fig. 2.4b, one of the three strings is a physical on-shell mass eigenstate. In the figure, we sketch a process $1 \to 1' + 2$ in which string state 2 is a mass eigenstate. 1 and $1'$ may or may not be mass eigenstates. The very concept of a mass eigenstate of the string is a quantum mechanical concept, not a classical concept, and if we restore Planck's constant in all formulas, a mass eigenstate has a width and mass squared that are of order $\hbar$. Thus, in the classical limit any mass eigenstate of the string is in some ways similar to a point particle. In the $1 \to 1'$ transition with emission of the on-shell state 2, the quantum state of $1'$ must be related to that of 1 by some linear transformation that depends on the state of string 2 – linear because we are discussing quantum mechanics. With 2 being pointlike, it is natural to suspect that string $1'$ is obtained from string 1 by the action of a local operator at the end of the string where 2 was emitted. The local operator is usually called V_2, the vertex operator for emission of the on-shell state 2. This heuristic argument leads us to suspect that we should associate to every on-shell physical state $|\phi\rangle$ a vertex operator V_ϕ with suitable properties, something we argued in a different way in chapter 1. Our goal in discussing vertex operators here is actually not to analyze interactions, which will be the subject of later chapters, but to develop a tool for analyzing the spectrum of physical states. The discussion will actually be a useful complement to what was said in chapter 1. It will be sufficient for our purposes here to concentrate on open strings, though vertex operators are certainly equally important in the theory of closed strings.

Consider a local operator $A(\sigma, \tau)$ in the open-string Hilbert space. We set $\sigma = 0$ (or $\sigma = \pi$) to study this operator at the string endpoint, and for brevity we refer to $A(0, \tau)$ as $A(\tau)$. As $L_0 - a$ is the string Hamiltonian, we have

$$A(\tau) = e^{i\tau L_0} A(0) e^{-i\tau L_0} \qquad (2.2.50)$$

We are interested in operators $A(\tau)$ that are transformed into themselves by the Virasoro algebra. $A(\tau)$ is defined to have conformal dimension J if and only if under an arbitrary change of variables $\tau \to \tau'(\tau)$, $A(\tau)$ transforms to

$$A'(\tau') = \left(\frac{d\tau}{d\tau'}\right)^J A(\tau). \qquad (2.2.51)$$

Equation (2.2.51) is equivalent to the definition used in §1.4.5 (and the footnote in §1.5.2), where the dimension of an operator was extracted from its two-point function. Considering the case of an infinitesimal transformation,

$$\tau \to \tau' = \tau + \epsilon(\tau), \tag{2.2.52}$$

the transformation law of a field of conformal dimension J is thus

$$\delta A(\tau) = -\epsilon \frac{dA}{d\tau} - JA\frac{d\epsilon}{d\tau}. \tag{2.2.53}$$

The L_m's generate transformations (2.2.52) with $\epsilon = ie^{im\tau}$, so the condition for A to have conformal dimension J is

$$[L_m, A(\tau)] = e^{im\tau}\left(-i\frac{d}{d\tau} + mJ\right)A(\tau). \tag{2.2.54}$$

If $A(\tau)$ has an expansion in Fourier modes of the form

$$A(\tau) = \sum_{m=-\infty}^{\infty} A_m e^{-im\tau}. \tag{2.2.55}$$

then this condition becomes for the Fourier modes

$$[L_m, A_n] = [m(J-1) - n]A_{m+n}. \tag{2.2.56}$$

This rule can easily be verified to be compatible with the Virasoro algebra and the Jacobi identity. By this definition, for example, the string coordinate $X^\mu(\tau)$ has $J = 0$ and the momentum operator $\dot{X}^\mu(\tau)$ has $J = 1$. (Later, when we discuss Faddeev–Popov ghosts, we will see that the ghost coordinate c has $J = -1$ and the antighost b has $J = 2$.) Operators that transform as in (2.2.54) for some definite J are said to have definite conformal dimension; they are the ones that transform 'nicely' under the Virasoro algebra.

Operators of definite conformal dimension are rather special. It is by no means true that every operator can be expanded as a linear combination of operators of definite conformal dimension.

For our present task of analyzing the physical states, the utility of introducing operators of definite conformal dimension is that these operators can be used to build new physical states from old ones. Indeed, if $|\phi\rangle$ is a physical state, $(L_n - a\delta_n)|\phi\rangle = 0$, $n \geq 0$, and $A(\tau)$ has conformal dimension $J = 1$, then it is easily seen that $[L_m, A_0] = 0$ and therefore that

$$|\phi'\rangle = A_0 |\phi\rangle \tag{2.2.57}$$

is also a physical state. Since the vertex operator associated with emission of a mass eigenstate 2 should certainly map a physical initial state 1 to a physical final state $1'$, (2.2.57) might suggest that open-string vertex operators should be operators of conformal dimension one. We indeed learned in another way in the introduction that open-string vertex operators are local operators of dimension one.

The vertex operator $V(k, 0, \tau) = V(k, \tau)$ for emission at time τ and $\sigma = 0$ of a physical state of momentum $-k^\mu$ or absorption of a physical state of momentum $+k^\mu$, must, among other things, change the momentum of whatever state it acts on by an amount k^μ. Thus its dependence on the center-of-mass coordinate of the string must consist of a factor $e^{ik \cdot x(\tau)}$, where

$$x^\mu(\tau) = x^\mu + p^\mu \tau \qquad (2.2.58)$$

is the center-of-mass position of the string at time τ. The obvious way to achieve this in the string context is to include in $V(k, \tau)$ a factor $\exp[ik \cdot X(0, \tau)]$. Indeed, it is natural that a string that absorbs a mass eigenstate of momentum k^μ at world-sheet position $(0, \tau)$ and space-time position $X^\mu(0, \tau)$ should have its wave function modified by a factor $\exp[ik \cdot X(0, \tau)]$. If the absorbed or emitted string state has no particular quantum numbers except for its momentum, then we might try simply $V(k, \tau) = \exp[ik \cdot X(0, \tau)]$ as a vertex operator. This expression actually requires normal ordering. Let us study the normal-ordered expression

$$V(k, \tau) =: e^{ik \cdot X(0, \tau)} :$$
$$= \exp\left(k \cdot \sum_{n=1}^\infty \frac{\alpha_{-n}}{n} e^{in\tau}\right) e^{ik \cdot x(\tau)} \exp\left(-k \cdot \sum_{n=1}^\infty \frac{\alpha_n}{n} e^{-in\tau}\right).$$
$$(2.2.59)$$

$x^\mu(\tau)$ was defined in (2.2.58). The exponent in (2.2.59) differs by the divergent sum $\alpha' k^2 \sum \frac{1}{n}$ from what it would be without normal ordering. (Thus normal ordering has no effect in the special case $k^2 = 0$.)

We would now like to compute the conformal dimension of $V(k, \tau)$. Since $X^\mu(\tau)$ has conformal dimension $J = 0$, one might expect that products such as $X^\mu(\tau) X^\nu(\tau)$ and more generally composite operators of the form $f(X^\mu(\tau))$ would also have conformal dimension zero. In fact, it follows naively from (2.2.56) that if $A_1(\tau)$ has conformal dimension J_1 and $A_2(\tau)$ has conformal dimension J_2, then $A_1(\tau) A_2(\tau)$ has conformal dimension $J_1 + J_2$. This is indeed true whenever $A_1(\tau) A_2(\tau)$ is unambiguously well-defined without any subtraction or renormalization other than what is required to define $A_1(\tau)$ and $A_2(\tau)$ separately, or in other words whenever there is no short-distance singularity in the operator product $A_1(\tau) A_2(\tau')$ as $\tau' \to \tau$. In fact, typical normal-ordered products such as

: $X^\mu(\tau)X_\mu(\tau)$: have no definite conformal dimension. On the other hand, $V(k,\tau)$ in (2.2.59) does have a definite conformal dimension.

The conformal dimension of $V(k,\tau)$ can be determined by straightforward oscillator manipulations, provided that one is careful in keeping track of ordering effects. In order to evaluate the commutator $[L_m, V(k,\tau)]$, we begin by noting that

$$[\alpha_p^\mu, e^{k \cdot \alpha_{-n}}] = p\delta_{p-n}k^\mu e^{k \cdot \alpha_{-n}}. \qquad (2.2.60)$$

Using the expression, $L_m = \frac{1}{2}\sum_q \alpha_{m-q} \cdot \alpha_q$, it is easy to show that

$$[L_m, e^{k \cdot \alpha_{-n}}] = \frac{1}{2}\sum_q \{\alpha_{m-q} \cdot [\alpha_q, e^{k \cdot \alpha_{-n}}] + [\alpha_{m-q}, e^{k \cdot \alpha_{-n}}] \cdot \alpha_q\}$$
$$= \frac{1}{2}n\{k \cdot \alpha_{m-n}, e^{k \cdot \alpha_{-n}}\}. \qquad (2.2.61)$$

In applying this to the evaluation of $[L_m, V]$ (with V given by (2.2.59)) suppose that $m > 0$ (the argument is the same for $m < 0$). Neglecting the issue of normal-ordering, the calculation is easily seen to give a result of the form of the (2.2.54) with $J = 0$ (and with A replaced by V). However, V is defined to be the normal-ordered expression : $\exp[ik \cdot X]$:, and this means that its derivative $dV/d\tau$ is also automatically normal-ordered. When one uses (2.2.61) to obtain an expression for $[L_m, V]$, one arrives at an expression that is not normal ordered. Of the infinite number of terms in the calculation (one for each term in the infinite product (2.2.59)), a finite number do not arise in a normal-ordered form. Those particular terms have lowering operators to the left of the raising operators in V and are given by

$$(\frac{1}{2}\sum_{n=1}^{m} k \cdot \alpha_{m-n}e^{in\tau})V(k,\tau). \qquad (2.2.62)$$

Normal ordering this expression gives the commutator contribution

$$[\frac{1}{2}\sum_{n=1}^{m} k \cdot \alpha_{m-n}e^{in\tau}, V(k,\tau)] = \frac{1}{2}\sum_{n=1}^{m} k^2 e^{im\tau}V(k,\tau)$$
$$= \frac{1}{2}mk^2 e^{im\tau}V(k,\tau). \qquad (2.2.63)$$

Thus, the result is

$$[L_m, V(k,\tau)] = e^{im\tau}\left(-i\frac{d}{d\tau} + \frac{1}{2}mk^2\right)V(k,\tau). \qquad (2.2.64)$$

As a result, comparing with (2.2.54), we deduce that $J = k^2/2$. In chapter 1 we computed the anomalous dimension of the operator $\exp[ik \cdot X]$

by computing its two-point function. We obtained $J = k^2/4$ for closed strings and $J = k^2/2$ for insertion on the boundary of an open string. It is reassuring to note the value $J = k^2/2$ that we obtain here by oscillator methods agrees with the latter open-string result.

The vertex $V_0(k)$ is a physical vertex operator with conformal dimension $J = 1$ provided that $k^2 = 2$. It actually is the proper vertex for emission of the ground-state tachyon, whose mass squared we have tentatively assigned as $M^2 = -2$.

The only case in which $V(k, \tau)$ requires no normal ordering is the case $k^2 = 0$; in this case the conformal dimension is $J = 0$. The condition $k^2 = 0$ is correct for a massless vector meson, but $J = 0$ is the wrong conformal dimension for a vertex operator. Since $dX^\mu/d\tau$ has conformal dimension one, one might try to interpret

$$V_\zeta(k, \tau) = \zeta \cdot \frac{dX}{d\tau} \exp[ik \cdot X] \qquad (2.2.65)$$

as the vertex operator for emission of a massless meson of polarization $\zeta^\mu(k)$. The operator product of $\zeta \cdot dX/d\tau$ and $\exp[ik \cdot X]$ in (2.2.65) is free of short-distance singularity, ensuring that V has $J = 1$, provided that $k \cdot \zeta = 0$. This restriction on the allowed polarization of a vector particle is familiar from electrodynamics and was encountered earlier in analyzing the physical state spectrum. Its appearance here is an illustration of the fact that vertex operators of conformal dimension one are in one to one correspondence with physical states. The vertex operators for the other states in the spectrum are more complicated. For a state with $\alpha' k^2 = -n$ it is generically of the form : $f(\dot{X}, \ddot{X}, \ldots) \exp[ik \cdot X]$:, where the total number of τ derivatives adds up to n. But there are additional restrictions needed to achieve $J = 1$.

Vertex operators for states of zero norm can be described as follows. Suppose $W(k, \tau)$ is an operator of conformal dimension 0 containing the factor : $\exp[ik \cdot X]$:. Then $V(k, \tau) = -idW(k, \tau)/d\tau = [L_0, W(k, \tau)]$ has $J = 1$, as the reader should verify. If we pick $W(k, \tau) = \exp[ik \cdot X]$ with $k^2 = 0$, for example, then $V(k, \tau)$ can be seen to be the vertex operator for emission of a massless vector meson with longitudinal polarization $\zeta^\mu = k^\mu$. Indeed, vertex operators of the form $V(\tau) = -idW/d\tau$, where W has $J = 0$, always describe emission of states of zero norm. At the risk of jumping rather ahead of our story, we might note that the reason that states of zero-norm decouple is that, with V being a total τ derivative, the zero-frequency component V_0, which as in (2.2.57) maps physical states to physical states, vanishes.

As a further illustration, consider emission vertices for the second excited level, states with $\alpha' k^2 = -1$. The factor $V_0(k)$ now has $J = -1$, and thus

$$\zeta^{\mu\nu} \dot{X}_\mu \dot{X}_\nu : \exp[ik \cdot X] : \qquad (2.2.66)$$

has $J = 1$ provided that there are no short-distance singularities in the operator product in (2.2.66). This is the case if $k_\mu \zeta^{\mu\nu} = \text{tr}\,\zeta = 0$. These are exactly the conditions for $\zeta^{\mu\nu}(k)$ to be the polarization tensor of a massive spin-two state, *i.e.*, a symmetric traceless tensor of $SO(D-1)$. One might suppose that it is possible at the same mass level to make vertex operators $Y_{k,\eta} = \eta_\mu \dot{X}^\mu \exp[ik \cdot X]$ for physical states of spin one and polarization η_μ. However, if $\eta_\mu k^\mu = 0$, the total τ derivative

$$-i\frac{d}{d\tau}(\eta \cdot \dot{X} : e^{ik \cdot X} :) = (-i\eta \cdot \ddot{X} + \eta \cdot \dot{X} k \cdot \dot{X}) : e^{ik \cdot X} :, \qquad (2.2.67)$$

describes emission of the zero-norm states $L_{-1}\eta \cdot \alpha_{-1} |0; k\rangle$ and accounts for $D-1$ of the possible components of Y. Therefore these components of Y differ from expressions of the form in (2.2.66) only by a total derivative that corresponds to a zero-norm emission. The remaining polarization corresponds to $\eta_\mu = k_\mu$. In the case of $D = 26$ this does not give a new physical emission vertex either, because it corresponds to a linear combination of the emission of a state of the form in (2.2.66) and the zero-norm state $(L_{-2} + \frac{3}{2} L_{-1}^2) |0; k\rangle$. Altogether, therefore, at the second excited mass level in 26 dimensions the only suitable open-string vertex operator is (2.2.66), which describes emission or absorption of a massive particle of 'spin two,' that is, a particle that transforms as a symmetric traceless second-rank tensor of $SO(25)$.

2.3   Light-Cone Gauge Quantization

In §2.2 we explored quantization of free bosonic strings in the covariant gauge $h_{\alpha\beta} = \eta_{\alpha\beta}$, imposing the Virasoro conditions as subsidiary constraints on physical states. However, as we pointed out, there is still a residual gauge symmetry that remains after setting $h_{\alpha\beta} = \eta_{\alpha\beta}$ and can be used to make further specific gauge choices. In fact, by making a particular noncovariant choice it becomes possible to solve the Virasoro constraint equations, and to describe the theory in a Fock space that describes physical degrees of freedom only. This formalism, which is somewhat analogous to unitary gauge in spontaneously broken gauge theories, was originally developed by Goddard, Goldstone, Rebbi, and Thorn in 1973.

The light-cone formalism, though not manifestly covariant, will be manifestly free of ghosts. By proving its equivalence to the covariant formalism, which is manifestly covariant though not manifestly free of ghosts, we will obtain a proof of the no-ghost theorem. There are many other reasons for describing the light-cone formalism here. Historically, it was light-cone quantization that first conclusively established that dual models were theories of strings. The light-cone picture is very 'physical' and also provides a useful framework for many calculations and for understanding the necessity of choosing $a = 1$ and $D = 26$.

2.3.1 Light-Cone Gauge and Lorentz Algebra

Recall that in the covariant gauge $h_{\alpha\beta} = \eta_{\alpha\beta}$ the string coordinates with open-string boundary conditions have mode expansions

$$X^\mu(\sigma, \tau) = x^\mu + p^\mu \tau + i \sum_{n \neq 0} \frac{1}{n} \alpha_n^\mu e^{-in\tau} \cos n\sigma, \qquad (2.3.1)$$

and satisfy the Virasoro subsidiary conditions $T_{++} = T_{--} = 0$. We have seen moreover in (2.1.44) and (2.1.45) that there is still a residual gauge symmetry. We would like to use this residual symmetry to impose an additional gauge condition, which will be noncovariant but quite convenient. We begin by introducing light-cone coordinates in space-time:

$$X^+ = (X^0 + X^{D-1})/\sqrt{2}, \quad X^- = (X^0 - X^{D-1})/\sqrt{2}. \qquad (2.3.2)$$

While these are rather similar to the light-cone coordinates $\sigma^\pm$ introduced earlier on the string world sheet, there is a big difference. In space-time, we have D coordinates in all, and (2.3.2) involves singling out two of them, namely X^0 and X^{D-1}, in an arbitrary and noncovariant way. On the world sheet, there are only two coordinates to begin with, and there is no arbitrary choice in defining $\sigma^\pm$. In a coordinate system in which the D space-time coordinates are $X^\pm$ and the remaining (transverse) space-like coordinates X^i, $i = 1 \ldots D - 2$, the nonzero components of the Minkowski metric are $\eta_{ij} = 1$, $\eta_{+-} = \eta_{-+} = -1$. In these coordinates, the components of a vector V^μ are

$$V^\pm = \frac{1}{\sqrt{2}}(V^0 \pm V^{D-1}) \qquad (2.3.3)$$

and V^i, $i = 1, \ldots, D - 2$. The inner product of the two vectors is

$$V \cdot W = V^i W^i - V^+ W^- - V^- W^+. \qquad (2.3.4)$$

Indices are raised and lowered by the rules $V^+ = -V_-$, $V^- = -V_+$, and

$V^i = V_i$.

What simplification can be achieved by using the residual gauge symmetry? In terms of $\sigma^{\pm}$, the residual invariance corresponds to the possibility of arbitrary reparametrizations

$$\sigma^+ \to \tilde{\sigma}^+(\sigma^+), \ \sigma^- \to \tilde{\sigma}^-(\sigma^-). \tag{2.3.5}$$

For closed strings σ^+ and σ^- are reparametrized independently, while for open strings they are linked by the boundary conditions. They transform $\tau = \frac{1}{2}(\sigma^+ + \sigma^-)$ and $\sigma = \frac{1}{2}(\sigma^+ - \sigma^-)$ into

$$\begin{aligned}
\tilde{\tau} &= \tfrac{1}{2}[\tilde{\sigma}^+(\tau + \sigma) + \tilde{\sigma}^-(\tau - \sigma)] \\
\tilde{\sigma} &= \tfrac{1}{2}[\tilde{\sigma}^+(\tau + \sigma) - \tilde{\sigma}^-(\tau - \sigma)].
\end{aligned} \tag{2.3.6}$$

The first equation in (2.3.6) asserts that $\tilde{\tau}$ may be an arbitrary solution of the free massless wave equation

$$\left(\frac{\partial^2}{\partial \sigma^2} - \frac{\partial^2}{\partial \tau^2} \right) \tilde{\tau} = 0. \tag{2.3.7}$$

On the other hand, once $\tilde{\tau}$ is chosen, $\tilde{\sigma}$ in (2.3.6) is completely determined (except for the possibility of making a rigid translation of σ in the case of closed strings). What is a natural way to choose the solution $\tilde{\tau}$ of the free massless wave equation? That is certainly an equation we have seen before; it is the equation obeyed (in conformal gauge) by the space-time coordinates $X^\mu(\sigma, \tau)$. Thus, our gauge freedom precisely corresponds to the fact that, if we wish, we can make a reparametrization so that $\tilde{\tau}$ will equal one of the X^μ. Light-cone gauge corresponds to the choice $\tilde{\tau} = X^+/p^+ + $ constant. This is usually expressed by saying that the light-cone gauge choice is

$$X^+(\sigma, \tau) = x^+ + p^+ \tau. \tag{2.3.8}$$

This corresponds in the classical description to setting the oscillator coefficients α_n^+ to zero for $n \neq 0$. The X^+ component of the string coordinates corresponds to the time coordinate as seen in a frame in which the string is traveling at infinite momentum. This choice of gauge has the conceptual advantage that every point on the string is at the same value of 'time' (since X^+ is independent of σ).

Having fixed $X^+(\sigma, \tau)$ according to (2.3.8), the Virasoro constraint equations $(\dot{X} \pm X')^2 = 0$ become

$$(\dot{X}^- \pm X'^-) = (\dot{X}^i \pm X'^i)^2/2p^+. \qquad (2.3.9)$$

This equation can be solved for X^- in terms of the X^i (with an unknown integration constant), so that actually in light-cone gauge both X^+ and X^- can be eliminated, leaving only the transverse oscillators X^i. Recalling the mode expansion of X^-, namely

$$X^- = x^- + p^- \tau + i \sum_{n \neq 0} \frac{1}{n} \alpha_n^- e^{-in\tau} \cos n\sigma, \qquad (2.3.10)$$

the explicit solution of (2.3.9) can be seen to be

$$\alpha_n^- = \frac{1}{p^+} \left(\frac{1}{2} \sum_{i=1}^{D-2} \sum_{m=-\infty}^{\infty} : \alpha_{n-m}^i \alpha_m^i : -a\delta_n \right). \qquad (2.3.11)$$

where, as in the covariant treatment, we introduce an unknown normal-ordering constant a in α_0^-. In light-cone gauge the identification of α_0^- with p^- is the mass-shell condition. Indeed, for $n = 0$ (2.3.11) is the formula

$$M^2 = (2p^+ p^- - p^i p^i) = 2(N - a), \qquad (2.3.12)$$

where

$$N = \sum_{n=1}^{\infty} \alpha_{-n}^i \alpha_n^i, \qquad (2.3.13)$$

This is the same mass-shell condition found in the covariant treatment, except that now only transverse oscillators contribute to N. The quantities $p^+ \alpha_n^-$ satisfy a Virasoro algebra

$$[p^+ \alpha_m^-, p^+ \alpha_n^-] = (m-n)p^+ \alpha_{m+n}^- + \left[\frac{D-2}{12}(m^3 - m) + 2am \right] \delta_{m+n}. \qquad (2.3.14)$$

The calculation is exactly analogous to our discussion of the Virasoro algebra in the covariant quantization.

These then are the basic formulas of light-cone quantization. We would now like to investigate whether the theory is really Lorentz invariant in this gauge. Naively it should be, since it was ostensibly obtained by gauge fixing of a Lorentz invariant underlying theory. If there is something wrong with the theory for some values of a and D, this will have the chance to show up as a failure of Lorentz invariance in light-cone gauge, where Lorentz invariance has not been explicitly maintained.

In the light-cone gauge all string excitations are generated by the transverse oscillators α_n^i. Thus, for example, the first excited state is given by $\alpha_{-1}^i |0; p\rangle$, which is a $(D-2)$-component vector representation of the transverse rotation group $SO(D-2)$. A transversely polarized vector, subjected to a Lorentz transformation, acquires a longitudinal polarization in general unless it is massless. This is just a statement of the well-known fact that the 'spin' of a massive particle is labeled by an irreducible representation of $SO(D-1)$, whereas a massless particle corresponds to an irreducible representation of $SO(D-2)$. (In the case of fermions one must use the covering groups spin$(D-1)$ and spin$(D-2)$.) Thus it is clear that the light-cone gauge cannot give a Lorentz-invariant string theory unless the vector state $\alpha_{-1}^i |0; p\rangle$ is massless, *i.e.*, the parameter a must equal 1.

Now we turn to the more difficult question of understanding the restriction on the space-time dimension D. First we consider a heuristic argument that uses the information just obtained that a must equal 1 for Lorentz invariance. Let us try to calculate the normal-ordering constant a directly. This normal-ordering constant arises from the formula

$$\frac{1}{2} \sum_{i=1}^{D-2} \sum_{n=-\infty}^{\infty} \alpha_{-n}^i \alpha_n^i = \frac{1}{2} \sum_{i=1}^{D-2} \sum_{n=-\infty}^{\infty} : \alpha_{-n}^i \alpha_n^i : + \frac{D-2}{2} \sum_{n=1}^{\infty} n. \quad (2.3.15)$$

Of course, the second sum in (2.3.15) diverges and must be regularized. One method of regularizing (2.3.15), which is used in similar normal-ordering problems in field theory, is the method of 'zeta function regularization'. One considers the more general sum

$$\sum_{n=1}^{\infty} n^{-s}. \quad (2.3.16)$$

For Re $s > 1$ this sum converges to a function known as the Riemann zeta function $\zeta(s)$. The zeta function has a unique analytic continuation to the point $s = -1$, where we get $\zeta(-1) = -1/12$. Inserting this 'value' of $\sum_{n=1}^{\infty} n$ in (2.3.15) gives the result

$$-\frac{D-2}{24} \quad (2.3.17)$$

for the normal-ordering constant in L_0. Since we know already that a should equal 1 for Lorentz invariance, this tells us that D should be 26. The zeta function regularization used in the above might appear rather formal, but in chapter 11 we will obtain the same answer from a more 'physical' method of regularizing the zero-point energy.

The above argument indicates, but perhaps rather heuristically, that Lorentz invariance of the light-cone formalism will require $D = 26$ as well as $a = 1$. We will now demonstrate rigorously that these conditions are necessary and sufficient for Lorentz invariance by a systematic study of the Lorentz generators $J^{\mu\nu}$. These were given earlier in the form

$$J^{\mu\nu} = l^{\mu\nu} + E^{\mu\nu}$$

$$l^{\mu\nu} = x^\mu p^\nu - x^\nu p^\mu$$

$$E^{\mu\nu} = -i \sum_{n=1}^{\infty} \frac{1}{n}(\alpha^\mu_{-n}\alpha^\nu_n - \alpha^\nu_{-n}\alpha^\mu_n).$$

(2.3.18)

It is clear that the effect of Lorentz transformations on the coordinates has to be quite subtle in the light-cone gauge since the choice of gauge is not Lorentz invariant. Certain of the transformations rotate the $+$ direction into the other directions so that it is necessary to perform a reparametrization (*i.e.*, a gauge transformation) in the transformed system in order to restore the gauge condition. We shall refer to this as a 'compensating' reparametrization. The transformations that affect X^+, and hence the gauge condition, are those generated by J^{+-} and J^{i-}. These are the transformations which may potentially have an anomaly; cancellation of this anomaly will give the restrictions on a and D. The remaining Lorentz generators are those associated with the transverse space which generate a $SO(D-2)$ subgroup that is a manifest symmetry of the light-cone gauge formalism.

Let us first consider the general expression for an infinitesimal Lorentz transformation on the coordinates in the classical theory, allowing for an arbitrary reparametrization, $\xi^\alpha(\sigma, \tau)$. This is

$$\delta X^\mu(\sigma, \tau) = a^\mu_{\ \nu} X^\nu(\sigma, \tau) + \xi^\alpha(\sigma, \tau)\partial_\alpha X^\mu(\sigma, \tau), \qquad (2.3.19)$$

where ξ^α is restricted to reparametrizations which are compatible with the gauge condition $h_{\alpha\beta} = \eta_{\alpha\beta}$. This means that ξ^α satisfies (2.1.44). On the other hand we want the gauge condition (2.3.8) to transform according to

$$\delta X^+ = a^+_{\ \nu} x^\nu + a^+_{\ \nu} p^\nu \tau, \qquad (2.3.20)$$

in order to preserve the gauge condition in the new coordinate frame. By considering the $+$ component of (2.3.19) and comparing with (2.3.20), we can determine the form of the compensating reparametrization ξ^α. This gives the condition

$$a^+_\nu X^\nu + \xi^0 p^+ = a^+_\nu (x^\nu + p^\nu \tau) = a^+_\nu x^\nu(\tau), \qquad (2.3.21)$$

so that

$$\xi^0 = \frac{a^+_\nu}{p^+} \left(x^\nu(\tau) - X^\nu(\sigma, \tau) \right). \qquad (2.3.22)$$

The two components of ξ^α are linked by (2.1.44) so that

$$\xi^1 = \int\limits_0^\sigma d\sigma' \frac{\partial \xi^0}{\partial \tau}. \qquad (2.3.23)$$

Substituting these expressions for ξ^α (which are linear in the coordinates) into (2.3.19) gives the form for the action of Lorentz transformations that takes into account the noncovariant gauge fixing. The important new feature is that for those transformations involving a^+_i the transformations act nonlinearly on the transverse coordinates, since there are terms on the right-hand side of (2.3.19) that are quadratic in the transverse coordinates. In the quantum theory such bilinear terms raise subtle issues to do with normal ordering, which are a potential source of anomalies in the Lorentz algebra.

It is therefore important to check that the operators in (2.3.18) really generate the Lorentz algebra (2.2.24), as they should. Most of the commutators can be checked straightforwardly, and give the correct answer for any D. However, as anticipated by our discussion in the previous paragraph, the J^{i-} transformations must be treated with care. In particular, the commutator $[J^{i-}, J^{j-}]$, which must vanish if Lorentz invariance is to hold, leads instead to an anomaly except under certain restrictions.

In light-cone gauge, we have $E^{\mu+} = E^{+\mu} = 0$. On the other hand E^{i-} is cubic in transverse oscillators when the light-cone-gauge expansion of α^-_n is substituted. As a result the commutator $[J^{i-}, J^{j-}]$ might contain terms quartic or quadratic in oscillators. (There cannot be a c number without any oscillators since $[J^{i-}, J^{j-}]$ transforms nontrivially under the transverse rotation group.) The terms in $[J^{i-}, J^{j-}]$ containing four oscillators are the same ones that would arise classically, and they cancel just as in the classical computation, which certainly has no anomaly. To check this explicitly is not too difficult. Thus if an anomaly is to arise, it must be quadratic in oscillators, and indeed it can only be of the form

$$[J^{i-}, J^{j-}] = -\frac{1}{(p^+)^2} \sum_{m=1}^\infty \Delta_m (\alpha^i_{-m} \alpha^j_m - \alpha^j_{-m} \alpha^i_m), \qquad (2.3.24)$$

where the coefficients Δ_m are c-numbers.

The calculation of the Δ_m goes as follows. Using the commutation relations

$$[x^-, 1/p^+] = i(p^+)^{-2} \qquad (2.3.25)$$

and defining $E^j = p^+ E^{j-}$

$$[x^i, E^j] = -iE^{ij}, \qquad (2.3.26)$$

gives

$$[J^{i-}, J^{j-}] = -(p^+)^{-2} C^{ij}, \qquad (2.3.27)$$

where

$$C^{ij} = 2ip^+ \alpha_0^- E^{ij} - [E^i, E^j] - iE^i p^j + iE^j p^i. \qquad (2.3.28)$$

Comparing with (2.3.24), we learn that the coefficients Δ_m can be determined by evaluating matrix elements of C^{ij} as follows

$$\langle 0| \alpha_m^k C^{ij} \alpha_{-m}^l |0\rangle = m^2 (\delta^{ik}\delta^{jl} - \delta^{jk}\delta^{il})\Delta_m. \qquad (2.3.29)$$

Using the oscillator commutation relations

$$[\alpha_m^i, \alpha_n^j] = m\delta^{ij}\delta_{m+n}, \qquad [\alpha_m^i, \alpha_n^-] = m\alpha_{m+n}^i/p^+, \qquad (2.3.30)$$

one finds that

$$\langle 0| \alpha_m^k C^{ij} \alpha_{-m}^l |0\rangle = \langle 0| \{2m(m-1)\delta^{ik}\delta^{jl} + mp^j p^k \delta^{il} - mp^j p^l \delta^{ik}$$

$$+ (m \sum_{n=1}^{m} \frac{1}{n} \alpha_{m-n}^k \alpha_n^i - \delta^{ik} p^+ \alpha_m^-)$$

$$\times (p^+ \alpha_{-m}^- \delta^{jl} - m \sum_{n=1}^{m} \frac{1}{n} \alpha_{-n}^j \alpha_{n-m}^l)\} |0\rangle - (i \leftrightarrow j). \qquad (2.3.31)$$

Now using

$$(p^+)^2 \langle 0| \alpha_m^- \alpha_{-m}^- |0\rangle = \frac{D-2}{12}(m^3 - m) + 2am, \qquad (2.3.32)$$

as follows from (2.3.14), and the identities

$$p^+ \langle 0| \alpha_m^- \sum_{n=1}^{m} \frac{1}{n} \alpha_{-n}^j \alpha_{n-m}^l |0\rangle = p^j p^l + \delta^{jl} m(m-1)/2 \qquad (2.3.33)$$

$$\langle 0| \sum_{n=1}^{m} \frac{1}{n}\alpha^k_{m-n}\alpha^i_n \sum_{p=1}^{m} \frac{1}{p}\alpha^j_{-p}\alpha^l_{p-m} |0\rangle - (i \leftrightarrow j) = (m-1)(\delta^{il}\delta^{jk} - \delta^{jl}\delta^{ik}),$$

$$(2.3.34)$$

one learns that

$$\Delta_m = m\left(\frac{26-D}{12}\right) + \frac{1}{m}\left(\frac{D-26}{12} + 2(1-a)\right). \qquad (2.3.35)$$

Requiring $\Delta_m = 0$ gives $D = 26$ and $a = 1$ as expected.

2.3.2 Construction of Transverse Physical States

At the classical level, the relation between the covariant discussion and light-cone gauge is quite transparent; light-cone gauge arises by fuller specification of the conformal gauge choice. At the quantum level the connection between the two formulations is far from evident; clarifying this will be our next task.

We return now to the covariant quantization of §2.2. In §2.2 we formulated the Virasoro conditions which should be obeyed by physical states, but we were not able to give a general description of the states which obey these conditions. Our goal here is to fill this gap and explicitly construct all the physical excited states. Among other things, this will enable us to make contact with light-cone gauge and (since that formalism is manifestly ghost free) to prove a no-ghost theorem for the covariant formalism.

The approach to be followed was pioneered by Del Giudice, Di Vecchia and Fubini (DDF). They constructed a set of operators that commute with the Virasoro operators, and which when applied successively to the ground state give all possible physical states. These operators form a closed algebra, called the 'spectrum generating algebra'. The DDF construction, which we describe below, provides 'spectrum-generating operators' A^i_n, where the index i runs over $D-2$ transverse dimensions of space-time and n is an arbitrary integer. These operators are in one-to-one correspondence with the transverse components of α^μ_n and describe the transverse modes of the string. The Virasoro constraints provide one restriction for each value of n so one might have expected that a spectrum generating algebra would have to contain $D-1$ operators for each value of n. The missing longitudinal operators, A^-_n, do enter in the theory and will be described in the next subsection.

Let $|0; p_0\rangle$ denote the tachyonic ground state of the bosonic open-string spectrum. We take $a = 1$, so this state has $p_0^2 = 2$. Suppose the tachyon is in the particular state of motion described by $p_0^+ = 1$, $p_0^- = -1$, and $p_0^i = 0$, a choice that satisfies the mass-shell condition $p_0^2 = 2$. Introduce a null vector k_0^μ with components $k_0^- = -1$ and $k_0^+ = k_0^i = 0$. Thus $k_0 \cdot p_0 = 1$. We will find it convenient to study only states with the property that if the mass is given by $\alpha' M^2 = N - 1$ then the momentum should be $p^\mu = p_0^\mu - N k_0^\mu$. We will call such states 'allowed' states. Any physical state can be Lorentz transformed into a configuration that obeys the condition just stated, so if we can gain a thorough understanding of the 'allowed' states, this will give us an understanding of all of the physical states with $p^i = 0$ and $p^+ = 1$. All other states (except those with $p^\mu = 0$) can be reached by a Lorentz transformation.

The massless vertex operator $V_\zeta(k, \tau)$, defined in (2.2.65), plays a crucial role in the construction of the spectrum generating algebra. It is a periodic function of τ with period 2π except for the factor $\exp(ik \cdot p\tau)$ arising from the term $p^\mu \tau$ in the expansion of $X^\mu(0, \tau)$. If we study the massless vector vertex only with $k^\mu = n k_0^\mu$ for integer n, then acting on 'allowed' states, $k \cdot p$ is an integer, and the factor $\exp(ik \cdot p\tau)$ is periodic in τ also. In such a situation, the vertex operator corresponding to transverse polarizations is

$$V^i(nk_0, \tau) = \dot{X}^i(\tau) e^{inX^+(\tau)}. \tag{2.3.36}$$

Because this is periodic in the 'allowed' subspace of Hilbert space, one can in this subspace define unambiguously the Fourier components

$$A_n^i = \frac{1}{2\pi} \int_0^{2\pi} V^i(nk_0, \tau) d\tau. \tag{2.3.37}$$

These are the DDF operators.

The DDF operators have two important properties. They commute with the L_n, and they obey a simple algebra. The proof of the first fact is very simple. The definition of conformal dimension implies that if $V(\tau)$ has $J = 1$, then

$$[L_m, V(\tau)] = -i \frac{d}{d\tau} \left(e^{im\tau} V(\tau) \right). \tag{2.3.38}$$

Thus $[L_m, A_n^i] = 0$, as desired, provided that the kinematic setup is restricted in the way described so that the integrand of (2.3.37) is periodic. A corollary obtained from the L_0 condition is that $[N, A_n^i] = -n A_n^i$. It follows that an arbitrary state of the form

$$A_{-n_1}^{i_1} A_{-n_2}^{i_2} \ldots A_{-n_m}^{i_m} |0; p_0\rangle \tag{2.3.39}$$

satisfies the Virasoro conditions and has $N = \sum_j n_j$. To determine the algebra of the A_n^i we require the commutators of the $\dot{X}^i(\tau)$ at *unequal* τ.

By use of the mode expansion[*]

$$\dot{X}^i(\tau) = \sum_{-\infty}^{\infty} \alpha_m^i e^{-im\tau}, \tag{2.3.40}$$

one finds

$$[\dot{X}^i(\tau_1), \dot{X}^j(\tau_2)] = 2\pi i \delta^{ij} \delta'(\tau_1 - \tau_2). \tag{2.3.41}$$

Now using (2.3.36) and (2.3.41), and noting that X^+ commutes with itself and with the X^i (even at unequal τ) we can readily compute the commutators of the A_n^i, giving

$$[A_m^i, A_n^j] = \frac{1}{(2\pi)^2} \int_0^{2\pi} d\tau_1 d\tau_2 [\dot{X}^i(\tau_1), \dot{X}^j(\tau_2)] \exp\left(im X^+(\tau_1) + in X^+(\tau_2)\right)$$

$$= \frac{m}{2\pi} \delta_{ij} \int_0^{2\pi} d\tau \dot{X}^+(\tau) \exp\left(i(m+n) X^+(\tau)\right) = m\delta_{ij}\delta_{m+n}. \tag{2.3.42}$$

The last step uses $p^+ = 1$. We see that the algebra is identical to that of the transverse oscillators α_m^i. The A_n^i also share with the transverse oscillators the reality property, $A_n^{i\dagger} = A_{-n}^i$, and the property that $A_n^i |0; p_0\rangle = 0$, $n > 0$. These facts ensure that the physical states (2.3.39) obtained by acting on the ground state with the DDF operators are all of positive metric, just like the states obtained in light-cone gauge by acting on the tachyon state with transverse oscillators. We shall call the states (2.3.39) spanned by the A_m^i operators 'DDF states'. Of course, we know that there are ghost states in the physical subspace for $D > 26$, so for general D it must be that the A_n^i do not generate the whole spectrum of physical states.

2.3.3 The No-Ghost Theorem and the Spectrum-Generating Algebra

The complete space of states created by the D-dimensional oscillator modes includes the undesirable ghost states. We have seen how to construct a subspace of 'transverse' states that satisfy the physical-state conditions and have positive norm – namely, the DDF states. From the isomorphism of the algebra of the A_m^i to that of transverse oscillators it is clear that the dimensionality of this subspace is that appropriate to

---

[*] Recall that for any operator A, the expression $A(\tau)$ is shorthand for $A(0, \tau)$.

$(D-2)$-dimensional oscillators. Our task now is to clarify the nature of the states spanned by the orthogonal complement to this subspace. We will in fact show that the DDF states essentially account for all the physical states in 26 dimensions when $a = 1$, thus proving the absence of physical negative-norm states in that case. This was first proved by Brower and by Goddard and Thorn; simplifications in the argument recently introduced by Thorn will be incorporated below. Absence of ghosts for $D < 26$ with $a \leq 1$ is a simple corollary of the result for $D = 26$, $a = 1$ (although it is somewhat more subtle to actually construct the complete basis of physical states).

We continue to consider the kinematical configuration used in the discussion of DDF operators; if there are no ghosts among the 'allowed' states, then Lorentz invariance of the covariant formalism ensures that there are no ghosts in the physical Hilbert space. Let F be the space of DDF states. A generic DDF state will be called $|f\rangle$. Let us define the operators

$$K_m = k_0 \cdot \alpha_m, \qquad (2.3.43)$$

(k_0 is the light-like vector that was introduced in the construction of the DDF states.) These operators are easily seen to obey

$$[K_m, L_n] = m K_{m+n}, \qquad [K_m, K_n] = 0. \qquad (2.3.44)$$

It is easily seen that if $|f\rangle$ is a DDF state then

$$K_n |f\rangle = 0, \qquad n > 0. \qquad (2.3.45)$$

Now, we would like to study the states made by acting on a DDF state $|f\rangle$ with a string of L_{-n} and K_{-m} operators. We define

$$|\{\lambda, \mu\}, f\rangle = L_{-1}^{\lambda_1} L_{-2}^{\lambda_2} \ldots L_{-m}^{\lambda_m} K_{-1}^{\mu_1} \ldots K_{-m}^{\mu_m} |f\rangle. \qquad (2.3.46)$$

It is convenient to define

$$\sum r \lambda_r + \sum s \mu_s = P. \qquad (2.3.47)$$

In (2.3.46) we have chosen to order the L_{-r} with r increasing from left to right. This is an arbitrary choice but it is important to keep to some convention since the L's do not commute. Similarly, we have chosen to put the K's to the right of the L's. We would like to show that, for any P, these states are linearly independent. Following the modern treatment by

Thorn, we consider the matrix of inner products of the states at a given value of P

$$\mathcal{M}^P_{\{\lambda,\mu\};\{\lambda',\mu'\}} = \langle f | K_n^{\mu_n} \dots K_1^{\mu_1} L_n^{\lambda_n} \dots L_1^{\lambda_1}$$
$$L_{-1}^{\lambda'_1} \dots L_{-m}^{\lambda'_m} K_{-1}^{\mu'_1} \dots K_{-m}^{\mu'_m} | f \rangle, \tag{2.3.48}$$

where $\sum r\lambda_r + \sum s\mu_s = \sum r\lambda'_r + \sum s\mu'_s = P$. This matrix is only a function of the K_0 and L_0 values of the state $|f\rangle$ (with $K_0 = k_0 \cdot \alpha_0 \neq 0$). If we can show that the determinant of the matrix (2.3.48) is nonzero, it follows that the states (2.3.46) of given P are linearly independent.

For $P = 1$, one computes

$$\mathcal{M}^1 = \begin{pmatrix} 2L_0 & K_0 \\ K_0 & 0 \end{pmatrix}, \tag{2.3.49}$$

which has determinant $-K_0^2 \neq 0$. The general proof that $\mathcal{M}^P$ has nonzero determinant for any P depends on the fact that there is a natural way of ordering the rows and columns so that the matrix always has a form with zeros below its minor diagonal (the diagonal from upper right to lower left corners) and nonzero elements along the minor diagonal. The determinant is then given (up to sign) by the product of these elements. Since this elementary point may be confusing, we write explicitly a general 4×4 matrix with zeros below the minor diagonal:

$$\begin{pmatrix} a_{11} & a_{12} & a_{13} & a_{14} \\ a_{21} & a_{22} & a_{23} & 0 \\ a_{31} & a_{32} & 0 & 0 \\ a_{41} & 0 & 0 & 0 \end{pmatrix}. \tag{2.3.50}$$

Clearly, the determinant of this matrix is the product of the elements on the minor diagonal, and so is nonzero if these are all nonzero. Thus, we can prove that the matrix $\mathcal{M}^P$ has nonzero determinant if we show that with suitable ordering of the states $\mathcal{M}^P$ has the general form (2.3.50). Let us illustrate the ordering that does the job for the matrix at level 2, i.e., $P = 2$. In this case the appropriate ordering of the states $|\{\lambda,\mu\},f\rangle$ is

$$L_{-1}^2, \quad L_{-2}, \quad L_{-1}K_{-1}, \quad K_{-2}, \quad K_{-1}^2. \tag{2.3.51}$$

In evaluating an inner product (2.3.48) we commute L's and K's past each other. However, the number of K's can never be reduced in this process.

To avoid the K's simply killing the conjugate end states there must be enough L's to turn all of the K's into factors of K_0 since a matrix element

$$\langle f' | K_{\mu_1} K_{\mu_2} \ldots K_{\mu_k} | f \rangle \tag{2.3.52}$$

(with $|f\rangle$ and $|f'\rangle$ being DDF states) vanishes unless $\mu_1 = \mu_2 = \ldots = \mu_k = 0$. It is easy to see that the arrangement of the states as in (2.3.51) gives a matrix of inner products of the general form (2.3.50). The generalization of (2.3.51) to higher mass levels is as follows. Recall that $\{\mu\}$ or $\{\lambda\}$ denotes a string of K's or L's, respectively. We first define an ordering between two strings of L's:

$$\{\lambda\} > \{\lambda'\} \quad \text{if } \sum r\lambda_r > \sum r\lambda'_r$$
$$\text{or } \sum r\lambda_r = \sum r\lambda'_r \text{ and } \lambda_1 > \lambda'_1.$$
$$\text{or } \sum r\lambda_r = \sum r\lambda'_r \text{ and } \lambda_1 = \lambda'_1, \lambda_2 > \lambda'_2, \text{ etc.} \tag{2.3.53}$$

(λ_1, λ_2, etc., are defined in (2.3.48).) Now we want to give a rule for combined strings $\{\lambda, \mu\}$ of L's and K's. The appropriate rule is

$$\{\lambda, \mu\} < \{\lambda', \mu'\} \quad \text{if } \{\lambda\} < \{\lambda'\}$$
$$\text{or } \{\lambda\} = \{\lambda'\} \text{ and } \{\mu\} > \{\mu'\}. \tag{2.3.54}$$

One may readily see that the examples given earlier are special cases of this rule, and that this rule always give $\mathcal{M}^P$ of the desired form, with zero elements everywhere below the minor diagonal and K_0^P along it.

This calculation is purely algebraic, making no reference to the representation of the L_{-n} or K_{-n}. The fact that $\mathcal{M}$ is nonsingular depends crucially on the presence of the K_{-n}. The corresponding matrix for states constructed simply from the L_{-n} leads to a determinant (the Kac determinant), which can be singular.

Now let $|f\rangle$ and $|g\rangle$ be two different DDF states, with $\langle f|g \rangle = 0$. We assume that $|f\rangle$ and $|g\rangle$ are eigenstates of L_0. Let $|\tilde{f}\rangle = |\{\lambda, \mu\}, f\rangle$ and $|\tilde{g}\rangle = |\{\lambda', \mu'\}, g\rangle$ be states obtained by acting on $|f\rangle$ and $|g\rangle$ with strings of L's and K's. Then we claim that

$$\langle \tilde{f} | \tilde{g} \rangle = 0. \tag{2.3.55}$$

Indeed, writing out $|\tilde{f}\rangle$ and $|\tilde{g}\rangle$ explicitly as in (2.3.48), and commuting L_{-n} and K_{-n} to the left and L_n and K_n to the right (for $n > 0$), the left-hand side of (2.3.55) reduces to a multiple of $\langle f|g \rangle$, which we assume to vanish.

We claim now that the states (2.3.46) with $|f\rangle$ running over all DDF
states and $\{\lambda, \mu\}$ running over all strings of L's and K's, are linearly
independent. This is the case because we have proved in (2.3.55) that the
towers of states based on orthogonal $|f\rangle$'s are orthogonal to each other,
and we have proved by studying the determinant of $\mathcal{M}$ that the tower of
states (2.3.46) based on a given $|f\rangle$ are linearly independent.

The statement that the states (2.3.46) are linearly independent may
seem rather technical, but it is an amazingly powerful tool. It follows from
(2.3.46) that every state in the boson string Fock space can be expressed
as a linear combination of states of the form (2.3.46). To see this is a
matter of simple counting. Any state in the Fock space can be written in
the form

$$\prod_{\rho=0}^{25}\prod_{n=1}^{\infty}\left(\alpha_{-n}^{\rho}\right)^{\epsilon_{n,\rho}}\cdot|0\rangle. \tag{2.3.56}$$

Although the total number of states in Fock space is infinite, there are
only finitely many states with a given eigenvalue of

$$N=\sum_{\rho=0}^{25}\sum_{n=1}^{\infty}\alpha_{-n}^{\rho}\alpha_{n\rho}. \tag{2.3.57}$$

The states in (2.3.56) are linearly independent, of course, and they are N
eigenstates, with eigenvalue

$$\langle N\rangle=\sum_{n,\rho}n\epsilon_{n,\rho}. \tag{2.3.58}$$

Now, explicitly, a general state (2.3.46) is of the form

$$\prod_{n=1}^{\infty}L_{-n}^{\lambda_n}\cdot K_{-n}^{\mu_n}\cdot\prod_{i=1}^{24}\left(A_{-n}^{i}\right)^{\beta_{n,i}}|0\rangle \tag{2.3.59}$$

for some λ_n, μ_n, and $\beta_{n,i}$. The states in (2.3.59) are again N eigenstates,
with eigenvalue

$$\langle N\rangle=\sum_{n=1}^{\infty}n\left(\lambda_n+\mu_n+\sum_{i}\beta_{n,i}\right). \tag{2.3.60}$$

Comparison of (2.3.58) and (2.3.60) reveals that the number of states of
the form (2.3.56) of given N is exactly the same as the number of states

(2.3.46) of the same N – since the combinatorics of 26 ϵ's is the same as the combinatorics of one λ, one μ, and 24 β's. Since the states (2.3.46) are linearly independent, and are as numerous (at each mass level) as states of the form (2.3.56), the states of the form (2.3.46) must furnish a basis for the Hilbert space.

We will now use this result to prove the no-ghost theorem – after assembling one or two more tools. Let S be the space of spurious states. According to our previous discussion, a spurious state is the same as a state $|s\rangle$ that can be written as

$$|s\rangle = L_{-1}|\chi_1\rangle + L_{-2}|\chi_2\rangle \qquad (2.3.61)$$

for some $|\chi_1\rangle$ and $|\chi_2\rangle$. Evidently, we could just as well write $|s\rangle$ as

$$|s\rangle = L_{-1}|\chi_1'\rangle + \tilde{L}_{-2}|\chi_2'\rangle, \qquad (2.3.62)$$

where

$$\tilde{L}_{-2} = L_{-2} + \frac{3}{2}L_{-1}^2. \qquad (2.3.63)$$

Recall that this is the combination that appeared in the construction of zero-norm states in (2.2.45). The utility of using $\tilde{L}_{-2}$ instead of L_{-2} will become apparent.

Let K be the space of all states of the form

$$\prod_{n=1}^{\infty} K_{-n}^{\mu_n} \cdot |f\rangle, \qquad (2.3.64)$$

where $|f\rangle$ is a DDF state. Every state of the type (2.3.46) either has some L's in its expansion, and so is spurious, or has no L's, and so belongs to K. Since the states (2.3.46) are a basis for the Fock space, we have learned that any state $|\phi\rangle$ in Fock space can be written

$$|\phi\rangle = |s\rangle + |k\rangle, \qquad (2.3.65)$$

where $|s\rangle$ is spurious and $|k\rangle$ is in K. Moreover, this representation is unique, since the states (2.3.46) are linearly independent. Therefore, if $|\phi\rangle$ is an L_0 eigenstate, then $|s\rangle$ and $|k\rangle$ are L_0 eigenstates with the same eigenvalue. In particular, if

$$(L_0 - 1)|\phi\rangle = 0, \tag{2.3.66}$$

then

$$(L_0 - 1)|s\rangle = (L_0 - 1)|k\rangle = 0. \tag{2.3.67}$$

Now, suppose that $|\phi\rangle$ is a physical state, and so obeys

$$L_m|\phi\rangle = 0, \quad m = 1, 2, 3 \ldots \tag{2.3.68}$$

as well as (2.3.66). Writing $|\phi\rangle$ as in (2.3.65) with $|s\rangle$ and $|k\rangle$ obeying (2.3.66), we will now show that $|s\rangle$ and $|k\rangle$ are physical states,

$$L_m|s\rangle = L_m|k\rangle = 0, \quad m = 1, 2, 3 \ldots. \tag{2.3.69}$$

It is enough to prove (2.3.69) for $m = 1$ and $m = 2$, since the L_m for $m > 2$ can be obtained as repeated commutators of L_1 and L_2. Also, we have

$$|s\rangle = L_{-1}|\chi_1\rangle + \tilde{L}_{-2}|\chi_2\rangle \tag{2.3.70}$$

for some $|\chi_1\rangle$ and $|\chi_2\rangle$. Equation (2.3.67) implies

$$0 = L_0|\chi_1\rangle = (L_0 + 1)|\chi_2\rangle. \tag{2.3.71}$$

To prove (2.3.69) for $m = 1$, we note from (2.3.65) and (2.3.68) that

$$0 = L_1|s\rangle + L_1|k\rangle. \tag{2.3.72}$$

We can write in more detail

$$L_1|s\rangle = L_1(L_{-1}|\chi_1\rangle + \tilde{L}_{-2}|\chi_2\rangle)) \tag{2.3.73}$$

and

$$L_1|k\rangle = L_1 \prod_n K^{\mu_n}_{-n}|f\rangle, \tag{2.3.74}$$

where $|\chi_1\rangle$ and $|\chi_2\rangle$ obey (2.3.71) and $|f\rangle$ is a DDF state. Using the commutation relations $[L_1, L_{-1}] = 2L_0$, $[L_1, L_{-2}] = 3L_{-1}$, and (2.3.71), we see from (2.3.73) that $L_1|s\rangle$ is a spurious state,

$$L_1|s\rangle = L_{-1}|\eta_1\rangle + \tilde{L}_{-2}|\eta_2\rangle \tag{2.3.75}$$

for some η_i. It also follows from (2.3.74), the fact that the DDF states are physical ($L_1|f\rangle = 0$) and the commutator $[L_m, K_n] = K_{m+n}$ that $L_1|k\rangle$

is in K. Equation (2.3.72) thus means that the sum of a spurious state $(L_1|s\rangle)$ and a state in K $(L_1|k\rangle)$ add up to zero. Because of the linear independence of the states (2.3.46), this is possible only if

$$L_1|s\rangle = L_1|k\rangle = 0, \qquad (2.3.76)$$

which is what we wished to show.

To complete the proof that $|s\rangle$ and $|k\rangle$ in (2.3.65) are physical if $|\phi\rangle$ is, we must repeat the argument that led to (2.3.76) to show under the same assumptions that

$$\tilde{L}_2|s\rangle = \tilde{L}_2|k\rangle = 0. \qquad (2.3.77)$$

The argument that gave (2.3.76) can be directly generalized to give (2.3.77), and we leave this as an exercise for the reader. We remark, however, that in the process, the reader will encounter the commutator $[L_2, L_{-2}]$, which is D dependent, because of the Virasoro anomaly. This is the one and only one place in which D dependence arises in the whole argument. The reader should be able to prove (2.3.77) in the case $D = 26$.

The no-ghost theorem is now in our grasp. In the decomposition (2.3.65) of the general physical state $|\phi\rangle$, we now know that $|s\rangle$ is physical as well as spurious. We know from our earlier discussion of spurious states that a state that is both physical and spurious is null and orthogonal to all physical states, so $\langle s|s\rangle = \langle k|s\rangle = 0$. Hence

$$\langle \phi|\phi\rangle = \langle k|k\rangle, \qquad (2.3.78)$$

but we can easily show that $|k\rangle$ has non-negative norm. Looking back to (2.3.64), we see that the general state in K can be written

$$|k\rangle = |f\rangle + \sum_\alpha \prod{}' K_{-n}^{\mu_n,\alpha}|f_\alpha\rangle, \qquad (2.3.79)$$

where the $|f_\alpha\rangle$ are DDF states and the $'$ in $\prod'$ means that the $\mu_{n,\alpha}$ are not all zero. Let us abbreviate (2.3.79) as

$$|k\rangle = |f\rangle + |\tilde{k}\rangle. \qquad (2.3.80)$$

It follows from elementary properties of the K's and the DDF states that $\langle \tilde{k}|\tilde{k}\rangle = \langle f|\tilde{k}\rangle = 0$. So $\langle k|k\rangle = \langle f|f\rangle \geq 0$ (with $\langle f|f\rangle = 0$ if and only if $|f\rangle = 0$). In view of (2.3.78) this completes the proof that the general physical state $|\phi\rangle$ has non-negative norm. There are no ghosts in the physical Hilbert space.

It is possible actually to prove a stronger statement. Using the commutator $[L_m, K_n] = -nK_{m+n}$ and the fact that the L_m for $m > 0$ annihilate the DDF states $|f\rangle$ and $|f_\alpha\rangle$, it is easy to show that if $|k\rangle$ in (2.3.79) is physical, then $|\tilde{k}\rangle = 0$ and $|k\rangle = |f\rangle$. Thus, we learn finally that the general physical state $|\phi\rangle$ can be written

$$|\phi\rangle = |f\rangle + |s\rangle, \tag{2.3.81}$$

where $|f\rangle$ is a DDF state and $|s\rangle$ is a spurious physical state. This is the strong statement of the no-ghost theorem. The appearance here of the spurious physical state $|s\rangle$ may at first sight seem like a nuisance, but actually it is closely related to the reason that string theory is interesting. The transformation $|f\rangle \to |f\rangle + |s\rangle$ is a string-theoretic analog of a gauge transformation.

The generalization of the no-ghost theorem to $D < 26$ is immediate since the physical space is simply a subspace of the 26-dimensional space in which

$$\alpha_m^i |\phi\rangle = 0, \qquad i = D, \ldots, 25 \tag{2.3.82}$$

for $m > 0$. It is only in 26 dimensions, however, that there are enough null states for the DDF states (*i.e.*, the physical $|f\rangle$ states) to form a complete basis for the physical states. For $D < 26$ it is possible to relax the condition on the mass of the ground state by allowing (as a formal trick) k_0^i to be nonzero when $i = D, \ldots, 25$. (Due to the absence of excitations in these dimensions, this does not alter the desirable features of k_0.) As a result the ground-state mass is shifted so that

$$a = 1 - \sum_{i=D}^{25} \left(k^i\right)^2 \leq 1. \tag{2.3.83}$$

The physical states in lower dimensions are not purely transverse.

We can, in fact, explicitly construct the operators that create these physical 'longitudinal' states from the transverse ones and see that they create zero-norm states when $D = 26$. This concrete representation of all $D - 1$ physical modes was developed by Brower in his presentation of the no-ghost theorem. We will now sketch this approach, though we will not enter into the degree of detail we considered above.

It is natural to try to extend the DDF operators to operators with D components, not just $D - 2$, by considering (2.3.36) with the transverse index replaced by $+$ or $-$. The $+$ component turns out to be trivial because

$$A_n^+ = \frac{1}{2\pi} \int\limits_0^{2\pi} \dot{X}^+ e^{inX^+} d\tau = \delta_n \qquad (2.3.84)$$

in the particular kinematical configuration that we have chosen to study. The $-$ component is the integral of $V^-(nk_0, \tau)$, where

$$V^-(nk_0, \tau) =: \dot{X}^- e^{inX^+} : \qquad (2.3.85)$$

does not have conformal dimension $J = 1$ because of the effect of normal ordering. Computation reveals instead that

$$[L_m, V^-(nk_0, \tau)] = e^{im\tau}\left(-i\frac{d}{d\tau} + m\right)V^-(nk_0, \tau) + \frac{1}{2}nm^2 e^{im\tau} e^{inX^+}. \qquad (2.3.86)$$

The second term in this expression has the same origin as the anomalous dimension $k^2/2$ for the vertex operator $: \exp(ik \cdot X) :$. Indeed, commutation of a normal-ordered expression with L_m gives an expression with some terms having the raising operators to the right of the corresponding lowering operators. Commuting them to normal-ordered form gives the last term in (2.3.86).

The first term in (2.3.86) has the correct form for an operator of conformal spin $J = 1$. The presence of the second term means that V^- cannot be used to create physical states. It is, however, possible to add a term to correct this. The required term turns out to involve a peculiar combination of oscillators that at first sight may look ill-defined, but actually is not. The zero-mode piece of $k \cdot X$ is $k \cdot p = nk_0 \cdot p = n$ for our assumed kinematic configuration. Therefore for $n \neq 0$ (which is the only case that we need), $\log(k \cdot \dot{X}) = \log(n + \sum_{q \neq 0} k \cdot \alpha_q) = \log n + \frac{1}{n}\sum_{q \neq 0} k \cdot \alpha_q + \ldots$ has a well-defined power series expansion of the general form

$$\log n + \text{oscillator terms.} \qquad (2.3.87)$$

The formula that enables us to cancel the last term in (2.3.86) is

$$[L_m, i\frac{d}{d\tau}\log(k \cdot \dot{X})] = e^{im\tau}\left(-i\frac{d}{d\tau} + m\right)i\frac{d}{d\tau}\log(k \cdot \dot{X}) - m^2 e^{im\tau}. \qquad (2.3.88)$$

Therefore the combination

$$\tilde{V}^\mu(k, \tau) =: \dot{X}^\mu e^{ik \cdot X} : +\frac{i}{2}k^\mu \frac{d}{d\tau}(\log k \cdot \dot{X})e^{ik \cdot X} \qquad (2.3.89)$$

has conformal dimension $J = 1$. Since the only nonzero component of k^μ

is $k^- = -n$, this reduces to (2.3.36) for $\mu = i$. For $\mu = -$ it gives

$$\tilde{V}^-(nk_0, \tau) =: \dot{X}^- e^{inX^+} : -\tfrac{1}{2}in\frac{d}{d\tau}(\log \dot{X}^+)e^{inX^+}, \qquad (2.3.90)$$

and hence

$$A_n^- = \frac{1}{2\pi}\int\limits_0^{2\pi} d\tau \tilde{V}^-(nk_0, \tau) \qquad (2.3.91)$$

can be used to generate physical states. Note that $A_0^- = p^-$.

These operators, taken together with the A_n^i, give $D-1$ operators for each value of n, and do in fact constitute a complete spectrum generating algebra. The algebra of these operators is

$$[A_m^i, A_n^j] = m\delta^{ij}\delta_{m+n} \qquad (2.3.92)$$

$$[A_m^-, A_n^i] = -nA_{m+n}^i \qquad (2.3.93)$$

$$[A_m^-, A_n^-] = (m-n)A_{m+n}^- + 2m^3\delta_{m+n}. \qquad (2.3.94)$$

Together with the Virasoro operators L_m, the spectrum generating operators A_m^i and A_m^- generate all the states in the string Fock space. The correspondence between the A_m and the α_m can be seen from the defining formulas, where it is evident that

$$A_{-n}^i = \alpha_{-n}^i + \dots, \qquad A_{-n}^- = k \cdot \alpha_{-n} + \dots, \qquad L_{-n} = p \cdot \alpha_{-n} + \dots,$$
$$(2.3.95)$$

where $+\dots$ denotes terms that involve more than one oscillator mode. This strongly suggests that at each mass level there is an invertible transformation from states made from spectrum generating and Virasoro operators to states made from the α_m^μ.

We can now consider the space of physical states explicitly to illustrate the fact that there are no negative-norm states. In fact, acting on physical states of positive norm, A_n^i gives physical states of positive norm. The delicate case is A_n^-. Instead of working with A_n^- we may as well work with

$$\tilde{A}_n^- = A_n^- - \frac{1}{2}\sum_{m=1}^{\infty}\sum_{i=1}^{D-2} : A_m^i A_{n-m}^i :, \qquad (2.3.96)$$

since the $\tilde{A}_n^-$ have the advantage that they commute with the A_m^i. Thus our proof that the A_n^- and A_k^i generate the physical spectrum implies that an arbitrary physical state $|\psi\rangle$ can be written in the form

$$|\psi\rangle = A^{i_1}_{k_1} A^{i_2}_{k_2} \ldots A^{i_r}_{k_r} \tilde{A}^-_{n_1} \tilde{A}^-_{n_2} \ldots \tilde{A}^-_{n_p} |0\rangle. \qquad (2.3.97)$$

The commutation relations of the A^i_n and $\tilde{A}_n$ can be used to arrange the spectrum-generating operators in the convenient order given in (2.3.97).

The $\tilde{A}^-_n$ obey a Virasoro algebra, which can be extracted from the algebra of the α^-_n,(2.3.14), together with (2.3.93) and (2.3.94):

$$[\tilde{A}^-_m, \tilde{A}^-_n] = (m - n)\tilde{A}^-_{m+n} + \frac{26 - D}{12}m^3\delta_{m+n}. \qquad (2.3.98)$$

If we can show that states of the form

$$|\chi\rangle = \tilde{A}^-_{n_1} \ldots \tilde{A}^-_{n_k} |0\rangle \qquad (2.3.99)$$

always have non-negative norm, then subsequent action of A^i_n as in (2.3.97) will always give states of non-negative norm, and this again will prove the no-ghost theorem. Since

$$\tilde{A}^-_n |0\rangle = 0, \quad n \geq 0 \qquad (2.3.100)$$

we can (after commuting the $\tilde{A}^-_n$ to the right for $n \geq 0$) assume that $n_i < 0$ in (2.3.99). To compute the norm of $|\chi\rangle$ we write

$$\langle\chi|\chi\rangle = \langle 0| \tilde{A}^-_{-n_k} \ldots \tilde{A}^-_{-n_1} \tilde{A}^-_{n_1} \ldots \tilde{A}^-_{n_k} |0\rangle. \qquad (2.3.101)$$

Using (2.3.98) to commute the $\tilde{A}^-_{-n_i}$ to the right where they eventually annihilate $|0\rangle$, we see that in $D = 26$ the states in (2.3.101) are all of zero norm (unless $k = 0$, in which case $|\chi\rangle = |0\rangle$, of norm 1). This argument is the basis of Brower's proof of the no-ghost theorem for $D = 26$. For $D \neq 26$, (2.3.101) can still be evaluated by use of (2.3.98) and (2.3.100). One finds that all states of the form $\langle\chi|$ have positive norm for $D < 26$ and some have negative norm for $D > 26$.

2.3.4 Analysis of the Spectrum

For $a = 1$ and $D = 26$ the ground state of the bosonic open string, $|0; p\rangle$, is a tachyon with $\alpha'M^2 = -1$, and the first excited state $\alpha^i_{-1}|0; p\rangle$ is a massless vector with 24 independent transverse polarizations. In order to get a better sense of what it means to have a string, it is instructive to study what happens at some of the higher levels as well.

The light-cone-gauge description of string states has the advantage of giving physical states only, described in a manifestly positive-definite Hilbert space. It has the disadvantage of presenting the states as multiplets of $SO(D-2)$, the transverse rotation group, even through the proof of Lorentz invariance, given above for $D = 26$, guarantees that the massive levels in fact fill out complete multiplets of $SO(D-1)$. One could determine how the $SO(24)$ multiplets fit together into $SO(25)$ multiplets by actually studying how they rotate into one another under suitable Lorentz transformations. This is somewhat tedious, however, and it is simpler (at least for the first few levels) to note that the $SO(24)$ multiplets that occur correspond to unique branchings of $SO(25)$ multiplets.

We consider first the open-string states. The only state with $\alpha' M^2 = -1$ is the ground state tachyon, and the only states of $\alpha' M^2 = 0$ are the 24 polarization states of the massless vector boson. With these states we are already quite familiar. The first states with positive M^2 occur at $\alpha' M^2 = 1$, and are given by

$$\alpha^i_{-2} |0; p\rangle \text{ and } \alpha^i_{-1}\alpha^j_{-1} |0; p\rangle , \qquad (2.3.102)$$

$D-2$ and $\frac{1}{2}(D-2)(D-1)$ states, respectively. The sum, $\frac{1}{2}(D-2)(D+1)$, is the dimensionality of a symmetric traceless representation, $\square$, of $SO(D-1)$, which must be the complete answer therefore. It is tempting to call such a representation 'spin two'. This is indeed the representation we found in §2.2.3 in trying to construct vertex operators at this mass level.

At the level $\alpha' M^2 = 2$ the possible states are

$$\alpha^i_{-3} |0\rangle , \quad \alpha^i_{-2}\alpha^j_{-1} |0\rangle , \text{ and } \alpha^i_{-1}\alpha^j_{-1}\alpha^k_{-1} |0\rangle , \qquad (2.3.103)$$

a total of $24 + 576 + 2600 = 3200$ states in all. These combine to give the $\square\square$ (**2900**) and $\boxminus$ (**300**) of $SO(25)$. Similarly, for $\alpha' M^2 = 3$ one obtains $\square\square\square$ (**20 150**), $\boxplus$ (**5175**), $\square\square$ (**324**), $\cdot$ (**1**), a total of 25,650 states in all. One fact worth noting is that maximum 'spin' for mass M is given by $n = \alpha' M^2 + 1$. The nth rank symmetric traceless tensor representation is built from portions of $\alpha^{i_1}_{-1} \cdots \alpha^{i_n}_{-1} |0\rangle$ plus other terms required to complete the $SO(25)$ multiplet. It is clear that if decomposed in terms of an $SO(3)$ subgroup it would give one term of spin n in the usual sense, and that this is the unique highest spin state at this mass level. Thus we have the inequality

$$J \leq \alpha' M^2 + 1. \qquad (2.3.104)$$

We already deduced this relation at the classical level in §2.1.3, and we encountered essentially the same fact in the introduction in the course of analyzing the poles in the Veneziano amplitude. This inequality has essentially the same form as one appearing in the theory of black holes, and is only one of a number of tantalizing similarities between properties of massive string states and black holes.

The spectrum of closed-string states is easily deduced from that of the open strings. Here at last we have the chance to enjoy the fruits of our labors, for it is in the closed string sector that we will find the graviton – the massless spin two state. Closed strings in the light-cone gauge are described by two sets of transverse oscillators, $\{\alpha_n^i\}$ and $\{\tilde{\alpha}_n^i\}$, corresponding to left and right movers. In addition, one has the restriction $L_0 = \tilde{L}_0$, mentioned earlier, which implied that there must be an equal amount of excitation of the left and right movers, i.e.,

$$\sum_{n=1}^{\infty} \alpha_{-n}^i \alpha_n^i = \sum_{n=1}^{\infty} \tilde{\alpha}_{-n}^i \tilde{\alpha}_n^i. \qquad (2.3.105)$$

Thus the closed-string multiplet with $\alpha' M^2 = 4(N-1)$ is given by tensor products of open-string states having $\alpha' M^2 = N - 1$ with themselves.

For example, the ground state is a scalar tachyon with $\alpha' M^2 = -4$. The next level is a set of massless states of the form

$$|\Omega^{ij}\rangle = \alpha_{-1}^i \tilde{\alpha}_{-1}^j |0\rangle \qquad (2.3.106)$$

with $SO(24)$ quantum numbers corresponding to the tensor product of a massless vector of $SO(24)$ from left-moving modes with a massless vector of $SO(24)$ from right-moving modes. The part of $|\Omega^{ij}\rangle$ which is symmetric and traceless in i and j transforms under $SO(24)$ as a massless spin two particle; this is the graviton. The trace term $\delta_{ij}|\Omega^{ij}\rangle$ is a massless scalar, usually called the dilaton. Finally, the antisymmetric part $|\Omega^{ij}\rangle - |\Omega^{ji}\rangle$ transforms under $SO(24)$ as an antisymmetric second rank tensor. They all have their counterparts in supersymmetric string theories, and each plays a fundamental role. We can also go on to describe closed-string states of positive mass squared by taking suitable tensor products of left- and right-moving open-string Hilbert space. For example, the level $\alpha' M^2 = 4$ has representations given by the decomposition of $\square \times \square$.

The spectrum described above is that of oriented closed strings – the spectrum of the 'extended Shapiro–Virasoro model'. It is possible, if one wishes, to restrict the spectrum to states corresponding to an unoriented string. Physically, an oriented string carries an intrinsic 'arrow', as in fig. 2.5a, while an unoriented string, as in fig. 2.5b, is directionless. Mathematically, an unoriented string is a string whose quantum wave

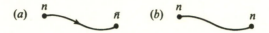

Figure 2.5. Oriented and unoriented open strings. An orientation is present (a) if the charges at the ends of the string are distinguishable (as in QCD with the usual $SU(n)$ gauge group), while there is no orientation (b) if the charges at the end of the string are indistinguishable, as in a gauge theory with $SO(n)$ or $Sp(n)$ gauge group. For closed strings the notion of the 'orientation' of the string is equally important though more abstract.

functional $\Psi(X^\mu(\sigma))$ is invariant under $\sigma \to -\sigma$. The quantum wave functional of an oriented closed string has no such restriction. As we have imposed no $\sigma \to -\sigma$ restriction on our string wave functionals, we have been discussing oriented closed strings, even though we have not said so explicitly. Let us now discuss the alternative case of unoriented closed strings. Since $\sigma \to -\sigma$ interchanges left- and right-moving oscillators α_m and $\tilde{\alpha}_m$, the state $|\Psi\rangle$ of an unoriented closed string must be symmetric under exchange of the two sets of oscillators. Thus only the symmetric product of the open-string multiplets should be used to describe unoriented closed strings. For example, at the massless level the antisymmetric tensor term should be dropped, whereas the graviton and dilaton terms should be kept. The spectrum of unoriented closed strings is the spectrum of the 'restricted Shapiro–Virasoro model'.

2.3.5 Asymptotic Formulas for Level Densities

We will now explore the asymptotic behavior of the level density for very highly excited states. This subsection is not entirely self-contained, and the reader may wish to return to it after reading the appendix to chapter 8 on modular functions.

The total number of open-string states with $\alpha' M^2 = n - 1$, denoted d_n, is conveniently described as the coefficient of w^n in

$$\text{tr}\, w^N, \tag{2.3.107}$$

where N is the number operator

$$N = \sum_{n=1}^{\infty} \alpha_{-n} \cdot \alpha_n. \tag{2.3.108}$$

Since we wish to count only physical states, the oscillators in (2.3.108) are transverse oscillators only, α_n^i, $i = 1, \ldots, 24$. More easily computed than the individual d_n is the generating function

$$G(w) = \sum_{n=0}^{\infty} d_n w^n = \text{tr } w^N. \qquad (2.3.109)$$

This can be evaluated by elementary methods of quantum statistical mechanics. In fact,

$$\text{tr } w^N = \prod_{n=1}^{\infty} \text{tr} w^{\alpha_{-n} \cdot \alpha_n} = \prod_{n=1}^{\infty} (1 - w^n)^{-24} = [f(w)]^{-24}, \qquad (2.3.110)$$

where

$$f(w) = \prod_{n=1}^{\infty} (1 - w^n) \qquad (2.3.111)$$

is known as the classical partition function; it enters in many problems in additive or combinatorial number theory.

In order to estimate the asymptotic density of states we shall require the behavior of the function $f(w)$ as $w \to 1$. This can be crudely estimated by using

$$f(w) = \exp\left(\sum_{n=1}^{\infty} \ln(1 - w^n)\right)$$

$$= \exp\left(-\sum_{m,n=1}^{\infty} \frac{w^{mn}}{m}\right) = \exp\left(-\sum_{m=1}^{\infty} \frac{w^m}{m(1 - w^m)}\right) \qquad (2.3.112)$$

$$\sim \exp\left(-\frac{1}{1-w} \sum_{m=1}^{\infty} \frac{1}{m^2}\right) \sim \exp\left(-\frac{\pi^2}{6(1-w)}\right).$$

A more precise estimate can be obtained by noting that upon replacing w by $e^{2\pi i \tau}$, the function $f(w)$ is closely related to the Dedekind eta function

$$\eta(\tau) = e^{i\pi\tau/12} \prod_{n=1}^{\infty} (1 - e^{2\pi i n \tau}). \qquad (2.3.113)$$

This function has the modular transformation formula

$$\eta(-1/\tau) = (-i\tau)^{1/2} \eta(\tau), \qquad (2.3.114)$$

(which will be derived in the appendix to chapter 8 in the context of the calculation of loop diagrams). Applied to $f(w)$ this gives the Hardy–Ramanujan formula

$$f(w) = \left(\frac{-2\pi}{\log w}\right)^{1/2} w^{-1/24} q^{1/12} f(q^2), \qquad (2.3.115)$$

where

$$q = \exp\left(\frac{2\pi^2}{\log w}\right). \qquad (2.3.116)$$

This relation enables one to deduce the asymptotic formula for $w \to 1$ (or $q \to 0$)

$$f(w) \sim A(1-w)^{-1/2} \exp\left(-\frac{\pi^2}{6(1-w)}\right). \qquad (2.3.117)$$

Having found an asymptotic formula for $f(w)$ and therefore for the generating function $G(w) = [f(w)]^{-24}$, we would now like to go back and determine the large n behavior of the d_n, which was our original interest. One can project out d_n from $G(w) = \sum d_n w^n$ by a contour integral on a small circle about the origin

$$d_n = \frac{1}{2\pi i} \oint \frac{G(w)}{w^{n+1}} dw. \qquad (2.3.118)$$

Using the asymptotic expansion of $f(w)$, this can be estimated for large n by a saddle point evaluation. $G(w)$ vanishes rapidly for $w \to 1$, while if n is very large, w^{n+1} is very small for $w < 1$. There is consequently for large n a sharply defined saddle point for w near 1. Indeed, the factor

$$\exp\left[-\frac{4\pi^2}{\ln w} - (n+1)\ln w\right] \qquad (2.3.119)$$

is stationary for $\ln w \sim -2\pi/\sqrt{n+1}$. Therefore one finds that as $n \to \infty$

$$d_n \sim (\text{const.}) n^{-27/4} \exp(4\pi\sqrt{n}). \qquad (2.3.120)$$

Using $n \sim \alpha' m^2$, the density of levels as a function of mass is asymptotically

$$\rho(m) \sim m^{-25/2} \exp(m/m_0), \qquad (2.3.121)$$

where

$$m_0 = \frac{1}{4\pi}(\alpha')^{-1/2}. \qquad (2.3.122)$$

The level density grows so rapidly with mass that the partition function $\text{tr}\, e^{-\beta H}$ of the free theory cannot be defined beyond a maximum temperature $T = m_0$. This has led to speculations about some kind of phase transition near this temperature. Such speculations cannot be pushed too far, however, because we are dealing with a system that includes gravity. The concept of temperature and the idea of a phase transition are really only valid in the limit of macroscopic systems of essentially infinite volume. Such a limit does not exist in a system that includes gravity, because in the presence of gravity any statistical ensemble of nonzero energy per unit volume is, in the infinite volume limit, inside its Schwarzschild radius and unstable against gravitational collapse. (Even nonrelativistically such a system always suffers from the Jeans instability, which is crucial in the theory of star and galaxy formation.) Thus there is no stable thermal ensemble, and the very concept of temperature and phase transitions is ill-defined in the presence of gravity. For most practical purposes, this does not worry us, since we deal with temperatures so low compared to the Planck scale that temperature and statistical mechanics are good approximate notions. At energy scales of order $(\alpha')^{-1/2}$, however, the notion of temperature has presumably lost its meaning, and whatever happens in string theory at that scale is unlikely to be understandable in terms of thermal physics.

2.4   Summary

Classical free string theory can be consistently formulated for any spacetime dimension, but quantization with a ghost-free spectrum requires $D \leq 26$. Also, the intercept of the leading Regge trajectory must satisfy $a \leq 1$. In the special case of $D = 26$ and $a = 1$ the spectrum is entirely transverse, with many decoupled zero-norm states. This is suggestive of an enormous underlying gauge invariance. These results were obtained using various covariant and noncovariant approaches, which will be useful for later developments.

Closed strings are described by a doubling of the degrees of freedom of the open strings. The mathematics of left- or right-moving modes on closed strings is essentially the same as that of the standing waves on open strings. The two sets of modes are independent except for the single relation $L_0 = \tilde{L}_0$. Closed strings are of particular importance, because the spectrum includes a massless graviton.

3. Modern Covariant Quantization

The propagation of a string in a background space-time is governed by a two-dimensional field theory, as we have discussed extensively in chapter 2. This two-dimensional theory has local reparametrization invariance. We have quantized this two-dimensional theory in chapter 2 by canonical methods – equal-time commutators, constraints and a Hilbert space of physical states. Path integrals offer an alternative approach to the quantization of field theories. Experience with ordinary gauge field theory in four space-time dimensions shows that path-integral quantization is particularly natural in the case of systems with local symmetries, and leads to many insights that are difficult to obtain otherwise. It is thus natural to consider the quantization of the free string using the same tools that are customarily used for quantization of any other gauge-invariant theory. This viewpoint was introduced by Polyakov in 1981, though path-integral approaches to gauge-fixed forms of the string action go back to the early days of string theory.

3.1 Covariant Path-Integral Quantization

The free field theory action

$$S_0[X] = -\frac{1}{2\pi} \int d^2\sigma \partial^\alpha X^\mu \partial_\alpha X_\mu, \qquad (3.1.1)$$

when supplemented with constraints, describes the propagation of a free string. It is a gauge-fixed form of

$$S[h, X] = -\frac{1}{2\pi} \int d^2\sigma \sqrt{h} h^{\alpha\beta} \partial_\alpha X^\mu \partial_\beta X_\mu. \qquad (3.1.2)$$

It is natural to attempt to quantize (3.1.2) by the same methods that are used for quantization of any other gauge-invariant theory, starting with the Euclidean path integral

$$Z = \int Dh(\sigma) DX(\sigma) e^{-S[h,X]}. \qquad (3.1.3)$$

We will apply to this expression the modern Faddeev–Popov techniques for interpreting the partition function of a gauge theory.

3.1.1 Faddeev–Popov Ghosts

The symbol $\int Dh(\sigma)$ denotes an integral over the three independent components $h_{++}(\sigma)$, $h_{--}(\sigma)$, $h_{+-}(\sigma)$. It is necessary to define a precise measure to be used in this integral, and anomalies potentially arise because there is no satisfactory way to define the measure so as to preserve all of the apparent symmetries of (3.1.3). Since there are three gauge invariances, the two reparametrizations and the Weyl scaling, one wishes to choose a gauge slice that makes a particular choice for each of the three functions in $h_{\alpha\beta}(\sigma)$. We would like to impose the usual gauge choice[*]

$$h_{\alpha\beta} = e^{\phi}\eta_{\alpha\beta}, \tag{3.1.4}$$

which in terms of light-cone coordinates means

$$0 = h_{++} = h_{--}. \tag{3.1.5}$$

Under a world-sheet reparametrization, $\sigma^+ \to \sigma^+ + \xi^+$, $\sigma^- \to \sigma^- + \xi^-$, the gauge conditions (3.1.5) transform as

$$\delta h_{++} = 2\nabla_+\xi_+, \qquad \delta h_{--} = 2\nabla_-\xi_-. \tag{3.1.6}$$

(This amounts to the formula – familiar in general relativity – for the transformation of the metric tensor under an infinitesimal coordinate transformation: $\delta h_{\alpha\beta} = \nabla_\alpha\xi_\beta + \nabla_\beta\xi_\alpha$. Here ∇ is the covariant derivative that includes the Christoffel connection, so $\nabla_\alpha\xi_\beta = \partial_\alpha\xi_\beta - \Gamma^\gamma{}_{\alpha\beta}\xi_\gamma$.) The procedure for imposing a gauge condition in a path integral such as (3.1.3) is well-known. Let G be the group of reparametrizations of the string world sheet Σ, and let Dg denote an integration over the group manifold. Let h^g denote the metric into which h is transformed by a reparametrization g. A basic tool in manipulating the path integral is then the identity

$$1 = \int Dg(\sigma)\,\delta(h^g_{++})\delta(h^g_{--})\det(\delta h^g_{++}/\delta g)\det(\delta h^g_{--}/\delta g). \tag{3.1.7}$$

The factors $\det(\delta h^g_{++}/\delta g)$, $\det(\delta h^g_{--}/\delta g)$ are the usual gauge-fixing determinants, which are needed so that the integral really equals one. The next

---

[*] This is a slight simplification; global properties will be discussed later.

step in gauge fixing of path integrals is to insert '1' in the path integral (3.1.3) in the form (3.1.7). This gives

$$Z = \int Dg(\sigma) \int Dh(\sigma) DX(\sigma) e^{-S[h,X]} \delta(h^g_{++}) \delta(h^g_{--})$$
$$\times \det(\delta h^g_{++}/\delta g) \det(\delta h^g_{--}/\delta g). \tag{3.1.8}$$

Since the action S is reparametrization invariant, $S[h,X] = S[h^g, X]$, the integrand in (3.1.8) depends on h and g only in the combination h^g. We therefore make a change of integration variables g and h to g and $h' = h^g$, and discard the $\int Dg$ integral, which now only contributes an infinite multiplicative factor. We thus arrive at the gauge fixed path integral

$$Z = \int Dh'(\sigma) DX(\sigma) \, e^{-S[h',X]} \, \delta(h'_{++}) \delta(h'_{--})$$
$$\times \det(\delta h'_{++}/\delta g) \det(\delta h'_{--}/\delta g). \tag{3.1.9}$$

The delta functions in (3.1.9) are easy to deal with; they mean that the integral $\int Dh'$ reduces to an integral over h'_{+-}, or equivalently over ϕ defined in (3.1.4). The determinants in (3.1.9) are more awkward. The usual way to deal with them is to represent the determinants as integrals over anticommuting 'ghosts' and 'antighosts'. The required formulas can be determined from the formula (3.1.6) for $\delta h'_{++}/\delta g = \delta h'_{++}/\delta \xi_+$ and $\delta h'_{--}/\delta g = \delta h'_{--}/\delta \xi_-$. Indeed,

$$\delta h'_{++}(\sigma)/\delta \xi_+(\sigma') = 2\nabla_+ \delta(\sigma - \sigma') \tag{3.1.10}$$

and likewise for $+ \rightarrow -$. The delta function in (3.1.10) is just the identity operator in coordinate space; it is the operator ∇_+ (and ∇_-) whose determinant is needed in (3.1.9). Therefore, to represent the first determinant in (3.1.9) we introduce an anticommuting 'ghost' c^- and an 'antighost' b_{--} and write (absorbing the factor of 2 in the normalization)[*]

$$\det(\delta h'_{++}/\delta g) = \int Dc^-(\sigma) Db_{--}(\sigma) \exp\left\{ -\frac{1}{\pi} \int d^2\sigma \, c^- \nabla_+ b_{--} \right\}. \tag{3.1.11}$$

The second determinant in (3.1.9) is likewise represented as an integral

---

[*] Integrals over anticommuting (Grassmann) coordinates, called Berezin integrals, are discussed in §4.1.2.

over a ghost c^+ and an antighost b_{++},

$$\det(\delta h'_{--}/\delta g) = \int Dc^+(\sigma)Db_{++}(\sigma)\exp\left\{-\frac{1}{\pi}\int d^2\sigma\, c^+\nabla_- b_{++}\right\}.$$
(3.1.12)

Using the delta functions in (3.1.9) to solve for h in terms of the conformal factor ϕ defined in (3.1.4), the gauge-fixed path integral becomes

$$Z = \int D\phi(\sigma)\int DX(\sigma)\,Dc(\sigma)\,Db(\sigma)\,\exp -S(X,b,c),$$
(3.1.13)

where now the action S includes the ghost terms defined in (3.1.11) and (3.1.12) in addition to the free-field action of (3.1.1).

The next issue is to discuss the $D\phi$ integral in (3.1.13). Formally, the integrand in (3.1.13) is independent of ϕ, so that the $D\phi$ integral would give merely an irrelevant overall infinite factor. In fact, because of problems with regularization, the decoupling of ϕ only holds in 26 dimensions, where the conformal anomaly cancels. The justification of this assertion, however, requires some tools we have not yet developed. For the time being we simply assume that the $D\phi$ integral can be discarded and study the resulting theory including the ghosts. We first show that precisely in 26 dimensions the c-number anomaly in the Virasoro algebra cancels if the ghost contributions are included. Then in §3.2.3 we show that this is equivalent to the decoupling of ϕ in 26 dimensions. It is conceivable that the integral in (3.1.13) is physically sensible even if the ϕ dependence does not cancel. This possibility has motivated some very ingenious suggestions, but remains uncertain. In any case, for superstring unification the critical dimension in which the ϕ dependence cancels is preferred, since in most suggestions about how to get outside of the critical dimension, one expects to lose the massless particles that are present in the critical dimension. We do not pursue the analysis of string theory outside of the critical dimension in this book.

3.1.2 Complex World-Sheet Tensor Calculus

Before trying to understand the ghosts, it is useful to work out the basic formulas of Riemannian geometry in a two-dimensional metric of the conformal form $h_{\alpha\beta} = e^\phi\eta_{\alpha\beta}$. It is convenient, though not essential, in this discussion to use a Euclidean language, discussing a string world sheet with a metric of Euclidean signature. The continuation of the formulas from Euclidean to Minkowskian signature and vice versa is straightforward. Though not really essential for our present purposes, the Euclidean picture is likely to be essential in some of the subsequent developments, since it leads to the theory of Riemann surfaces and complex analysis. We will present the mathematical setting for this subject in a deeper way in chapter 15.

On a world sheet whose metric (at least locally) has been put in the form $ds^2 = e^\phi(d\sigma^2 + d\tau^2)$, it is natural to introduce a complex coordinate $z = \tau + i\sigma$ and its conjugate $\bar{z} = \tau - i\sigma$. When using a world sheet of Minkowski signature, we have previously referred to z and $\bar{z}$ as $\sigma^\pm$. In the $z, \bar{z}$ coordinate system, the components of a vector are

$$t_\pm = t_0 \pm it_1. \tag{3.1.14}$$

Furthermore, the components of the gradient $\partial/\partial\sigma^\alpha$ are now

$$\partial_\pm = \frac{1}{2}\left(\frac{\partial}{\partial\tau} \mp i\frac{\partial}{\partial\sigma}\right). \tag{3.1.15}$$

The metric components are $h_{++} = h_{--} = 0$ and $h_{+-} = h_{-+} = \frac{1}{2}e^\phi$. The invariant line element is given by

$$ds^2 = e^\phi dz\, d\bar{z}, \tag{3.1.16}$$

and indices are raised and lowered according to the rule

$$t_+ = \frac{1}{2}e^\phi t^- \tag{3.1.17}$$

$$t_- = \frac{1}{2}e^\phi t^+. \tag{3.1.18}$$

A change of coordinates $z \to z' = f(z)$, where f is a holomorphic function of z, preserves the conformally flat form of the metric. It simply sends

$$\rho \to \rho' = \left|\frac{dz'}{dz}\right|^{-2}\rho, \tag{3.1.19}$$

where $\rho = e^\phi$. In general, a tensor with n_u upper and n_l lower holomorphic indices and $\bar{n}_u$ upper and $\bar{n}_l$ lower antiholomorphic indices transforms according to

$$t \to t' = \left(\frac{dz'}{dz}\right)^{n_u - n_l}\left(\frac{d\bar{z}'}{d\bar{z}}\right)^{\bar{n}_u - \bar{n}_l} t. \tag{3.1.20}$$

The quantities $n_l - n_u$ ($\bar{n}_l - \bar{n}_u$) are referred to as the holomorphic (anti-holomorphic) conformal dimension of the tensor t. The qualifiers 'holomorphic' and 'antiholomorphic' can usually be dropped without confusion.

We define covariant derivatives of tensors using the Christoffel connection

$$\Gamma^{\gamma}{}_{\alpha\beta} = \tfrac{1}{2}h^{\gamma\delta}(\partial_{\alpha}h_{\beta\delta} + \partial_{\beta}h_{\alpha\delta} - \partial_{\delta}h_{\alpha\beta}) \qquad (3.1.21)$$

in the usual manner. The Riemann curvature tensor is conventionally defined as

$$R^{\gamma}{}_{\alpha\beta\rho} = \partial_{\rho}\Gamma^{\gamma}{}_{\alpha\beta} + \Gamma^{\epsilon}{}_{\alpha\beta}\Gamma^{\gamma}{}_{\rho\epsilon} - (\beta \leftrightarrow \rho). \qquad (3.1.22)$$

The only nonzero components of the Christoffel connection for the conformally flat metric are

$$\Gamma^{+}{}_{++} = \partial_{+}\phi \qquad (3.1.23)$$

$$\Gamma^{-}{}_{--} = \partial_{-}\phi. \qquad (3.1.24)$$

Therefore, for example, a tensor with n lower or upper $+$ indices has

$$\nabla_{-}t_{++\cdots+} = \partial_{-}t_{++\cdots+} \qquad (3.1.25)$$

$$\nabla_{+}t_{++\cdots+} = (\partial_{+} - n\partial_{+}\phi)t_{++\cdots+} \qquad (3.1.26)$$

$$\nabla_{-}t^{++\cdots+} = \partial_{-}t^{++\cdots+} \qquad (3.1.27)$$

$$\nabla_{+}t^{++\cdots+} = (\partial_{+} + n\partial_{+}\phi)t^{++\cdots+}. \qquad (3.1.28)$$

It follows that

$$[\nabla_{-}, \nabla_{+}]t_{++\cdots+} = [\partial_{-}, \partial_{+} - n\partial_{+}\log\rho]t_{++\cdots+}$$

$$= -n(\partial_{+}\partial_{-}\log\rho)t_{++\cdots+} \qquad (3.1.29)$$

$$= n\rho R^{(2)}t_{++\cdots+}.$$

From this we read off the two-dimensional scalar curvature for a conformally flat metric:

$$R^{(2)} = -\frac{1}{\rho}\partial_{+}\partial_{-}\phi. \qquad (3.1.30)$$

Now, let us reconsider the Faddeev–Popov ghost action for the ghosts and antighosts c and $/b$ that were introduced in the preceding section. Let us write this action in a way that makes sense for any world-sheet metric, not necessarily in the conformal gauge. The ghost fields c^{+} and c^{-} may be interpreted as the components of a vector field c^{α}. The antighost fields

b_{++} and b_{--} can be interpreted as the components of a symmetric, trace-less tensor $b_{\alpha\beta}$. (Saying this in reverse, a general symmetric tensor has components b_{++}, b_{--} and b_{+-}, but b_{+-} vanishes for a *traceless* symmetric tensor.) The ghost action can now be written

$$S_g = -\frac{i}{2\pi} \int d^2\sigma \sqrt{h} h^{\alpha\beta} c^\gamma \nabla_\alpha b_{\beta\gamma}, \qquad (3.1.31)$$

where the ghost field c^α is a contravariant vector and the antighost field $b_{\alpha\beta}$ is a covariant symmetric traceless tensor. The ghost fields b and c are anticommuting quantities (*i.e.*, Grassmann valued).

As discussed in §1.3.3, the world-sheet energy–momentum tensor is defined by

$$T_{\alpha\beta} = -\frac{2\pi}{\sqrt{h}} \frac{\delta S}{\delta h^{\alpha\beta}}. \qquad (3.1.32)$$

In using this formula to deduce the ghost contribution one must be careful to include the contribution from the Christoffel connection in (3.1.31). Also, the tracelessness of $b_{\beta\gamma}$ must be taken into account. Doing this carefully, one finds for the ghost contribution to the energy–momentum tensor

$$T_{\alpha\beta}^{(c)} = -i\left[\tfrac{1}{2} c^\gamma \nabla_{(\alpha} b_{\beta)\gamma} + (\nabla_{(\alpha} c^\gamma) b_{\beta)\gamma} - \text{trace}\right]. \qquad (3.1.33)$$

The parentheses represent symmetrization of the enclosed subscripts.

Being a traceless symmetric tensor, the only components of $T_{\alpha\beta}$ in the complex basis described above are T_{++} and T_{--}. The same applies to $b_{\alpha\beta}$. For example

$$T_{++}^{(c)} = -i\left[\tfrac{1}{2} c^+ \partial_+ b_{++} + (\partial_+ c^+) b_{++}\right]. \qquad (3.1.34)$$

We also note that in the conformal gauge S_g simplifies to

$$S_{FP} = \frac{i}{\pi} \int (c^+ \partial_- b_{++} + c^- \partial_+ b_{--}) d^2\sigma. \qquad (3.1.35)$$

This indeed coincides with the Faddeev–Popov ghost action defined in the preceding section. (In checking this the reader must note the rules (3.1.17) and (3.1.18) for raising and lowering indices.)

3.1.3 Quantization of the Ghosts

The ghost action S_{FP} of equation (3.1.35) implies that b and c are conjugate degrees of freedom with simple canonical anticommutation relations

$$\{b_{++}(\sigma, \tau), c^+(\sigma', \tau)\} = 2\pi\delta(\sigma - \sigma') \tag{3.1.36}$$

$$\{b_{--}(\sigma, \tau), c^-(\sigma', \tau)\} = 2\pi\delta(\sigma - \sigma'). \tag{3.1.37}$$

In the conformal gauge their equations of motion are

$$\partial_- c^+ = \partial_- b_{++} = 0 \tag{3.1.38}$$

$$\partial_+ c^- = \partial_+ b_{--} = 0, \tag{3.1.39}$$

with b_{++} conjugate to c^+ and b_{--} conjugate to c^-. Open-string boundary conditions imply that $c^+ = c^-$ at the ends of the string, so that

$$c^+ = \sum_{-\infty}^{\infty} c_n e^{-in(\tau+\sigma)} \tag{3.1.40}$$

$$c^- = \sum_{-\infty}^{\infty} c_n e^{-in(\tau-\sigma)}. \tag{3.1.41}$$

Similarly they require $b_{++} = b_{--}$ at the ends so that

$$b_{++} = \sum_{-\infty}^{\infty} b_n e^{-in(\tau+\sigma)} \tag{3.1.42}$$

$$b_{--} = \sum_{-\infty}^{\infty} b_n e^{-in(\tau-\sigma)}. \tag{3.1.43}$$

(The symbol b_n for the ghost modes should not be confused with the symbol for anticommuting modes of the bosonic sector of superstrings introduced in the next chapter.) In terms of the modes the canonical anticommutation relations are

$$\{c_m, b_n\} = \delta_{m+n} \tag{3.1.44}$$

$$\{c_m, c_n\} = \{b_m, b_n\} = 0. \tag{3.1.45}$$

For closed strings the boundary condition is just periodicity in σ, so that c^+ and c^- have independent mode expansions (as in the case of left-

and right-moving X coordinates)

$$c^+ = \sqrt{2} \sum_{-\infty}^{\infty} c_n e^{-2in(\tau+\sigma)} \tag{3.1.46}$$

$$c^- = \sqrt{2} \sum_{-\infty}^{\infty} \tilde{c}_n e^{-2in(\tau-\sigma)}. \tag{3.1.47}$$

Similarly, the coordinates b_{++} and b_{--} involve modes b_n and $\tilde{b}_n$.

In these formulas b and c enter symmetrically, despite the asymmetrical looking tensor structures such as c^- and b_{--}. They enter symmetrically because on a *flat* world sheet the ghost Lagrangian treats b and c symmetrically. This is not so on a curved world sheet, as is evident from our formula (3.1.31) for the ghost action on a curved world sheet. Likewise, b and c do not enter symmetrically in the world-sheet energy–momentum tensor, since this is derived by varying with respect to the world-sheet metric; even on a flat world sheet, the energy–momentum tensor treats b and c differently because these would propagate differently on a curved world sheet. In all deeper aspects of the theory, b and c enter quite differently. Indeed, in (3.1.33) and (3.1.34) we determined the form of the world-sheet energy–momentum tensor:

$$T_{++}^{(c)} = -i\left[\tfrac{1}{2}c^+\partial_+ b_{++} + \partial_+ c^+ b_{++}\right]$$
$$\tag{3.1.48}$$
$$T_{--}^{(c)} = -i\left[\tfrac{1}{2}c^-\partial_- b_{--} + \partial_- c^- b_{--}\right].$$

Inserting the mode expansions in T_{++} and T_{--} and extracting Fourier modes $L_m = \frac{1}{\pi}\int_{-\pi}^{\pi} d\sigma\, e^{im\sigma} T_{++}$ at $\tau = 0$ (for the open string) gives the Virasoro generators

$$L_m^{(c)} = \sum_{n=-\infty}^{\infty} [m(J-1)-n]b_{m+n}c_{-n}, \tag{3.1.49}$$

where $J = 2$ is the conformal dimension of the antighost (whereas the ghost, c, has conformal dimension $J = -1$). We have included a free parameter J here, rather than setting $J = 2$, because we will analyze later a system in which b and c are replaced by anticommuting fields of dimensions J and $1 - J$. As usual, normal ordering is required in (3.1.49) in the case of $m = 0$. Of course, for closed strings there is also a second set of ghost Virasoro generators. The $L_m^{(c)}$ satisfy the usual Virasoro algebra

$$[L_m^{(c)}, L_n^{(c)}] = (m - n)L_{m+n}^{(c)} + A^c(m)\delta_{m+n} \qquad (3.1.50)$$

with an anomaly term

$$A^c(m) = \frac{1}{12}[1 - 3k^2]m^3 + \frac{1}{6}m, \qquad (3.1.51)$$

where $k = 2J - 1$, which for $J = 2$ gives

$$A^c(m) = \frac{1}{6}(m - 13m^3). \qquad (3.1.52)$$

Just as in chapter 2, the easiest and safest way to determine the anomaly is by evaluating specific matrix elements. For example,

$$A^c(1) = \langle 0|\,[L_1^{(c)}, L_{-1}^{(c)}]\,|0\rangle = \langle 0|\,L_1^{(c)}L_{-1}^{(c)}\,|0\rangle$$

$$= \langle 0|\,(b_1c_0 + 2b_0c_1)(-b_{-1}c_0 - 2b_0c_{-1})\,|0\rangle \qquad (3.1.53)$$

$$= -2\,\langle 0|\,(c_0b_0 + b_0c_0)\,|0\rangle = -2.$$

Given the mode expansion (3.1.49) of the Virasoro operators, we can straightforwardly compute the commutation relations of the $L_m^{(c)}$ with b and c and quantify the extent to which the ghosts and antighosts are different. For example, for open strings the $\sigma = 0$ expressions

$$c(\tau) = \sum_{-\infty}^{\infty} c_n e^{-in\tau} \qquad (3.1.54)$$

and

$$b(\tau) = \sum_{-\infty}^{\infty} b_n e^{-in\tau} \qquad (3.1.55)$$

have conformal dimension $J = -1$ and $J = +2$, respectively, since

$$[L_m^{(c)}, b_n] = (m - n)b_{m+n} \qquad (3.1.56)$$

$$[L_m^{(c)}, c_n] = -(2m + n)c_{m+n}. \qquad (3.1.57)$$

The conformal dimensions of c and b agree with the assignments that we would have made on the basis of their holomorphic tensor types as defined earlier. This is indeed a general rule.

Let us define the complete Virasoro generators corresponding to $S_0 + S_{gh}$ by

$$L_m = L_m^{(\alpha)} + L_m^{(c)} - a\delta_m. \qquad (3.1.58)$$

Notice that we have shifted the earlier definition of L_0 so that the zeroth constraint is now $L_0 = 0$. The anomaly – adding ghost and matter contributions – is

$$A(m) = \frac{D}{12}(m^3 - m) + \frac{1}{6}(m - 13m^3) + 2am. \qquad (3.1.59)$$

This vanishes if and only if $D = 26$ and $a = 1$, providing yet one more determination of these magic values. Only for these values is the theory really conformally invariant.

3.2 BRST Quantization

Incorporation of the ghosts has given us Virasoro generators without anomaly, obeying the naive algebra $[L_m, L_n] = (m-n)L_{m+n}$. The absence of the anomaly means, for example, that we could impose the constraint equations in a far more straightforward way; it makes sense to look for states $|\chi\rangle$ obeying $L_n |\chi\rangle = 0$ for all n. However, incorporation of ghosts brings about a new conundrum. Now that we are working in a much larger Fock space that contains ghost and antighost excitations as well as excitations of the coordinates X^μ, how are we to identify the physical states? The answer to this question is provided by BRST quantization.

BRST quantization was first introduced in quantization of Yang–Mills theory as a useful device for proving the renormalizability of nonabelian gauge theories in four dimensions. It was found that a global fermionic symmetry is present after Yang–Mills gauge fixing, and this unbroken symmetry is a useful tool in analyzing the structure of possible counterterms. We will see that BRST quantization is useful in a somewhat similar way in string theory as a tool for facilitating the understanding of the gauge-fixed action.

For our purposes in this section, this is enough motivation for the study of the BRST quantization, but it is useful to point out that BRST quantization actually seems to have a much more far-reaching significance in the case of string theory. In fact, in the case of string theory the very phrase 'BRST quantization' must be taken with a grain of salt. Yang–Mills theory is, like almost every physical theory, a system in which after one obtains the linear Schrödinger equation one is finished at the level of principle – all that remains is to study this equation. String theory is rather different, as we discussed at length in chapter 1. In string theory after quantizing the free string the goal is to study interactions of strings – or, in a sense, to introduce nonlinearities in the linear Schrödinger equation. While other gauge theories are studied as ends unto themselves, the quantum theory of the free string is a tool toward the invention – still in progress – of something much richer and more far-reaching. But such lofty thoughts need not concern us here. We have on our hands a subtle technical problem of identifying the physical states in the 'large' Fock space containing both X^μ and the ghosts and antighosts; we will see that BRST quantization gives an elegant solution of this problem.

3.2.1 Construction of the BRST Charge

Consider any physical system with symmetry operators K_i that form a closed Lie algebra G,

$$[K_i, K_j] = f_{ij}{}^k K_k, \tag{3.2.1}$$

with $f_{ij}{}^k$ being the structure constants of G. BRST quantization involves the introduction of 'antighosts' b_i, which transform in the adjoint representation of G, and 'ghosts' c^i, which transform in the dual of the adjoint representation.[*] They obey canonical anticommutation relations,

$$\{c^i, b_j\} = \delta^i_j. \tag{3.2.2}$$

One defines the ghost number U as

$$U = \sum_i c^i b_i. \tag{3.2.3}$$

(In the case of infinite-dimensional Lie algebras, it is necessary to subtract a normal-ordering constant to make sense out of U. We temporarily avoid this step in order to agree with the mathematical literature. Note that

---

[*] In the familiar case of compact Lie algebras, the adjoint representation and its dual are the same thing. In general, this is not so. Saying that the ghosts c^i transform in the dual of the adjoint representation is a fancy way to say that they carry a contravariant Lie algebra index, while the b_j carry a covariant one. The Virasoro algebra is a typical case in which there is no invariant way to raise and lower Lie algebra indices, so that c^i and b_j transform differently under the Virasoro algebra; we have seen that they have conformal dimension -1 and 2, respectively. In general J is dual to $1 - J$.

the eigenvalues of U are integers running from 0 to n, with n being the dimension of the Lie algebra G.) One then introduces the operator

$$Q = c^i K_i - \tfrac{1}{2} f_{ij}{}^k c^i c^j b_k. \tag{3.2.4}$$

To physicists this is known as the BRST operator; to mathematicians it is the operator that computes the cohomology of the Lie algebra G, with values in the representation defined by the K_i. The basic property of Q is that

$$Q^2 = 0. \tag{3.2.5}$$

To prove this one must use the commutation relations (3.2.1) as well as the identity

$$f_{ij}{}^m f_{mk}{}^l + f_{jk}{}^m f_{mi}{}^l + f_{ki}{}^m f_{mj}{}^l = 0, \tag{3.2.6}$$

which follows from (3.2.1) via the Jacobi identity.

Let C^k be the Hilbert space of states of ghost number $U = k$. A state χ in C^k is said to be BRST invariant if

$$Q\chi = 0. \tag{3.2.7}$$

There is a trivial way to find BRST-invariant states; any state of the form $\chi = Q\lambda$ is invariant, in view of (3.2.5). λ would necessarily have ghost number $k-1$, since the form of Q shows that it changes the ghost number of any state it acts on by $+1$. The interesting solutions of (3.2.7) are the ones that cannot be written in the form $\chi = Q\lambda$. Given two solutions χ and χ' of (3.2.7), we consider them to be equivalent if $\chi - \chi'$ is a trivial solution of (3.2.7) in the sense that

$$\chi - \chi' = Q\lambda \tag{3.2.8}$$

for some λ. The equivalence classes of solutions of (3.2.7) of ghost number k, with two solutions considered equivalent under the conditions just stated, form what in the mathematical literature would be called the kth cohomology group of the Lie algebra G, with values in the representation R determined by the matrices K_i. It would commonly be denoted $H^k(G; R)$. The equivalence classes are called cohomology classes.

Of special interest are the BRST-invariant states of ghost number zero. The form of the ghost number operator U shows that a state χ of ghost number zero must be annihilated by all of the b_k, so acting on such a state the second term in Q vanishes; indeed for such a state

$$Q\chi = \sum_i c^i K_i \chi. \qquad (3.2.9)$$

A state annihilated by the b_j cannot be annihilated by any of the c^i, so the condition $Q\chi = 0$ is equivalent to

$$K_i \chi = 0, \quad i = 1, \ldots, n. \qquad (3.2.10)$$

Thus, a state χ of ghost number zero is BRST invariant if and only if it is G invariant. On the other hand a state χ of ghost number zero cannot possibly be written as $\chi = Q\lambda$, since there are no states of ghost number -1. Hence, states of ghost number zero that obey (3.2.10) are the same thing as cohomology classes of ghost number zero. Hence the cohomology group $H^0(G; R)$ is the same as the space of G-invariant states of ghost number zero. Of course, a state of ghost number zero is a state that is annihilated by ghost annihilation operators b_i and hence does not contain any ghosts.

This is more or less what we want. Asking for BRST-invariant states of ghost number zero is a way to isolate the G invariant states that do not contain ghosts. While this way of looking at things could be carried out in physical problems with a finite number of degrees of freedom, it would be an excessively heavy-handed approach, to say the least. In string theory it is actually a useful approach.

Most of what we have said above can be carried out in the case of an infinite-dimensional Lie algebra such as the Virasoro algebra. There are a few differences. The equation $Q^2 = 0$ might be afflicted with an anomaly, so this has to be checked carefully. Also, the ghost number U contains a normal-ordering constant. Because of this, while it is reasonable to expect that the physical states of the string are BRST cohomology classes χ (modulo gauge transformations $\chi \to \chi + Q\lambda$) of some definite ghost number, it would be naive to assume that the required ghost number is zero; the ghost number of the physical states is a normal-ordering constant that depends on which physical system one chooses to consider.

We now choose G to be the Virasoro algebra and attempt to carry out the BRST program. Corresponding to the Virasoro generators L_m (m being an arbitrary integer) we must introduce ghosts c_m and antighosts b_n – these are precisely the Fourier modes considered earlier in the process of quantizing the ghosts. The BRST operator would be

$$Q = \sum_{-\infty}^{\infty} L_{-m}^{(\alpha)} c_m - \frac{1}{2} \sum_{-\infty}^{\infty} (m - n) : c_{-m} c_{-n} b_{m+n} : - a c_0. \qquad (3.2.11)$$

where we have used the explicit form of the Virasoro structure constants. Comparing to the form of the L_m derived earlier for both matter and ghost fields, we note that we can write

$$Q = \sum_{-\infty}^{\infty} : \left(L_{-m}^{(\alpha)} + \frac{1}{2} L_{-m}^{(c)} - a\delta_m \right) c_m :, \qquad (3.2.12)$$

Similarly, the ghost number is

$$U = \sum_{-\infty}^{\infty} : c_{-m} b_m :, \qquad (3.2.13)$$

where now normal ordering is necessary. Of course, for closed strings we would add to these formulas the contribution of a second set of left-moving ghosts and oscillators.[*]

An important property in the physical system, which is not guaranteed by the general introductory discussion, is that both the BRST operator Q and the ghost number U can be obtained as integrals of conserved charge densities. The BRST current is defined as (ignoring normal ordering)

$$J_+^B = 2c^+ (T_{++}^{(\alpha)} + \frac{1}{2} T_{++}^{(c)}) \qquad (3.2.14)$$

with J_-^B obtained via $- \leftrightarrow +$. $T_{++}^{(c)}$ is given in (3.1.48) and $T_{++}^{(\alpha)} = (\partial_+ X)^2$ was obtained in §2.1.3. The ghost-number current is defined by

$$J_+ = c^+ b_{++} \qquad (3.2.15)$$

with, again, J_- defined via $- \leftrightarrow +$. It is easily seen (from the equations of motion of b and c and the law of energy–momentum conservation on the world sheet) that these currents are conserved,

$$\partial_- J_+^B = \partial_- J_+ = 0 \qquad (3.2.16)$$

and that the corresponding conserved charges are indeed the BRST charge

$$Q = \frac{1}{2\pi} \int_0^{\pi} d\sigma (J_+^B + J_-^B) \qquad (3.2.17)$$

---

[*] The factor of 1/2 multiplying the $L^{(c)}$ in (3.2.12) may appear surprising, but is a consequence of evaluating (3.2.11). This factor of 1/2 is needed not only for $Q^2 = 0$, but also for many other formulas in the theory such as $L_m = \{Q, b_m\}$, with $L_m = L_m^{(\alpha)} + L_m^{(c)} - a\delta_m$.

and the ghost number

$$U = \frac{1}{2\pi} \int_0^\pi d\sigma (J_+ + J_-).$$ (3.2.18)

To be more precise, these formulas reduce exactly to the previous ones in the case of open strings, while in the case of closed strings we have added both left- and right-moving modes in writing (3.2.17) and (3.2.18).

We now include normal ordering in the definition of $L_0^{(\alpha)}$ and the second term of Q. The resulting ambiguities can be absorbed in a term linear in c_0 with a free coefficient a. The previous discussion ensures that $Q^2 = 0$ in the classical sense, but we now wish to ask whether this is true at the quantum level. To investigate this, we note it follows from (3.2.11) and (3.2.12) that

$$Q^2 = \tfrac{1}{2}\{Q, Q\} = \tfrac{1}{2} \sum_{-\infty}^{\infty} ([L_m, L_n] - (m-n)L_{m+n})c_{-m}c_{-n},$$ (3.2.19)

with L_m as given in (3.1.58). Therefore $Q^2 = 0$ for $D = 26$ and $a = 1$ as a consequence of the vanishing of the anomaly $A(m)$ in (3.1.59).

The converse of the last result, namely that $Q^2 = 0$ implies that the Virasoro algebra has no anomaly, can also be demonstrated. To show this we note first that the complete Virasoro generators of (3.1.58) are given by

$$L_m = \{Q, b_m\},$$ (3.2.20)

as the reader should verify. Note that

$$[L_m, Q] = [\{Q, b_m\}, Q] = 0,$$ (3.2.21)

as a consequence of $Q^2 = 0$. It follows in a similar manner that the algebra of the complete Virasoro generators L_m closes without an anomaly term:

$$[L_m, L_n] = [L_m, \{Q, b_n\}] = \{Q, [L_m, b_n]\}$$
$$= (m-n)\{Q, b_{m+n}\} = (m-n)L_{m+n}.$$ (3.2.22)

Let us define the BRST transformation of an arbitrary physical quantity Y as

$$\delta Y = [\lambda Q, Y],$$ (3.2.23)

where λ is a constant Grassmann parameter. One can show that the coordinates satisfy

$$\delta X^\mu = \lambda c^+ \partial_+ X^\mu + \lambda c^- \partial_- X^\mu$$ (3.2.24)

$$\delta c^+ = \lambda c^+ \partial_+ c^+ \qquad (3.2.25)$$

$$\delta b_{++} = 2i\lambda T_{++} \qquad (3.2.26)$$

$$\delta T_{++} = 0, \qquad (3.2.27)$$

and similarly with $+ \leftrightarrow -$. $T_{++} = T_{++}^{(\alpha)} + T_{++}^{(c)}$ is the complete energy-momentum tensor. It is easily seen that the square of this transformation is zero and that it corresponds to an invariance of the gauge-fixed action.

Now let us study the normal ordering of the ghost number operator:

$$U = \frac{1}{2}(c_0 b_0 - b_0 c_0) + \sum_{n=1}^{\infty}(c_{-n}b_n - b_{-n}c_n). \qquad (3.2.28)$$

We have separated out the ghost and antighost zero modes c_0 and b_0 for special treatment, since there is no obviously natural way to normal order them.

In fact, c_0 and b_0 both commute with the Hamiltonian, so the ground state has a degeneracy that arises from the fact that it must furnish a representation of these operators. Actually, c_0 and b_0 have the anticommutation relations $c_0^2 = b_0^2 = 0$, $\{c_0, b_0\} = 1$. The irreducible representation of these commutation relations requires two states, which we may call $|\uparrow\rangle$ and $|\downarrow\rangle$; we may choose them to be annihilated, respectively, by c_0 and b_0. They must then obey

$$c_0 |\downarrow\rangle = |\uparrow\rangle, \; b_0 |\uparrow\rangle = |\downarrow\rangle. \qquad (3.2.29)$$

What are the 'ghost numbers' $U_\uparrow$ and $U_\downarrow$ of $|\uparrow\rangle$ and $|\downarrow\rangle$? It is clear from (3.2.29) that $U_\uparrow = U_\downarrow + 1$, but this does not fix the separate values of $U_\uparrow$ and $U_\downarrow$, which indeed depend on a normal-ordering constant in the definition of U. The most symmetrical choice is $U_\downarrow = -1/2$, $U_\uparrow = +1/2$. This choice corresponds to the precise normal-ordering prescription in (3.2.28). It is in some ways a rather peculiar prescription since it means that all eigenvalues of the ghost-number operator are half-integral. Nevertheless, it seems to be the most natural choice; for instance, it is the choice that makes gauge-invariant string field theory as simple as possible, though we do not develop this subject in this book.

Our hope is now that the physical states can be characterized as BRST cohomology classes of some definite ghost number. Since we expect that physical states need not contain ghost excitations, it should be possible (after a possible transformation $\psi \rightarrow \psi + Q\lambda$) to put a physical state ψ in a form in which the ghost wave function is proportional to one of the two ground states $|\uparrow\rangle$ and $|\downarrow\rangle$. Consequently, the possible choices for the ghost number of a physical state are $\pm 1/2$. The choice between these two options is not simply a matter of convention, since the ghost and antighost fields do not enter symmetrically in the theory. They have conformal dimension -1 and 2, for example. The correct choice turns out to be that physical states have ghost number $-1/2$. Indeed, let χ be a state that is annihilated by the ghost and antighost annihilation operators

$$c_n |\chi\rangle = b_n |\chi\rangle = 0, \qquad n > 0. \tag{3.2.30}$$

We may think of such a state as 'containing no ghosts or antighosts'. Let us suppose in addition that χ has ghost number $-1/2$ and so is annihilated by b_0. Acting on a state of this form, the condition of BRST invariance reduces to

$$0 = Q |\chi\rangle = \left(c_0(L_0^{(\alpha)} - 1) + \sum_{n>0} c_{-n} L_n^{(\alpha)}\right) |\chi\rangle, \tag{3.2.31}$$

so that the single condition $Q |\chi\rangle = 0$ reproduces all of the physical state conditions of the older covariant quantization. If instead we chose $|\chi\rangle$ to have ghost number $+1/2$ (and so to be annihilated by c_0 rather than by b_0) the first term on the right-hand side of (3.2.31) would drop out, and we would not quite get all of the physical state conditions.[*] If we find a state χ that obeys (3.2.30) as well as (3.2.31), can χ be written as $\chi = Q\lambda$ for some λ? It is easily seen that this would require

$$\chi = \sum_{n>0} L_{-n}^{(\alpha)} |\lambda_n\rangle \tag{3.2.32}$$

for some states $|\lambda_n\rangle$. This, in turn, implies that the state χ is null, since

$$\langle \chi | \chi \rangle = \langle \sum_{n>0} L_{-n} \lambda_n | \chi \rangle = \sum_{n>0} \langle \lambda_n | L_n |\chi\rangle = 0 \tag{3.2.33}$$

by virtue of (3.2.31). In fact, such states are precisely the physical spurious states discussed in chapter 2. Thus, we have shown that states obeying the traditional physical-state conditions of the bosonic string theory give rise to BRST cohomology classes of ghost number $-1/2$, and that a physical state is trivial as a cohomology class (can be written as $Q\lambda$ for some λ) if

---

[*] In the unphysical case of finite space-time volume, but only in that case, the extra condition is recovered when one considers the invariance $\chi \to \chi + Q |\lambda\rangle$.

and only if it is a null state in the older language. Thus, we conclude that *physical states in the bosonic string theory are BRST cohomology classes of ghost number* $-1/2$. Actually, to complete the proof of this statement we must establish the converse to this statement. Given a state χ of ghost number $-1/2$ that is BRST invariant, we would like to show that it can be written in the form $\chi = \chi' + Q\lambda$, where χ' obeys (3.2.30) and therefore corresponds to a physical state of the old language, embedded in the enlarged Fock space in the way that we have just described. Though it is rather clear that this is true, a really complete and economical proof does not seem to have appeared as of this writing, and we will not try to prove it here.

3.2.2 Covariant Calculation of The Virasoro Anomaly

In this section we describe an alternative computation of the Virasoro anomaly that leads to many useful insights. We recall the general form of the Virasoro commutation relations:

$$[L_m, L_n] = (m - n)L_{m+n} + (am^3 + bm)\delta_{m+n}. \qquad (3.2.34)$$

Of the two anomaly coefficients a and b, it is really only a that has an invariant meaning; b can be absorbed in shifting the normal-ordering constant that appears in L_0. We would now like to describe how to calculate a using what might be called world-sheet methods, as opposed to the mode expansions employed in chapter 2.

We consider a conformally invariant free bosonic field theory on a world sheet, which we take to be the whole complex plane:

$$S_B = -\frac{1}{2\pi} \int d^2\sigma \partial_\alpha \phi \partial^\alpha \phi. \qquad (3.2.35)$$

In our work hitherto the bosonic fields of interest have always been space-time coordinates $X^\mu(\sigma)$, but with an eye to later developments we give the bosonic field in (3.2.35) the neutral name ϕ. For the same reason, we also work out a few formulas that are not strictly needed until later sections. The propagator of ϕ is

$$\langle \phi(\sigma)\phi(\sigma') \rangle = \pi \int \frac{d^2k}{4\pi^2} \frac{e^{ik \cdot (\sigma-\sigma')}}{k^2} = -\tfrac{1}{2}\ln(|\sigma - \sigma'|\mu) \qquad (3.2.36)$$

with μ an infrared cutoff that cancels out of all relevant formulas. This expression arose in §1.4.4. Introducing $\sigma^\pm = \tau \pm \sigma$, ϕ obeys the free wave equation

$$0 = \partial_+ \partial_- \phi \qquad (3.2.37)$$

This means that ϕ splits according to

$$\phi(\sigma^+, \sigma^-) = \frac{1}{2}\phi^+(\sigma^+) + \frac{1}{2}\phi^-(\sigma^-) \qquad (3.2.38)$$

There is a slight ambiguity in the splitting in (3.2.38): we could add a constant to ϕ^+ while subtracting the same constant from ϕ^-. In our present discussion we formulate the quantum field theory in (3.2.35) on a plane; in the infinite volume limit the zero-mode ambiguity in (3.2.38) is inconsequential. It will be important shortly when we discuss bosonization in finite volume. Explicit formulas for ϕ^+ and ϕ^- can easily be given. Indeed,

$$\phi^+(\sigma, \tau) = \phi(\sigma, \tau) - \int_\sigma^\infty d\sigma' \frac{\partial \phi}{\partial \tau}$$

$$\phi^-(\sigma, \tau) = \phi(\sigma, \tau) + \int_\sigma^\infty d\sigma' \frac{\partial \phi}{\partial \tau}. \qquad (3.2.39)$$

Equation (3.2.39) evidently obeys (3.2.38), while $(\partial_\tau \mp \partial_\sigma)\phi^\pm = 0$ is easily checked using the equation of motion $(\partial_\tau^2 - \partial_\sigma^2)\phi = 0$.

As ϕ^+ is a function of σ^+ only, and ϕ^- is a function of σ^- only, the two-point function $\langle \phi^+ \phi^- \rangle$ must vanish, while $\langle \phi^+ \phi^+ \rangle$ must be a function only of σ^+ and $\langle \phi^- \phi^- \rangle$ must be a function only of σ^-. Writing out (3.2.36) in the form

$$\langle (\phi^+(\sigma^+) + \phi^-(\sigma^-)) \cdot (\phi^+(\sigma'^+) + \phi^-(\sigma'^-)) \rangle$$
$$= -\ln((\sigma^+ - \sigma'^+)(\sigma^- - \sigma'^-)\mu^2) \qquad (3.2.40)$$

we have enough information to disentangle the separate pieces:

$$\langle \phi^+(\sigma^+) \phi^+(\sigma'^+) \rangle = -\ln[(\sigma^+ - \sigma'^+)\mu]$$
$$\langle \phi^-(\sigma^-) \phi^-(\sigma'^-) \rangle = -\ln[(\sigma^- - \sigma'^-)\mu]. \qquad (3.2.41)$$

Of course, this can also be checked using (3.2.39).

Now we turn to our real interest, the world-sheet energy–momentum tensor

$$T_{++}(\sigma^+) = \partial_+\phi\partial_+\phi = \frac{1}{4}\partial_+\phi^+\partial_+\phi^+ \qquad (3.2.42)$$

with an analogous formula for T_{--}. The energy–momentum tensor obeys

$$0 = \partial_- T_{++}. \qquad (3.2.43)$$

To evaluate the Virasoro anomaly, we borrow a trick that is well-known in current algebra. Consider the time-ordered (or rather τ-ordered) two-point function $\langle T(T_{++}(\sigma^+)T_{++}(\sigma'^+))\rangle$. It is not conserved, but rather obeys the Ward-like identity

$$\partial_-\langle T(T_{++}(\sigma,\tau)T_{++}(\sigma',\tau'))\rangle = \tfrac{1}{2}\delta(\tau - \tau')\langle [T_{++}(\sigma,\tau), T_{++}(\sigma',\tau')]\rangle, \qquad (3.2.44)$$

which arises (as is customary in current algebra) because in pulling ∂_- inside the T-product, one picks up an equal-time commutator.

$$\langle T_{++} \, T_{++} \rangle = \quad$$

Figure 3.1. In free field theory, the two-point function of the energy–momentum tensor is given by a simple one-loop diagram.

On the right-hand side of (3.2.44), the expectation value of the commutator $[T_{++}(\sigma,\tau), T_{++}(\sigma',\tau)]$ extracts the c-number piece of this commutator, which is the Virasoro anomaly. Therefore, we can evaluate the Virasoro anomaly by evaluating the left-hand side of (3.2.44). In free field theory, the two-point function of the energy–momentum tensor is given by the simple one-loop diagram of fig. 3.1. There is no need to do any integral; to evaluate fig. 3.1 in coordinate space, one merely takes the product of the various propagators. Using (3.2.41) and (3.2.42) we see that

$$\langle T(T_{++}(\sigma^+)\, T_{++}(\sigma'^+))\rangle = \frac{1}{8}(\sigma^+ - \sigma'^+)^{-4}. \qquad (3.2.45)$$

The left-hand side of (3.2.44) involves

$$\partial_-[(\sigma^+ - \sigma'^+)^{-4}]. \qquad (3.2.46)$$

At first sight this may appear to vanish, but in fact we must note that

$$\partial_-\frac{1}{\sigma^+} = i\pi\delta(\sigma)\delta(\tau) \equiv i\pi\delta^2(\sigma), \tag{3.2.47}$$

whence

$$\partial_-(\sigma^+-\sigma'^+)^{-4} = -\frac{1}{6}\frac{\partial^3}{\partial\sigma^3}\partial_-(\sigma^+-\sigma'^+)^{-1} = -\frac{i\pi}{6}\frac{\partial^3}{\partial\sigma^3}\delta^2(\sigma-\sigma'). \tag{3.2.48}$$

Thus, (3.2.44) and (3.2.45) correspond to an anomalous part of the energy–momentum commutator at equal τ:

$$[T_{++}(\sigma,\tau),T_{++}(\sigma',\tau)]_A = -i\frac{\pi}{24}\delta'''(\sigma-\sigma'). \tag{3.2.49}$$

The subscript A in $[\,,]_A$ means that we have here evaluated only the anomalous, c-number part of the commutator.

In this evaluation, we have formulated the free field theory on the plane, but the anomaly (3.2.49) is determined only by the short distance behavior of the free field theory, and so (3.2.49) remains equally valid if the theory is formulated on the world sheet of, say, a closed string. In that case we define the Virasoro generators as Fourier moments of T_{++},

$$L_n = \frac{1}{2\pi}\int_0^\pi d\sigma\, e^{2in\sigma}T_{++}(\sigma) \tag{3.2.50}$$

and (3.2.49) gives a formula for the crucial coefficient a in (3.2.34), namely

$$a = 1/12 \tag{3.2.51}$$

in agreement with our previous evaluation.

One of the virtues of this calculation is that it shows, in a sense, the inevitability of the Virasoro anomaly. The $(\sigma^+ - \sigma'^+)^{-4}$ dependence of (3.2.45) is completely determined by scale invariance and holds in any conformally invariant theory in $1+1$ dimensions. Only the coefficient of $(\sigma^+ - \sigma'^+)^{-4}$ might be different in another conformally invariant theory. This coefficient must be positive in any field theory with physical degrees of freedom only, since the two-point function of the hermitian operator T_{++} must be positive. Only ghosts might cancel the Virasoro anomaly.

To see how this works, we now consider the conformally invariant field theory describing the ghosts. It is enough to focus on left-moving modes:

$$S = \frac{i}{\pi} \int d^2\sigma c^+ \partial_- b_{++}. \tag{3.2.52}$$

The two-point function is (for $\sigma^+ \sim \sigma'^+$)

$$\langle c^+(\sigma) b_{++}(\sigma') \rangle = \frac{1}{2\pi} \int d^2 k e^{ik \cdot (\sigma - \sigma')} / k_-$$

$$= -\frac{i}{\pi} \partial_+ \int d^2 k e^{ik \cdot (\sigma^+ - \sigma'^+)} / k^2 = -\frac{i}{\sigma^+ - \sigma'^+}. \tag{3.2.53}$$

The energy–momentum tensor cannot be uniquely determined from the form of the flat world-sheet action (3.2.52), since in fact

$$T_{++}^k = -\frac{i}{4}[(\partial_+ c^+ \cdot b_{++} - c^+ \partial_+ b_{++}) + k\partial_+(c^+ b_{++})] \tag{3.2.54}$$

is conserved for any k. (The energy and momentum operators constructed from T_{++} are independent of k since the k-dependent term in (3.2.54) is a total derivative.) Our study of the ghost action on a curved world sheet showed that $k = 3$ is the correct value, but we leave k as a free parameter, since in later work we will encounter systems with other values of k. The Fourier modes of (3.2.54) coincide with (3.1.49) if one sets

$$J = \frac{k+1}{2}. \tag{3.2.55}$$

Treating b and c symmetrically would amount to taking $k = 0$. In this case b and c transform under scaling and conformal transformations as conventional fermion fields of conformal dimension $1/2$. Introducing a nonzero k shifts the commutator of any field Z with T_{++} by an amount linear in k. We have already computed that, for $k = 3$, b_{++} has conformal dimension 2 and c has conformal dimension -1. More generally, in view of the obvious linearity in k the conformal dimension of b is $(1+k)/2$ and that of c is $(1 - k)/2$. This is a special case of a still more general statement. We recall that the ghost-number current is $J_+ = c^+ b_{++}$. The k-dependent piece in (3.2.54) is precisely the derivative of the ghost-number current, and because of this, the k-dependent part of the conformal dimension of any physical field Z depends only on the ghost number of Z. Thus, let

$d(Z)$ be the conformal dimension of Z at $k = 0$, and let $g(Z)$ be its ghost number. The conformal dimension $d_k(Z)$ of Z at general k is then

$$d_k(Z) = d(Z) - kg(Z)/2. \qquad (3.2.56)$$

The enterprising reader is urged to check this assertion using the definition of conformal dimension and the fact that the k-dependent term in (3.2.54) is the derivative of the ghost current. Equation (3.2.56) is not essential in our present discussion, but plays a role in the construction of the covariant fermion vertex operator in chapter 7.

Notice that in (3.2.54) b_{++} and c^+ are *distinct* anticommuting variables. If one sets $b_{++} = c^+$ to get a system with one anticommuting field only, the k-independent part of (3.2.54) would still make sense, but the k-dependent term would vanish by Fermi statistics.

Returning to (3.2.54), it is straightforward to use the form of the propagator to compute the two-point function of the energy–momentum tensor. This involves evaluating a one-loop diagram similar to that in fig. 3.1. One obtains

$$\langle T \left(T^k_{++}(\sigma^+) T^k_{++}(\sigma'^+) \right) \rangle = \frac{1}{8}(1 - 3k^2)(\sigma^+ - \sigma'^+)^{-4}. \qquad (3.2.57)$$

Setting $k = 3$ and comparing with (3.2.45), we see that the ghosts can cancel the Virasoro anomaly of 26 bosons. This is why the critical dimension of the Veneziano model is 26. The k dependence of (3.2.57) coincides with that of (3.1.51) if we set $k = 2J - 1$. The reason for this agreement will become clearer in the next subsection.

The superconformal ghosts that we will encounter in the next chapter have $k = 2$, so (3.2.57) and (3.2.45) indicate at first sight that they make the same contribution to the anomaly as -11 bosons. Actually, these ghosts are commuting fields unlike the anticommuting fields b and c; the statistics give a minus sign, and the superconformal ghosts will be shown have the same Virasoro anomaly as $+11$ bosons.

3.2.3 Virasoro, Conformal and Gravitational Anomalies

At the end of §3.1.1, in analyzing the $D\phi$ integral in (3.1.13), we made the assertion that this integral can be dropped in $D = 26$ since invariance under Weyl rescaling of the metric is valid in that case. We will establish this assertion in this section, and show that Weyl rescaling of the metric is valid precisely when the Virasoro anomaly cancels – something that we know happens precisely in $D = 26$. The key ideas here involve general properties of conformally invariant theories in two dimensions, and for illustrative purposes we will consider the theory of free fermions. In the course of our investigation, we will also meet a new kind of anomaly, the gravitational anomaly, which will interest us further in chapters 10 and 13.

Consider a real right-moving Majorana fermion ψ_+ with action

$$S = \frac{i}{\pi} \int d^2\sigma \psi_+ \partial_- \psi_+. \qquad (3.2.58)$$

The two-point function of the energy–momentum tensor is given by the expression in (3.2.57), except that now we set $k = 0$ and divide by a factor of two, since we have only a single mode ψ_+ instead of the pair b, c:[*]

$$\langle T_{++}(\sigma^+) T_{++}(\sigma'^+) \rangle = \frac{1}{16}(\sigma^+ - \sigma'^+)^{-4}. \qquad (3.2.59)$$

In this section, it is more useful to work in momentum space. At first sight it might appear messy to take the Fourier transform of the right-hand side of (3.2.59). This is, however, easily done using (3.2.48) above if one notes that in momentum space $\partial_- = ip_-, \partial_+ = ip_+$ and that the Fourier transform of $\delta^2(\sigma - \sigma')$ is 1. Hence the momentum-space counterpart of (3.2.59) is

$$\langle T_{++}(p) T_{++}(-p) \rangle = -\frac{\pi}{96} \frac{p_+^3}{p_-}. \qquad (3.2.60)$$

Previously we considered (3.2.59) on a flat world sheet, and interpreted it as evidence for a c-number anomaly in the commutator of T_{++} with itself. Now we consider the significance of (3.2.59) and (3.2.60) for fermions propagating on a curved world sheet. It is adequate to work to first order in the deviation from a flat world-sheet metric, so we set

$$h_{\alpha\beta} = \eta_{\alpha\beta} + f_{\alpha\beta}, \qquad (3.2.61)$$

where $f_{\alpha\beta}$ is the disturbance in the metric, which we will treat to lowest order. This is easily done. The interaction of matter with a gravitational

---

[*] This section uses a Euclidean signature on the world sheet, so we drop the T-ordering symbol.

field is given by

$$\Delta I = \frac{1}{2\pi} \int d^2\sigma f^{\alpha\beta} T_{\alpha\beta}. \tag{3.2.62}$$

For the simple system (3.2.58), the only nonzero component of $T_{\alpha\beta}$ would seem to be T_{++}, so the coupling is

$$f^{++}T_{++}/2\pi. \tag{3.2.63}$$

We would like to compute the expectation value of the induced fermion energy–momentum in a gravitational field. In view of (3.2.63), this can be read off from (3.2.60) and is

$$\langle T_{++}(p)\rangle = -\frac{1}{96}\frac{p_+^3}{p_-}f^{++}(p). \tag{3.2.64}$$

Now, we would like to verify the fundamental physical principle of conservation of energy–momentum. In a background gravitational field one expects

$$\langle D^\alpha T_{\alpha\beta}\rangle = 0. \tag{3.2.65}$$

In the case at hand, since the only nonzero component of T is T_{++}, and since to lowest order in the gravitational field we can replace covariant derivatives with ordinary ones, (3.2.65) reduces to $\langle\partial_-T_{++}\rangle = 0$. In momentum space this amounts to $p_-\langle T_{++}\rangle = 0$, which is obviously not true. Rather we have the anomalous formula

$$p_-\langle T_{++}\rangle = -\frac{1}{96}p_+^3 f^{++}(p) = -\frac{1}{24}p_+^3 f_{--}. \tag{3.2.66}$$

Again we see the near inevitability of anomalies in two dimensions. The formal reasoning that says that the left-hand side of (3.2.66) should vanish cannot be correct unless $\langle T_{++}\rangle = 0$, which would mean that there is no coupling at all to gravity. The breakdown (3.2.66) of energy–momentum conservation in the coupling of a chiral fermion to gravity is called a gravitational anomaly. It means that in two dimensions the coupling of a chiral fermion to gravity is nonsensical, unless extra degrees of freedom are added to cancel the anomaly.

   Notice that the right-hand side of (3.2.66) is a polynomial in momentum and so exhibits $\langle\partial_-T_{++}\rangle$ as a local functional of f_{--}. This is a general property of anomalies; the anomaly can be understood as an ultraviolet effect (though there are other ways to look at it as well), so it must be given by a local formula. Though (3.2.66) is local, (3.2.64) from which it was derived is not (because of the $1/p_-$ singularity). So there is no way to add a local term to (3.2.64) to eliminate the anomaly in (3.2.66). It is only upon verifying this that one can be sure that there is an anomaly. Were it possible to add a local term to (3.2.64) to achieve energy–momentum conservation, it would be physically sensible to do so.

Now, let us consider a theory with left-moving fermions ψ_- as well as the right mover ψ_+. The action of ψ_- is

$$S' = \frac{i}{\pi} \int d^2\sigma \psi_- \partial_+ \psi_-. \tag{3.2.67}$$

The analog of (3.2.64) is now

$$\langle T_{++} \rangle = -\frac{1}{24} \frac{p_+^3}{p_-} f_{--}$$
$$\langle T_{--} \rangle = -\frac{1}{24} \frac{p_-^3}{p_+} f_{++}. \tag{3.2.68}$$

At first, it seems that we are no better off than before, since (3.2.66) becomes

$$\langle p_- T_{++} \rangle = -\frac{1}{24} p_+^3 f_{--}$$
$$\langle p_+ T_{--} \rangle = -\frac{1}{24} p_-^3 f_{++}. \tag{3.2.69}$$

There is, however, a world of difference between (3.2.69) and (3.2.66). In the case of (3.2.69), the anomalous violation of energy–momentum conservation can be removed by adding local counterterms to (3.2.68) and writing

$$\langle T_{++} \rangle = -\frac{1}{24} \frac{p_+}{p_-} \left(p_+^2 f_{--} - 2p_+ p_- f_{+-} + p_-^2 f_{++} \right)$$
$$\langle T_{+-} \rangle = \frac{1}{24} \left(p_+^2 f_{--} - 2p_+ p_- f_{+-} + p_-^2 f_{++} \right) \tag{3.2.70}$$
$$\langle T_{--} \rangle = -\frac{1}{24} \frac{p_-}{p_+} \left(p_+^2 f_{--} - 2p_+ p_- f_{+-} + p_-^2 f_{++} \right).$$

Energy–momentum conservation is now obeyed:

$$\langle p_- T_{++} \rangle + \langle p_+ T_{+-} \rangle = \langle p_+ T_{--} \rangle + \langle p_- T_{+-} \rangle = 0. \tag{3.2.71}$$

Consequently, the theory described by the sum of (3.2.58) and (3.2.67) in which there are both left- and right-moving fermions can be consistently coupled to gravity. However, the form of (3.2.70) has a fateful consequence. Although formally it appears that (3.2.58) and (3.2.67) are invariant under Weyl rescaling of the metric, corresponding to a theory with $T_{+-} = 0$, in fact by the time we manage to respect energy–momentum conservation (a more fundamental physical principle than tracelessness of the energy–momentum tensor), we learn that the energy–momentum tensor is not traceless but has a trace given to lowest order in f by the second equation in (3.2.70). That equation is actually an approximation to a formula

$$T_{+-} = -\frac{1}{3}R^{(2)}, \tag{3.2.72}$$

where $R^{(2)}$ is the curvature scalar of the string world sheet. We conclude then that a two-dimensional theory with massless fermions coupled to gravity makes sense (energy–momentum conservation is valid) but does not possess at the quantum level the Weyl invariance that seems to be present in the classical Lagrangian.

The lesson is really much more general than the special case that we have considered. Consider a general two-dimensional theory that is scale invariant on a flat world sheet; thus $T_{+-} = 0$ and

$$\langle T_{++}(p)T_{++}(-p)\rangle = -\frac{c}{96}\frac{p_+^3}{p_-}$$
$$\langle T_{--}(p)T_{--}(-p)\rangle = -\frac{d}{96}\frac{p_-^3}{p_+} \tag{3.2.73}$$

for some constants c and d. By virtue of our discussion in the preceding section we know that c and d can be interpreted as the Virasoro anomaly of right- and left-moving modes, respectively. Upon coupling this theory to a curved world sheet, energy–momentum conservation breaks down unless $c = d$. If $c = d$, then Weyl invariance is lost, with an anomaly of the form (3.2.72), unless $c = d = 0$.

Applying these considerations to the Veneziano model, it has (with ghosts included) $c = d = (D - 26)$ in any space-time dimension D, so the world-sheet energy–momentum tensor is conserved even on a curved world sheet for any D. The world-sheet energy–momentum tensor, however, is traceless only for $D = 26$ when $c = d = 0$, and it is precisely in this case that the Weyl invariance used to eliminate the ϕ integral in (3.1.13) is valid.

3.2.4 Bosonization of Ghost Coordinates

The ghost equation of motion derived from (3.2.52) is

$$\partial_- c^+ = \partial_- b_{++} = 0. \tag{3.2.74}$$

This is the same equation as that obeyed by ϕ^+,

$$\partial_- \phi^+ = 0. \tag{3.2.75}$$

This raises the question of whether it is possible to express anticommuting variables such as c^+ and b_{++} in terms of the right-moving boson ϕ^+.

Free fermion field theory is largely specified by the two-point function

$$\langle c^+(\sigma^+) b_{++}(\sigma'^+) \rangle = \frac{1}{\sigma^+ - \sigma'^+}. \tag{3.2.76}$$

Can we find operators in the bosonic theory that reproduce this two-point function? Let

$$D_t(\sigma^+) = \mu^{t^2/2} : e^{it\phi^+(\sigma^+)} :, \tag{3.2.77}$$

where μ is the infrared cutoff present in (3.2.41). Using (3.2.41) and the sort of reasoning that was used in the introduction to calculate the expectation value of a product of tachyon vertex operators, the two-point function of the D_t is

$$\langle D_t(\sigma^+) D_{-t}(\sigma'^+) \rangle = (\sigma^+ - \sigma'^+)^{-t^2}. \tag{3.2.78}$$

The infrared cutoff μ cancels out of this formula. Comparison of (3.2.78) with (3.2.76) suggests the tentative identification

$$c^+(\sigma^+) \sim : e^{i\phi^+(\sigma^+)} :, \quad b_{++}(\sigma^+) \sim : e^{-i\phi^+(\sigma^+)} : \tag{3.2.79}$$

for which we will give further evidence shortly.[*] If we denote the conformal dimension of D_t as d_t, then from (3.2.78) we can read off the fact that

$$d_t + d_{-t} = t^2. \tag{3.2.80}$$

In chapter 7 we will have need for a generalization of (3.2.78), namely

$$\langle D_{t_1}(\sigma_1^+) \ldots D_{t_n}(\sigma_n^+) \rangle = \mu^{(\sum_{j=1}^n t_j)^2/2} \prod_{1 \le i < j \le n} (\sigma_i^+ - \sigma_j^+)^{t_i t_j}. \tag{3.2.81}$$

This vanishes as $\mu \to 0$ unless $\sum t_j = 0$. This reflects the fact that the left-hand side of (3.2.81) is invariant under $\phi^+ \to \phi^+ +$constant only if

---

[*] The ghosts obey $\langle c\,c \rangle = \langle b\,b \rangle = 0$. The counterpart of this statement is that $\lim_{\mu \to 0} \langle D_1 D_1 \rangle = \lim_{\mu \to 0} \langle D_{-1} D_{-1} \rangle = 0$.

$\sum t_j = 0$; as $\phi^+ \to \phi^+ +$constant is a continuous symmetry of the free boson field theory, and continuous symmetries cannot be spontaneously broken in quantum field theory in $1+1$ dimensions, the symmetry predicts the vanishing of (3.2.81) except when $\sum t_j = 0$. We have kept the infrared cutoff μ in the formulas until this point in order to explain why observables that are not invariant under adding a constant to ϕ^+ all vanish. The cutoff μ always cancels out of such observables, while other observables all vanish as $\mu \to 0$. In what follows, we restrict ourselves to observables that are invariant under a constant shift in ϕ^+, and we correspondingly delete μ from all equations. The tentative identification (3.2.79) shows that under $\phi^+ \to \phi^+ + a$, $c \to ce^{ia}$ while $b \to be^{-ia}$, suggesting that the symmetry of shifting ϕ^+ can be interpreted as fermion number. This will be confirmed by the discussion that follows.

To establish (3.2.79) conclusively we would like to establish that the bosonic operators indicated in (3.2.79) obey the correct fermion anticommutation relations. A typical relation that we would like to establish is the equal τ anticommutator

$$\{c^+(\sigma, \tau), c^+(\sigma', \tau)\} = 0. \tag{3.2.82}$$

To establish this, we study the product

$$D_1(\sigma, \tau)D_1(\sigma', \tau) = e^{i\phi(\sigma,\tau)} \exp(-i \int_\sigma^\infty d\tilde{\sigma}\partial_\tau\phi)\, e^{i\phi(\sigma',\tau)} \exp(-i \int_{\sigma'}^\infty d\tilde{\sigma}'\partial_\tau\phi), \tag{3.2.83}$$

where we have used the explicit formula for ϕ^+ given in (3.2.39). Rearranging (3.2.83) to move ϕ to the left and $\partial_\tau\phi$ to the right by use of the well-known formula $e^A e^B = e^B e^A e^{[A,B]}$, which holds when $[A, B]$ is a c-number, and using the canonical commutation relations of ϕ and $\partial_\tau\phi$, we get

$$D_1(\sigma, \tau)D_1(\sigma', \tau) = e^{i\pi\theta(\sigma'-\sigma)}e^{i\phi(\sigma,\tau)}e^{i\phi(\sigma',\tau)}$$

$$\times \exp(-i \int_\sigma^\infty d\tilde{\sigma}\partial_\tau\phi)\exp(-i \int_{\sigma'}^\infty d\tilde{\sigma}'\partial_\tau\phi). \tag{3.2.84}$$

Here $\theta(x)$ is $+1$ for positive x and 0 otherwise. The anticommutator $\{D_1(\sigma, \tau), D_1(\sigma', \tau)\}$ vanishes, since the phase factor on the right-hand side of (3.2.84) is an odd function of $\sigma - \sigma'$. We leave it to the enterprising reader to study the other anticommutators that are implicit in the identification (3.2.79).

We would now like to find a bosonic formula for the ghost-number current. As we recall the ghost-number current is $J_+ = c^+ b_{++}$, or more precisely

$$J_+(\sigma^+) = \lim_{\sigma'^+ \to \sigma^+} \left[c^+(\sigma^+) b_{++}(\sigma'^+) + \frac{i}{\sigma^+ - \sigma'^+} \right]. \qquad (3.2.85)$$

In the bosonic language this becomes

$$J_+(\sigma^+) = \lim_{\sigma'^+ \to \sigma^+} \left[: e^{i\phi^+(\sigma^+)} : : e^{-i\phi^+(\sigma'^+)} : + \frac{i}{\sigma^+ - \sigma'^+} \right]. \qquad (3.2.86)$$

To take the limit $\sigma'^+ \to \sigma^+$, we expand

$$e^{-i\phi^+(\sigma'^+)} = e^{-i\phi^+(\sigma^+)}(1 - i(\sigma'^+ - \sigma^+)\partial_+\phi^+). \qquad (3.2.87)$$

Inserted in (3.2.86), the first term on the right side of (3.2.87) gives a c-number that is discarded upon normal ordering. The second term on the right side of (3.2.87) would at first sight appear to vanish as $\sigma' \to \sigma$, but this is precisely canceled by the fact that $: e^{i\phi^+(\sigma^+)} : : e^{-i\phi^+(\sigma'^+)} :$ has a short-distance singularity proportional to $(\sigma^+ - \sigma'^+)^{-1}$. The limit in (3.2.86) thus actually gives

$$J_+ = \partial_+\phi^+, \qquad (3.2.88)$$

and this is the bosonized expression for the ghost-number current. Using the canonical commutation relations, it can be seen that

$$[J_+(\sigma, \tau), \phi^+(\sigma', \tau)] = -i\pi\delta(\sigma - \sigma'), \qquad (3.2.89)$$

and this confirms the earlier assertion that the symmetry under constant shift in ϕ^+ corresponds to ghost number.

Now we turn to the study of the energy–momentum tensor in the bosonized language. The ghost theory has (on a flat world sheet only) a 'ghost conjugation' symmetry $b \leftrightarrow c$. The ghost current $J_+ = c^+ b_{++}$ is odd under ghost conjugation. In view of (3.2.88) (or for that matter (3.2.79)), we must therefore interpret ghost conjugation as $\phi \to -\phi$. In the preceding section we formulated a one-parameter family of energy–momentum tensors (3.2.54), which differed from one another by addition of the derivative of the ghost current. The corresponding one-parameter family of energy–momentum tensors in the bosonic language is

$$T^k_{++} = \frac{1}{4}\partial_+\phi^+\partial_+\phi^+ - \frac{1}{8}ik\partial_+^2\phi^+. \qquad (3.2.90)$$

In particular, $k = 0$ is the unique choice that is invariant under ghost conjugation. Using formulas in the preceding section, the reader can easily verify that the Virasoro anomaly of T^k_{++} is proportional to $(1 - 3k^2)$, just as in the fermionic language.

At $k = 0$, the operators D_t and D_{-t} are related by ghost conjugation, and must have the same conformal dimension. Using (3.2.80) we can thus see that the conformal dimension of D_t at $k = 0$ is $t^2/2$. What about $k \neq 0$? D_t is an operator of ghost number t, in view of the identification that we have made of the ghost number in bosonic terms, so the conformal dimension of D_t at general k is determined in terms of the value at $k = 0$ by (3.2.56):

$$d_t(k) = t^2/2 - kt/2. \tag{3.2.91}$$

This formula plays a crucial role in constructing fermion vertex operators for superstrings.

Just as in the fermionic description, the existence of a one-parameter family of energy–momentum tensors ought, logically, to correspond to the existence of a one-parameter family of couplings of the free field ϕ on a *curved* world sheet. The relevant family is

$$S_k = -\frac{1}{2\pi} \int d^2\sigma \sqrt{h} \left(\partial_\alpha \phi \partial^\alpha \phi - \frac{i}{2} k R^{(2)} \phi \right), \tag{3.2.92}$$

with $R^{(2)}$ being the scalar curvature of the string world sheet. On a flat world sheet, (3.2.92) is independent of k. However, varying (3.2.92) with respect to the world-sheet metric to derive the energy–momentum tensor and *then* setting the metric to $\eta_{\alpha\beta}$, one derives the k-dependent energy–momentum tensor (3.2.90), as the reader is invited to check. A significant feature of (3.2.92) is that ghost number is conserved ((3.2.92) is invariant under $\phi \to \phi+$ constant) only if $k = 0$. The violation of ghost number on a curved world sheet can be straightforwardly read off from (3.2.92); in the fermionic description it is far less straightforward, and requires discussion of anomalies in one-loop diagrams on a curved world sheet. Alternatively, in the fermionic description the violation of ghost number of a curved world sheet shows up in the fact that on a compact world sheet the b and c fields have different numbers of normalizable zero modes, something that we will discuss for the case $k = 3$ in the next section.

We have been discussing the bosonization of fermions for (3.2.92) formulated on a $(1+1)$-dimensional world of infinite volume. For many applications, however, it is essential to study the bosonization of fermions that are propagating on a circle. If we study the free theory (3.2.92) on a finite one-dimensional world with $0 \leq \sigma \leq 2\pi$ and our familiar periodic boundary conditions,[*] then ϕ^+ has the traditional normal mode expansion that we have discussed extensively:

$$\phi^+(\sigma) = \phi_0 + \sigma p_0 + i \sum_{n \neq 0} \frac{1}{n} \phi_n e^{-in\sigma}. \qquad (3.2.93)$$

Here $[p_0, \phi_0] = -i$, and $[\phi_n, \phi_m] = n\delta_{n+m}$. Notice that p_0 is the ghost-number operator that shifts ϕ by a constant. One may ask whether bosonization of fermions is valid not just in infinite volume but also on a circle. The answer turns out to be rather subtle and full of consequences. We try to define as in infinite volume

$$c^+(\sigma) =: e^{i\phi^+(\sigma)} :, \, b_{++}(\sigma) =: e^{-i\phi^+(\sigma)} : . \qquad (3.2.94)$$

We must decide precisely what the normal ordering is supposed to mean. A correct prescription turns out to be

$$c^+(\sigma) = \exp(-\sum_{n<0} \frac{1}{n} e^{-in\sigma} \phi_n) \, e^{i\phi_0} \, e^{i\sigma(p_0+1/2)} \, \exp(-\sum_{n>0} \frac{1}{n} e^{-in\sigma} \phi_n)$$

$$(3.2.95)$$

$$b_{++}(\sigma) = \exp(\sum_{n<0} \frac{1}{n} e^{-in\sigma} \phi_n) \, e^{-i\phi_0} \, e^{-i\sigma(p_0+1/2)} \, \exp(\sum_{n>0} \frac{1}{n} e^{-in\sigma} \phi_n).$$

$$(3.2.96)$$

The patient reader should be able, with mode by mode use of the identity $e^A c^B = e^B e^A e^{[A,B]}$, to verify that c^+ and b_{++} defined in those equations do obey the correct anticommutation relations. The fact that $p_0 + 1/2$, rather than p_0, appears in (3.2.95) and (3.2.96) may seem rather surprising, but it is needed for a correct Bose–Fermi correspondence. Thus, the periodicity $c^+(\sigma + 2\pi) = c^+(\sigma)$, $b_{++}(\sigma + 2\pi) = b_{++}(\sigma)$ requires that the ghost number operator p_0 in (3.2.95) and (3.2.96) should have half-integral eigenvalues, and this is in agreement with our earlier determination that the ghost number of the states of the bosonic open string is indeed half-integral.

The requirement that p_0 should have half-integral eigenvalues has an interesting 'physical' interpretation. Since p_0 is canonically conjugate to the zero-mode coordinate ϕ_0 (thus $p_0 = -i\partial_{\phi_0}$), saying that p_0 only has half-integral eigenvalues means that ϕ_0 is an angular variable, ϕ_0 and $\phi_0 +$

---

[*] In this book the closed-string periodicity is taken to be π, but for open strings after doubling of the interval the periodicity is effectively 2π. We adopt the latter convention here.

2π being physically equivalent, with a quantum wave functional Ψ that (suppressing other variables) obeys $\Psi(\phi_0 + 2\pi) = -\Psi(\phi_0)$. Actually, the form of the mode expansion (3.2.93) shows that when we add a constant to ϕ_0, the whole quantum field $\phi(\sigma)$ is shifted by that constant, so we could express this behavior without making any mode expansion by saying that the exact quantum wave functional obeys $\Psi(\phi(\sigma)) = -\Psi(\phi(\sigma) + 2\pi)$.

The crucial point here is that p_0 takes discrete values; this is more fundamental than the fact that those values are half integers instead of integers. When in finite volume one bosonizes a fermion theory, the resulting Bose field is an angular variable, living on a circle of circumference 2π in terms of the above conventions. Actually, when one bosonizes not just a single pair of anticommuting fields b_{++} and c^+ but n such pairs, one must introduce n Bose fields $\phi^a, a = 1, \ldots, n$, and some very important generalizations of the above statement become possible, as we will discuss in chapter 6.

Although less basic, the peculiar minus sign in the condition $\Psi(\phi(\sigma) + 2\pi) = -\Psi(\phi(\sigma))$ also merits comment. The minus sign originated from requiring half-integral eigenvalues for p_0 in order that $c^+(\sigma)$ and $b_{++}(\sigma)$ should obey the appropriate boundary conditions for the ghost coordinates, namely $c^+(\sigma + 2\pi) = c^+(\sigma)$, $b_{++}(\sigma + 2\pi) = b_{++}(\sigma)$. If we choose to study fermion fields that obey the opposite boundary conditions $c^+(\sigma + 2\pi) = -c^+(\sigma)$, $b_{++}(\sigma + 2\pi) = -b_{++}(\sigma)$, then we would in (3.2.96) and (3.2.95) require integral eigenvalues for p_0, so the bosonized wave functional would obey the opposite condition $\Psi(\phi(\sigma) + 2\pi) = +\Psi(\phi(\sigma))$.

The ability to bosonize fermions on a circle has some interesting consequences. In terms of ϕ^+, the Hamiltonian $H = \int_0^{2\pi} d\sigma T_{++}$ of the free field theory is

$$H = p_0^2/2 + \sum_{n=1}^{\infty} \phi_{-n}\phi_n - 1/24, \qquad (3.2.97)$$

where we have included the normal-ordering constant that was one of the fruits of our study of string quantization. The ghost number operator is

$$U = p_0. \qquad (3.2.98)$$

One knows from elementary quantum statistical mechanics how to write down the partition function of such a free field theory:

$$\operatorname{tr} e^{-\beta H} = e^{\beta/24} \sum_{n=-\infty}^{\infty} e^{-\beta(n+1/2)^2/2} \prod_{m=1}^{\infty} \frac{1}{(1 - e^{-\beta m})}. \qquad (3.2.99)$$

It is quite useful to calculate not just $\operatorname{tr} e^{-\beta H}$ but the more general trace tr $e^{-\beta H} e^{i\theta U}$. Letting $q = e^{-\beta}$, this can be expressed

$$\mathrm{tr}\, q^H e^{i\theta U} = q^{-1/24} \sum_{n=-\infty}^{\infty} q^{(n+1/2)^2/2} e^{i\theta(n+1/2)} \prod_{m=1}^{\infty} \frac{1}{1-q^m}. \qquad (3.2.100)$$

What happens if we express H and U in terms of fermions rather than bosons? In this case,

$$H = \sum_{n=1}^{\infty} n(b_{-n}c_n + c_{-n}b_n) + x \qquad (3.2.101)$$

(of course, the zero modes b_0 and c_0 do not appear in H), where x is an unknown normal-ordering constant, and as in (3.2.28)

$$U = \frac{1}{2}(c_0 b_0 - b_0 c_0) + \sum_{n=1}^{\infty}(c_{-n}b_n - b_{-n}c_n). \qquad (3.2.102)$$

Using standard methods of quantum statistical mechanics, (3.2.101) and (3.2.102) give

$$\mathrm{tr}\, q^H e^{i\theta U} = q^x 2\cos(\theta/2) \prod_{n=1}^{\infty}(1+q^n e^{i\theta})(1+q^n e^{-i\theta}). \qquad (3.2.103)$$

It is not obvious that (3.2.100) is equal to (3.2.103) for any value of x, but it is a theorem due to Jacobi that these identities coincide if $x = 1/12$. Indeed, Jacobi's triple product formula

$$\sum_{n=-\infty}^{+\infty} q^{(n+1/2)^2/2} e^{i(n+1/2)\theta}$$

$$= 2q^{1/8} \cos(\theta/2) \prod_{n=1}^{\infty}(1-q^n)(1+q^n e^{i\theta})(1+q^n e^{-i\theta}) \qquad (3.2.104)$$

is tantamount to the equivalence of (3.2.100) and (3.2.103) for $x = 1/12$. We will find use for (3.2.104), originally proved by Jacobi by the study of elliptic functions, in appendix 8.A.

Another interesting result of this investigation is the following. We have learned that if p_0 is a half-integer, the bosonic Hamiltonian (3.2.97) is equivalent to a Hamiltonian for two anticommuting degrees of freedom b and c obeying *periodic* boundary conditions $b(\sigma + 2\pi) = b(\sigma)$, $c(\sigma + 2\pi) = c(\sigma)$. The ground-state energy of (3.2.97) with half-integral p_0 is

$(1/2)^3 - 1/24 = 1/12$. So we see that the normal-ordering constant for two anticommuting fermions is $1/12$, so that the normal-ordering constant for a single anticommuting degree of freedom (a Majorana fermion) with periodic boundary conditions must be

$$\epsilon^F_+ = +1/24. \qquad (3.2.105)$$

On the other hand, if we take p_0 in (3.2.97) to be integral, then (3.2.97) is, from the above analysis, equivalent to a theory of two anticommuting fields b and c obeying *antiperiodic* boundary conditions, $b(\sigma+2\pi) = -b(\sigma)$, $c(\sigma + 2\pi) = -c(\sigma)$. With integer p_0, the ground state energy of (3.2.97) is $-1/24$, so we conclude that the normal-ordering constant for fermions that obey antiperiodic boundary conditions is

$$\epsilon^F_- = -1/48 \qquad (3.2.106)$$

for each (Majorana) field. As for bosons, we know from chapter 2 that the normal-ordering constant for bosons with periodic boundary conditions is

$$\epsilon^B_+ = -1/24. \qquad (3.2.107)$$

More general boundary conditions can be considered, but these are the cases for which we have the most need in this book. The above considerations may seem somewhat heuristic. We will recover the same results by methods that are perhaps more rigorous in the next chapter. The above results can also be obtained by considering zeta functions, as was done for ϵ^B_+ in §2.3.1.

3.3   Global Aspects of the String World Sheet

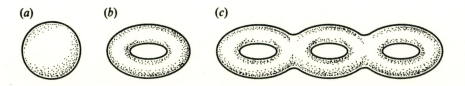

(a)   *(b)*   *(c)*

Figure 3.2. World-sheet diagrams for a theory of closed, oriented strings: (*a*) tree, (*b*) one loop, (*c*) multiloop.

In all of our discussion of the bosonic string, we have used the fact that by combined use of reparametrization invariance and Weyl invariance, we can locally put the world-sheet metric $h_{\alpha\beta}$ in any prescribed form. Indeed, a world-sheet reparametrization depends on two arbitrary functions, and a Weyl rescaling of the metric depends on one; together these are enough to gauge away the three independent components of the world-sheet metric. This simple counting of degrees of freedom has been enough for our purposes so far, but in deeper aspects of the theory the global understanding of the string world sheet plays an essential role; we give here an elementary exposition of some of these matters.

We concentrate on the case of oriented closed strings; this leads to the most elegant analysis. The world sheet of an n-loop closed-string diagram is, as in fig. 3.2, a sphere with n handles. Tree-level diagrams correspond to a world sheet that has no handles and is topologically an ordinary sphere, as in fig. 3.2a, while one-loop diagrams correspond to a torus, as in fig. 3.2b, and multiloop diagrams give rise to a surface with n handles, as in fig. 3.2c. A sphere with n handles is known as a Riemann surface of genus n.

In the tree-level case (genus 0), there is a theorem due to Riemann stating that the metric can be put *globally* in a standard form by a diffeomorphism plus Weyl rescaling. Given, say, the round metric h_0 on the sphere S^2, *any* metric h is up to a diffeomorphism of the form $h = e^\phi h_0$. This may be proved by use of the Riemann–Roch theorem (or the Atiyah–Singer index theorem), but we will not do so in this book. Two metrics on a Riemann surface that can be related to each other by a reparametrization plus Weyl rescaling are said to be conformally equivalent; Riemann's theorem says that any two metrics on S^2 are conformally equivalent.

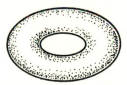 

Figure 3.3. Two tori that are conformally inequivalent.

Next, we move on to the one-loop case, genus one. Here life is much more interesting. A genus one surface is an ordinary torus. It is no longer true that globally any two metrics on the torus are equivalent up to a diffeomorphism plus Weyl rescaling. The two tori sketched in fig. 3.3 cannot be related in this way. To describe the difference analytically

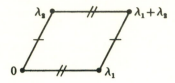

Figure 3.4. A torus can be made by identifying the marked line segments in the complex z plane.

(but only heuristically), note that a torus can be constructed as follows. Starting with the complex z plane, pick two complex numbers λ_1 and λ_2 such that

$$\tau = \lambda_2/\lambda_1 \qquad (3.3.1)$$

is not real.[*] By exchanging λ_1 with λ_2, if necessary, we may suppose that $\mathrm{Im}\,\tau > 0$, so that τ defines a point in the upper half plane. Then we define a torus by making the identifications

$$z \approx z + n\lambda_1 + m\lambda_2 \qquad (3.3.2)$$

for arbitrary integers n and m, as shown in fig. 3.4. This torus inherits from the z plane a flat metric, and we wish to ask whether tori defined using different values of λ_1 and λ_2 are equivalent under diffeomorphisms plus Weyl rescalings.

It is fairly clear that only the ratio $\tau = \lambda_2/\lambda_1$ can be a diffeomorphism plus Weyl invariant – usually called a conformal invariant. Indeed, a complex rescaling of z,

$$z \to z' = kz \qquad (3.3.3)$$

k being a nonzero complex number, changes the metric of the torus only by the constant $|k|^2$, which can be absorbed in a conformal rescaling. The transformation (3.3.3) rescales λ_1 and λ_2, while leaving fixed the ratio τ. It is thus only τ that might be a conformal invariant.

It is almost but not quite true that τ is a conformal invariant that cannot be changed by diffeomorphisms plus Weyl rescalings. Let a, b, c and d be four integers with $ad - bc = 1$, or in other words such that the

---

[*] The name τ for this variable is conventional, and we will follow this convention, hoping the parameter τ defined in (3.3.1) will not be confused with the τ coordinate on the string world sheet.

matrix

$$\begin{pmatrix} a & b \\ c & d \end{pmatrix} \qquad (3.3.4)$$

has determinant one. Such integer-valued matrices of determinant one form a group called the modular group $SL(2, Z)$. In claiming that these matrices form a group, the slightly nontrivial assertion is the claim that the matrix in (3.3.4) has an integer-valued inverse, namely

$$\begin{pmatrix} d & -b \\ -c & a \end{pmatrix}. \qquad (3.3.5)$$

Suppose that we transform λ_1 and λ_2 by an element of $SL(2, Z)$ to

$$\begin{pmatrix} \lambda_2' \\ \lambda_1' \end{pmatrix} = \begin{pmatrix} a & b \\ c & d \end{pmatrix} \cdot \begin{pmatrix} \lambda_2 \\ \lambda_1 \end{pmatrix}. \qquad (3.3.6)$$

Then the torus defined by

$$z \approx z + n'\lambda_1' + m'\lambda_2' \qquad (3.3.7)$$

is exactly the same as that defined by (3.3.2); indeed, (3.3.2) can be converted into (3.3.7) by transforming n and m by the matrix (3.3.5).

The fact that (3.3.2) and (3.3.7) define equivalent tori means that the conformal structure of a torus is invariant under a certain action of the modular group on τ; by comparing (3.3.2) to (3.3.7) we can determine this action to be

$$\tau \to (a\tau + b)/(c\tau + d). \qquad (3.3.8)$$

The complex number τ, subject to the equivalence relation (3.3.8), is the only feature of the metric of a torus that cannot be absorbed in a diffeomorphism plus Weyl transformation, though a full proof of this assertion would go beyond the scope of this book.

We will now discuss more briefly the case of a surface of genus $g > 1$. Consider adding a handle to a surface Σ of genus g to make a surface of genus $g + 1$. Roughly speaking, one must first make two punctures in Σ, as depicted in fig. 3.5a. One then joins the two punctures by a tube shown in fig. 3.5b, which may have an arbitrary length and may be twisted by an arbitrary angle, as in fig. 3.5c. To specify the position of either of the two punctures requires two real parameters or one complex parameter. Altogether, in going from genus g to genus $g+1$, we introduce

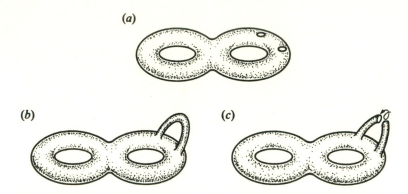

Figure 3.5. Adding a handle to a surface of genus g to make a surface of genus $g+1$ can be done in stages. One makes two punctures, as in (a), and then connects them with a tube of given length and 'twist' angle, as in (b). The purpose of (c) is to underscore the appearance of the twist angle. We have glued in two half-tubes in the two punctures; to complete the construction, we must glue together the ends of the tubes, but there is an ambiguous relative angle in doing this.

six new real parameters (four from the positions of the two punctures, as well as a length and an angle) or three complex parameters. If we denote the number of conformally invariant complex parameters that enter in describing a surface Σ of genus g as B_g, then the argument indicates heuristically that

$$B_{g+1} = B_g + 3. \tag{3.3.9}$$

In fact, (3.3.9) is only valid for $g \geq 2$, because surfaces of genus zero and one have continuous symmetries and for $g = 0$ or $g = 1$ there is no invariant significance in the choice of the positions of the punctures in fig. 3.5. The actual value of B_g is $(3g - 3)$ for $g \geq 2$. We can heuristically motivate this value as follows. A surface of genus zero has enough symmetry so that the positions of both punctures in fig. 3.5 are irrelevant, so that there is only one complex parameter (or two real parameters: a length and twist) added in going from genus zero to genus one. Hence $B_1 = B_0 + 1 = 1$. A surface of genus one has (in the presentation of it given above) only the rigid translations of the z plane for continuous symmetries; this is enough to shift the position of one puncture to, say, the origin, but the position of the second puncture (as well as the length and twist parameters) has invariant meaning, so $B_2 = B_1 + 2 = 3$. Beyond this point, the surface of genus g has no continuous symmetries, so (3.3.9) takes over, giving $B_g = 3g - 3$.

We will discuss here one other aspect of the global geometry of the string world sheet. In performing the gauge-fixed path integral of (3.1.13) including the ghosts, one encounters the question of whether c^+ and b_{++} may have normalizable zero modes on the string world sheet. The equations are

$$\nabla_- c^+ = 0, \qquad \nabla_+ c^- = 0 \qquad (3.3.10)$$

and

$$\nabla_- b_{++} = 0, \qquad \nabla_+ b_{--} = 0. \qquad (3.3.11)$$

In discussing the significance of these equations, the key element is to recall the transformation law of the metric under infinitesimal coordinate reparametrizations $\sigma^\alpha \rightarrow \sigma^\alpha + \xi^\alpha$; this is as in (3.1.6)

$$\delta h_{++} = 2\nabla_+ \xi_+, \qquad \delta h_{--} = 2\nabla_- \xi_-. \qquad (3.3.12)$$

Comparing (3.3.12) to (3.3.10), we see that a zero mode of c is the generator of a 'conformal symmetry' – a world-sheet reparametrization that changes the metric only by a multiple of itself. Such a change in the metric can of course be absorbed in a Weyl rescaling.

At tree level, the world sheet being a sphere, we can make a stereographic projection onto the complex plane. If we set $z = \tau + i\sigma, \bar{z} = \tau - i\sigma$, then the equation for, say, c^+ becomes

$$\frac{\partial}{\partial \bar{z}} c^+ = 0 \qquad (3.3.13)$$

so that c^+ must be an analytic function of z. For reasons that were explained at the end of §1.4.3, in order for the conformal symmetry generated by $c^+(z)\partial_z$ to have no pole at infinity, $c^+(z)$ must grow at infinity at most like z^2. There are thus three acceptable solutions of (3.3.13), namely $c^+ = 1$, $c^+ = z$, and $c^+ = z^2$. The corresponding conformal transformations $c^+(z)\partial_z$ generate a closed Lie algebra, as the reader should verify. The Lie algebra in question is that of the three parameter group $SL(2, C)$.

To analyze ghost zero modes for higher genus surfaces, we note that the equation for c^+ in (3.3.10) implies

$$0 = \nabla_+ \nabla_- c^+ = \frac{1}{2}(\nabla_+ \nabla_- + \nabla_- \nabla_+)c^+ + \frac{R^{(2)}}{2}c^+, \qquad (3.3.14)$$

with $R^{(2)}$ being the scalar curvature of the surface. Multiplying by c^{+*}

and integrating over the world sheet Σ, this implies

$$
\begin{aligned}
0 &= \int_\Sigma c^{+*}(\nabla_+\nabla_- + \nabla_-\nabla_+ + R^{(2)})c^+ \\
&= -\int_\Sigma \left(|\nabla_-c^+|^2 + |\nabla_+c^+|^2 - R^{(2)}|c^+|^2 \right).
\end{aligned}
\tag{3.3.15}
$$

In the case of the torus, with its flat metric, the last term drops out of (3.3.15), which then implies that c^+ is covariantly constant and thus (as the torus is flat) actually constant. There is thus precisely one normalizable ghost zero mode on a genus-one surface, namely $c^+ = 1$. Correspondingly, the only conformal symmetry of a torus is the rigid translation $z \to z + a$, with complex a. For genus greater than one, it can be shown that the surface Σ admits a metric of everywhere negative scalar curvature, and consequently (3.3.15) implies in this case that $c^+ = 0$; there are no normalizable ghost zero modes on a surface of genus greater than one. If we denote the number of ghost zero modes on a genus g surface as C_g, then $C_0 = 3$, $C_1 = 1$, and $C_g = 0$ for $g \geq 2$.

Now we turn to antighost zero modes. As a preliminary to discussing the significance of (3.3.11), let us consider the following. We have discussed qualitatively above the fact that on a surface of genus g for $g > 0$ there are conformally inequivalent metrics. Let us now put this on a quantitative basis. Picking a background metric $h_{\alpha\beta}$, let us ask whether a general perturbation $\delta h_{\alpha\beta}$ of h can be absorbed in a reparametrization plus Weyl rescaling. Working in a local coordinate system on the world sheet Σ in which $h_{++} = h_{--} = 0$, $h_{+-} = e^\phi$, it is clear that δh_{+-} can be absorbed in a Weyl rescaling; the real question is whether δh_{++} and δh_{--} can be absorbed in a diffeomorphism. As δh_{--} is the complex conjugate of δh_{++}, we may as well study the latter. Examination of (3.3.12) shows that δh_{++} can be absorbed in a diffeomorphism if and only if there is some globally defined ξ_+ with $\delta h_{++} = 2\nabla_+\xi_+$. If this is *not* true, then

$$
S = \int_\Sigma |\delta h_{++} - 2\nabla_+\xi_+|^2
\tag{3.3.16}
$$

is nonzero for all ξ_+. Even if we cannot choose ξ_+ to make (3.3.16) vanish, we can certainly expect to choose ξ_+ to minimize (3.3.16). The variational equation for minimizing (3.3.16) with respect to ξ_+ is that

$$
\nabla_-(\delta h_{++} - 2\nabla_+\xi_+) = 0.
\tag{3.3.17}
$$

With $b_{++} = \delta h_{++} - 2\nabla_+\xi_+$, we can recognize in (3.3.17) the equation (3.3.11) for antighost zero modes! In fact, antighost zero modes are in one-to-one correspondence with choices of δh_{++} for which (3.3.16) cannot be made to vanish, or in other words with deformations of the metric of Σ that cannot be absorbed in reparametrization plus Weyl rescaling. Thus, the number of antighost zero modes on a surface of genus g is the number that we previously called B_g, namely $B_0 = 0$, $B_1 = 1$, $B_g = 3g - 3$, $g \geq 2$. These values for $g = 0$ and $g = 1$ can be checked by studying (3.3.11) explicitly. Indeed, for $g = 0$, upon making the stereographic projection to the plane, (3.3.11) becomes

$$\frac{\partial}{\partial z} b_{++} = 0 \qquad (3.3.18)$$

so that b_{++} must be an analytic function of z. However, study of (3.3.11) shows that we must require that $b_{++} \to 0$ as $z \to \infty$, which is impossible for an analytic function, so b_{++} has no normalizable zero modes. For genus one, (3.3.11) shows that b_{++} (like c^+) must be covariantly constant and hence constant, so that there is precisely one antighost zero mode for genus one.

While the behavior of B_g and C_g as a function of g is somewhat irregular for small g, it is noteworthy that the difference

$$\Delta_g = C_g - B_g = -3(g - 1) \qquad (3.3.19)$$

behaves more smoothly. This difference is indeed given by a classical theorem known as the Riemann–Roch theorem, whose modern generalization, the Atiyah–Singer index theorem, will be discussed in chapter 14.

The conformally invariant parameters that enter in specifying the metric of a Riemann surface of genus g are known as moduli of the Riemann surface. The space of these parameters is called moduli space. Loop integrals over string world sheets include integrals over moduli space, which enter in one way or another depending on the formalism. In formalisms in which the ghosts are present, their zero modes play an important role in obtaining the correct integration measure in integrating over moduli space, since as we have seen the b and c zero modes are related to infinitesimal deformations of the moduli and to unbroken symmetries left by the gauge-fixing procedure, respectively. The modular group $SL(2, Z)$ that we defined at the one-loop level has a multiloop generalization.

Without taking account of the discrete $SL(2, Z)$ equivalences, the conformal structure of a genus-one surface would be specified by a point τ in the upper half of the complex plane. The upper half plane is called the Teichmüller space of the genus-one surface, while its quotient by $SL(2, Z)$ (identifying values of τ that are related by $SL(2, Z)$) is the actual moduli space. This has an analog for multiloop surfaces; there is a comparatively simple Teichmüller space that describes the conformal structure of the surface up to certain discrete equivalences, while the moduli space, which takes account of these discrete equivalences, is more subtle to describe. In many approaches to loop diagrams, an integration over Teichmüller space appears naturally, and modular invariance is not self-evident. Modular invariance is needed for an acceptable theory, since (in the one-loop case, for example) surfaces described by values of τ that are related by $SL(2, Z)$ are really the 'same' surface up to reparametrization, and modular invariance is one aspect of reparametrization invariance. We will describe one-loop integrals in this book in detail, but the most efficient approach to multiloop integrals is still under active investigation, and this subject will not be presented here.

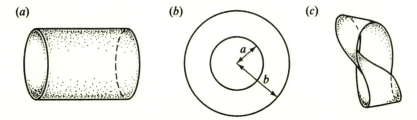

(a)   *(b)*   *(c)*

Figure 3.6. The simplest one-loop open-string world sheet is the cylinder of (a). It can always be conformally mapped onto a standard annulus in the complex plane, as in (b). The other one-loop string world sheet is the twisted cylinder, or Möbius strip, of (c), which leads to similar considerations.

Our considerations have analogs for open strings and world sheets with boundary. For example, at one-loop level for open strings we meet a world sheet that is topologically a cylinder, as in fig. 3.6a. According to classical theorems, any metric on the cylinder is conformally equivalent to the standard flat metric on the annulus $a \leq |z| \leq b$ in the complex z plane, for some a and b as shown in fig. 3.6b. Which boundary of the cylinder corresponds to the outer boundary of the annulus and which to the inner boundary is a matter of convention, since the two boundaries of the annulus are exchanged by the conformal mapping of the complex plane $z \rightarrow ab/z$ (which changes the metric only by an irrelevant Weyl rescaling). By scaling $z \rightarrow z/b$, we can set $b = 1$, so the conformal structure of the annulus is described by the ratio $x = a/b$, which is a real parameter that ranges from 0 to 1; an integral over this parameter must appear in one-loop integrals for open strings. A similar parameter appears in the other one-loop open-string world sheet of fig. 3.6c.

3.4   Strings in Background Fields

3.4.1 Introduction of a Background Space-Time Metric

Until now, we have discussed the propagation of a string in flat 26-dimensional Minkowski space. The action is

$$S_0 = -\frac{1}{2\pi} \int d^2\sigma \sqrt{h} h^{\alpha\beta} \partial_\alpha X^\mu \partial_\beta X^\nu \eta_{\mu\nu}, \qquad (3.4.1)$$

where $h_{\alpha\beta}$ is the world-sheet metric, which is regarded as a dynamical variable, and $\eta_{\mu\nu}$ is the Minkowski metric in space-time. We would like in this section to consider string propagation not on flat Minkowski space but on some more general 26-dimensional manifold **M** with metric tensor $g_{\mu\nu}$. The obvious generalization of (3.4.1) is to replace the Minkowski metric $\eta_{\mu\nu}$ with $g_{\mu\nu}$, giving

$$S = -\frac{1}{2\pi} \int d^2\sigma \sqrt{h} h^{\alpha\beta} \partial_\alpha X^\mu \partial_\beta X^\nu g_{\mu\nu}(X^\rho). \qquad (3.4.2)$$

Equation (3.4.2) is such a natural generalization of (3.4.1) that in a sense it is hardly necessary to derive it, but it is instructive to consider the following. Suppose that the space-time metric is

$$g_{\mu\nu}(X^\rho) = \eta_{\mu\nu} + f_{\mu\nu}(X^\rho), \qquad (3.4.3)$$

where f – which we treat as a perturbation – represents the deviation from Minkowski space. The world-sheet path integral derived from (3.4.1) is

$$Z_0 = \int DX^\mu Dh_{\alpha\beta} e^{-S_0}, \qquad (3.4.4)$$

while that derived from (3.4.2) is

$$\begin{aligned}
Z &= \int DX^\mu Dh_{\alpha\beta} \, e^{-S} \\
&= \int DX^\mu Dh_{\alpha\beta} \, e^{-S_0} \left(1 + \frac{1}{2\pi} \int d^2\sigma \sqrt{h} h^{\alpha\beta} \partial_\alpha X^\mu \partial_\beta X^\nu f_{\mu\nu}(X^\rho) \right. \\
&\left. + \frac{1}{2} \left[\frac{1}{2\pi} \int d^2\sigma \sqrt{h} h^{\alpha\beta} \partial_\alpha X^\mu \partial_\beta X^\nu f_{\mu\nu}(X^\rho) \right]^2 + \ldots \right).
\end{aligned}$$

$$(3.4.5)$$

Here

$$\frac{1}{2\pi} \int d^2\sigma \sqrt{h} h^{\alpha\beta} \partial_\alpha X^\mu \partial_\beta X^\nu f_{\mu\nu}(X^\rho) \qquad (3.4.6)$$

is the vertex operator V for emission of a graviton of wave function $f_{\mu\nu}(X^\rho)$. (We have hitherto generally considered gravitons whose wave function is a plane wave $f_{\mu\nu}(X^\rho) = \zeta_{\mu\nu}e^{ik\cdot X}$, but there is no reason not to consider instead a wave function that is a general superposition of plane waves.) An insertion of V in the Minkowskian path integral Z_0 of (3.4.4) would accommodate the interaction of strings with an external graviton of wavefunction $f_{\mu\nu}$. The insertion of e^V in (3.4.4) would describe interaction with a coherent state of gravitons in this wave function, and this corresponds precisely to string propagation in the metric $g_{\mu\nu} = \eta_{\mu\nu} + f_{\mu\nu}$.

We would now like to consider some of the simple properties of (3.4.2). The actions (3.4.1) and (3.4.2) are both two-dimensional quantum field theories, but there is an essential difference. The action (3.4.1) becomes a free field theory in the conformal gauge

$$h_{\alpha\beta} = \eta_{\alpha\beta}, \qquad (3.4.7)$$

but (3.4.2) does not. The action (3.4.2) reduces in this gauge to

$$S' = -\frac{1}{4\pi\alpha'} \int d^2\sigma \partial_\alpha X^\mu \partial^\alpha X^\nu g_{\mu\nu}(X^\rho). \qquad (3.4.8)$$

This is a nontrivial quantum field theory, which is known as the nonlinear sigma model. We have restored the α' dependence; the previous formulas correspond to the usual choice $\alpha' = 1/2$.

Of course, just as in the case of string propagation in flat Minkowski space, (3.4.8) must be supplemented by the Virasoro conditions

$$T_{\alpha\beta} = 0, \qquad (3.4.9)$$

which are conjugate to the gauge choice (3.4.7). Formally, (3.4.8) is invariant under rescalings or conformal mappings of the σ's (this being a reflection of the underlying Weyl invariance in (3.4.2)), so at the classical level we have

$$T_{+-} = 0, \qquad (3.4.10)$$

just as in our study of string propagation in flat space. Equation (3.4.10) leaves the two sets of Virasoro conditions

$$T_{++} = T_{--} = 0, \qquad (3.4.11)$$

and these (in the critical dimension) are just enough to eliminate the modes of negative norm while leaving an interesting theory. If there were an anomaly in the classical statement (3.4.10), then (3.4.11) would have to supplemented with the extra Virasoro condition $T_{+-} = 0$. This condition, which has no counterpart in the flat-space limit, would certainly lead to inconsistencies.

Even in the case of flat Minkowski space ($g_{\mu\nu} = \eta_{\mu\nu}$), where (3.4.8) reduces to a free field theory, we have found that there can be an anomaly in T_{+-} if the string world sheet is curved. This anomaly, which arises in $D \neq 26$, was found to be of the form $T_{+-} \sim R^{(2)}$, with $R^{(2)}$ being the curvature scalar of the world sheet. This is a relatively gentle (though unacceptable) sort of anomaly; if (3.4.8) is formulated on a world sheet with a given geometry, then $R^{(2)}$ is a definite c-number function of the world-sheet coordinates σ and τ, so that the conformal anomaly encountered in §3.2.2 is a c number only. In the case $g_{\mu\nu} \neq \eta_{\mu\nu}$, with (3.4.8) describing an interacting nonlinear theory, we will meet a more drastic, q-number anomaly in T_{+-}.

3.4.2 Weyl Invariance

Depending on the form of $g_{\mu\nu}$, scale invariance breaks down in (3.4.8), because there is no way to regularize (3.4.8) while preserving world-sheet scale or conformal invariance. Pauli–Villars regularization (subtracting from loops the contribution of a massive regulator field) certainly violates scale invariance. Equation (3.4.8) can be regularized by dimensional regularization, but this violates scale invariance because the two-dimensional sigma model (3.4.8) is only scale invariant in two dimensions. We will use dimensional regularization.

The breakdown of scale invariance in a quantum field theory is usually described in terms of the β function. One way or another, depending on the formalism that is used in defining the β function, the nonzero beta function arises from ultraviolet divergences in Feynman diagrams. In string theory, the beta function and ultraviolet divergences are not the really basic issues; the basic issue is whether (3.4.8) is Weyl invariant if formulated on a curved world sheet. These questions are intimately related, however. Weyl invariance implies global scale invariance, which in turn implies vanishing beta function and thus ultraviolet finiteness (modulo possible wave function renormalization). Historically, the Callan-Symanzik equation was originally derived by studying the Ward identity associated with the trace of the energy–momentum tensor or in other words the Ward identity associated with Weyl transformations.

The two calculations — finiteness and Weyl invariance — are essentially equivalent. We will see that demanding Weyl invariance on a curved world-sheet necessarily implies the vanishing of the renormalization group β function and hence finiteness.

First, let us discuss what the expansion parameter is in our one-loop calculations. Inspection of (3.4.8) shows that in the limit of very small α' the action is large and quantum corrections are small. Quantum-mechanical perturbation theory is an expansion in powers of α'. There is another equivalent way to think of this. Rescaling the space-time metric

$$g_{\mu\nu} \rightarrow t^2 g_{\mu\nu} \tag{3.4.12}$$

in (3.4.8) shows that large t is equivalent to small α'. Since all lengths on the manifold $\mathbf{M}$ are rescaled by a factor t under (3.4.12), large t is the limit in which the size of $\mathbf{M}$ is very large in units of α'. The dimensionless expansion parameter is $\sqrt{\alpha'}/r$, with r being the characteristic length or 'radius' of $\mathbf{M}$.

We shall choose the gauge

$$h_{\alpha\beta} = e^{2\phi}\eta_{\alpha\beta}. \tag{3.4.13}$$

The calculation of the possible breakdown of Weyl symmetry requires regularization, which may be achieved by working in $2 + \epsilon$ dimensions. Inserting (3.4.13) in (3.4.2) we get

$$\tilde{S} = -\frac{1}{2\pi} \int d^{2+\epsilon}\sigma e^{\epsilon\phi} \partial_\alpha X^\mu \partial^\alpha X^\nu g_{\mu\nu}(X^\rho). \tag{3.4.14}$$

We want to investigate the question as to whether the ϕ dependence vanishes in the limit $\epsilon \rightarrow 0$. In the process we will see (at least at the one-loop level) that this condition is connected to the condition for the theory to be ultraviolet finite. In a complete treatment we should also study the possible breakdown of two-dimensional reparametrization invariance due to two-dimensional gravitational anomalies (as arose in the treatment of strings in flat space-time in §3.2.3) since the regularization (3.4.13) is not coordinate invariant.

Treating (3.4.8) as a quantum field theory in which the quantum field is $X^\mu(\sigma, \tau)$, the first step is to pick a vacuum expectation value — call it X_0^μ — and expand the quantum field around this value

$$X^\mu(\sigma, \tau) = X_0^\mu + x^\mu(\sigma, \tau), \tag{3.4.15}$$

where x^μ is the quantum fluctuation. In more general applications of this 'background field method' the classical background X_0^μ is taken to be any function of σ and τ that satisfies the classical field equations rather than the constant solution that we are choosing here. We shall expand the metric around $X^\mu = X_0^\mu$. The general form of the expansion is rather complicated and unwieldy unless we note that (3.4.8) is a 'geometrical' expansion, invariant under a redefinition of the field variables

$$X^\mu \to \tilde{X}^\mu(X^\rho) \tag{3.4.16}$$

accompanied by a suitable transformation of the space-time metric tensor $g_{\mu\nu}$. By making, if necessary, such a redefinition of the field variables, we can assume that on the space-time manifold **M** the coordinates X^μ are locally inertial coordinates at the point X_0^μ. (The Jacobian for this transformation does not affect the following results). In such coordinates the metric $g_{\mu\nu}(X^\rho)$ equals the Minkowski metric $\eta_{\mu\nu}$ at $X^\mu = X_0^\mu$, and differs from it only in order $(x^\mu)^2$. The higher-order terms cannot be set to zero by a field redefinition, but they can be simplified by a choice of coordinates known as Riemann normal coordinates. In such coordinates there is an expansion

$$g_{\mu\nu}(X^\rho) = \eta_{\mu\nu} - \frac{1}{3}R_{\mu\lambda\nu\kappa}(X_0^\rho)x^\lambda x^\kappa - \frac{1}{6}D_\rho R_{\mu\lambda\nu\kappa}x^\rho x^\lambda x^\kappa + O((x^\mu)^4), \tag{3.4.17}$$

with $R_{\mu\lambda\nu\kappa}$ being the Riemann tensor of **M** at the point X_0^ρ. With this choice of field variables and expanding $e^{\epsilon\phi} = 1 + \epsilon\phi + \dots$, (3.4.14) takes the form

$$\tilde{S} = -\frac{1}{2\pi}\int d^{2+\epsilon}\sigma[(\partial_\alpha x^\mu \partial^\alpha x^\nu)(1 + \epsilon\phi)\eta_{\mu\nu}$$
$$- \frac{1}{3}R_{\mu\lambda\nu\kappa}(X_0^\rho)x^\lambda x^\kappa \partial_\alpha x^\mu \partial^\alpha x^\nu(1 + \epsilon\phi) + O(x^5)]. \tag{3.4.18}$$

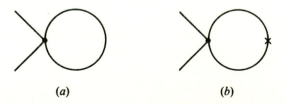

(a)   (b)

Figure 3.7. The one-loop counterterm in the nonlinear sigma model that contributes to wave-function renormalization comes from (a). In checking Weyl invariance of the effective action there is also a contribution of the form of diagram (b), where the cross represents the insertion of a kinetic term with coefficient $\epsilon\phi$.

The expansion in powers of $\sqrt{\alpha'}/r$ is equivalent to an expansion in powers of x in (3.4.18), since, for instance, the curvature tensor of **M** is of order $1/r^2$. Thus, the lowest-order counterterm is just obtained by contracting two of the x^μ's that appear in the quartic term in (3.4.18). The relevant Feynman diagram is sketched in fig. 3.7a. In dimensional regularization poles arise only from logarithmically divergent integrals. The contraction $\langle x^\lambda x^\kappa \rangle$ gives a logarithmically divergent integral, while a contraction $\langle \partial x^\mu \partial x^\nu \rangle$ gives a quadratically divergent integral, which is discarded in dimensional regularization.[*] The logarithmic divergence in $2 + \epsilon$ dimensions is

$$\langle x^\lambda(\sigma) x^\kappa(\sigma') \rangle_{\sigma \to \sigma'} = \pi \eta^{\lambda\kappa} \lim_{\sigma \to \sigma'} \int \frac{d^{2+\epsilon}k}{(2\pi)^{2+\epsilon}} \frac{e^{ik\cdot(\sigma-\sigma')}}{k^2} \sim \frac{\eta^{\lambda\kappa}}{2\epsilon}. \qquad (3.4.19)$$

There are therefore ϵ poles in the one-loop effective action derived from the theory defined by (3.4.18). As well as giving rise to nonzero β functions such poles can lead to ϕ dependence that survives in the limit $\epsilon \to 0$. For instance, matrix elements of the operator

$$W = x^\lambda x^\kappa \partial_\alpha x^\mu \partial^\alpha x^\nu R_{\mu\lambda\nu\kappa}(X_0^\rho) \qquad (3.4.20)$$

have a pole as $\epsilon \to 0$. The pole comes from the $\langle x^\lambda x^\kappa \rangle$ contraction of (3.4.19), giving

$$W \to W' = W - \frac{1}{2\epsilon} \partial_\alpha x^\mu \partial^\alpha x^\nu R_{\mu\nu}(X_0^\rho) \qquad (3.4.21)$$

as $\epsilon \to 0$. Here $R_{\mu\nu}(X_0^\rho)$ is the Ricci tensor of the manifold **M**, defined in terms of the Riemann tensor by $R_{\mu\nu} = R^\lambda{}_{\mu\lambda\nu}$. Substituting (3.4.21) in (3.4.18) gives a finite ϕ-dependent term as $\epsilon \to 0$:

$$-\frac{1}{12\pi} \int d^2\sigma \phi(\sigma) \partial_\alpha x^\mu \partial^\alpha x^\nu R_{\mu\nu}(X_0^\rho). \qquad (3.4.22)$$

However, the kinetic term in (3.4.18) also gives rise to other ϕ-dependent terms in the one-loop effective action in the limit $\epsilon \to 0$. For example, it is easy to see that there is another one-loop contribution to the effective action that is quadratic in x^μ, as shown in fig. 3.7b. This again contributes a

---

[*] This quadratically divergent integral has a physical significance that can be understood by a careful treatment of the infrared divergences. It is related to the possibility of adding nonderivative couplings to (3.4.8), corresponding to an expectation value of the tachyon field.

finite ϕ dependence due to a $1/\epsilon$ pole that just cancels with the factor of ϵ in the ϕ-dependent kinetic insertion. The ϕ dependence of the sum of the two contributions of fig. 3.7 vanishes. In addition there are diagrams like fig. 3.7b with external $x^\mu \partial_\alpha x^\nu$ and $\partial_\alpha x^\mu \partial^\alpha x^\nu$ that are proportional to $\partial^\alpha \phi$. After integration by parts (and dropping terms proportional to $\partial_\alpha \partial^\alpha x^\mu$, which vanish by the lowest-order equations of motion) these terms lead to a net ϕ dependence of the unrenormalized effective action that is again proportional to (3.4.22). In order to obtain the correct $\epsilon \to 0$ limit we must still renormalize the ϕ-independent ϵ pole terms that contribute to the effective action. Such pole terms arise at one loop from fig. 3.7a as well as from the one-loop diagram similar to fig. 3.7a but with four external x^μ fields. The latter contributes to the renormalization of the $R_{\mu\lambda\nu\kappa}(X_0^\rho)$ coupling in (3.4.18). It is an important feature of nonlinear sigma models that these two infinities can be absorbed into a wave function renormalization

$$x^\mu \to x^\mu + \frac{1}{6\epsilon} R^\mu{}_\nu(X_0^\rho) x^\nu + O(x^2), \qquad (3.4.23)$$

and a renormalization of the space-time metric

$$g_{\mu\nu} \to g_{\mu\nu} - \frac{1}{2\epsilon} R_{\mu\nu}(X_0^\rho). \qquad (3.4.24)$$

Making these substitutions into (3.4.18) resurrects the terms with coefficient $\epsilon\phi$ due to the cancellation of ϵ factors by the poles[*]. Combining all the ϕ-dependent contributions gives an effective action that can be written at this order (and when $D = 26$) as

$$\hat{S} = -\frac{1}{4\pi} \int d^2\sigma \, \phi \, R_{\mu\nu}(X^\rho) \partial_\alpha X^\mu \partial^\alpha X^\nu, \qquad (3.4.25)$$

where X^μ is given by (3.4.15). Thus, to this order (3.4.2) leads to a Weyl-invariant (ϕ-independent) quantum theory if and only if

$$R_{\mu\nu}(X^\rho) = 0. \qquad (3.4.26)$$

The renormalization of the metric in (3.4.24) means that there is a one-loop beta functional given by

$$\beta_{\mu\nu}(X^\rho) = -\frac{1}{2\pi} R_{\mu\nu}(X^\rho). \qquad (3.4.27)$$

The notion of a beta *functional* may be unfamiliar, but it really is not

---

[*] Furthermore, local counterterms should be added to cancel any gravitational anomalies that arise at one loop. Such anomalies are calculated, much as in §3.2.3, by considering the one-loop corrections to $\langle h^{\mu\nu} h^{\rho\sigma} \rangle$. They result in background-independent terms that are cancelled in 26 dimensions by the ghost contributions.

that strange. It is well-known that in a theory with n coupling constants there are n beta functions, one for each coupling. These n beta functions are indexed by whatever label indexes the couplings. Similarly, in a theory which has a coupling *functional* $g_{\mu\nu}(X^\rho)$, which is a continuously infinite number of couplings, there will be a beta functional $\beta_{\mu\nu}(X^\rho)$ depending on the same degrees of freedom that enter in the coupling. From (3.4.27), the condition for the vanishing of the one-loop beta functional — or equivalently the condition for the vanishing of the one-loop ϕ-independent counterterms ((3.4.23) and (3.4.24)) in the effective action — is $R_{\mu\nu}(X^\rho) = 0$. As stated earlier this is the same condition as is required by the Weyl invariance (ϕ independence) of the effective action. The two statements are related by the fact that the β function is the trace of the energy–momentum tensor.

Although our one-loop calculation may not make it altogether obvious, this relationship between Weyl invariance and the vanishing of beta functions holds much more generally. A more systematic exploration of it, suitable for higher loop calculations, would make use of the background field method, generalized to allow X_0^μ to be a general non-constant solution of the classical field equations.[†]

3.4.3 Conformal Invariance and the Equations of Motion

The equations we have found, namely $R_{\mu\nu} = 0$, are the familiar Einstein equations in vacuum. Is it merely an accident for the Einstein equations to arise in this way?

In investigating a physical theory, the only equations that we are entitled to impose are the equations of motion – or, at the quantum level, the equations for minimizing a quantum effective potential. On the other hand, Weyl invariance or vanishing beta function is needed if (3.4.2) is to make sense in string theory. The condition for vanishing β function is certainly needed in string theory (at the classical level, as we will discuss shortly), and this condition must coincide with the equations of motion if it is to have any sensible physical interpretation. Therefore, in hindsight, we should breathe a sigh of relief that the condition (3.4.26) for vanishing of the lowest-order β function has a sensible interpretation as a long-wavelength approximation to the equation of motion of the gravita-

---

[†] For a much more complete treatment of this subject see the papers by Fradkin and Tseytlin listed in the bibliography or, for example, Hull, C.M. and Townsend, P.K. (1986), 'Finiteness and conformal invariance in non-linear sigma models', *Nucl. Phys.* **B274**, 349.

tional field. Equation (3.4.26) must have such an interpretation if string theory is to make sense.

The beta function, or breakdown of Weyl invariance, which we have computed depends only on the short-distance behavior of the quantum field theory (3.4.2). The same short-distance behavior would occur on a Riemann surface of any topology. When Weyl invariance holds in (3.4.2), we are able to compute the path integral derived from (3.4.2) on the Riemann sphere – corresponding to tree diagrams or the classical approximation to string theory – and also on surfaces of higher genus – corresponding to quantum corrections. This suggests that Weyl invariance of (3.4.2) is to be interpreted as the condition for finding a classical solution of string theory (a Weyl-invariant path integral on the Riemann sphere) around which we then expand to compute quantum corrections.

Let us now attempt to argue more directly that Weyl invariance of (3.4.2) corresponds to finding a classical solution. Consider any physical theory with fields Φ^k, $k = 1, \ldots$. We describe a vacuum state (in perturbation theory, at any rate) by choosing vacuum expectation values

$$\Phi_0^k = \langle \Phi^k \rangle \tag{3.4.28}$$

and writing

$$\Phi^k = \Phi_0^k + \phi^k, \tag{3.4.29}$$

where ϕ^k is the quantum fluctuation. One then computes vacuum expectation values of products of the ϕ^k to describe scattering amplitudes:

$$A_n = \langle \phi^{k_1} \phi^{k_2} \ldots \phi^{k_n} \rangle. \tag{3.4.30}$$

In string theory, there is a vertex operator V^k corresponding to each field ϕ^k; properties of these operators were discussed in chapter 1 and in §2.2.3. The analog of (3.4.30) in string theory is then

$$\hat{A}_n = \langle V^{k_1} V^{k_2} \ldots V^{k_n} \rangle. \tag{3.4.31}$$

Despite the formal similarity between (3.4.30) and (3.4.31), there is a big difference between them: the expectation value in (3.4.30) is computed in space-time while that in (3.4.31) is computed on the string world sheet.

For $n \geq 4$, (3.4.30) and (3.4.31) describe scattering amplitudes; for $n = 3$ they describe vertex corrections; for $n = 2$ they describe mass shifts. What about $n = 1$? In field theory, the expectation value A_n for

$n = 1$ is of fundamental importance. Its vanishing

$$\langle \phi^k \rangle = 0 \quad k = 1, 2, \ldots \tag{3.4.32}$$

is the statement that the candidate vacuum state around which we are expanding, with expectation values Φ_0^k for the quantum fields Φ^k, is a solution of the classical field equations (or, at the quantum level, an extremum of the effective potential). The parallel between (3.4.30) and (3.4.31) shows what the corresponding statement must be in string theory. The condition for a classical solution in string theory (or an extremum of the effective potential) must be that

$$\langle V \rangle = 0, \tag{3.4.33}$$

where V is the vertex operator corresponding to any physical state.

We will now attempt to explain why the Einstein equations appear in (3.4.26) and more generally why world-sheet conformal invariance is related to the equations of motion. In particular, we will show that at tree level in string theory (3.4.33) is a consequence of world-sheet conformal invariance. The argument uses ingredients that are valid at tree level of string theory only and therefore it will show specifically that a conformally invariant nonlinear sigma model (3.4.2) or (3.4.8) corresponds to a solution of string theory at the *classical* level. The idea is very simple. Considering closed strings, for example, the world sheet at the classical level is a sphere that can be stereographically projected to the $x - y$ plane. In evaluating (3.4.33) we can assume that the vertex operator V is inserted on the plane at $x = y = 0$. Conformal invariance of (3.4.8) means, in particular, invariance under the scaling transformation

$$x \to \lambda x, \quad y \to \lambda y. \tag{3.4.34}$$

A physical closed-string vertex operator V has dimension two (as we learned in §1.4.5), and so transforms under (3.4.34) as

$$V \to \lambda^{-2} V. \tag{3.4.35}$$

Invariance under (3.4.34) thus implies that

$$\langle V \rangle = \langle \lambda^{-2} V \rangle, \tag{3.4.36}$$

from which it follows that $\langle V \rangle = 0$, as we wished to show.

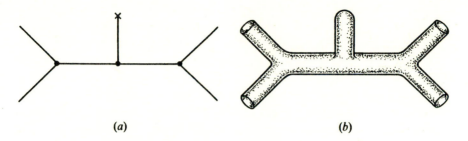

Figure 3.8. A 'tadpole' insertion on a Feynman diagram is sketched in (a). In string theory such tadpoles are automatically included in any calculation, as shown in (b), since adding a tadpole insertion to a propagating string does not change the topology.

A few points in this argument deserve comment. First of all, the vertex operators in (3.4.31) are all vertex operators of on-shell physical states. It seems that in string theory there is no really natural off-shell continuation of (3.4.31) – none, that is, of comparable elegance and simplicity to the on-shell formula. In (3.4.32) and its putative string theory analog (3.4.33), the operators ϕ^k and V are certainly to be evaluated at zero momentum. Zero momentum is on shell only for massless particles. It may seem thus that in (3.4.33) we have found a sensible way of checking the equations of motion in string theory only in the case of massless external states.

The resolution of this issue is instructive. What happens if we attempt in quantum field theory to expand around a misidentified vacuum in which (3.4.32) is *not* obeyed? In this case, tadpole insertions in Feynman diagrams, as sketched in fig. 3.8a, are nonzero. The tadpoles represent a shift of the expectation values of the fields Φ^k, and summation of the tadpoles shifts the invalid vacuum state to a valid one, if there is a nearby valid vacuum state to shift to. A tadpole of a particle Φ of mass M is proportional to

$$\frac{g}{M^2},\qquad(3.4.37)$$

where g is the coupling constant for emission of a ϕ particle and the factor $1/M^2$ comes from the ϕ propagator $1/(k^2 + M^2)$ evaluated at $k^\mu = 0$, the appropriate value for tadpoles. If the coupling is weak (and for strong coupling the expansion around a classical solution is not too useful anyway) the tadpoles are small for any nonzero M. As long as they are small, tadpoles, though a nuisance, are a harmless nuisance, which simply bring about a small shift to a valid vacuum state.

What happens in string theory? In any calculation, one automatically sums all possible tadpoles, since (as in fig. 3.8b) adding a tadpole to a diagram does not change the topology of the diagram. As we have just seen, tadpoles of massive particles are small (in the weakly coupled regime in which the perturbation expansion makes sense). It is really not necessary in string theory to check the equations of motion of massive states, since if these equations are not obeyed, there merely results a small, harmless shift in the vacuum due to tadpoles that are in any case automatically included in any calculation. String theory has its roots in S matrix considerations, and it is in keeping with this that there is no simple way to ask a question whose answer is not needed, like the question of whether the equations of motion of massive particles are obeyed. On the other hand, tadpoles of massless particles are always dangerous, and accordingly there is a good way, namely (3.4.33), to probe for such tadpoles.

Why is the above argument limited to *tree level* of string theory? Essential in the argument is the existence on the Riemann sphere or the $x-y$ plane of the transformation (3.4.34), which is a conformal transformation (the change it induces in the world-sheet metric can be absorbed in a Weyl rescaling) but not an isometry. For closed-string diagrams other than the sphere, there is no analog of this. On the torus, for example, or on RP^2 (which arises in the case of unoriented strings and will be discussed in chapter 8) the only conformal transformations are rigid motions that do not give the nontrivial scaling law (3.4.35) that was used in the argument. For surfaces of higher genus there are no conformal transformations at all.

What happens when open strings are included? The tree-level world sheet for open strings is a disk, which can be conformally mapped to the upper half of the $x-y$ plane. An open-string vertex operator is inserted on the boundary of the upper half plane, say at $x=y=0$. In a theory of open strings only, the scaling argument based on (3.4.34) shows, just as in the closed-string case, that conformal invariance implies the vanishing of tadpoles, the only change in the argument being that since open-string vertex operators have conformal dimension one, the λ^{-2} in (3.4.35) and (3.4.36) is replaced by λ^{-1}.

Things are different if one couples open and closed strings. For example, a closed-string vertex operator would be inserted on the upper half plane at an interior point, say $x=0$, $y \neq 0$. For coupling of a closed-string vertex operator to open strings (the upper half plane), rather than closed strings (the whole plane), (3.4.36) is replaced by

$$\langle V(\lambda y)\rangle = \langle \lambda^{-2}V(y)\rangle. \qquad (3.4.38)$$

This does not imply the vanishing of the tadpole; it only implies that $\langle V(y)\rangle$ is proportional to y^{-2}. Thus, closed-string tadpoles can be nonzero on the upper half plane, the torus, RP^2, and in fact on any world sheet except the plane (or Riemann sphere). We will meet such tadpoles extensively in chapters 8–10.

One further comment is called for. The graviton and dilaton vertex operators are both of the form

$$V = \zeta_{\mu\nu}\partial_\alpha X^\mu \partial^\alpha X^\nu e^{ik\cdot X}, \qquad (3.4.39)$$

where $k^2 = 0$. The polarization tensor $\zeta_{\mu\nu}$ is symmetric and satisfies

$$k^\mu \zeta_{\mu\nu} = 0 \qquad (3.4.40)$$

in order that (3.4.39) have the correct conformal dimension. In the case of the graviton $\zeta_{\mu\nu}$ is traceless, since the trace describes spin zero, as we discuss below. Note that the polarization tensor $\zeta_{\mu\nu}$ is defined up to a transformation

$$\zeta_{\mu\nu} \rightarrow \zeta_{\mu\nu} + \epsilon_\mu k_\nu + \epsilon_\nu k_\mu, \qquad (3.4.41)$$

where ϵ_μ is an arbitrary vector satisfying $\epsilon \cdot k = 0$, which preserves (3.4.40). The change corresponds to a longitudinally polarized graviton, which decouples from physical processes (as explained in §7.2.2), as a consequence of on-shell gauge invariance.

In the case of the dilaton, one might suppose that $\zeta_{\mu\nu} \sim \eta_{\mu\nu}$, but this does not satisfy (3.4.40). This can be overcome by choosing

$$\zeta_{\mu\nu} = \eta_{\mu\nu} - k_\mu \overline{k}_\nu - \overline{k}_\mu k_\nu, \qquad (3.4.42)$$

where $\overline{k}_\mu$ is an arbitrary vector satisfying $\overline{k} \cdot \overline{k} = 0$ and $k \cdot \overline{k} = 1$. The last two terms in (3.4.42) correspond to longitudinal pieces that decouple in physical processes. Except at $k^\mu = 0$, the graviton and dilaton conditions are distinct, and so are the corresponding vertex operators. However, in (3.4.33) we are working precisely at $k^\mu = 0$, where (3.4.40) and (3.4.42) are compatible, so that we do not have enough vertex operators to probe independently the equations of motion of all components of the gravitational and dilaton fields. As a result, we have in (3.4.33) essentially checked for tadpoles of all massless fields except one. It can be shown by use of a Bianchi identity that the one 'missing' equation is equivalent to the requirement that the Virasoro anomaly c should have the correct value.

3.4.4 String-Theoretic Corrections to General Relativity

It is now straightforward, at least conceptually, to derive string-theoretic corrections to general relativity. The Einstein equations $R_{\mu\nu} = 0$ correspond to vanishing of the one-loop beta function (3.4.27), and corrections to the Einstein equation can be found by computing corrections to the one-loop beta function. Including one and two-loop contributions, the β function is

$$\beta_{\mu\nu}(X^\rho) = -\frac{1}{4\pi}\left(R_{\mu\nu} + \frac{\alpha'}{2}R_{\mu\kappa\lambda\tau}R_\nu{}^{\kappa\lambda\tau}\right). \qquad (3.4.43)$$

The calculation of the second term in (3.4.43) is nontrivial. However, it should be evident that since the coupling constants in (3.4.8) are proportional (in Riemann normal coordinates) to the Riemann tensor and its derivatives, a two-loop beta function is given by an expression involving terms roughly of the form seen in (3.4.43).

The second term is thus a 'stringy' correction to general relativity that vanishes for $\alpha' \to 0$ or, equivalently, when the radius of M becomes very large compared to the square root of α'.

3.4.5 Inclusion of Other Modes

We now proceed to a more systematic treatment of the bosonic string in a background field, including all of the massless states of the closed string (and not only the graviton) as part of the background. The relevant closed-string fields are the antisymmetric tensor $B_{\mu\nu}(X^\rho)$ and the dilaton $\Phi(X^\rho)$, as well as the gravitational field $g_{\mu\nu}(X^\rho)$.

Let us write down the most general action for the field $X^\mu(\sigma, \tau)$ that is invariant under reparametrizations of the string world sheet and also renormalizable by power counting; the latter condition means that there must be precisely two world-sheet derivatives in each term in the action. One possible term is the one that we have been studying:

$$S_1 = -\frac{1}{4\pi\alpha'}\int d^2\sigma\,\sqrt{h}h^{\alpha\beta}\partial_\alpha X^\mu \partial_\beta X^\nu g_{\mu\nu}(X^\rho). \qquad (3.4.44)$$

It incorporates the effects of 26-dimensional gravity. A second term

$$S_2 = -\frac{1}{4\pi\alpha'}\int d^2\sigma\epsilon^{\alpha\beta}\partial_\alpha X^\mu \partial_\beta X^\nu B_{\mu\nu}(X^\rho), \qquad (3.4.45)$$

which makes use of the world-sheet antisymmetric tensor $\epsilon^{\alpha\beta}$, [*] gives a way to incorporate the effects of the antisymmetric tensor field $B_{\mu\nu}$ in σ

---

[*] $\epsilon^{\alpha\beta}$ takes the values $\epsilon^{01} = -\epsilon^{10} = 1$, $\epsilon^{00} = \epsilon^{11} = 0$. It is actually a tensor density, since $\epsilon^{\alpha\beta}/\sqrt{h}$ transforms as a tensor.

models. The factor of α', which we usually set equal to $1/2$, is needed to make (3.4.44) and (3.4.45) dimensionless, since X^μ has dimensions of length. Note that the integrand in (3.4.45) changes by a total divergence under the gauge transformation

$$\delta B_{\mu\nu} = \partial_\mu \Lambda_\nu - \partial_\nu \Lambda_\mu. \tag{3.4.46}$$

This point will be explored critically, in a global context, in chapter 14. We still must find a way to incorporate in the σ model the dilaton field Φ. At first sight there is no obvious operator, independent of (3.4.44), which can be used. The proper construction is perhaps rather surprising.

The world-sheet Ricci scalar $R^{(2)}$ contains two derivatives of the world-sheet metric h, so at first sight the two-dimensional Einstein–Hilbert action

$$\chi = \frac{1}{4\pi} \int d^2\sigma \sqrt{h} R^{(2)} \tag{3.4.47}$$

seems to be a renormalizable and reparametrization-invariant term that could be considered in the world-sheet theory. This is illusory, however, since (3.4.47) is actually a topological invariant that gives no dynamics to the two-dimensional metric h. To see that (3.4.47) is a topological invariant, note first that in any dimension the Riemann tensor $R_{\alpha\beta\gamma\delta}$ obeys

$$R_{\alpha\beta\gamma\delta} = -R_{\beta\alpha\gamma\delta} = -R_{\alpha\beta\delta\gamma}. \tag{3.4.48}$$

In two dimensions, a second-rank antisymmetric tensor must be proportional to $\epsilon_{\alpha\beta}$, so that $R^{(2)}_{\alpha\beta\gamma\delta}$ is proportional to $R^{(2)}$. It follows that

$$R^{(2)}_{\alpha\beta\gamma\delta} = (h_{\alpha\gamma}h_{\beta\delta} - h_{\beta\gamma}h_{\alpha\delta})R^{(2)}/2. \tag{3.4.49}$$

By contracting (3.4.49) with $h^{\beta\delta}$ one learns that in two dimensions

$$R^{(2)}_{\alpha\gamma} - \frac{1}{2}h_{\alpha\gamma}R^{(2)} = 0. \tag{3.4.50}$$

On the other hand, in any number of dimensions the variation of (3.4.47) under an infinitesimal variation in the metric tensor is

$$\int d^n\sigma \sqrt{h} \delta h^{\alpha\beta}(R_{\alpha\beta} - \frac{1}{2}h_{\alpha\beta}R), \tag{3.4.51}$$

which in view of (3.4.50) vanishes in two dimensions. This does not mean that (3.4.47) vanishes in two dimensions. It means that (3.4.47), being invariant under arbitrary variations in the world-sheet metric, depends only on the topology of the string world sheet. A standard result, which we will derive in §12.5.3, shows that if the string world sheet is a compact Riemann surface of genus g, then

$$\chi = 2(1 - g). \tag{3.4.52}$$

The quantity χ is known as the Euler characteristic of the two-dimensional manifold.

As a topological invariant of the world sheet, (3.4.47) does not really contribute to the dynamics of the σ model. However, by using the fact that scalar fields are dimensionless in two dimensions, we can generalize (3.4.47) to the more general renormalizable interaction

$$S_3 = \frac{1}{4\pi} \int d^2\sigma \sqrt{h} \Phi(X^\rho) R^{(2)}. \tag{3.4.53}$$

This proves to be a correct way to include the 26-dimensional dilaton field in the sigma model.

We now consider the sigma model with action $S = S_1 + S_2 + S_3$ and probe for conformal invariance. Thus, we take the world-sheet metric to be of the form

$$h_{\alpha\beta} = e^\phi \eta_{\alpha\beta}. \tag{3.4.54}$$

Working in $2 + \epsilon$ dimensions, we compute the ϕ dependence of the effective action, and ask whether this vanishes in the limit $\epsilon \to 0$. The conditions for Weyl invariance to hold in two dimensions in the lowest nontrivial approximation in α' turn out to be

$$0 = R_{\mu\nu} + \frac{1}{4} H_\mu{}^{\lambda\rho} H_{\nu\lambda\rho} - 2D_\mu D_\nu \Phi$$

$$0 = D_\lambda H^\lambda{}_{\mu\nu} - 2(D_\lambda \Phi) H^\lambda{}_{\mu\nu} \tag{3.4.55}$$

$$0 = 4(D_\mu \Phi)^2 - 4D_\mu D^\mu \Phi + R + \frac{1}{12} H_{\mu\nu\rho} H^{\mu\nu\rho},$$

where

$$H_{\mu\nu\rho} = \partial_\mu B_{\nu\rho} + \partial_\rho B_{\mu\nu} + \partial_\nu B_{\rho\mu} \tag{3.4.56}$$

is a third-rank antisymmetric tensor field strength that is invariant under 'gauge transformations' $\delta B_{\mu\nu} = \partial_\mu \Lambda_\nu - \partial_\nu \Lambda_\mu$. (Covariant derivatives in D dimensions are denoted D_μ, whereas on the world sheet they are ∇_α.) If the dimension is not restricted to $D = 26$, the third equation in (3.4.55) has an additional term $(D - 26)/3\alpha'$.

We will not describe the calculations that are needed to obtain (3.4.55); they are similar to those we described earlier in the purely gravitational case though much more elaborate. It is appropriate, however, to draw attention to one subtlety. While S_1 and S_2 as defined above are Weyl invariant at the classical level, this is not true of S_3. How, then, can it make sense to add S_3 to S_1 and S_2 in discussing Weyl invariance? At least at a technical level, the reason that it does make sense is that for dimensional reasons S_1 and S_2 are proportional to $1/\alpha'$, while this is not so in the case of S_3. Since the perturbation expansion is really an expansion in powers of α', S_3 should be considered as an order α' correction compared to S_1 and S_2. This being so, it is logical to compare a classical effect coming from S_3 to quantum effects coming from S_1 and S_2. It is necessary to do this to derive the first equation in (3.4.55). The last equation in (3.4.55) receives contributions from one-loop diagrams constructed from S_3 and *two*-loop diagrams constructed from S_1 and S_2; they are all of the same order in α'.

A crucial test of all this is that the equations (3.4.55) must have a sensible physical interpretation. In fact, it can easily be seen that they are the Euler-Lagrange equations coming from the 26-dimensional action

$$S_{26} = -\frac{1}{2\kappa^2} \int d^{26}x \sqrt{g} e^{-2\Phi} \left(R - 4D_\mu \Phi D^\mu \Phi + \frac{1}{12} H_{\mu\nu\rho} H^{\mu\nu\rho} \right). \quad (3.4.57)$$

Equation (3.4.57) thus describes the long-wavelength limit of the interactions of the massless modes of the bosonic closed string. If desired, the gravitational action in (3.4.57) can be put in the form $\int d^{26}x \sqrt{g} R$, rather than $\int d^{26}x \sqrt{g} e^{-2\Phi} R$, by absorbing a suitable power of $e^{-\Phi}$ in the definition of the space-time metric $g_{\mu\nu}$. String-theoretic corrections to (3.4.57) can be computed, just as in the purely gravitational case, by calculating higher-order corrections in sigma-model perturbation theory.

We have discussed sigma-model interaction terms with two world-sheet derivatives, corresponding to vertex operators of massless fields. It is also possible to try to include other operators in the sigma-model Lagrangian. A particularly simple possibility is to include in the Lagrangian a non-derivative interaction $S(X^\mu)$, with S being some scalar function. Since the tachyon vertex operator $e^{ik \cdot X}$ is nonderivative, the inclusion of a non-derivative interaction corresponds to giving an expectation value to the tachyon field. It is very natural to do so, in the bosonic sigma model, and in fact it is possibly unnatural not to. In computing the one-loop sigma-model beta function, even in the purely gravitational case, we encountered quadratic divergences that we simply ignored on the grounds that they are irrelevant in dimensional regularization. The quadratic divergences actually reflect the fact that it would be possible to include a dimension-zero nonderivative interaction in the sigma model.

3.4.6 The Dilaton Expectation Value and the String Coupling Constant

The formalism for studying string propagation in background fields that we have been developing in this section gives many insights that are difficult to obtain in other ways. Here we give just one example of an insight that can be conveniently obtained by thinking in terms of sigma models.

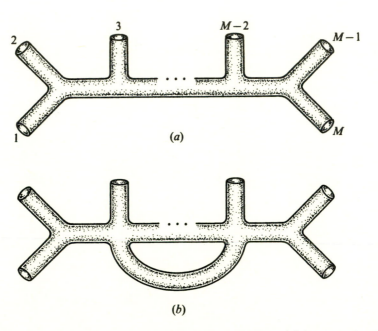

Figure 3.9. A tree-level process for scattering of M gravitons (a) is proportional to κ^{M-2}. Each loop adds an extra factor of κ^2, as shown in (b).

If we consider graviton scattering in the bosonic closed-string theory (or for that matter if we consider the scattering of any closed-string modes), we require at each interaction vertex a factor of the gravitational coupling constant κ. A tree diagram with M external gravitons has $M - 2$ interaction vertices, as in fig. 3.9a, while each loop adds two more, as shown in fig. 3.9b. Thus, a general loop diagram is proportional to

$$\kappa^M \cdot \kappa^{-2(1-g)}, \tag{3.4.58}$$

where M is the number of external vertex operators, and g, the genus of the Riemann surface, is the number of loops. The factor of κ^M can be absorbed in the normalization of external vertex operators; we simply use κV rather than V as the vertex operator for emission of a string state. We now focus on the interpretation of the loop-dependent factor $\kappa^{-2(1-g)}$.

Apart from κ, which in 26 dimensions has dimensions of (length)^{12}, the usual construction of the bosonic string theory involves a parameter α', with dimensions of (length)^2. It therefore seems at first sight that the closed bosonic string theory contains an arbitrary fundamental dimensionless parameter $\alpha'/\kappa^{1/6}$. It would be disappointing if reconciling quantum mechanics with gravity involved the introduction of a new fundamental dimensionless constant. Happily, this is not the case. Let us return to the sigma model of the preceding subsection with its action $S = S_1 + S_2 + S_3$. The path integral on a surface of genus g involves the partition function

$$Z_g = \kappa^{-2(1-g)} \int DX^\mu Dh \, e^{-S}. \tag{3.4.59}$$

Here, for emphasis, we include the factor $\kappa^{-2(1-g)}$, which seems to be required by (3.4.58).

To understand why there is no fundamental dimensionless parameter in the bosonic closed-string theory, the key is to note the relatively simple dependence of (3.4.59) on the dilaton field Φ. Looking back to (3.4.47), (3.4.52) and (3.4.53), we see that under

$$\Phi \to \Phi + a \tag{3.4.60}$$

(with a being an arbitrary constant), the sigma-model action changes by

$$S \to S + 2a(1-g). \tag{3.4.61}$$

In (3.4.59) the effect of this is equivalent to the effect of a redefinition of the gravitational coupling

$$\kappa \to e^{-a}\kappa. \tag{3.4.62}$$

Thus, the value of κ can be absorbed in a shift in the vacuum expectation value of Φ; κ is not a fundamental parameter of the theory.

The result given above can also be checked in the low-energy effective action (3.4.57). While (3.4.57) appears to contain the gravitational coupling constant κ as a free parameter, in fact the value of κ can be absorbed in a shift in the value of Φ. (This does not change the value of α', since α' does not appear in S_3.) Instead of a one-parameter family of theories labeled by a fundamental dimensionless parameter $\alpha'/\kappa^{1/6}$, the bosonic string theory is a single theory that has at tree level a one-parameter family of vacuum states labeled by the arbitrary expectation value of the massless scalar field Φ.

Though we will not do so in this book, the considerations of this section can be generalized to supersymmetric string theories by considering supersymmetric nonlinear sigma models instead of the bosonic sigma model considered here. Superstring theories also have no fundamental dimensionless parameter – as we will see from the low-energy point of view in chapter 13. In chapter 14 we will discuss briefly an interesting example of a physical question (involving axions and world-sheet instantons) in which the sigma-model approach to string theory gives significant results that would be rather obscure from other viewpoints. There are many other such examples.

3.5   Summary

The covariant path-integral method of quantization utilizing Faddeev–Popov ghost fields in a gauge-fixed action with BRST symmetry was developed originally in the study of Yang–Mills theories. In this chapter we have seen that these same techniques can be effectively and fruitfully applied to string theory. In particular, we have obtained new perspectives on many subjects first introduced in chapter 2. In the study of superstrings in the next chapter we will learn that analogous constructions are possible for them as well provided that $D = 10$. This formal framework provides the basis for modern studies that attempt to formulate the theory in terms of an action based on functional string fields with an enormous gauge symmetry.

We have also discussed the dynamics of a string in the presence of background fields, including a nontrivial space-time geometry. By requiring that the two-dimensional world-sheet theory be conformally invariant, even at the quantum level where anomalies in the trace of the two-dimensional energy–momentum tensor must vanish, we learned that the background fields must satisfy the field equations of the string theory itself. This implies a subtle self-consistency between the background fields that determine the dynamics of the string and the dynamics of the string whose solutions determine the possible background fields. The understanding of these connections, which are surely of fundamental importance, is still being developed.

4. World-Sheet Supersymmetry in String Theory

The bosonic string theory described in chapters 2 and 3, despite all its beautiful features, has a number of shortcomings. The most obvious of these are the absence of fermions and the presence of tachyons. It is conceivable that the latter feature merely indicates that the vacuum has been incorrectly identified, and that (as in a Higgs theory) there is some other stable vacuum that does not give rise to tachyons. Despite considerable effort over the years, this remains a conjecture. Another route, which has proved more fruitful, is to try to formulate another string theory instead. Progress in this direction has involved the introduction of internal degrees of freedom propagating along the string.

The particular string theory described in this chapter is based on the introduction of a world-sheet supersymmetry that relates the space-time coordinates $X^\mu(\sigma, \tau)$ to fermionic partners $\psi^\mu(\sigma, \tau)$. The latter are two-component world-sheet spinors. We will demonstrate that an action principle with $N = 1$ supersymmetry gives rise to a consistent string theory with critical dimension $D = 10$. Truncating the spectrum in a manner proposed by Gliozzi, Scherk and Olive gives supersymmetry in the $D = 10$ space-time sense, as well, with one or two Majorana–Weyl supercharges ($N = 1$ or $N = 2$) depending on the choice of boundary conditions.

In view of the enormously rich and fascinating structure that results from the introduction of $N = 1$ world-sheet supersymmetry, it is natural to consider generalizations based on extended supersymmetry. This is briefly explored in §4.5, where it is shown that $N = 2$ world-sheet supersymmetry leads to a string theory with critical dimension $D = 2$. Starting with $N = 4$ supersymmetry leads to an even more disappointing result, a negative critical dimension!

There is another possibility for inventing new string theories arising from the observation that left-moving and right-moving modes can be introduced independently in consistent closed-string theories. This makes it possible to use different schemes for each of them. In particular, using $D = 10$ superstring modes for right movers and $D = 26$ bosonic string modes for left movers gives the heterotic string. This will be explored in detail in chapter 6.

4.1   The Classical Theory

Let us recall that the action for the bosonic string, in conformal gauge, is

$$S = -\frac{1}{2\pi} \int d^2\sigma \partial_\alpha X_\mu \partial^\alpha X^\mu. \qquad (4.1.1)$$

This is a free field theory in two dimensions. $X^\mu(\sigma, \tau)$, $\mu = 0, 1, \ldots, D-1$, are coordinates for a string that is propagating in D space-time dimensions. If we wish to generalize the bosonic string theory, we can replace (4.1.1) with some more general two-dimensional field theory. The simplest possibility is that this more general theory might be again a free field theory.[*] Thus we are led to the idea of introducing additional free fields in (4.1.1). These would correspond physically to internal degrees of freedom that are free to propagate along the string. Various choices can be contemplated. One might, for instance, try to introduce a free fermion field $\psi_A(\sigma, \tau)$. (Capital letters A, B, C denote world-sheet spinor indices. In two dimensions the spinor index A takes two values, if both chiralities are included.) If one pursues that path one must decide whether ψ is to be a Dirac fermion or a Majorana fermion and whether it is to carry additional quantum numbers. There are surprisingly few choices that lead to interesting theories, but one that does is to introduce a D-plet of Majorana fermions $\psi_A^\mu(\sigma, \tau)$ transforming in the vector representation of the Lorentz group $SO(D-1, 1)$. We thus consider the Lagrangian

$$S = -\frac{1}{2\pi} \int d^2\sigma \{\partial_\alpha X^\mu \partial^\alpha X_\mu - i\overline{\psi}^\mu \rho^\alpha \partial_\alpha \psi_\mu\}. \qquad (4.1.2)$$

Here the symbol ρ^α represents two-dimensional Dirac matrices. (The symbols γ^μ and Γ^μ are reserved for four-dimensional and D-dimensional space-time gamma matrices.) A convenient basis is

$$\rho^0 = \begin{pmatrix} 0 & -i \\ i & 0 \end{pmatrix}, \quad \rho^1 = \begin{pmatrix} 0 & i \\ i & 0 \end{pmatrix}. \qquad (4.1.3)$$

These matrices obey

$$\{\rho^\alpha, \rho^\beta\} = -2\eta^{\alpha\beta}. \qquad (4.1.4)$$

---

[*] It may be surprising that we are trying to generalize (4.1.1) rather than trying to generalize the reparametrization invariant Lagrangian that it can be derived from by fixing conformal gauge. It turns out that the generalization of (4.1.1) is much easier to guess. After understanding the key features of this generalization, we will be able to go back and find the appropriate reparametrization invariant Lagrangian.

We refer to the components of ψ in this basis as $\psi_\pm$:

$$\psi = \begin{pmatrix} \psi_- \\ \psi_+ \end{pmatrix}. \tag{4.1.5}$$

We have chosen the ρ^α to be to be purely imaginary, so the Dirac operator $i\rho^\alpha \partial_\alpha$ is real and it makes sense in this representation of the Dirac algebra to demand that the components of the world-sheet spinor ψ^μ should be real. Such a two-component real spinor is known as a Majorana spinor. The symbol $\overline{\psi}$ indicates $\psi^\dagger \rho^0$ as usual.

Majorana spinors obey a number of important identities that do not hold for Dirac spinors. For example, for Majorana fermions, $\overline{\chi}$ is simply $\chi^T \rho^0$; there is no need to take the complex conjugate (or hermitian conjugate) of χ, since it is real anyway. Therefore $\overline{\chi}\psi$ is the same as $\rho^0_{AB}\chi_A\psi_B$. Since ρ^0 is an antisymmetric matrix, that last expression is symmetric in χ and ψ if they are anticommuting variables, so in this case

$$\overline{\chi}\psi = \overline{\psi}\chi. \tag{4.1.6}$$

This manipulation is typical of those that arise in working with Majorana fermions in two dimensions.

Several things about (4.1.2) may be disturbing. First of all, it should seem somewhat counterintuitive to introduce an anticommuting field ψ^μ that transforms as a *vector* – a bosonic representation – of $SO(D-1,1)$. This choice means that ψ maps bosons to bosons and fermions to fermions, in the space-time sense. While this may be counterintuitive, there is no clash with the spin-statistics theorem. On the contrary, (4.1.2) is a *two*-dimensional field theory, not a field theory in space-time, and ψ^μ_A transforms as a spinor under transformations of the two-dimensional world sheet, in perfect agreement with the usual spin-statistics relation.[†] The Lorentz group $SO(D-1,1)$ is merely an internal symmetry from the world-sheet point of view, and the spin and statistics theorem says nothing about whether anticommuting fields should transform as vectors or spinors under an internal symmetry. Although not paradoxical, the assignment of Lorentz quantum numbers to ψ is nonetheless surprising and is a matter we will re-examine later.

[†] Properly speaking, in $1+1$ dimensions there is no such thing as spin, but there is a two-dimensional Lorentz group (or local Lorentz group, in the case of a generally covariant theory), and it makes sense to ask how a two-dimensional field transforms under this group. The spin-statistics theorem says that in local quantum field theory in two dimensions an anticommuting field must have half-integral Lorentz quantum numbers.

There is actually a much more urgent problem that faces us in trying to understand (4.1.2) as a two-dimensional quantum field theory. We recall that the equal τ commutation relations of the bosonic coordinates are $[X^\mu(\sigma), \dot{X}^\nu(\sigma')] = i\pi\eta^{\mu\nu}\delta(\sigma - \sigma')$. The Lorentz metric $\eta^{\mu\nu}$ is not positive, and for this reason the $X^0(\sigma)$ oscillators in (4.1.1) create modes of wrong metric, or 'ghosts'. Happily, (4.1.1) has an infinite-dimensional symmetry algebra, the Virasoro algebra, which can be used to eliminate the ghosts in the critical dimension $D = 26$. To make sense of (4.1.2) we have to face the precisely analogous question for the fermions. From (4.1.2) we can deduce the equal τ commutation relations of the fermions,

$$\{\psi_A^\mu(\sigma), \psi_B^\nu(\sigma')\} = \pi\eta^{\mu\nu}\delta_{AB}\delta(\sigma - \sigma'). \qquad (4.1.7)$$

This anticommutation relation is the quantum version of the Poisson bracket for Grassmann variables. The familiar problem appears in a new guise. Since $\eta^{00} = -1$, the 'timelike' fermions $\psi_A^0(\sigma)$ create wrong metric states, just like the 'timelike' bosons $X^0(\sigma)$. The Virasoro conditions (4.1.2) may, as in the purely bosonic model, suffice for eliminating the wrong metric modes created by $X^0(\sigma)$, but to solve the analogous problem for $\psi_A^0(\sigma)$ we have to find a new symmetry and new constraints that can do for fermions what the Virasoro conditions do for bosons. It turns out that this step can be taken. The new symmetry is supersymmetry or more precisely superconformal symmetry. It proves to have many ramifications.

4.1.1 Global World-Sheet Supersymmetry

The new symmetry of the free field theory (4.1.2) may be surprising at first sight, but it is not difficult to demonstrate. Let ϵ represent a constant (*i.e.*, independent of σ and τ) anticommuting infinitesimal Majorana spinor. The action S of (4.1.2) is invariant under the infinitesimal transformations

$$\begin{aligned} \delta X^\mu &= \bar{\epsilon}\psi^\mu \\ \delta\psi^\mu &= -i\rho^\alpha\partial_\alpha X^\mu\epsilon, \end{aligned} \qquad (4.1.8)$$

with ϵ a constant anticommuting spinor. These transformations mix bosonic and fermionic coordinates and have come to be known as supersymmetry transformations.

A basic algebraic fact about supersymmetry is that the commutator of two supersymmetry transformations gives a spatial translation. In the present context a 'spatial translation' means a translation of the string world sheet. To see this explicitly consider

$$[\delta_1, \delta_2]X^\mu = \delta_1(\bar{\epsilon}_2\psi^\mu) - (1 \leftrightarrow 2) = a^\alpha\partial_\alpha X^\mu, \qquad (4.1.9)$$

where

$$a^\alpha = 2i\bar{\epsilon}_1\rho^\alpha\epsilon_2. \qquad (4.1.10)$$

It is important here that for Majorana fermions in $1 + 1$ dimensions, $\bar{\epsilon}_1\rho^\alpha\epsilon_2 = -\bar{\epsilon}_2\rho^\alpha\epsilon_1$. The reader is urged to check likewise that

$$[\delta_1, \delta_2]\psi^\mu = a^\alpha\partial_\alpha\psi^\mu. \qquad (4.1.11)$$

Here it is necessary to use the fact that ψ obeys the Dirac equation derived from (4.1.2), namely $\rho^\alpha\partial_\alpha\psi = 0$. In the next subsection we will reformulate the theory in a form for which the supersymmetry algebra closes without use of equations of motion.

Using (4.1.2) and the subsequent formula (4.1.8) for the supersymmetry transformation law, one can derive a formula for the supercurrent and energy–momentum tensor. An efficient way to do so makes use of the Noether method, described in §2.1.3, as follows. Consider, for instance, the supersymmetry transformation (4.1.8). If ϵ is a constant, it leaves the action S invariant. If ϵ is *not* a constant, then (4.1.8) does not leave the action invariant, but its variation is of the general form

$$\delta S = \frac{2}{\pi}\int d^2\sigma(\partial_\alpha\bar{\epsilon})J^\alpha. \qquad (4.1.12)$$

J^α is then the conserved Noether current, as was explained in §2.1.3. Applied to (4.1.8), this procedure gives the formula for the supercurrent

$$J_\alpha = \frac{1}{2}\rho^\beta\rho_\alpha\psi^\mu\partial_\beta X_\mu, \qquad (4.1.13)$$

(where the normalization has been defined for later convenience). Applied to the translation $\delta\sigma^\alpha$ = constant, it gives the formula for the energy–momentum tensor:

$$T_{\alpha\beta} = \partial_\alpha X^\mu\partial_\beta X_\mu + \frac{i}{4}\bar{\psi}^\mu\rho_\alpha\partial_\beta\psi_\mu + \frac{i}{4}\bar{\psi}^\mu\rho_\beta\partial_\alpha\psi_\mu - (\text{trace}), \qquad (4.1.14)$$

The reader is urged to check explicitly from the equations of motion that (4.1.13) and (4.1.14) are conserved. The energy–momentum tensor is traceless, just as in the purely bosonic theory, so, in terms of light-cone components, $T_{+-} = T_{-+} = 0$. Some components of the supercurrent also vanish, because it obeys the analogous restriction

$$\rho^\alpha J_\alpha = 0, \qquad (4.1.15)$$

as a consequence of the two-dimensional identity $\rho^\alpha \rho^\beta \rho_\alpha = 0$.

4.1.2 Superspace

Equation (4.1.2) is a two-dimensional field theory formulated on an ordinary two-dimensional space Σ, the string world sheet. Supersymmetry can be made manifest by formulating the theory in a two-dimensional superspace $\hat{\Sigma}$, in which the world-sheet coordinates σ^α are supplemented by two Grassmann coordinates θ^A forming a two-component Majorana spinor. Anticommuting coordinates may seem strange at first, but they are actually not difficult to use. A general function Y^μ in superspace would be a function

$$Y^\mu(\sigma, \theta) = X^\mu(\sigma) + \bar{\theta}\psi^\mu(\sigma) + \frac{1}{2}\bar{\theta}\theta B^\mu(\sigma), \qquad (4.1.16)$$

which depends in a general way on both the bosonic and fermionic coordinates of superspace. Such a function is called a superfield. Equation (4.1.16) is the complete power series expansion in powers of θ, because the anticommutation properties of the θ imply that any product of more than two of them vanishes. The superfield Y^μ unites X^μ and ψ^μ with a new field B^μ whose utility may not be apparent at first.

Let us begin by explaining how superspace makes supersymmetry manifest. Supersymmetry is represented on superspace by the generator

$$Q_A = \frac{\partial}{\partial\bar{\theta}^A} + i(\rho^\alpha\theta)_A \partial_\alpha. \qquad (4.1.17)$$

It is frequently convenient to introduce an arbitrary anticommuting parameter ϵ_A, the infinitesimal parameter of a supersymmetry transformation, and to work not with Q_A but with $\bar{\epsilon}Q$. The latter generates the transformation

$$\delta\theta^A = [\bar{\epsilon}Q, \theta^A] = \epsilon^A,$$

$$\delta\sigma^\alpha = [\bar{\epsilon}Q, \sigma^\alpha] = i\bar{\epsilon}\rho^\alpha\theta \qquad (4.1.18)$$

of the superspace coordinates. In this way supersymmetry is realized in superspace as a 'geometrical' transformation. The supercharge Q can also be used to define transformations of the coordinates according to

$$\delta Y^\mu = [\bar\epsilon Q, Y^\mu] = \bar\epsilon Q Y^\mu. \tag{4.1.19}$$

Since

$$[\bar\epsilon_1 Q, \bar\epsilon_2 Q] = -2i\bar\epsilon_1 \rho^\alpha \epsilon_2 \partial_\alpha, \tag{4.1.20}$$

it is evident that

$$[\delta_1, \delta_2]Y^\mu = -a^\alpha \partial_\alpha Y^\mu, \tag{4.1.21}$$

with a^α as given in (4.1.10), without using the equations of motion. Expanding (4.1.19) in components and using the two-dimensional Fierz relation

$$\theta_A \bar\theta_B = -\frac{1}{2}\delta_{AB}\bar\theta_C \theta_C, \tag{4.1.22}$$

gives

$$\delta X^\mu = \bar\epsilon \psi^\mu$$

$$\delta \psi^\mu = -i\rho^\alpha \epsilon \partial_\alpha X^\mu + B^\mu \epsilon \tag{4.1.23}$$

$$\delta B^\mu = -i\bar\epsilon \rho^\alpha \partial_\alpha \psi^\mu.$$

This reduces to the previous transformation formulas (4.1.8) if one sets $B^\mu = \rho^\alpha \partial_\alpha \psi^\mu = 0$, corresponding to the fact that in our initial formulation B^μ was absent, and $\rho^\alpha \partial_\alpha \psi^\mu$ was zero by a field equation. Because of the role of the auxiliary field B^μ, closure of the supersymmetry algebra is now achieved without use of equations of motion.

Now, let $Y_1, \ldots, Y_k$ be some superfields. Their transformation law under supersymmetry is $\delta Y_k = \bar\epsilon Q Y_k$. Any product of such superfields transforms in the same way. For instance

$$\delta(Y_1 Y_2) = \bar\epsilon Q(Y_1 Y_2). \tag{4.1.24}$$

This is so because, as $\bar\epsilon Q$ is a first-order differential operator in superspace, it obeys the Leibniz rule

$$\bar\epsilon Q(Y_1 Y_2) = \bar\epsilon Q(Y_1)Y_2 + Y_1\bar\epsilon Q(Y_2) \tag{4.1.25}$$

that is characteristic of such first-order operators. This ensures that the product $Y_1 Y_2$ transforms like a superfield, as in (4.1.24), if Y_1 and Y_2 do. This result should not seem surprising. A superfield is simply a function in superspace, and naturally the product of two such functions Y_1 and Y_2 is again such a function.

We would like to learn to use superspace to write Lagrangians invariant under supersymmetry transformations (4.1.23). To this end, we need first of all a derivative operator that is invariant under supersymmetry. It is

$$D = \frac{\partial}{\partial \overline{\theta}} - i\rho^\alpha \theta \partial_\alpha. \qquad (4.1.26)$$

This is known as the superspace covariant derivative. Its basic properties are that it anticommutes with Q,

$$\{D_A, Q_B\} = 0, \qquad (4.1.27)$$

and obeys

$$\{D_A, \overline{D}_B\} = 2i(\rho^\alpha)_{AB}\partial_\alpha \qquad (4.1.28)$$

or

$$\{D_A, D_B\} = 2i(\rho^\alpha \rho^0)_{AB}\partial_\alpha. \qquad (4.1.29)$$

The superspace covariant derivative is useful in constructing Lagrangians for the following reason. Equation (4.1.27) ensures that if an object Y transforms under supersymmetry as $\delta Y = \overline{\epsilon}QY$, then its covariant derivative $D_A Y$ transforms in the same way. Thus, the covariant derivative of a superfield is again a superfield. This enables us to write supersymmetric Lagrangians that contain derivatives – as, of course, all interesting Lagrangians do.

Another essential ingredient to formulate actions that are invariant under supersymmetry is that one must have an integration measure in superspace. The natural choice is the integral over 'all' of superspace,

$$\int d^2\sigma d^2\theta, \qquad (4.1.30)$$

where the fermion integration $d^2\theta$ is the standard Berezin integral for fermions. The Berezin integral is defined by saying that the integral over anticommuting coordinates of a general function is[*]

$$\int d^2\theta(a + \theta^1 b_1 + \theta^2 b_2 + \theta^1 \theta^2 c) = c. \qquad (4.1.31)$$

In other words, the integration picks out the coefficient of $\theta^1 \theta^2$. It follows

---

[*] In discussing the path integral over anticommuting ghost coordinates in §3.1.1, we were actually using infinite-dimensional Berezin integrals.

from this that (since $\overline{\theta}\theta = \theta\rho^0\theta = -2i\theta^1\theta^2$)

$$\int d^2\theta\overline{\theta}\theta = -2i. \qquad (4.1.32)$$

Like ordinary bosonic integrals, the Berezin integral has the property that one can integrate by parts:

$$\int d^2\theta\frac{\partial V}{\partial\theta^A} = 0 \qquad (4.1.33)$$

for any V. Equation (4.1.33) holds because the theta derivative removes one of the two θ's from the integrand, leaving an integral that vanishes by the Berezin rules.

The basic property of (4.1.30) is invariance under supersymmetry in the following sense. Let Y be any superfield, and let

$$S = \int d^2\sigma d^2\theta Y. \qquad (4.1.34)$$

Then S is invariant under supersymmetry transformations $\delta Y = \overline{\epsilon}QY$. The reason for this is that writing

$$\delta S = \int d^2\sigma d^2\theta\overline{\epsilon}QY \qquad (4.1.35)$$

and writing out the explicit form of Q, we find that (4.1.35) vanishes upon integrating by parts. By 'integrating by parts', we mean of course (4.1.33) as well as ordinary bosonic integration by parts. Thus we have learned how to write supersymmetric Lagrangians. For any superfield Y, the action S is invariant under supersymmetry. Y may in turn be constructed as an arbitrary product of elementary superfields and their covariant derivatives.

We now return to our original problem in which the elementary superfields are a D-tuplet transforming in the vector representation of $SO(D - 1, 1)$. We now know how to construct an infinity of supersymmetric Lagrangians for this field. One of special interest is

$$S = \frac{i}{4\pi}\int d^2\sigma d^2\theta\overline{D}Y^\mu DY_\mu. \qquad (4.1.36)$$

To evaluate the θ integrals explicitly, we first note that

$$DY^\mu = \psi^\mu + \theta B^\mu - i\rho^\alpha\theta\partial_\alpha X^\mu + \frac{i}{2}\overline{\theta}\theta\rho^\alpha\partial_\alpha\psi^\mu \qquad (4.1.37)$$

$$\overline{D}Y^\mu = \overline{\psi}^\mu + B^\mu\overline{\theta} + i\partial_\alpha X^\mu\overline{\theta}\rho^\alpha - \frac{i}{2}\overline{\theta}\theta\partial_\alpha\overline{\psi}^\mu\rho^\alpha, \qquad (4.1.38)$$

where we have used (4.1.22). Thus $\overline{D}Y^\mu DY_\mu$ contains the following terms quadratic in θ

$$\partial_\alpha X^\mu \partial_\beta X_\mu \overline{\theta} \rho^\alpha \rho^\beta \theta + B^\mu B_\mu \overline{\theta} \theta$$
$$+ \frac{i}{2} (\overline{\psi}^\mu \rho^\alpha \partial_\alpha \psi_\mu - \partial_\alpha \overline{\psi}^\mu \rho^\alpha \psi_\mu) \overline{\theta} \theta \qquad (4.1.39)$$
$$= (-\partial^\alpha X^\mu \partial_\alpha X_\mu + i \overline{\psi}^\mu \rho^\alpha \partial_\alpha \psi_\mu + B^\mu B_\mu) \overline{\theta} \theta.$$

Using the integration rule in (4.1.32), the action can be expanded in components to give

$$S'_0 = -\frac{1}{2\pi} \int d^2\sigma (\partial_\alpha X^\mu \partial^\alpha X_\mu - i \overline{\psi}^\mu \rho^\alpha \partial_\alpha \psi_\mu - B^\mu B_\mu). \qquad (4.1.40)$$

In this action, the field equations say that $B^\mu = 0$, so it is legitimate to simply set B^μ to zero and forget it. In this way we retrieve our starting point (4.1.2), but with a major difference: now we have a much better understanding of why (4.1.2) is invariant under supersymmetry.

Let us now recall the original objective in our search for a new symmetry. If (4.1.2), with its timelike fermions ψ^0 of wrong metric, is to lead to an acceptable theory, it must contain an enlarged symmetry that makes it possible to eliminate the unwanted modes. The supersymmetry of (4.1.2) or (4.1.40) is certainly a step in the right direction, but it will take an infinite-component symmetry algebra to do the job.

4.1.3 Constraint Equations

The required infinite-component symmetry algebra is not hard to find. The only symmetries that we have discussed so far are global supersymmetry transformations, given in (4.1.8) with a constant supersymmetry parameter ϵ. Constant translations of the world-sheet coordinates are implicitly present in the discussion, as well, since the commutator of two global supersymmetries Q_A is a translation. The translations in question are translations of the world-sheet coordinates σ and τ. But in the bosonic string theory, the translations of σ and τ are generated by L_0 and $\tilde{L}_0$, two of the Virasoro generators. Just as L_0 and $\tilde{L}_0$ generate a tiny subalgebra of the infinite-dimensional symmetry algebra of the bosonic string theory, we must extend the Q_A to an infinite-component 'supersymmetry'.

The fermion equation of motion derived from (4.1.2) is simply the two-dimensional Dirac equation $\rho^\alpha \partial_\alpha \psi = 0$, which must be supplemented by boundary conditions that will be discussed later. In the basis for ρ^α given in (4.1.3) this decomposes into a pair of decoupled equations for the upper and lower components of ψ^μ

$$\left(\frac{\partial}{\partial\sigma} + \frac{\partial}{\partial\tau}\right)\psi^\mu_- = 0$$

$$\left(\frac{\partial}{\partial\sigma} - \frac{\partial}{\partial\tau}\right)\psi^\mu_+ = 0. \qquad (4.1.41)$$

Thus ψ_- and ψ_+ describe right- and left-moving modes, respectively. The two-dimensional Dirac action can be written in a form that makes manifest the decoupling of ψ_+ and ψ_-, as well as the form (4.1.41) of the equations of motion, by introducing light-cone coordinates on the world sheet $\sigma^\pm = \tau \pm \sigma$ and $\partial_\pm = \frac{1}{2}(\partial_\tau \pm \partial_\sigma)$. In terms of these variables the fermion part of the action (4.1.2) becomes

$$S_F = \frac{i}{\pi}\int d^2\sigma\,(\psi_-\partial_+\psi_- + \psi_+\partial_-\psi_+) \qquad (4.1.42)$$

(where the index μ has been suppressed). Equation (4.1.42) also makes it apparent that we could, say, set ψ_+ to zero and discuss a two-dimensional Lagrangian with right-moving fermions only. The two-dimensional chirality operator $\bar{\rho} = \rho^0\rho^1$ actually has $\psi_\pm$ for its eigenstates (to be precise, $\bar{\rho}\psi_\pm = \mp\psi_\pm$), so setting $\psi_+ = 0$ amounts to working with a spinor field of positive chirality only.[*]

Equation (4.1.41) and the boson equation $0 = \partial^2 X^\mu/\partial\sigma^\alpha\partial\sigma_\alpha$ can be written in a way that makes it much more transparent why there can be a symmetry between bosons and fermions:

$$0 = \partial_+\psi^\mu_- = \partial_+(\partial_-X^\mu)$$
$$0 = \partial_-\psi^\mu_+ = \partial_-(\partial_+X^\mu). \qquad (4.1.43)$$

Thus, ψ^μ_- and ∂_-X^μ are both functions of σ^- only, while ψ^μ_+ and ∂_+X^μ are both functions of σ^+ only. Supersymmetry is the symmetry between ψ^μ_- and ∂_-X^μ (or between ψ^μ_+ and ∂_+X^μ), which obey the same equation.

In view of the decoupling of positive- and negative-chirality modes, the world-sheet supersymmetry current and energy–momentum tensor are bound to simplify if written in terms of positive- and negative-chirality

---

[*] A rough analogy to this would be the two-component neutrino in four dimensions, but there is a crucial difference. In four dimensions the CPT conjugate of the neutrino is the antineutrino, of opposite chirality, but in two dimensions, CPT maps ψ_- to itself and permits a theory in which there simply is no fermion of opposite chirality.

modes. Indeed, if we write the supercurrent $J_{\alpha A}$ of (4.1.13) in terms of its light-cone components J_{+A} and J_{-A} (with $\pm$ labeling the light-cone vector components and A being the spinor index), then, *a priori*, J_{+A} and J_{-A} are both two-component spinors. In fact, however, because of (4.1.15), only the positive-chirality spinor component of J_{+A} or the negative-chirality spinor component of J_{-A} are nonzero. It is convenient to call the nonzero spinor components of J_{+A} and J_{-A} simply J_+ and J_-. These are

$$J_+ = \psi_+^\mu \partial_+ X_\mu$$

$$J_- = \psi_-^\mu \partial_- X_\mu. \tag{4.1.44}$$

They are obviously conserved,

$$0 = \partial_- J_+ = \partial_+ J_-, \tag{4.1.45}$$

but what algebra do they generate? Using the equal τ (anti)commutators

$$\{\psi_+^\mu(\sigma), \psi_+^\nu(\sigma')\} = \{\psi_-^\mu(\sigma), \psi_-^\nu(\sigma')\} = \pi \eta^{\mu\nu} \delta(\sigma - \sigma')$$

$$[\partial_\pm X^\mu(\sigma), \partial_\pm X^\nu(\sigma')] = \pm i \frac{\pi}{2} \eta^{\mu\nu} \delta'(\sigma - \sigma') \tag{4.1.46}$$

$$\{\psi_+^\mu, \psi_-^\nu\} = [\partial_+ X^\mu, \partial_- X^\nu] = 0,$$

one can readily calculate the algebra

$$\{J_+(\sigma), J_+(\sigma')\} = \pi \delta(\sigma - \sigma') T_{++}(\sigma)$$

$$\{J_-(\sigma), J_-(\sigma')\} = \pi \delta(\sigma - \sigma') T_{--}(\sigma) \tag{4.1.47}$$

$$\{J_+(\sigma), J_-(\sigma')\} = 0.$$

More precisely, (4.1.47) emerges from formal manipulations or Poisson brackets; quantum mechanically, there is an anomaly term, which we will consider later. Here T_{++} and T_{--} are the light-cone components of the energy–momentum tensor

$$T_{++} = \partial_+ X^\mu \partial_+ X_\mu + \frac{i}{2} \psi_+^\mu \partial_+ \psi_{+\mu}$$

$$T_{--} = \partial_- X^\mu \partial_- X_\mu + \frac{i}{2} \psi_-^\mu \partial_- \psi_{-\mu}. \tag{4.1.48}$$

We would now like to formulate constraint equations that can eliminate the timelike components of ψ^μ and X^μ alike. We recall that in the bosonic case, the timelike components of X^μ were eliminated (in 26 dimensions!) with the use of the Virasoro constraints $T_{++} = T_{--} = 0$. It is natural to aspire to repeat this success here. But in view of the algebra (4.1.47), we can hardly expect to set T_{++} and T_{--} to zero without setting J_+ and J_- to zero as well. We are thus led to pin our hopes on the super-Virasoro constraints,

$$0 = J_+ = J_- = T_{++} = T_{--}. \qquad (4.1.49)$$

To be sure, there is a big difference between this guess and our more systematic discussion in the case of the bosonic theory. Indeed, in the bosonic case we deduced the Virasoro conditions as constraint equations that arise by gauge fixing of a gauge-invariant Lagrangian. In the supersymmetric case we have just postulated (4.1.49). The super-Virasoro constraints (4.1.49) can be systematically derived by gauge fixing of a suitable two-dimensional supergravity Lagrangian, but the required discussion is much more intricate than in the bosonic case. The reader who would like to see a systematic derivation of (4.1.49) now is urged to jump ahead to §4.3.5 (which does not depend on the intervening material). Here, however, we accept (4.1.49) as plausible and analyze the content of the resulting theory.

4.1.4 Boundary Conditions and Mode Expansions

Our first task is to analyze the possible boundary conditions and determine the content of the *unconstrained* theory. This too is somewhat more involved than the analogous discussion in the purely bosonic case.

The space-time coordinate X^μ satisfies the same free wave equation as in the bosonic string theory. The possible boundary conditions correspond to open or closed strings, and the resulting normal-mode expansions are completely unchanged from before and therefore do not need to be repeated here. In the case of the Fermi coordinates, the surface terms arise in the variation of the Lagrangian to obtain the Euler-Lagrange equations. Vanishing of these surface terms requires that $\psi_+ \delta\psi_+ - \psi_- \delta\psi_-$ should vanish at each end of an open string. This is satisfied by making $\psi_+ = \pm\psi_-$ (and hence $\delta\psi_+ = \pm\delta\psi_-$) at each end. The overall relative sign between ψ_- and ψ_+ is a matter of convention, and so without loss of generality we set

$$\psi_+^\mu(0, \tau) = \psi_-^\mu(0, \tau). \qquad (4.1.50)$$

The relative sign at the other end now becomes meaningful and there are two cases to be considered. In the first case (Ramond (R) boundary conditions)

$$\psi^\mu_+(\pi,\tau) = \psi^\mu_-(\pi,\tau), \tag{4.1.51}$$

and the mode expansion of the Dirac equation becomes

$$\psi^\mu_-(\sigma,\tau) = \frac{1}{\sqrt{2}} \sum_{n \in Z} d^\mu_n e^{-in(\tau-\sigma)} \tag{4.1.52}$$

$$\psi^\mu_+(\sigma,\tau) = \frac{1}{\sqrt{2}} \sum_{n \in Z} d^\mu_n e^{-in(\tau+\sigma)}, \tag{4.1.53}$$

where the sums run over all integers n. In the second case (Neveu–Schwarz (NS) boundary conditions), one chooses

$$\psi^\mu_+(\pi,\tau) = -\psi^\mu_-(\pi,\tau) \tag{4.1.54}$$

so that the mode expansions become

$$\psi^\mu_-(\sigma,\tau) = \frac{1}{\sqrt{2}} \sum_{r \in Z+1/2} b^\mu_r e^{-ir(\tau-\sigma)} \tag{4.1.55}$$

$$\psi^\mu_+(\sigma,\tau) = \frac{1}{\sqrt{2}} \sum_{r \in Z+1/2} b^\mu_r e^{-ir(\tau+\sigma)}, \tag{4.1.56}$$

where now the sums run over half-integer modes r. We always use the symbol m or n for integers and r or s for half integers. Thus $r - 1/2$ or $s - 1/2$ are integers. As we will explain in the next section, the boundary condition (4.1.51) and integer modes are appropriate to the description of string states which are space-time fermions, whereas the boundary condition (4.1.54) and half-integers give bosonic states. These bosonic states are different, of course, from those of the bosonic string theory in chapter 2.

For closed strings the surface terms vanish when the boundary conditions are periodicity or antiperiodicity for each component of ψ separately. Thus we can have

$$\psi^\mu_- = \sum d^\mu_n e^{-2in(\tau-\sigma)} \tag{4.1.57}$$

or

$$\psi^\mu_- = \sum b^\mu_r e^{-2ir(\tau-\sigma)}, \tag{4.1.58}$$

and

$$\psi_+^\mu = \sum \tilde{d}_n^\mu e^{-2in(\tau+\sigma)} \qquad (4.1.59)$$

or

$$\psi_+^\mu = \sum \tilde{b}_r^\mu e^{-2ir(\tau+\sigma)}. \qquad (4.1.60)$$

Corresponding to the different pairings of left-moving and right-moving modes, there are four distinct closed-string sectors that can be referred to as NS–NS, NS–R, R–NS, R–R. As is explained below, the first and last cases describe closed-string states that are bosons and the other two describe fermions.

The super-Virasoro operators are given by the modes of $T_{\alpha\beta}$ and J_α. For open strings, there is one independent set of L_m's defined, just as in chapter 2, by the linear combination,

$$L_m = \frac{1}{\pi} \int_0^\pi d\sigma \{ e^{im\sigma} T_{++} + e^{-im\sigma} T_{--} \} = \frac{1}{\pi} \int_{-\pi}^\pi d\sigma e^{im\sigma} T_{++}. \qquad (4.1.61)$$

For the fermionic generators of the algebra we define

$$F_m = \frac{\sqrt{2}}{\pi} \int_0^\pi d\sigma \{ e^{im\sigma} J_+ + e^{-im\sigma} J_- \} = \frac{\sqrt{2}}{\pi} \int_{-\pi}^\pi d\sigma e^{im\sigma} J_+ \qquad (4.1.62)$$

in the case of R boundary conditions or

$$G_r = \frac{\sqrt{2}}{\pi} \int_0^\pi d\sigma \{ e^{ir\sigma} J_+ + e^{-ir\sigma} J_- \} = \frac{\sqrt{2}}{\pi} \int_{-\pi}^\pi d\sigma e^{ir\sigma} J_+ \qquad (4.1.63)$$

in the case of NS boundary conditions. In the case of the closed strings there are two sets of super-Virasoro generators. One is given by the modes of T_{++} and J_+ and the other by the modes of T_{--} and J_-. In the classical string theory all these expressions are simply required to vanish. In the quantum theory there are various options for dealing with them, just as there were in the bosonic string theory. We discuss the options in the following sections.

4.2 Quantization – The Old Covariant Approach

In this section we describe the quantization of the superstring using the techniques described in §2.2 for the bosonic string. The super-Virasoro constraints on physical states are implemented and analyzed in essentially the same way as the Virasoro constraints were previously. One new feature is the existence of two sectors, bosonic and fermionic, that need to be studied separately. Eventually, the spectrum should be truncated by the GSO conditions, which we have not yet explained, and the two sectors become related by space-time supersymmetry. That fact is deeply hidden and highly nontrivial to explain in the formalism described in this chapter.

4.2.1 Commutation Relations and Mode Expansions

In the covariant gauge (explained in §4.3.5) the dynamics of the coordinates $X^\mu(\sigma, \tau)$ and $\psi^\mu(\sigma, \tau)$ are given by a free two-dimensional Klein-Gordon equation and a free Dirac equation supplemented by certain constraints. The quantization of these coordinates is just that of free two-dimensional field theory. The analysis of the X^μ coordinates is unchanged from chapter 2, where we found

$$[\dot{X}^\mu(\sigma, \tau), X^\nu(\sigma', \tau)] = -i\pi\delta(\sigma - \sigma')\eta^{\mu\nu}. \tag{4.2.1}$$

This resulted in commutation relations for the Fourier coefficients

$$[\alpha_m^\mu, \alpha_n^\nu] = m\delta_{m+n}\eta^{\mu\nu}. \tag{4.2.2}$$

The α_m^μ represent coefficients in either an open- or closed-string mode expansion. In the latter case there is a second set denoted $\tilde{\alpha}_m^\mu$.

The quantization of the fermionic coordinates is equally easy. The canonical anticommutation relations for the coordinates $\psi_A^\mu(\sigma, \tau)$ are

$$\{\psi_A^\mu(\sigma, \tau), \psi_B^\nu(\sigma', \tau)\} = \pi\delta(\sigma - \sigma')\eta^{\mu\nu}\delta_{AB}. \tag{4.2.3}$$

This implies that the modes b_r^μ or d_n^μ introduced in §4.1.4 satisfy

$$\{b_r^\mu, b_s^\nu\} = \eta^{\mu\nu}\delta_{r+s} \tag{4.2.4}$$

$$\{d_m^\mu, d_n^\nu\} = \eta^{\mu\nu}\delta_{m+n}. \tag{4.2.5}$$

In the subsequent discussion we describe a single set of modes α_m^μ and b_r^μ or α_m^μ and d_m^μ, as appropriate, for the description of open strings or the right-moving sector of closed strings. It is easy enough to include oscillators with tildes for left-moving closed-string modes, but we do not bother to say so explicitly in each case, as that would be rather tedious.

Recall that there are two distinct types of string states corresponding to $\psi_1(\pi, \tau) = \pm \psi_2(\pi, \tau)$ in the open-string case. The plus sign gives the integrally moded d oscillators and the minus sign gives the half-integrally moded b oscillators. The zero-frequency part of the Virasoro constraint gives the mass-shell condition

$$\alpha' M^2 = N + \text{constant}. \tag{4.2.6}$$

The constant is a normal-ordering effect to be described later, and

$$N = N^\alpha + N^d \tag{4.2.7}$$

or

$$N = N^\alpha + N^b, \tag{4.2.8}$$

where

$$N^\alpha = \sum_{m=1}^{\infty} \alpha_{-m} \cdot \alpha_m \tag{4.2.9}$$

$$N^d = \sum_{m=1}^{\infty} m d_{-m} \cdot d_m \tag{4.2.10}$$

$$N^b = \sum_{r=1/2}^{\infty} r b_{-r} \cdot b_r. \tag{4.2.11}$$

The state of lowest mass-squared corresponds to the Fock-space ground state

$$\alpha_m^\mu |0\rangle = d_m^\mu |0\rangle = 0 \quad m > 0 \tag{4.2.12}$$

or

$$\alpha_m^\mu |0\rangle = b_r^\mu |0\rangle = 0 \quad m, r > 0. \tag{4.2.13}$$

An excitation by a raising operator α_{-m}^μ or d_{-m}^μ increases the eigenvalue of $\alpha' M^2$ by m units. Similarly, b_{-r}^μ increases $\alpha' M^2$ by r units.

Note that in the case with half-integer modes it is possible to choose a unique nondegenerate ground state, which may therefore be identified as a spin zero state. In the case with integer modes this is not possible because of the oscillators d_0^μ. The d_0^μ have the algebra

$$\{d_0^\mu, d_0^\nu\} = \eta^{\mu\nu} \tag{4.2.14}$$

and commute with the M^2 operator. This algebra is just the Dirac algebra, so up to normalization the zero modes d_0^μ are Dirac matrices. If we require the Dirac matrices to obey $\{\Gamma^\mu, \Gamma^\nu\} = -2\eta^{\mu\nu}$, then

$$\Gamma^\mu = i\sqrt{2}d_0^\mu \qquad (4.2.15)$$

for our sign and metric conventions.

The states at each mass level must furnish representations of (4.2.14), and this must, in particular, be true for the ground state of the Fock space. The ground state, in fact, should be an irreducible representation of (4.2.14), since there are no other zero modes that would cause any degeneracy. The representation theory of the Clifford algebra (4.2.14) is well-known. The irreducible representations correspond to spinors of $SO(1,9)$. Therefore the boundary conditions that give integrally moded d oscillators must give fermionic strings, as we asserted earlier. The sectors with integrally or half-integrally moded world-sheet spinors are usually called, respectively, the fermionic (or R) and bosonic (or NS) sectors. The bosonic sector of superstrings must not be confused with the bosonic strings of chapters 2 and 3, of course. The former involve b oscillators in their description, whereas the latter do not.

As in the bosonic case, the densities of momentum and angular momentum along the string can be described as Noether currents associated with the global symmetries $X^\mu \to a^\mu_\nu X^\nu + b^\mu$, $\psi^\mu \to a^\mu_\nu \psi^\nu$. One finds (for $T = 1/\pi$)

$$P^\mu_\alpha = \frac{1}{\pi}\partial_\alpha X^\mu$$
$$J^{\mu\nu}_\alpha = \frac{1}{\pi}(X^\mu \partial_\alpha X^\nu - X^\nu \partial_\alpha X^\mu + i\overline{\psi}^\mu \rho_\alpha \psi^\nu). \qquad (4.2.16)$$

Explicit formulas for the Lorentz generators in terms of mode expansions can be given by inserting mode expansions for $X^\mu(\sigma,\tau)$ and $\psi^\mu(\sigma,\tau)$ in (4.2.16). The conserved charges are given by

$$J^{\mu\nu} = \int_0^\pi J^{\mu\nu}_\tau \, d\sigma = l^{\mu\nu} + E^{\mu\nu} + K^{\mu\nu}, \qquad (4.2.17)$$

where

$$l^{\mu\nu} = x^\mu p^\nu - x^\nu p^\mu \qquad (4.2.18)$$

and

$$E^{\mu\nu} = -i\sum_{n=1}^\infty \frac{1}{n}(\alpha^\mu_{-n}\alpha^\nu_n - \alpha^\nu_{-n}\alpha^\mu_n) \qquad (4.2.19)$$

as in the bosonic string theory. There is also a contribution from the fermionic modes

$$K^{\mu\nu} = -i \sum_{r=1/2}^{\infty} (b^{\mu}_{-r} b^{\nu}_r - b^{\nu}_{-r} b^{\mu}_r) \quad \text{(NS)} \tag{4.2.20}$$

$$K^{\mu\nu} = -\frac{i}{2}[d^{\mu}_0, d^{\nu}_0] - i \sum_{n=1}^{\infty} (d^{\mu}_{-n} d^{\nu}_n - d^{\nu}_{-n} d^{\mu}_n) \quad \text{(R)}. \tag{4.2.21}$$

In forming $J^{\mu\nu}$ we are simply adding together a number of commuting representations of the Lorentz algebra, so the sum obviously satisfies the same algebra.

4.2.2 Super-Virasoro Algebra and Physical States

In §4.1.4 we defined generalized Virasoro constraint conditions L_m and F_m or G_r as Fourier modes of T_{++} and J_+, respectively. It is straightforward to insert the mode expansions given for $X^{\mu}(\sigma, \tau)$ and $\psi^{\mu}(\sigma, \tau)$ to obtain explicit expressions in terms of the various oscillators. The Virasoro operators become

$$L_m = L_m^{(\alpha)} + L_m^{(b)} \quad \text{(NS)} \tag{4.2.22}$$

$$L_m = L_m^{(\alpha)} + L_m^{(d)}, \quad \text{(R)} \tag{4.2.23}$$

where

$$L_m^{(\alpha)} = \tfrac{1}{2} \sum_{n=-\infty}^{\infty} : \alpha_{-n} \cdot \alpha_{m+n} : \tag{4.2.24}$$

as before, and

$$L_m^{(b)} = \tfrac{1}{2} \sum_{r=-\infty}^{\infty} (r + \tfrac{1}{2}m) : b_{-r} \cdot b_{m+r} : \tag{4.2.25}$$

$$L_m^{(d)} = \tfrac{1}{2} \sum_{n=-\infty}^{\infty} (n + \tfrac{1}{2}m) : d_{-n} \cdot d_{m+n} : . \tag{4.2.26}$$

In each case the normal ordering is only required for $m = 0$. For the fermionic generators one finds

$$G_r = \sum_{n=-\infty}^{\infty} \alpha_{-n} \cdot b_{r+n} \quad \text{(NS)} \tag{4.2.27}$$

$$F_m = \sum_{n=-\infty}^{\infty} \alpha_{-n} \cdot d_{m+n} \quad \text{(R)}. \qquad (4.2.28)$$

The super-Virasoro algebra in the bosonic (or NS) sector is

$$[L_m, L_n] = (m - n)L_{m+n} + A(m)\delta_{m+n}$$

$$[L_m, G_r] = (\tfrac{1}{2}m - r)G_{m+r} \qquad (4.2.29)$$

$$\{G_r, G_s\} = 2L_{r+s} + B(r)\delta_{r+s}.$$

Here $A(m)$ and $B(r)$ are c-number anomaly terms, analogous to those that arise in the bosonic theory. Except for the anomaly terms, (4.2.29) is easily computed canonically; the last equation in (4.2.29), for example, corresponds simply to the Fourier modes of (4.1.47). Arguments similar to those in the bosonic case show that the anomaly terms must be c-numbers. The anomaly terms are most easily determined, as in the bosonic case, by evaluating expectation values in the Fock-space ground state. One finds that

$$A(m) = \tfrac{1}{8}D(m^3 - m)$$

$$B(r) = \tfrac{1}{2}D(r^2 - \tfrac{1}{4}). \qquad (4.2.30)$$

The anomaly $A(m)$ receives two-thirds of its contribution from the α oscillators and one-third from b oscillators. The fermionic (R) sector has a very similar algebra

$$[L_m, L_n] = (m - n)L_{m+n} + A(m)\delta_{m+n}$$

$$[L_m, F_n] = (\tfrac{1}{2}m - n)F_{m+n} \qquad (4.2.31)$$

$$\{F_m, F_n\} = 2L_{m+n} + B(m)\delta_{m+n},$$

where now the anomalies are

$$A(m) = \tfrac{1}{8}Dm^3$$

$$B(m) = \tfrac{1}{2}Dm^2. \qquad (4.2.32)$$

The discrepancy between the expressions in (4.2.30) and in (4.2.32) could be removed by redefining L_0 by a constant, if one so wished. We find the conventions used here to be convenient.

Note that in the NS sector the five generators L_1, L_0, L_{-1}, $G_{1/2}$, $G_{-1/2}$ form a closed superalgebra. This algebra is known as $OSp(1|2)$ and will be further discussed later. The situation is strikingly different in the fermionic sector. As soon as one adjoins F_0 to L_1, L_0, L_{-1}, the entire infinite algebra follows.

In the old covariant approach the constraint equations are incorporated into the quantum theory by requiring that their positive-frequency components annihilate physical states. Thus, proceeding in analogy to §2.2.2, we require that a physical bosonic state $|\phi\rangle$ satisfy

$$G_r |\phi\rangle = 0 \quad r > 0 \tag{4.2.33}$$

$$L_n |\phi\rangle = 0 \quad n > 0 \tag{4.2.34}$$

$$(L_0 - a) |\phi\rangle = 0, \tag{4.2.35}$$

where a is a constant to be determined. The infinite number of conditions in (4.2.33) and (4.2.34) all follow from the particular two $G_{1/2} |\phi\rangle = G_{3/2} |\phi\rangle = 0$ as a consequence of the algebra (4.2.29).

Now we would like to determine the critical values of a and D in the theory. We recall from chapter 2 the structure of the ghost question in the bosonic string theory. For a certain range of a and D there are ghosts; in another range there are no ghosts. At the boundary between the two regions there are extra zero-norm states (which are on the verge of becoming ghosts). Finding extra zero-norm states that occur for special values of a and D can help us delineate the ghost-free region; even more important, these extra zero-norm states are of great physical importance, because they are related to gauge invariances, and the borderline values of a and D at which these extra zero-norm states appear are the really interesting values.

Let us therefore begin the study of the physical states by searching, as in §2.2.2, for special values of a and D that produce zero-norm states that are not present generically. The ground state $|0; k\rangle$ is on shell for $k^2/2 = a$, and therefore the excited state $G_{-1/2} |0; k\rangle$ is on shell for $k^2/2 = a - 1/2$. On the other hand if $a = 1/2$ this state satisfies the physical-state condition $G_{1/2} |\phi\rangle = 0$ in which case it is also a state of zero norm, $\langle \phi | \phi \rangle = 0$. It has negative norm if a is bigger than one-half. Thus $a = 1/2$ is the preferred value of a, analogous to $a = 1$ in the bosonic string theory. This again has as a consequence that the ground state is a scalar tachyon and the first excited state $(b^\mu_{-1/2} |0\rangle)$ is a massless vector. For the choice $a = 1/2$ there is an infinite family of zero-norm physical states given by $G_{-1/2} |\tilde{\phi}\rangle$, where $G_{1/2} |\tilde{\phi}\rangle = G_{3/2} |\tilde{\phi}\rangle = L_0 |\tilde{\phi}\rangle = 0$.

Now let us try to discover the critical dimension by constructing a second family of zero-norm states of the form

$$|\phi\rangle = (G_{-3/2} + \lambda G_{-1/2}L_{-1})|\tilde{\phi}\rangle, \tag{4.2.36}$$

where

$$G_{1/2}|\tilde{\phi}\rangle = G_{3/2}|\tilde{\phi}\rangle = (L_0 + 1)|\tilde{\phi}\rangle = 0. \tag{4.2.37}$$

Using the super-Virasoro algebra it is easy to show that

$$G_{1/2}|\phi\rangle = (2 - \lambda)L_{-1}|\tilde{\phi}\rangle \tag{4.2.38}$$

$$G_{3/2}|\phi\rangle = (D - 2 - 4\lambda)|\tilde{\phi}\rangle. \tag{4.2.39}$$

Thus $|\phi\rangle$ is a zero-norm state for $\lambda = 2$ and $D = 10$. Ten is therefore the critical dimension. This dimension will be shown to play a role entirely analogous to $D = 26$ for the bosonic string theory. In particular, the physical states that propagate can be put in correspondence with states in the Fock space built from the eight transverse components of α_m^μ and b_r^μ.

Let us turn our attention now to the fermionic sector. Once again a physical state $|\psi\rangle$ is required to be annihilated by the positive-frequency components of the constraint conditions

$$F_n|\psi\rangle = L_n|\psi\rangle = 0, \quad n > 0. \tag{4.2.40}$$

In addition, the zero-mode condition gives a wave equation

$$(F_0 - \mu)|\psi\rangle = 0. \tag{4.2.41}$$

We have introduced an arbitrary constant μ into this generalized Dirac equation. It is proportional to the mass of the fermionic ground state. Since $F_0^2 = L_0$, (4.2.41) implies that

$$(L_0 - \mu^2)|\psi\rangle = 0. \tag{4.2.42}$$

The definition of F_0 does not have a normal-ordering ambiguity in passing from the classical to the quantum theory, so it is hard to see how a nonzero number μ could arise from any sensible alteration in the meaning of the fermionic operator F_0. Indeed, since F_0 is an anticommuting operator, it would seem quite unnatural to add to it a (commuting) c-number. Our search for zero-norm states immediately confirms that this option is not favored, because $|\psi\rangle = F_0|\tilde{\psi}\rangle$ is zero norm for $L_0|\tilde{\psi}\rangle = F_1|\tilde{\psi}\rangle = 0$ (L_0 and

F_1 generate an infinite tower of conditions), but this is only on shell for $\mu = 0$. Therefore this is a first family of zero-norm fermionic states. A second set can be written in the form

$$|\psi\rangle = F_0 F_{-1} |\tilde{\psi}\rangle, \qquad (4.2.43)$$

where

$$F_1 |\tilde{\psi}\rangle = (L_0 + 1)|\tilde{\psi}\rangle = 0. \qquad (4.2.44)$$

This obviously satisfies $F_0 |\psi\rangle = 0$. It is a zero-norm physical state if it is also annihilated by L_1. A short calculation gives

$$L_1 |\psi\rangle = \left(\tfrac{1}{4}D - \tfrac{5}{2}\right)|\tilde{\psi}\rangle, \qquad (4.2.45)$$

and thus $D = 10$ is the critical dimension for the fermionic sector as well.

4.2.3 Boson-Emission Vertex Operators

Vertex operators can be used to construct the physical-state operators of a spectrum-generating algebra for superstrings in much the same way as for bosonic strings. (They are also useful for explicit constructions of amplitudes, of course.) There are three cases that can be considered. The first is emission of a specific on-shell bosonic state from a bosonic string, and the second is emission of such a state from a fermionic string. The third case is emission of a physical on-shell fermionic state from a bosonic string that turns into a fermionic string or vice versa. The last case involves much more complicated issues and mathematics than the first two. Also it is not required for the construction of a spectrum-generating algebra for use in proving a no-ghost theorem. The reason is that to prove a no-ghost theorem, it is adequate to prove that ghosts are absent separately in the bosonic and fermionic sectors; there is no need for an operator that turns bosons into fermions or vice versa. An operator for emission of fermions would have this property. The fermion-emission vertex is crucial for other purposes, however, and we will discuss it in chapter 7.

Let us begin by considering boson emission from a bosonic state. In §2.2.3 we learned that a physical vertex operator must have conformal dimension $J = 1$ in order that its zero-momentum component should map physical states to physical states. This result applies without change to the present problem, but it is no longer the whole story. To map physical states to physical states, an operator must commute correctly with the G_r, as well.

Given a candidate vertex operator $V \equiv V(\tau = 0)$, suppose that there exists another operator W such that for every $r \in Z + 1/2$

$$V(0) = [G_r, W(0)]. \qquad (4.2.46)$$

Note that we restrict the allowed choices of W by requiring that the operator V is independent of r. This condition is suitable when W is a bosonic operator (on the world sheet), but when W is fermionic it needs to be replaced by an anticommutator. Since

$$G_r^2 = L_{2r}, \qquad (4.2.47)$$

it follows that

$$\{G_r, V(0)\} = [L_{2r}, W(0)]. \qquad (4.2.48)$$

Using the definitions

$$V(\tau) = e^{iL_0\tau} V(0) e^{-iL_0\tau} \qquad (4.2.49)$$

and the definition of an operator of conformal spin J,

$$[L_m, V(\tau)] = e^{im\tau}(-i\frac{d}{d\tau} + mJ)V(\tau), \qquad (4.2.50)$$

it is an easy exercise to show that V has conformal dimension $J = 1$ if and only if W has conformal dimension $J = 1/2$.

As a simple first example consider

$$W(0) =: e^{ik \cdot X(0)} : \qquad (4.2.51)$$

This expression has conformal dimension $J = 1/2$ for $k^2 = 1$, which is the mass-shell condition for the ground-state tachyon in the bosonic sector. The associated vertex operator is

$$V(0) = [G_r, W(0)] = k \cdot \psi(0) : e^{ik \cdot X(0)} :, \qquad (4.2.52)$$

or

$$V(\tau) = k \cdot \psi(\tau) : e^{ik \cdot X(\tau)} :,$$

where

$$\psi^\mu(\tau) = \frac{1}{\sqrt{2}} \sum_{-\infty}^{\infty} b_r^\mu e^{-ir\tau}. \qquad (4.2.53)$$

The vertex operator in (4.2.52) obviously has $J = 1$, since each factor has $J = 1/2$ and they commute. Thus V is a suitable vertex operator for emission of the tachyonic ground state (called the 'pion' in the original papers on the subject).

The first excited state is a massless vector (of polarization ζ^μ and momentum k^μ) described by the Fock-space state $\zeta \cdot b_{-1/2} |0; k\rangle$. The $G_{1/2}$ subsidiary condition implies that $\zeta \cdot k = 0$ for this state be physical. To construct the vertex operator for emission of this state consider

$$W_1(0) = \zeta \cdot \psi(0) e^{ik \cdot X(0)}. \tag{4.2.54}$$

For $k^2 = 0$, this operator has $J = 1/2$ as required. The corresponding vertex operator

$$V_1 = \{G_r, W_1\} = \left(\zeta \cdot \dot{X}(0) - \zeta \cdot \psi(0) k \cdot \psi(0)\right) e^{ik \cdot X(0)} \tag{4.2.55}$$

is independent of r, as required. The operator products in this expression are well-defined because $\zeta \cdot k = 0$. Therefore V_1 has $J = 1$ and satisfies all the requirements for a physical emission vertex operator. The gauge invariance of V_1 is demonstrated by replacing ζ^μ by k^μ and noting that the ψ terms then drop out leaving a total derivative.

Two types of states and vertices can be distinguished according to whether they involve an even or odd number of ψ excitations. If W is a bosonic operator on the world sheet (even in ψ), then the commutator $[G_r, W]$ gives a fermionic vertex operator V, as in (4.2.52). If W is a fermionic operator on the world sheet, on the other hand, then the anticommutator $\{G_r, W\}$ gives a bosonic V, as in (4.2.55). The string states emitted or absorbed by fermionic vertices V correspond to Fock-space states with an even number of b oscillator excitations and are called states of 'odd G parity'. Ones associated with bosonic vertices V correspond to Fock-space states with an odd number of b oscillator excitations and are called states of 'even G parity'. (These names originated in the hadronic interpretation of the theory.) We will show in chapter 7 that G is a multiplicatively conserved quantum number for bosonic tree amplitudes. We will also show there that the present Fock-space description of the states (called F_2) can be replaced by a different one (called F_1) in which the correspondence between G parity and the number of b oscillator excitations is reversed to the more natural one.

It may seem strange that boson emissions can be described by operators that are fermionic. However, the operators in question are fermionic only in a two-dimensional world-sheet sense and not in the D-dimensional space-time sense. In fact, for reasons to be described later, the odd G parity states do lead to inconsistencies, and we are forced to truncate the spectrum to the even G parity sector. In particular, this has the virtue of eliminating the tachyon from the spectrum.

Bosonic emissions from a fermionic (R) string can be described by essentially the same equations as their emissions from a bosonic (NS) string. The change in boundary conditions that distinguishes an R string from an NS one cannot influence the local form of the vertex operator. These boundary conditions only affect the question of whether $\psi(\tau)$ is to be expanded in half-integral modes, as in (4.2.53) or in integral modes according to

$$\psi^\mu(\tau) = \frac{1}{\sqrt{2}} \sum_{-\infty}^{\infty} d_n^\mu e^{-in\tau}. \qquad (4.2.56)$$

Of course, in the formulas relating V to W, the operators G_r of the bosonic sector must also be replaced by ones F_m for the fermionic sector. In chapter 7 we will see that tree amplitudes describing a fermionic string with a set of emitted bosons can be calculated either using a propagator L_0^{-1} (generalizing a Klein–Gordon propagator) and vertices of the V type, or a propagator F_0^{-1} (generalizing a Dirac propagator) and vertices of the W type. For purely bosonic trees there is no analog of the second case, so it is always necessary to use V-type vertices.

4.3   Light-Cone Gauge Quantization

Here we wish to repeat for superstrings the analysis given for bosonic strings in §2.3. First we quantize the theory in the manifestly ghost-free (but not manifestly covariant) light-cone gauge, and show that Lorentz invariance is valid if $D = 10$ and $a = 1/2$. Then we show that the manifestly covariant (but not manifestly ghost-free) formalism is in fact equivalent to the light-cone formalism and therefore ghost-free. This again involves use of the massless boson-emission vertex and certain longitudinal operators to generate the physical spectrum. We then move on to matters that do not have a bosonic analog; the first evidence for space-time supersymmetry will be described in §4.3.3.

4.3.1 The Light-Cone Gauge

In the bosonic theory of chapter 2, we saw that even in the covariant gauge $h_{\alpha\beta} = \eta_{\alpha\beta}$ there are residual gauge invariances that allow further choices, such as light-cone gauge. The RNS model as we have been discussing it so far is really a gauge fixed version of a model with local world-sheet supersymmetry, which will be formulated in §4.3.5. In chapter 2, we saw that reparametrization invariance of the bosonic string theory permits the imposition of an extra gauge condition, light cone gauge. Here we will discuss the analogous light cone gauge formulation of the RNS model. This will involve a bit of guesswork, since we are postponing the systematic treatment of local world-sheet supersymmetry to §4.3.5.

Recall that in the bosonic string theory the residual reparametrization invariance that preserves the covariant gauge choice was just sufficient to be able to gauge away the $+$ components of all the nonzero mode oscillators, so that

$$X^+(\sigma, \tau) = x^+ + p^+ \tau. \tag{4.3.1}$$

That reasoning applies so long as X^+ satisfies the two-dimensional wave equation, so it applies equally well in the present context. However, there is now also the freedom of applying local supersymmetry transformations that preserve the gauge choices. They turn out to be just sufficient to gauge away ψ^+ completely so that we may make the gauge choice

$$\psi^+ = 0. \tag{4.3.2}$$

As a consistency check we note that under a global supersymmetry transformation

$$\delta X^+ = \bar{\epsilon}\psi^+ = 0, \tag{4.3.3}$$

since $\psi^+ = 0$, so that the X^+ gauge choice is not altered in the process.

Let us concentrate on the bosonic open-string sector to begin with and define $\psi^\mu(\sigma^\pm) = \psi^\mu_+(\sigma^+)$ for $\sigma > 0$ and $= \psi^\mu_-(\sigma^-)$ for $\sigma < 0$ (recalling that $\sigma^\pm = \tau \pm \sigma$ and $\partial_\pm = (\partial_\tau \pm \partial_\sigma)/2$). The mode expansion of $\psi^\mu(\tau)$ is given by (4.2.53). The corresponding expansions for closed strings involve modes with doubled exponents. The X^μ expansions were presented in §2.1.2 (in particular, $\partial_+ X^\mu = \frac{1}{2}\sum_{-\infty}^{\infty} \alpha_n^\mu e^{-in(\tau+\sigma)}$). The subsidiary constraints implied by the vanishing of J_α and $T_{\alpha\beta}$ take the form

$$\psi \cdot \partial_+ X = 0 \tag{4.3.4}$$

and

$$(\partial_+ X)^2 + \frac{i}{2}\psi \cdot \partial_+ \psi = 0. \tag{4.3.5}$$

Given the gauge choices $\partial_+ X^+ = \frac{1}{2}p^+$ and $\psi^+ = 0$, these equations can be solved for the light-cone components ψ^- and $\partial_+ X^-$

$$\partial_+ X^- = \frac{1}{p^+}(\partial_+ X^i \partial_+ X^i + \frac{i}{2}\psi^i \partial_+ \psi^i) \tag{4.3.6}$$

$$\psi^- = \frac{2}{p^+}\psi^i \partial_+ X^i. \tag{4.3.7}$$

In terms of the Fourier modes this gives

$$\alpha_n^- = \frac{1}{2p^+} \sum_{i=1}^{D-2} \left(\sum_{m=-\infty}^{\infty} : \alpha_{n-m}^i \alpha_m^i : \right.$$

$$\left. + \sum_{r=-\infty}^{\infty} (r - n/2) : b_{n-r}^i b_r^i : \right) - \frac{a\delta_n}{2p^+} \tag{4.3.8}$$

$$b_r^- = \frac{1}{p^+} \sum_{i=1}^{D-2} \sum_{s=-\infty}^{\infty} \alpha_{r-s}^i b_s^i. \tag{4.3.9}$$

Normal ordering is required for α_0^-, as usual.

Comparing with formulas in §4.2.2 it is evident that $p^+\alpha_n^-$ and $p^+b_r^-$ satisfy a super-Virasoro algebra with anomaly terms

$$A(m) = \frac{D-2}{8}(m^3 - m) + 2am \tag{4.3.10}$$

$$B(r) = \frac{D-2}{2}(r^2 - 1/4) + 2a. \tag{4.3.11}$$

The light-cone gauge is not consistent with the quantum structure of the theory in general. In fact, we can deduce restrictions on the space-time dimension D and the parameter a, just as in the bosonic string theory, by requiring that the Lorentz algebra be satisfied in the light-cone gauge. Substituting for $\alpha_n^\pm$ and $b_r^\pm$ in the covariant expressions of (4.2.19) and (4.2.20) gives

$$E^{+\mu} = K^{+\mu} = 0 \tag{4.3.12}$$

$$E^{i-} = -i\sum_{n=1}^{\infty} \frac{1}{n}(\alpha_{-n}^i \alpha_n^- - \alpha_{-n}^- \alpha_n^i) \tag{4.3.13}$$

$$K^{i-} = \frac{1}{p^+} \sum_{n=-\infty}^{\infty} \sum_{j=1}^{D-2} K_{-n}^{ij} \alpha_n^j, \tag{4.3.14}$$

where

$$K_m^{ij} = -\frac{i}{2} \sum_{-\infty}^{\infty} (b_{m-r}^i b_r^j - b_{m-r}^j b_r^i) \tag{4.3.15}$$

and α_n^- is given by (4.3.8). As in the bosonic string theory, it is easy to verify that the Lorentz generators

$$J^{\mu\nu} = l^{\mu\nu} + E^{\mu\nu} + K^{\mu\nu} \tag{4.3.16}$$

satisfy the usual Lorentz algebra, except for the case of $[J^{i-}, J^{j-}]$, which is supposed to vanish.

The proof that $[J^{i-}, J^{j-}] = 0$ requires basically the same methods as were employed in §2.3.1 for the bosonic string theory in the light-cone gauge. Again, terms quartic in oscillators are guaranteed to cancel as a consequence of the Lorentz invariance of the classical theory. Only terms quadratic in oscillators need to be evaluated carefully. Defining $L^{\mu\nu} = l^{\mu\nu} + E^{\mu\nu}$, we have $J^{i-} = L^{i-} + K^{i-}$. Manipulations identical to those of §2.3.1 now give

$$[L^{i-}, L^{j-}] = -(p^+)^{-2} \sum_{m=1}^{\infty} \Delta_m (\alpha^i_{-m}\alpha^j_m - \alpha^j_{-m}\alpha^i_m), \tag{4.3.17}$$

where

$$\Delta_m = m\left(\frac{D-2}{8}\right) + \frac{1}{m}\left(2a - \frac{D-2}{8}\right). \tag{4.3.18}$$

The difference from the previous result is attributable to the changed anomaly term in the commutator $[\alpha^-_m, \alpha^-_n]$. To complete the calculation, we use the identities

$$[K^{ij}_m, \alpha^-_n] = mK^{ij}_{m+n} \tag{4.3.19}$$

and

$$[K^{ij}_m, K^{kl}_n] = -i(K^{il}_{m+n}\delta^{jk} - K^{jl}_{m+n}\delta^{ik} - K^{ik}_{m+n}\delta^{jl} + K^{jk}_{m+n}\delta^{il})$$
$$+ m(\delta^{ik}\delta^{jl} - \delta^{il}\delta^{jk})\delta_{m+n} \tag{4.3.20}$$

to deduce that

$$[K^{i-}, K^{j-}] + [L^{i-}, K^{j-}] + [K^{i-}, L^{j-}]$$
$$= (p^+)^{-2} \sum_{m=1}^{\infty} m(\alpha^i_{-m}\alpha^j_m - \alpha^j_{-m}\alpha^i_m). \tag{4.3.21}$$

Combining these results we see that the Lorentz algebra is satisfied only for the choices $D = 10$ and $a = 1/2$.

4.3.2 No-Ghost Theorem and the Spectrum-Generating Algebra

The proof of the absence of ghosts is based on an extension of the proof in the bosonic theory. Since the proof of the theorem is almost exactly identical in the bosonic and fermionic sectors, we only describe the bosonic sector in order not to be overly repetitious. As in our discussion of the bosonic theory in chapter 2, we also construct the operators that generate the spectrum of the theory.

As in §2.3.2 we begin by constructing the DDF operators, which describe physical transverse excitations. There we considered a massless vector vertex operator for a transverse polarization, where the emitted momentum is $k^\mu = nk_0^\mu$ and the vertex operator is restricted to act on states of momentum $p^\mu = p_0^\mu - Nk_0^\mu$. The momenta k_0 and p_0 are given by $k_0^- = p_0^- = -1$, $p_0^+ = 1$, $k_0^+ = k_0^i = p_0^i = 0$. Also, n and N are integers. In view of (4.2.55), the appropriate operator is

$$V^i(nk_0, \tau) = [\dot{X}^i(\tau) - \psi^i(\tau)k \cdot \psi(\tau)]e^{inX^+(\tau)}. \qquad (4.3.22)$$

Then, as before, the operators

$$A_n^i = \langle V^i(nk_0, \tau) \rangle = \frac{1}{2\pi} \int\limits_0^{2\pi} d\tau\, V^i(nk_0, \tau) \qquad (4.3.23)$$

satisfy $[L_m, A_n^i] = 0$, when restricted to act on states of momentum p^μ, where they are well-defined. In the case of superstrings it is also necessary that A_n^i commute with G_r if it is to map physical states to physical states so that it can be included in a spectrum-generating algebra. This is established using (4.2.48) and the fact that W has conformal dimension $1/2$, which gives

$$[G_r, V^i(nk_0, \tau)] = e^{-ir\tau}[L_{2r}, W^i(nk_0, \tau)]$$
$$= e^{ir\tau}\left(-i\frac{d}{d\tau} + r\right)W^i(nk_0, \tau). \qquad (4.3.24)$$

It follows that

$$[G_r, A_n^i] = -i\langle \frac{d}{d\tau}\left(e^{ir\tau}W^i(nk_0, \tau)\right)\rangle = 0, \qquad (4.3.25)$$

so that A_n^i can belong to the spectrum-generating algebra. These operators are in one-to-one correspondence with the transverse oscillators α_n^i as in the bosonic string theory. However, now we need additional transverse fermionic operators B_r^i to correspond to the oscillators b_r^i.

It is possible to construct transverse operators B_r^i that are suitable for inclusion in a spectrum-generating algebra even though there is no corresponding vertex operator. The formulas that work are

$$Z^i(k) = -2\psi^i k \cdot \psi(k \cdot \dot{X})^{-1/2} e^{ik \cdot X} \tag{4.3.26}$$

$$
\begin{aligned}
Y^i(k) &= [G_r, Z^i(k)] \\
&= \{\psi^i(k \cdot \dot{X})^{1/2} - \dot{X}^i k \cdot \psi(k \cdot \dot{X})^{-1/2} \\
&\quad - \frac{i}{4}\psi^i k \cdot \psi k \cdot \dot{\psi}(k \cdot \dot{X})^{-3/2}\} e^{ik \cdot X},
\end{aligned}
\tag{4.3.27}
$$

$$B_r^i = \langle Y^i(rk_0, \tau) \rangle. \tag{4.3.28}$$

The peculiar-looking fractional powers give well-defined expressions in the present context for the reasons explained in §2.3.3. This operator satisfies

$$\{G_r, B_s^i\} = 0, \tag{4.3.29}$$

as required, since $Z^i(k)$ has $J = 1/2$ for $k^2 = 0$.

The algebra of the operators A_n^i and B_r^i can be worked out by the same sorts of manipulations as shown in §2.3.2. One finds that

$$[A_m^i, A_n^j] = m\delta^{ij}\delta_{m+n}$$

$$[A_n^i, B_r^j] = 0 \tag{4.3.30}$$

$$\{B_r^i, B_s^j\} = \delta^{ij}\delta_{r+s}.$$

Thus the algebra is isomorphic to that of the transverse oscillators α_n^i and b_r^i and gives a manifestly positive-definite Fock space.

The DDF states are states created by the DDF operators acting on a ground state of an arbitrary level. The states that form an orthogonal basis in the space of DDF states, denoted $|f\rangle$, satisfy the same conditions as in the bosonic theory, namely,

$$L_m|f\rangle = 0, \qquad K_m|f\rangle = 0, \qquad m > 0, \tag{4.3.31}$$

(where $K_m = k_0 \cdot \alpha_m$) as well as the fermionic conditions

$$G_r|f\rangle = 0, \qquad H_r|f\rangle = 0. \tag{4.3.32}$$

The operators H_r are defined by

$$H_r = k_0 \cdot b_m \sim b_m^+. \tag{4.3.33}$$

The fact that H_m annihilates the DDF states is obvious from the fact that the DDF operators do not contain any powers of b_m^-.

We can now describe the complete basis of states in the combined Fock space of all the bosonic and fermionic oscillators of the NS sector in terms of the basis formed by the bosonic and fermionic DDF states together with their orthogonal complement. This consists of the states formed by acting with arbitrary powers of L_{-m}, K_{-m} (defined in §2.3.3), G_{-r} and H_{-r}. An arbitrary state of this type has the form

$$
\begin{aligned}
G_{-1/2}^{\epsilon_{1/2}} G_{-3/2}^{\epsilon_{3/2}} \cdots G_{-r}^{\epsilon_r} L_{-1}^{\lambda_1} \cdots L_{-n}^{\lambda_n} \\
H_{-1/2}^{\delta_{1/2}} \cdots H_{-s}^{\delta_s} K_{-1}^{\mu_1} \cdots K_{-m}^{\mu_m} |f\rangle,
\end{aligned}
\tag{4.3.34}
$$

where $|f\rangle$ is an arbitrary DDF state. The fermionic occupation numbers ϵ_r and δ_s are equal to zero or one and

$$
\sum i\epsilon_i + \sum j\lambda_j + \sum k\delta_k + \sum l\mu_l = P > 0.
\tag{4.3.35}
$$

P is the level of the state above the level of the DDF state, $|f\rangle$. As in §2.3.3 such states are linearly independent for any choice of $|f\rangle$ if the matrix, $\mathcal{M}^P$, of their inner products has nonzero determinant for any choice of P. This can again be shown by a simple extension of the argument given for the bosonic theory. As a result, a general state in the Hilbert space can be written in the form

$$
|\phi\rangle = |s\rangle + |k\rangle,
\tag{4.3.36}
$$

where $|s\rangle$ is a spurious state and $|k\rangle$ is a state in the space of states of the form

$$
|k\rangle = \prod_{s=\frac{1}{2}}^{\infty} H_{-s}^{\delta_s} \prod_{m=1}^{\infty} K_{-m}^{\lambda_m} |f\rangle.
\tag{4.3.37}
$$

The next step in the proof of the no-ghost theorem is to demonstrate that in the critical dimension an arbitrary spurious state $|s\rangle$, satisfying $L_0 = 1/2$, is mapped into another spurious state by the action of the L_m and G_r gauges ($m, r > 0$). The argument again parallels that of §2.3.3, starting this time by considering the action of $G_{1/2}$ and $\tilde{G}_{3/2} = G_{3/2} + 2L_1 G_{1/2}$ on the spurious state. Only if $L_0 = 1/2$ and if $D = 10$ do these operations on a spurious state result in another spurious state. Since all the higher modes G_r, as well as the L_m modes with $m > 0$, can be formed by iterating $G_{1/2}$ and $G_{3/2}$, it follows that $|s\rangle$ is null and that $|k\rangle$ must be a DDF state. The no-ghost theorem follows as in §2.3.3.

To complete the description of the spectrum-generating algebra it is still necessary to construct longitudinal operators A_m^- and B_r^-. The construction of the longitudinal operators of the spectrum generating algebra involves the same principles we have used above and in §2.3.3. However, the algebra is tedious and the expressions are lengthy, so we only quote the results.

The complete spectrum-generating algebra is given by (4.3.30) and

$$[A_m^-, A_n^i] = -nA_{m+n}^i \tag{4.3.38}$$

$$[A_m^-, B_r^i] = -(\tfrac{1}{2}m + r)B_{m+r}^i \tag{4.3.39}$$

$$[A_m^-, A_n^-] = (m - n)A_{m+n}^- + m^3\delta_{m+n} \tag{4.3.40}$$

$$[B_r^-, A_n^i] = -nB_{r+n}^i \tag{4.3.41}$$

$$\{B_r^-, B_s^i\} = A_{r+s}^i \tag{4.3.42}$$

$$\{B_r^-, B_s^-\} = 2A_{r+s}^- + 4r^2\delta_{r+s} \tag{4.3.43}$$

$$[A_m^-, B_r^-] = (\tfrac{1}{2}m - r)B_{m+r}^-. \tag{4.3.44}$$

These operators can be proved to generate the entire physical spectrum by the same reasoning used in §2.3.3. The important fact about this algebra is that A_m^- and B_r^- have commutation relations analogous to L_m and G_r, except for the form of the anomaly terms. Note that this algebra is the same as that of the light-cone gauge operators α_m^i, $p^+\alpha_m^-$, b_r^i, $p^+b_r^-$, that the anomalies match those in (4.3.10) and (4.3.11) for $D = 10$ and $a = 1/2$, the same conditions that were required for Lorentz invariance. If one scales the + components of of k^μ and p^μ by $1/\gamma$ and the − components by γ, corresponding to a longitudinal boost, the spectrum-generating algebra is unchanged. In the limit $\gamma \to 0$ the operators A_m^i, A_m^-/p^+, B_r^i, B_r^-/p^+ reduce to the oscillator expressions α_m^i, α_m^-, b_r^i, b_r^- given in §4.3.1. This explains why the anomalies agree when the Lorentz algebra is satisfied.

Just as in the bosonic theory it is convenient to define the operators

$$\tilde{A}_m^- = A_m^- - \tfrac{1}{2}\sum_{i=1}^{8}\sum_{n=-\infty}^{\infty} : A_{m-n}^i A_n^i :$$

$$- \tfrac{1}{2}\sum_{i=1}^{8}\sum_{r=-\infty}^{\infty}(r - \tfrac{1}{2}m) : B_{m-r}^i B_r^i : + \tfrac{1}{2}\delta_m \tag{4.3.45}$$

$$\tilde{B}_r^- = B_r^- - \sum_{i=1}^{8}\sum_{n=-\infty}^{\infty} B_{r-n}^i A_n^i, \tag{4.3.46}$$

which (anti)commute with the transverse DDF operators. The physical subspace of the Fock space can then be identified with a subspace of a manifestly positive-definite Hilbert space for $a = 1/2$ and $D \leq 10$. When $D = 10$, the operators $\tilde{A}_m^-$ and $\tilde{B}_r^-$ create physical states of zero norm that are orthogonal to every state in the physical spectrum. Therefore they describe states that decouple from the physical spectrum, which is entirely built up by the transverse DDF operators A_m^i and B_r^i.

The spectrum does contain ghosts when $D > 10$. For example, the pair of states $B_{-3/2}^- |0; p_0\rangle$ and $\sum_{i=1}^{D-2} A_{-1}^i B_{-1/2}^i |0; p_0\rangle$ has a matrix of inner products $\begin{pmatrix} 8 & D-2 \\ D-2 & D-2 \end{pmatrix}$ with determinant $(D-2)(10-D)$. Thus, for $D > 10$ one linear combination describes a ghost.

4.3.3 The GSO Conditions

The RNS model, as described so far, is an inconsistent quantum theory, even for $D = 10$ and $a = 1/2$ (or $a = 0$ in the fermionic sector) unless further conditions are imposed. The spectrum must be truncated in a very specific manner first proposed by Gliozzi, Scherk and Olive (GSO).

There are several arguments that suggest that a truncation of the spectrum is required. First of all, the theory has a tachyon, and we would like to eliminate it. To impose an extra restriction on the states that would eliminate some of the states, including the tachyon, while keeping the massless particles that really interest us, would make the theory far more attractive.

Second, even though there is no actual conflict with the spin-statistics theorem, it is unnerving to have anticommuting operators ψ^μ that map bosons to bosons. Thus, let $|\phi\rangle$ be a bosonic state that we consider 'good', perhaps a massless vector meson. Then $\psi^\mu |\phi\rangle$ is a state of integer spin that nonetheless is obtained from $|\phi\rangle$ by acting with an anticommuting operator – a state of affairs that feels unnatural. More generally, consider the state

$$\psi^{\mu_1}(\sigma_1)\psi^{\mu_2}(\sigma_2)\dots\psi^{\mu_n}(\sigma_n)|\phi\rangle. \tag{4.3.47}$$

This state is bosonic for any n, since the ψ^μ are all bosonic. For even n, there is nothing peculiar here, since the product of n anticommuting operators ψ^{μ_k} is commuting. But for odd n contemplation of the states in (4.3.47) may give us the feeling that all is not well. We are tempted to propose discarding the states in (4.3.47) that have odd n and keeping those of even n. This can be stated formally by introducing a quantum number called $(-1)^F$ (historically it was called G parity in the effort to apply string theory to strong interactions) under which the Fermi fields ψ^μ

are odd and the Bose fields X^μ are even. That property only characterizes the operator $(-1)^F$ up to sign. We fix the sign by saying that the massless vector has $(-1)^F = +1$. Then the general state in (4.3.47) has $(-1)^F = (-1)^n$. Keeping only states of even n amounts to keeping only states of $(-1)^F = +1$. This is the GSO projection.

The third great virtue of the GSO projection is that it gives a supersymmetric theory (in the ten-dimensional sense, to be contrasted with the two-dimensional supersymmetry that is already present). We will see evidence for this here, and a complete proof in the next chapter. Space-time supersymmetry is an elegant and attractive thing, and on these grounds alone the GSO projection is attractive. Moreover, in our study of the spectrum we will learn that the theory contains a massless spin 3/2 particle. Consistency of the theory can hardly be expected at the interacting level unless this massless spin 3/2 particle couples to a conserved current. The corresponding conserved charge would have spin 1/2 and so would be a supersymmetry charge. Thus, it must be that at the interacting level the GSO projection (or some other modification to either eliminate the massless spin 3/2 particle or give space-time supersymmetry) is needed for consistency. We will learn why in chapter 9.

We devote the remainder of this subsection to developing circumstantial evidence that the RNS model with the GSO projection is a supersymmetric theory in the ten-dimensional sense. It is necessary to first discuss some background.

We have learned that the massless open-string states consist of a vector and a spinor. The vector is described by the Fock-space state $b^\mu_{-1/2} |0; k\rangle$. The massless spinor is the lowest-mass solution of the condition $F_0 |\psi\rangle = 0$. Such a state is described by $|a; k\rangle u^a(k)$, where $u^a(k)$ is a spinor satisfying the massless Dirac equation. (Since the ground state in the R sector is a spinor, we denote it as $|a; k\rangle$ where a is a spinor index and k is the momentum.) A necessary condition for supersymmetry of the theory is that this pair of states should form a supersymmetry multiplet. (Unbroken supersymmetry requires that each mass level must form a separate supermultiplet.) A vector field A^μ in $D = 10$ has ten components, but only the eight transverse components describe independent propagating modes. Thus supersymmetry requires that the massless spinor should also have eight propagating modes. In general, a spinor in ten dimensions might be expected to have $2^{D/2} = 32$ complex components.[*] However, we

---

[*] A detailed description of the construction of the spinor representation of $SO(D)$ will be given in the appendix 5.A, showing in particular that for even D this representation has dimension $2^{D/2}$ for even D.

show below that it is possible to impose simultaneous Majorana and Weyl constraints, each of which reduces the number of components by a factor of two thereby reducing the total to 16 real components. These remaining 16 components must still satisfy the Dirac equation $\Gamma \cdot \partial \chi = 0$ to describe physical propagating degrees of freedom. This linear equation relates half of the components to the other half, which in turn satisfy a Klein–Gordon equation. Thus, as always, the number of propagating modes described by a spinor satisfying a Dirac equation is just half the number of components that the spinor contains. It follows that a Majorana–Weyl spinor in $D = 10$ describes eight propagating modes, the number required to form a supermultiplet together with the vector A_μ.

Let us first consider the Majorana condition. A Majorana condition is simply a reality condition on the fermion field. The massless Dirac equation is $0 = \Gamma^\mu \partial_\mu \psi$. If it is possible to choose the Dirac matrices Γ^μ to be all real or all imaginary, then it makes sense to impose a condition that ψ should be real. A representation of gamma matrices in which they are all real or all imaginary is called a Majorana representation, and real spinors are called Majorana spinors.

We will construct a representation of the $D = 10$ Dirac algebra in which the gamma matrices are all imaginary. In such a representation, necessarily Γ_0 is antisymmetric and the other nine Γ matrices are symmetric. Since only the case $D = 10$ is important in this book, it is the only one we discuss in detail. However, we note without proof that Majorana fermions are possible whenever the dimension of space-time is $D = 2$, 3, or 4 (mod 8).

Of the ten matrices Γ_μ we may single out eight associated with 'transverse' directions. They form a Clifford algebra for an $SO(8)$ subalgebra of $SO(1,9)$, the $D = 10$ Lorentz algebra. The Clifford algebra for $SO(8)$ is constructed in detail in appendix 5.B, which is self-contained and can be consulted at this point. There we construct eight real symmetric 16×16 matrices γ^i satisfying $\{\gamma^i, \gamma^j\} = 2\delta^{ij}$. We now wish to utilize these in the construction of ten imaginary 32×32 matrices satisfying $\{\Gamma^\mu, \Gamma^\nu\} = -2\eta^{\mu\nu}$. (We denote 16×16 gamma matrices of $SO(8)$ by a small letter γ and 32×32 gamma matrices of $SO(1,9)$ by capital Γ.) We construct 32×32 matrices as tensor products of 2×2 Pauli matrices and 16×16 matrices. Specifically we set

$$
\begin{aligned}
\Gamma^0 &= \sigma_2 \otimes 1_{16} \\
\Gamma^i &= i\sigma_1 \otimes \gamma^i \qquad i = 1, \dots, 8 \\
\Gamma^9 &= i\sigma_3 \otimes 1_{16}.
\end{aligned}
\tag{4.3.48}
$$

Here 1_{16} is the 16×16 identity matrix. In the usual representation in which σ_1 and σ_3 are real and σ_2 is imaginary, it is evident that the Γ^μ are imaginary as desired. This shows that a Majorana condition is possible in ten dimensions.

Let us now turn to the second important condition that can be imposed on spinors – the Weyl condition. Whenever D is even one can define a matrix, analogous to γ_5 in four dimensions, that can be used to define chirality of spinors. In $D = 10$, we introduce

$$\Gamma_{11} = \Gamma^0 \Gamma^1 \cdots \Gamma^9, \tag{4.3.49}$$

which satisfies $\{\Gamma_{11}, \Gamma^\mu\} = 0$ and $(\Gamma_{11})^2 = 1$. To verify the latter equation requires a careful counting of minus signs. Nine arise from squaring spatial gamma matrices and 45 arise from anticommuting matrices to bring like ones together. Thus the total is even. Spinors ψ with $\Gamma_{11}\psi = +\psi$ or $\Gamma_{11}\psi = -\psi$ are called spinors of positive and negative chirality, respectively. The operators $(1 \pm \Gamma_{11})/2$ are chirality projection operators; they project on spinors of definite chirality. A spinor of definite chirality is called a Weyl spinor. The restriction to spinors of one chirality or the other is called a Weyl condition.

In the Majorana representation given above, with the ten Γ^μ imaginary, it is clear that Γ_{11} is real. Therefore given a real spinor χ, the two pieces of definite chirality $\frac{1}{2}(1 \pm \Gamma_{11})\chi$ are also real. This means that the Majorana and Weyl conditions are compatible. It is possible to require that a spinor field ψ be a Weyl spinor, of, say, positive chirality, and also be real (Majorana). This is to be contrasted with the case in $D = 4$ where both Majorana and Weyl spinors can be defined, but the conditions are not compatible because γ_5 is imaginary in a Majorana representation. In general, the Majorana and Weyl conditions are compatible for dimensions $D = 2 \pmod 8$. A spinor of definite chirality that also obeys a Majorana condition is called a Majorana–Weyl spinor.

A Majorana–Weyl spinor in $D = 10$ has 16 real components. (We started with 32 complex components, the Majorana condition makes them real, and the Weyl condition eliminates half of them.) As we have already explained, the Dirac equation $\Gamma \cdot \partial \chi = 0$ implies that eight of them can be related to the other eight. Thus there are just eight independent propagating degrees of freedom for χ, just right to form a supermultiplet with the vector A^μ. Attaching indices that span the adjoint representation of a semisimple Lie group to each of them gives the $D = 10$ super Yang–Mills multiplet. The corresponding Lagrangian field theory is described in appendix 4.A.

We have just learned that the massless fermions in $D = 10$ must be simultaneously Majorana and Weyl. This ensures that the massless sector forms a supersymmetry multiplet. The Weyl condition means that the ground-state spinor is an eigenstate of Γ_{11}. The generalization of this condition to an arbitrary fermion mass level requires the operator

$$\overline{\Gamma} = \Gamma_{11}(-1)^{\sum_{n=1}^{\infty} d_{-n} \cdot d_n}, \qquad (4.3.50)$$

which has the property

$$\{\overline{\Gamma}, d_n^{\mu}\} = 0, \qquad (4.3.51)$$

since the factor Γ_{11} anticommutes with $d_0^{\mu} \sim \Gamma^{\mu}$, and the other factor in $\overline{\Gamma}$ anticommutes with the d_n^{μ} modes having $n \neq 0$. Since ψ^{μ} is linear in the d_n^{μ} it follows that

$$\{\overline{\Gamma}, \psi^{\mu}(\sigma, \tau)\} = 0. \qquad (4.3.52)$$

It is the operator $\overline{\Gamma}$ that plays in the R sector the role of $(-1)^F$, which we defined with less effort in the NS sector. The GSO condition is thus

$$\overline{\Gamma}\,|\psi\rangle = |\psi\rangle \qquad (4.3.53)$$

for a physical fermion state. This is the counterpart of the even G parity restriction

$$G\,|\phi\rangle = |\phi\rangle\,, \qquad (4.3.54)$$

where

$$G = -(-1)^{\sum_{r=1/2}^{\infty} b_{-r} \cdot b_r}, \qquad (4.3.55)$$

for physical bosons. The operators $\overline{\Gamma}$ and G represent $(-1)^F$ in the fermionic and bosonic sectors, respectively.

For the fermionic ground state, $(-1)^F$ reduces to Γ_{11}, the chirality operator, and the GSO projection means keeping only massless fermions of positive chirality in the ten-dimensional sense. Any state in the Ramond sector can be built as

$$d_{-m_1}^{i_1} d_{-m_2}^{i_2} \cdots |\alpha\rangle\,, \qquad (4.3.56)$$

a suitable product of mode operators (possibly including zero-mode operators) acting on a positive-chirality massless ground state, and with $\overline{\Gamma} = (-1)^F$, such a state has $(-1)^F$ equal to $(-1)^n$, with n being the number of fermion mode operators in (4.3.56).

The condition imposed on massive fermions at first sight seems to imply that they too are Weyl, which is impossible. The massive spinor representations of the Lorentz group do not admit Weyl spinors, since Γ_{11} does not anticommute with $i\Gamma \cdot \partial + m$. However, in string theory $\{\overline{\Gamma}, F_0\} = 0$ is an exact statement, not an approximation valid only for massless states, so the GSO projection can be made even for massive levels. To see in detail how this works let us consider the first excited fermionic level. In light-cone gauge the possible states are $\alpha^i_{-1}|0\rangle\, u_1$ and $d^i_{-1}|0\rangle\, u_2$, where the $\overline{\Gamma}$ condition implies that u_1 and u_2 are Majorana–Weyl spinors of opposite handedness

$$\Gamma_{11} u_1 = u_1 \qquad\qquad (4.3.57)$$

$$\Gamma_{11} u_2 = -u_2. \qquad\qquad (4.3.58)$$

These two spinors combine to give a Majorana spinor that forms an irreducible massive representation of the Lorentz group.

A necessary condition for supersymmetry is that there should be an equal number of boson and fermion states at each mass level. For massless states, this follows from (4.3.53). The Weyl projection leaves eight physical massless fermion states, the same as the number of propagating modes of the massless vector. Let us now examine how many bosonic and fermionic states there are at each of the excited mass levels. This is a combinatoric problem analogous to the one worked out for bosonic strings in §2.3.5. We saw there that the degeneracies were given by $\mathrm{tr}\,w^N$. Essentially the same is true now, with only minor modifications. In the bosonic sector we need to take account of the G parity projection and the fact that the massless level involves $1/2$ unit of excitation. As a result the number of states with $\alpha' M^2 = n$ is given by $d_{\mathrm{NS}}(n)$, where

$$f_{\mathrm{NS}}(w) = \sum_{n=0}^{\infty} d_{\mathrm{NS}}(n)w^n = \frac{1}{\sqrt{w}}\mathrm{tr}\left[\frac{1}{2}(1+G)w^N\right] \qquad (4.3.59)$$

$$N = \sum_{n=1}^{\infty}\alpha^i_{-n}\alpha^i_n + \sum_{r=1/2}^{\infty} r b^i_{-r} b^i_r. \qquad (4.3.60)$$

(Recall that G is the same as $(-1)^F$.) The evaluation of the trace is a straightforward generalization of the example treated in §2.3.5 for the bosonic theory. The new feature is the factor of the trace over the b^i_r modes, $\mathrm{tr}\,w^{\sum r b^i_{-r} b^i_r}$. Since each fermionic state is either occupied or unoccupied this factor simply gives $(1+w^r)$ for each mode. The presence of G in $\mathrm{tr}\,Gw^N$ changes the signs of the occupied states, giving $(1-w^r)$ at each level. The result for the complete trace is therefore

$$f_{\text{NS}}(w) = \frac{1}{2\sqrt{w}} \left[\prod_{m=1}^{\infty} \left(\frac{1 + w^{m-1/2}}{1 - w^m} \right)^8 - \prod_{m=1}^{\infty} \left(\frac{1 - w^{m-1/2}}{1 - w^m} \right)^8 \right].$$

$$(4.3.61)$$

The degeneracy of the fermionic levels is given in similar manner by

$$f_{\text{R}}(w) = \sum_{n=1}^{\infty} d_{\text{R}}(n) w^n = \text{tr}\frac{1}{2}(1 + \overline{\Gamma}) w^N = 8 \, \text{tr} \, w^N \qquad (4.3.62)$$

with

$$N = \sum_{n=1}^{\infty} (\alpha_{-n}^i \alpha_n^i + n d_{-n}^i d_n^i). \qquad (4.3.63)$$

The factor of 8 represents the degeneracy of the ground state and takes account of the $\overline{\Gamma} = 1$ projection for every mass level. (It simply removes half the states that would be present otherwise.) There is no factor of $\sqrt{w}$ in the formula for $f_{\text{R}}(w)$ because the massless fermion ground state corresponds to $N = 0$. Performing the trace gives

$$f_{\text{R}}(w) = 8 \prod_{m=1}^{\infty} \left(\frac{1 + w^m}{1 - w^m} \right)^8. \qquad (4.3.64)$$

Now we can address the question of whether there really are an equal number bosons and fermions at each mass level. The required identity would be

$$f_{\text{NS}}(w) = f_{\text{R}}(w) = F(w). \qquad (4.3.65)$$

Remarkably, this identity was proved in 1829 by Jacobi, who referred to it as 'a rather obscure formula' (*aequatio identica satis abstrusa*). The reader should verify that each expression has the expansion

$$F(w) = 8(1 + 16w + 144w^2 + \cdots). \qquad (4.3.66)$$

This observation certainly does not constitute a proof of supersymmetry, but it is a very encouraging indication. We will give a proof in the next chapter.

4.3.4 Locally Supersymmetric Form of the Action

By now we have formulated the superstring in a formalism with world-sheet supersymmetry, but we would like to put all of this on a sounder logical foundation. Instead of arbitrarily postulating the super-Virasoro conditions, we would like to derive them as constraints that arise upon gauge fixing of a gauge-invariant Lagrangian. For the bosonic string, the Virasoro constraints arose by gauge fixing a Lagrangian with two-dimensional general covariance; for supersymmetric strings, the super-Virasoro constraints have to arise by gauge fixing a two-dimensional Lagrangian with local supersymmetry as well as general covariance. This means we must overcome a variety of technical obstacles. We must learn how to introduce spinors on a curved world sheet, and we must formulate two-dimensional supergravity.

It is not straightforward to couple fermions to general relativity. The reason for this is that general coordinate transformations (in D dimensions) transform Bose fields by $GL(D, R)$ transformations ($GL(D, R)$ being the group of invertible $D \times D$ real-valued matrices), and $GL(D, R)$ does not have any finite-dimensional spinor representations. There is a spinor representation of the Lorentz group, so it is clear how spinors are supposed to transform under Lorentz transformations, but it is not straightforward to describe how they should transform under general coordinate transformations (also called diffeomorphisms) that are not Lorentz transformations. This subject is discussed in textbooks on general relativity, and we will have more to say about it in chapter 12, where the notion of spinors in general relativity is used to give an introduction to certain geometric and topological ideas. Here we attempt to give a pragmatic introduction adequate for our needs in this section.

Consider spinors on a D-dimensional manifold M. The equivalence principle implies the existence of an inertial frame at each point of the manifold. In this frame Lorentz transformations can act in a meaningful way. They can be regarded as acting on the (flat) tangent space to the manifold at the point in question. In the locally inertial frame, we introduce a basis e_μ^m, $m = 0, \ldots, D-1$ of orthonormal tangent vectors. Thus, for each m, e_μ^m is a tangent vector; μ is the vector index (tangent to M) of this vector while m is the name of the vector. Of course, such a choice is somewhat arbitrary; if we act with a Lorentz transformation on the m index, the 'vielbein' e_μ^m is converted into a new but equally good basis of tangent vectors. These matters are discussed in a more thorough way in chapter 12. The index μ of e_μ^m transforms like any other vector index under diffeomorphisms of M. The Lorentz index m, since it is merely the name of the tangent vector e_μ^m, transforms like a scalar under diffeomorphisms of M. But since the introduction of the e_μ^m involves arbitrary choices at each space-time point, we are free to make local $SO(D-1,1)$ transformations on the 'local Lorentz index' m.

The advantage of introducing the vielbein is that the Lorentz group $SO(D-1,1)$ does admit spinor representations. Thus spinor indices must be regarded as local Lorentz indices. Saying that the e_μ^m are *orthonormal* tangent vectors means that the metric tensor $g_{\mu\nu}$ of the manifold and the Minkowski metric η_{mn} of the tangent space are related by

$$g_{\mu\nu} = \eta_{mn} e_\mu^m e_\nu^n$$

$$\eta^{mn} = g^{\mu\nu} e_\mu^m e_\nu^n.$$

(4.3.67)

We use the symbol e_m^μ for the inverse vielbein and $e = \sqrt{g}$ for the determinant of the vielbein. The vielbein has $\frac{1}{2}D(D-1)$ more components that $g_{\mu\nu}$, corresponding to its antisymmetric part. However, this many additional local symmetries – the local Lorentz transformations – are introduced at the same time, so the net number of propagating modes is unaltered.

The local Lorentz symmetry is very similar to an ordinary Yang–Mills symmetry. Analogous to the Yang–Mills potential (or connection) A_μ, one must introduce the spin connection ω_μ^{mn}, the gauge field for local Lorentz transformations. Under an infinitesimal Lorentz transformation described by parameter $\Theta^{mn} = -\Theta^{nm}$, the variation of the spin connection is

$$\delta\omega_\mu^{mn} = \partial_\mu\Theta^{mn} + [\omega_\mu, \Theta]^{mn} = (D_\mu\Theta)^{mn}.$$

(4.3.68)

We introduce gamma matrices Γ^m, which obey the standard Dirac algebra $\{\Gamma^m, \Gamma^n\} = -2\eta^{mn}$. The rules for the local Lorentz transformation of a spinor ψ are given by

$$\delta\psi = -\frac{1}{4}\Theta^{mn}\Gamma_{mn}\psi,$$

(4.3.69)

where we use the notation $\Gamma_{mnp\ldots}$ for a totally antisymmetrized product of $\Gamma_m\Gamma_n\Gamma_p\ldots$. In particular, $\Gamma_{mn} = \frac{1}{2}[\Gamma_m, \Gamma_n]$. The covariant derivative of the spinor is given by

$$D_\mu\psi = (\partial_\mu + \frac{1}{4}\omega_\mu^{mn}\Gamma_{mn})\psi.$$

(4.3.70)

The basic significance of a covariant derivative, such as this, is that the transformation law of $D_\mu\psi$ does not involve derivatives of the parameter Θ, and thus it transforms as a tensor of the indicated type.

In curved space, we introduce gamma matrices by

$$\Gamma^\mu(x) = e^\mu_m(x)\Gamma^m. \qquad (4.3.71)$$

The action for a spinor (Dirac or Majorana) in a gravitational background is

$$S_\psi = \frac{i}{2} \int dx\, e\overline{\psi}\Gamma^\mu D_\mu\psi. \qquad (4.3.72)$$

It is an instructive exercise to show that this is invariant under both general coordinate and local Lorentz transformations.

So far we have not said what the spin connection is supposed to be, except that it transforms like a gauge field of the Lorentz group. If desired, one could consider a theory in which the spin connection is simply a new propagating field. However, if we wish to discuss standard general relativity, the spin connection should not be an arbitrary new construct but should be determined (up to a local Lorentz transformation) by the metric. How should it be determined? In general relativity the metric tensor is covariantly constant, and (4.3.67) shows that the vielbein is a sort of square root of the metric, so it is natural to think that we should require the vielbein to be covariantly constant. Indeed, the covariant derivative of the vielbein, if not zero, is a quantity that does not have an analog in standard general relativity. Hence we require

$$D_\mu e^m_\nu = \partial_\mu e^m_\nu + \omega^m_{\mu\ n} e^n_\nu - \Gamma^\rho_{\mu\nu} e^m_\rho = 0. \qquad (4.3.73)$$

By counting equations and unknowns, one can see that this equation uniquely determines the spin connection. The requirement of orthonormality or equivalently (4.3.67) likewise determines the vielbein up to a local Lorentz transformation, so introducing the vielbein and the spin connection does not change the content of general relativity. The Riemann curvature tensor can be defined as the Yang–Mills field strength formed from the spin connection, *i.e.*,

$$R^{mn}_{\mu\nu} = \partial_\mu \omega^{mn}_\nu - \partial_\nu \omega^{mn}_\mu + [\omega_\mu, \omega_\nu]^{mn}. \qquad (4.3.74)$$

Indeed, the right-hand side of (4.3.74) is a tensor determined by at most two derivatives of the metric (since the spin connection is implicitly determined in terms of the metric by (4.3.73)), so it must be the Riemann tensor. This can be checked directly, of course.

As a nontrivial example of the application of the vielbein formalism, let us briefly review $N = 1$ supergravity in four dimensions. This theory contains in addition to the vierbein and spin connection a Rarita–Schwinger field $\chi_{A\mu}$ with a spinor index A and a vector index μ. The action is given by

$$S = \int d^4xe\{-\frac{1}{2\kappa^2}R - \frac{i}{2}\overline{\chi}_\mu\gamma^{\mu\nu\rho}D_\nu\chi_\rho\}, \qquad (4.3.75)$$

where $R = e_m^\mu e_n^\nu R_{\mu\nu}^{mn}$ is the curvature scalar and κ is the gravitational coupling constant (Planck length) whose square is proportional to Newton's constant. The Dirac matrices in four dimensions are written lower case (γ^μ, etc.). The above action has manifest general coordinate invariance and local Lorentz invariance. Less obvious is the fact that it also possesses local supersymmetry. The supersymmetry transformations that leave it invariant are

$$\delta\chi_\mu = \frac{1}{\kappa}D_\mu\epsilon \qquad (4.3.76)$$

$$\delta e_\mu^m = -\frac{i}{2}\kappa\overline{\epsilon}\gamma^m\chi_\mu. \qquad (4.3.77)$$

It is understood that ω_μ^{mn} is to be determined as the solution of its classical field equation (which is just algebraic). This gives a contribution bilinear in χ_μ in addition to the usual expression obtained by solving (4.3.73). It is not necessary to know how ω transforms to verify supersymmetry, since its variation multiplies its field equation, which is set equal to zero.

The theory just described was the first supergravity theory to be invented; its discovery opened up a new field of study, which – among other things – has played an important role in the development of superstring theory. The study of supergravity is by now a vast subject with many different facets. Those that are not directly relevant to superstring theory (and even some that are) will not be reviewed here.

4.3.5 Superstring Action and Its Symmetries

In order to derive the constraints that should accompany the gauge-fixed action in (4.1.2), we need to formulate a more fundamental action principle. By analogy with the bosonic case it is clear that invariance under reparametrizations is desirable. However, to have a framework in which the super-Virasoro conditions can emerge as gauge conditions requires local supersymmetry at the same time. The necessity of this follows from the fact that the commutator of two supersymmetry transformations gives a world-sheet translation in the gauge-fixed action. This means that we must incorporate a 'zweibein' $e_\alpha^a(\sigma,\tau)$ and a Rarita–Schwinger field $\chi_{A\alpha}(\sigma,\tau)$ (a two-component Majorana spinor and a world-sheet vector) in addition to the physical coordinates $X^\mu(\sigma,\tau)$ and $\psi^\mu(\sigma,\tau)$.

Reparametrization invariance on the world sheet (two-dimensional general coordinate invariance) and two-dimensional local Lorentz invariance are implemented by replacing (4.1.2) by

$$S_2 = -\frac{1}{2\pi} \int d^2\sigma e \{ h^{\alpha\beta} \partial_\alpha X^\mu \partial_\beta X_\mu - i\overline{\psi}^\mu \rho^\alpha \nabla_\alpha \psi_\mu \}. \tag{4.3.78}$$

By use of Fermi statistics, it can be shown that the spin connection does not contribute to the spinor term, so ∇_α can be replaced by ∂_α. This is a special feature of Majorana spinors in two dimensions. S_2 does not yet have local supersymmetry. Under the transformations of (4.1.23) with ϵ an unrestricted function of σ and τ, one obtains a variation proportional to $\int (\nabla_\alpha \overline{\epsilon}) J^\alpha d^2\sigma$ where J^α is the supercurrent introduced in §4.1.1:

$$J^\alpha = \frac{1}{2} \rho^\beta \rho^\alpha \psi^\mu \partial_\beta X_\mu. \tag{4.3.79}$$

The construction now proceeds according to the 'Noether method'. Introduce the supersymmetry gauge field χ_α (the gravitino) with supersymmetry transformation

$$\delta \chi_\alpha = \nabla_\alpha \epsilon. \tag{4.3.80}$$

(Here α is a vector index; χ_α carries an extra spinor index, which is being suppressed.) The variation obtained above can be balanced by varying χ_α in the term

$$S_3 = -\frac{1}{\pi} \int d^2\sigma e \, \overline{\chi}_\alpha \rho^\beta \rho^\alpha \psi^\mu \partial_\beta X_\mu. \tag{4.3.81}$$

However, the variation of X_μ in this expression gives an additional $\nabla \epsilon$ term of the form

$$\overline{\chi}_\alpha \rho^\beta \rho^\alpha \psi^\mu \overline{\psi}_\mu \nabla_\beta \epsilon = -\frac{1}{2} \overline{\psi}_\mu \psi^\mu \overline{\chi}_\alpha \rho^\beta \rho^\alpha \nabla_\beta \epsilon. \tag{4.3.82}$$

This can be canceled by adding another term to the action of the form

$$S_4 = -\frac{1}{4\pi} \int d^2\sigma e \, \overline{\psi}_\mu \psi^\mu \overline{\chi}_\alpha \rho^\beta \rho^\alpha \chi_\beta. \tag{4.3.83}$$

The complete action $S = S_2 + S_3 + S_4$ is now invariant under the local supersymmetry transformations

$$\delta X^\mu = \overline{\epsilon} \psi^\mu, \qquad \delta \psi^\mu = -i\rho^\alpha \epsilon (\partial_\alpha X^\mu - \overline{\psi}^\mu \chi_\alpha)$$

$$\delta e_\alpha^a = -2i\overline{\epsilon} \rho^a \chi_\alpha, \qquad \delta \chi_\alpha = \nabla_\alpha \epsilon. \tag{4.3.84}$$

There are several points worth making about the result that has been obtained. First of all, unlike supergravity theories for $D > 2$, there are no kinetic terms for the e_α^a and χ_α fields. The Einstein–Hilbert action $\sim \int eR d^2\sigma$ is a topological invariant (proportional to the Euler characteristic) that could be added. However, it does not affect the classical analysis. There is no kinetic term for a gravitino at all. The expression $\overline{\chi}_\alpha \rho^{\alpha\beta\gamma} \nabla_\beta \chi_\gamma$ vanishes simply because there is no third-rank antisymmetric tensor $\rho^{\alpha\beta\gamma}$ in two dimensions. The commutator of two supersymmetry transformations gives a combination of the other local symmetries with field-dependent coefficients, as is customary in supergravity theories. Even this requires using the equations of motion for the Fermi fields. It is possible to extend the superspace formulation described in §4.1.1 to this case, so as to achieve off-shell closure of the algebra. The description of how that works would take us too far afield, however, especially as it is not required for the rest of our discussion.

In addition to the symmetries described above, there are two further local symmetries of S. One is the extension of the local Weyl (or conformal) symmetry already encountered for the bosonic string. The Weyl transformations that leave S invariant are

$$\delta X^\mu = 0, \quad \delta\psi^\mu = -\frac{1}{2}\Lambda\psi^\mu$$

$$\delta e_\alpha^a = \Lambda e_\alpha^a, \quad \delta\chi_\alpha = \frac{1}{2}\Lambda\chi_\alpha. \tag{4.3.85}$$

There is also another local fermionic symmetry given by

$$\delta\chi_\alpha = i\rho_\alpha\eta$$

$$\delta e_\alpha^a = \delta\psi^\mu = \delta X^\mu = 0, \tag{4.3.86}$$

with η an arbitrary Majorana spinor. The proof requires the identity $\rho^\alpha \rho_\beta \rho_\alpha = 0$, which is true in two dimensions. Altogether, these symmetries imply that S is a 'superconformal' theory. The conformal symmetry of the bosonic string theory resulted in the Virasoro symmetry constraints after covariant gauge fixing. In this case we obtain a supersymmetric extension of that algebra.

Conformal and superconformal supergravity theories have been constructed in four dimensions. There the action inevitably contains R^2 terms that lead to a host of pathologies such as propagating ghosts and related effects. Fortunately these problems do not arise for $D = 2$ conformally invariant world-sheet theories. (They also do not occur in the ten-dimensional physics!)

We can now finally give a proper derivation of the super-Virasoro constraint equations by formulating covariant gauge choices that lead to simple field equations and constraint conditions, just as we did for the bosonic string theory in §2.1.3. The analysis is presented first for the classical theory, and then it is reconsidered in the next section in the quantum context.

There are altogether four local bosonic symmetries: two world-sheet reparametrizations, one local Lorentz, and one Weyl scaling. Locally, these can be used to gauge the four components of the zweibein into the standard form $e_\alpha^a = \delta_\alpha^a$. Similarly, the two supersymmetries (ϵ) and two superconformal symmetries (η) can be used locally to set the four components of $\chi_\alpha = 0$. In this gauge the action simplifies to the globally supersymmetric one of (4.1.2) and the equations of motion are just

$$\partial^\alpha \partial_\alpha X^\mu = 0 \tag{4.3.87}$$

$$\rho^\alpha \partial_\alpha \psi^\mu = 0. \tag{4.3.88}$$

These must be supplemented by the equations of motion of the fields e_α^a and χ_α evaluated in the gauge $e_\alpha^a = \delta_\alpha^a$, $\chi_\alpha = 0$. The resulting equations are the vanishing of the energy momentum tensor and supercurrent on the world sheet:

$$J_\alpha \equiv -\frac{\pi}{2e}\frac{\delta S}{\delta \chi^\alpha} = \frac{1}{2}\rho^\beta \rho_\alpha \psi^\mu \partial_\beta X_\mu = 0 \tag{4.3.89}$$

$$T_{\alpha\beta} = \partial_\alpha X^\mu \partial_\beta X_\mu + \frac{i}{2}\overline{\psi}^\mu \rho_{(\alpha}\partial_{\beta)}\psi_\mu - (\text{trace}) = 0. \tag{4.3.90}$$

These are the super-Virasoro constraint equations, finally derived from an underlying gauge invariance.

4.4   Modern Covariant Quantization

The RNS version of superstrings can be quantized by the same techniques that were described for bosonic strings in §2.4. This time there are spinorial bosonic ghosts associated with the local world-sheet supersymmetry in addition to the ones obtained previously. When correctly included the entire super-Virasoro algebra becomes anomaly-free for the correct choice of dimension and ground-state mass. Since the analysis is so similar to that which we have already presented in the bosonic case, we will be brief.

4.4.1 Faddeev–Popov Ghosts

In §3.1.1 we showed that the path integration over metrics $h_{\alpha\beta}$ could be replaced by one over conformal factors ϕ and reparametrization coordinates ξ^α. This was possible whenever any given metric could be reached from a fixed reference metric by local scale and reparametrization transformations. For world sheets of genus $g > 0$ it is not possible to reach all metrics from just one in this way. Instead they form equivalence classes, characterized by a finite number of Teichmüller parameters. This complication may be ignored if the purpose is only to develop the formalism of ghost coordinates and BRST symmetry and determine the critical dimension rather then to actually evaluate the integral for genus $g > 0$.

In chapter 3, gauge fixing of the the local reparametrization invariance gave rise to anticommuting ghost coordinates $b_{\alpha\beta}$ and c^β, having conformal dimension 2 and -1, respectively. These coordinates are present for the superstring as well, but now there are additional ones due to the local world-sheet supersymmetry. We will derive the ghost action in a slightly different way from the path that we followed in the earlier discussion. The classical statement that χ_α can be gauged away means that it can always be expressed in the form

$$\chi_\alpha = i\rho_\alpha \eta + \nabla_\alpha \epsilon, \tag{4.4.1}$$

and with reasonable boundary conditions this expression is unique. So we can change variables in the path integral from χ_α to η and ϵ. After the change of variables, the integral over η and ϵ can be dropped, since they are symmetry parameters and the action does not depend on them. (This is slightly oversimplified since some fields other than χ_α transform nontrivially under the symmetries generated by η and ϵ; the proper treatment leads, however, to the same conclusion.) This does not mean that the presence of the gravitino χ_α in the two-dimensional supergravity action is without moment. The change of variables from χ_α to η and ϵ gives rise to a Jacobian, which can again be represented as a ghost path integral.

Identifying the proper ghosts requires a slight digression about the 'spin' of a field in $1 + 1$ dimensions. The Lorentz group is $SO(1,1)$, or $SO(2)$ in the case of a Euclidean world sheet; for definiteness we adopt the latter terminology. $SO(2)$ has a single generator, which we might call W, and an $SO(2)$ representation is specified by giving the eigenvalue of W, which is the 'spin'. For our present discussion, the interesting representations have integer or half-integer values of W. For example, the gravitino field $\chi_{\alpha A}$ has a vector index α, corresponding to spin ± 1, and a spinor index A (which was suppressed in many formulas above) that carries spin $\pm 1/2$. Altogether, $\chi_{\alpha A}$ has four components of spin $\pm 1 \pm 1/2$, i.e., $3/2$, $1/2$, $-1/2$ and $-3/2$, respectively. The gauge parameters η and ϵ in (4.4.1) are spinors with two components each of spin $\pm 1/2$. The derivative operator ∇_α is a vector with components of spin ± 1. It is convenient to adopt the following temporary convention. Given a field V with components of various spin, let us denote the spin q component of V as V_q. Thus, the derivative ∇_α, for example, has components ∇_1 and ∇_{-1}. Then (4.4.1) can be written in more detail as

$$
\begin{aligned}
\delta\chi_{3/2} &= \nabla_1 \epsilon_{1/2} \\
\delta\chi_{1/2} &= \nabla_1 \epsilon_{-1/2} + \eta_{1/2} \\
\delta\chi_{-1/2} &= \nabla_{-1} \epsilon_{1/2} + \eta_{-1/2} \\
\delta\chi_{-3/2} &= \nabla_{-1} \epsilon_{-1/2}.
\end{aligned}
\tag{4.4.2}
$$

Notice that η appears only in the second and third equations and in a nonderivative fashion. The change of variables from $\chi_{\pm 1/2}$ to $\eta_{\pm 1/2}$ introduces no nontrivial Jacobian. We may as well discard the second and third equations in (4.4.2) since the real interest focuses on the first and last equations. In changing variables from $\chi_{3/2}$ to $\epsilon_{1/2}$ the Jacobian that arises is

$$
J_{3/2} = \det{}^{-1} \nabla_1^{1/2 \to 3/2},
\tag{4.4.3}
$$

where the superscript on $\nabla_1^{1/2 \to 3/2}$ means that ∇_1 is to be viewed as an operator mapping spin $1/2$ to spin $3/2$. To represent it as a determinant we introduce fields $\gamma_{1/2}$ and $\beta_{-3/2}$ of the indicated spins and write

$$
J_{3/2} = \int D\gamma_{1/2} D\beta_{-3/2} \exp - \left(\frac{1}{\pi} \int d^2\sigma\, \beta_{-3/2} \nabla_1 \gamma_{1/2} \right).
\tag{4.4.4}
$$

γ and β must be commuting fields since they are ghosts for an anticommuting symmetry, or alternatively because it is $\det^{-1}$ that appears in (4.4.3). (However, this -1 power is present, again, because χ and ϵ are anticommuting.) Likewise, the change of variables from $\chi_{-3/2}$ to $\epsilon_{-1/2}$ gives a Jacobian that can be represented as

$$
J_{-3/2} = \int D\gamma_{-1/2} D\beta_{3/2} \exp - \left(\frac{1}{\pi} \int d^2\sigma\, \beta_{3/2} \nabla_{-1} \gamma_{-1/2} \right),
\tag{4.4.5}
$$

with new commuting ghosts $\gamma_{-1/2}$ and $\beta_{3/2}$. The fields γ and β are called superconformal ghosts.

The two components of $\gamma_{\pm 1/2}$ make up a spinor γ_A. The two components of $\beta_{\pm 3/2}$ make up a vector-spinor $\beta_{\alpha A}$, which is subject to the constraints

$$\rho^{\alpha AB}\beta_{\alpha B} = 0. \tag{4.4.6}$$

(Indeed, the left-hand side of (4.4.6) is a spinor with components of spin $\pm 1/2$, so this equation just sets to zero the spin $\pm 1/2$ terms of β.) The ghost action indicated in (4.4.5) and (4.4.4) can be re-expressed in the covariant form

$$S_{FP} = -\frac{i}{2\pi}\int d^2\sigma e h^{\alpha\beta}\bar{\gamma}\partial_\alpha\beta_\beta. \tag{4.4.7}$$

In the gauge $h_{\alpha\beta} = \delta_{\alpha\beta}$ the equations of motion imply that $\beta_{3/2}$ and $\gamma_{-1/2}$ are right-moving while the other components are left-moving.

Focusing on right-moving components and suppressing the 3/2 and $-1/2$ in $\beta_{3/2}$ and $\gamma_{-1/2}$, we get (by varying (4.4.7) with respect to the world-sheet metric) the energy–momentum tensor of superconformal ghosts:

$$T_{++} = \frac{i}{2}\gamma\partial_+\beta + \frac{3i}{2}\beta\partial_+\gamma. \tag{4.4.8}$$

The ghost current of the $\beta - \gamma$ system is

$$J_+^{gh} = \beta\gamma. \tag{4.4.9}$$

Equation (4.4.8) corresponds to $k = 2$ in §3.2.2.

In terms of mode expansions

$$\gamma(\tau) = \frac{1}{\sqrt{2}}\sum_{-\infty}^{\infty}\gamma_n e^{-2in\tau} \tag{4.4.10}$$

$$\beta(\tau) = \frac{1}{\sqrt{2}}\sum_{-\infty}^{\infty}\beta_n e^{-2in\tau} \tag{4.4.11}$$

the commutation relations implied by S_{FP} are

$$[\gamma_m, \beta_n] = \delta_{m+n} \tag{4.4.12}$$

$$[\gamma_m, \gamma_n] = [\beta_m, \beta_n] = 0. \tag{4.4.13}$$

As defined here, γ is hermitian and β is antihermitian. Of course, β could be redefined by a factor of i if one wished. We have written the expansion of γ and β with integrally moded coefficients γ_m and β_m. This is the choice that is appropriate to the fermionic sector of theory. In the bosonic sector it is necessary to use half-integer modes γ_r and β_r.

The contribution of the ghost coordinates to the super-Virasoro generators follow from the formula for T_{++} and J_+. Altogether the ghost contributions are

$$L_m^{gh} = \sum (m+n) : b_{m-n} c_n : + \sum (\tfrac{1}{2}m + n) : \beta_{m-n} \gamma_n : \qquad (4.4.14)$$

and

$$F_m^{gh} = -2 \sum b_{-n} \gamma_{m+n} + \sum (\tfrac{1}{2}n - m) c_{-n} \beta_{m+n}. \qquad (4.4.15)$$

It is easy to verify that L_m^{gh} does in fact imply that c and b have conformal dimension -1 and 2 and γ and β have conformal dimension $-1/2$ and $3/2$. The anticommutator

$$\{F_m^{gh}, F_n^{gh}\} = 2L_{m+n}^{gh} + B^{gh}(m)\delta_{m+n} \qquad (4.4.16)$$

has an anomaly

$$B^{gh}(m) = -5m^2. \qquad (4.4.17)$$

In §4.2.2 we found that the contribution of the α and d oscillators is

$$B^{\alpha,d}(m) = \tfrac{1}{2}Dm^2 + 2a. \qquad (4.4.18)$$

Therefore the anomaly cancels for $D = 10$ and $a = 0$. It also cancels in $[L_m, L_n]$ as a consequence of the Jacobi identity. In the bosonic sector one has

$$B^{gh}(r) = \tfrac{1}{4} - 5r^2 \qquad (4.4.19)$$

$$B^{\alpha,b}(r) = \tfrac{1}{2}D(r^2 - \tfrac{1}{4}) + 2a, \qquad (4.4.20)$$

so that the sum vanishes for $D = 10$ and $a = 1/2$.

4.4.2 BRST Symmetry

The general prescription for writing down a nilpotent BRST charge in terms of the structure constants in the algebra of constraints, described in

§3.2.1, is applicable for graded algebras as well. Applying the prescription
to the super-Virasoro algebra gives

$$Q = \sum (L_{-n}^{(\alpha,d)} c_n + F_{-n}^{(\alpha,d)} \gamma_n)$$
$$- \frac{1}{2} \sum (m - n) : c_{-m} c_{-n} b_{m+n} :$$
$$+ \sum (\tfrac{3}{2} n + m) : c_{-n} \beta_{-m} \gamma_{m+n} :$$
$$- \sum \gamma_{-m} \gamma_{-n} b_{m+n} - a c_0. \qquad (4.4.21)$$

As before, the term $a c_0$ characterizes the ambiguity in the classical prescription arising from the need for normal ordering.

A somewhat tedious calculation, equivalent to ones we have described
previously, shows that $Q^2 = 0$ provided that $D = 10$ and $a = 0$. As a
simple check of this fact we note that

$$\{Q, b_n\} = L_n = L_n^{(\alpha,d)} + L_n^{gh} \qquad (4.4.22)$$

$$[Q, \beta_n] = F_n = F_n^{(\alpha,d)} + F_n^{gh}. \qquad (4.4.23)$$

The absence of anomalies in the algebra of L_n and F_n is consistent with the
nilpotency of Q. The same analysis applies for the bosonic sector with the
obvious replacements $F_m \to G_r$, $\gamma_m \to \gamma_r$, $\beta_m \to \beta_r$, $a = 0 \to a = 1/2$.
There is a complication in the fermionic sector not shared by the bosonic
one. In this sector the γ and β oscillators have zero modes, giving an
infinite degeneracy to the Fock-space ground state. This degeneracy has
important implications, whose elucidation unfortunately goes beyond the
scope of this book.

4.4.3 Covariant Computation of the Virasoro Anomaly

In this section we describe the covariant computation of the Virasoro
anomaly, analogous to the discussion of the bosonic case in §3.2.2. The
discussion is easy because the necessary formulas were all worked out in
that section and the following one on bosonization.

In the theory at hand we have ten bosons X^μ, ten fermions ψ^μ, conformal ghosts b and c and superconformal ghosts γ and β. In §3.2.2 we
noted that given a pair of anticommuting degrees of freedom such as b
and c, a one parameter family of energy–momentum tensors is possible,
parametrized by a label that we called k. A pair of anticommuting degrees of freedom have the same Weyl anomaly as $(1 - 3k^2)$ of the bosons
X^μ. The ψ^μ are ten anticommuting degrees of freedom and (as they

all have conformal spin 1/2, or by comparing their energy–momentum tensor to those discussed in §3.2.2) they have $k = 0$. Hence they contribute the same Virasoro anomaly that five X^μ would contribute. On the other hand, b and c have $k = 3$, and contribute the Virasoro anomaly of -26 bosons. What about the superconformal ghosts γ and β? Comparing their energy–momentum tensor to the discussion in §3.2.2 (or thinking about their conformal dimensions) they have $k = 2$. At first sight, therefore, it might seem that they would have the conformal anomaly of $(1 - 3 \cdot 2^2) = -11$ ordinary bosons. However, we must remember that the superconformal ghosts are *commuting* degrees of freedom, as opposed to the anticommuting degrees of freedom discussed in §3.2.2. Therefore, in computing the one-loop diagram for $< T_{++}T_{++} >$ that gives the Virasoro anomaly, they lack a minus sign from Fermi statistics that plays a part in the computations in §3.2.2, so they give the contribution of $+11$ rather than -11 bosons. The total conformal anomaly is thus $10 + 5 - 26 + 11 = 0$, showing the cancellation in ten dimensions.

4.5 Extended World-Sheet Supersymmetry

The generalization of the bosonic string world-sheet action to include $N = 1$ supersymmetry on the world sheet has provided us with string theories with space-time supersymmetry in $D = 10$. Given this success, it is natural to wonder what can be achieved starting from an action with extended world-sheet supersymmetry. For a string theory we must actually require superconformal symmetry in order that all the gravity-like fields introduced to implement the local gauge symmetries drop out of the classical equations of motion in a suitable gauge. The possibilities for superconformal symmetries are quite limited. Thus, whereas there are very many $D = 2$ supersymmetric theories that can be deduced by truncation of theories in $D > 2$, the vast majority do not have the requisite conformal symmetries. There are only two remaining possibilities that possess the conformal symmetry required of a string theory. One has $N = 2$ and the other has $N = 4$ on the world sheet. In this section we take a brief look at each. The conclusion will be that they do not seem to give physically interesting string theories, even though they are mathematically interesting structures that may prove to have some significance. We will not be as detailed in describing these theories as we have been in describing the models whose physical significance is more apparent, but it seems appropriate to alert the reader to their existence.

4.5.1 The N = 2 Theory

The $N = 1$ superconformal symmetry group was described by an infinite algebra. That algebra contained an $OSp(1,2)$ subalgebra in the left- and right-moving sectors for bosonic boundary conditions. The generalization to $N = 2$ replaces this subalgebra by $OSp(2,2)$. This implies that the bosonic generators include an $SO(2) \approx U(1)$ in addition to $Sp(2) \approx SU(1,1)$. The fermionic generators form an $SO(2)$ doublet, or complex representation of $U(1)$.

The appropriate gauge-fixed action with $N = 2$ supersymmetry contains a bosonic coordinate $Y^\mu(\sigma, \tau)$ in addition to $X^\mu(\sigma, \tau)$ and an $SO(2)$ doublet of Fermi coordinates $\psi_i^\mu(\sigma, \tau)$, $i = 1, 2$

$$S = -\frac{1}{2\pi} \int d^2\sigma \{ \partial_\alpha X^\mu \partial^\alpha X_\mu + \partial_\alpha Y^\mu \partial^\alpha Y_\mu - i\overline{\psi}_i^\mu \rho^\alpha \partial_\alpha \psi_{\mu i} \}. \quad (4.5.1)$$

This obviously has an $SO(2)$ symmetry that rotates ψ_1 and ψ_2 into one another while leaving X and Y inert. The supersymmetry charge also is an $SO(2)$ doublet, as one sees from the transformation formulas

$$\delta X = \overline{\epsilon}_i \psi_i \quad (4.5.2)$$

$$\delta Y = \epsilon_{ij} \overline{\epsilon}_i \psi_j \quad (4.5.3)$$

$$\delta \psi_i = -i\rho^\alpha \partial_\alpha X \epsilon_i + i\epsilon_{ij} \rho^\alpha \partial_\alpha Y \epsilon_j. \quad (4.5.4)$$

Here ϵ_{ij} is the antisymmetric symbol with $\epsilon_{12} = 1$. This action must be supplemented by appropriate constraints in the usual manner.

The index $\mu = 0, \ldots, D - 1$ labels the space-time directions, and $X^\mu(\sigma, \tau)$ is regarded as describing the embedding of the world sheet in space-time. However, Y^μ enters the mathematics in essentially the same way, so what is its interpretation? The Lorentz symmetry of S is just $SO(D - 1, 1)$, so Y certainly does not describe additional dimensions in the usual sense. It may be that crucial subtleties in this theory have not yet been unraveled.

Let us set aside these interpretation issues and proceed with the mathematical analysis. The action (4.5.1) can be written more succinctly in terms of complex coordinates $Z^\mu = X^\mu + iY^\mu$ and $\psi^\mu = \psi_1^\mu + i\psi_2^\mu$. The latter is just a Dirac spinor in $D = 2$. Now

$$S = -\frac{1}{2\pi} \int d^2\sigma \{ \partial_\alpha Z \partial^\alpha \overline{Z} - i\overline{\psi} \rho^\alpha \partial_\alpha \psi \}, \quad (4.5.5)$$

where we have dropped the space-time index μ. In terms of a Dirac supersymmetry parameter $\epsilon = \epsilon_1 + i\epsilon_2$, the global $N = 2$ supersymmetry transformations are

$$\delta Z = \bar{\epsilon}\psi$$

$$\delta\psi = -i\rho^\alpha \epsilon \partial_\alpha Z. \qquad (4.5.6)$$

Next we wish to re-express the action in reparametrization-invariant form. This requires a multiplet of supergravity coordinates consisting of the zweibein e_α^a, a Dirac gravitino χ_α, and a vector gauge field A_α required to make the $SO(2)$ symmetry local. The result is

$$S = -\frac{1}{2\pi}\int d^2\sigma e\{h^{\alpha\beta}\partial_\alpha Z\partial_\beta\overline{Z} - i\overline{\psi}\rho^\alpha \overset{\leftrightarrow}{\partial}_\alpha\psi + A_\alpha\overline{\psi}\rho^\alpha\psi$$

$$+ (\partial_\alpha Z - \tfrac{1}{2}\overline{\chi}_\alpha\psi)\overline{\psi}\rho^\beta\rho^\alpha\chi_\beta + (\partial_\alpha\overline{Z} - \tfrac{1}{2}\overline{\psi}\chi_\alpha)\overline{\chi}_\beta\rho^\alpha\rho^\beta\psi\}. \qquad (4.5.7)$$

Note that there is no kinetic term for the A_α field, just as there is none for its partners e_α^a and χ_α. In fact, the only appearance of A_α is in the one term where it is shown explicitly. Its equation of motion, $\overline{\psi}\rho^\alpha\psi = 0$, expresses the vanishing of the $SO(2)$ current, which supplements the conditions $T_{\alpha\beta} = J_{\alpha i} = 0$ that follow from the e_α^a and χ_α field equations.

In addition to local reparametrization invariance, Lorentz invariance, and supersymmetry, the action (4.5.7) has a number of other local symmetries. There is the usual Weyl scaling symmetry,

$$\delta e_\alpha^a = \Lambda e_\alpha^a \qquad\qquad \delta\psi = -\tfrac{1}{2}\Lambda\psi$$

$$\delta Z^\mu = \delta A_\alpha = 0 \qquad\qquad \delta\chi_\alpha = \tfrac{1}{2}\Lambda\chi_\alpha \qquad (4.5.8)$$

and an $N = 2$ superconformal symmetry

$$\delta\chi_\alpha = i\rho_\alpha\eta$$

$$\delta A_\alpha = \tfrac{1}{2}(\overline{\chi}_\beta\rho_\alpha\rho^\beta\eta + \overline{\eta}\rho^\beta\rho_\alpha\chi_\beta). \qquad (4.5.9)$$

There is also the local $SO(2)$ symmetry gauged by the A_α field

$$\delta\psi = i\Sigma\psi \qquad\qquad \delta A_\alpha = \partial_\alpha\Sigma$$

$$\delta\chi_\alpha = i\Sigma\chi_\alpha \qquad\qquad \delta e_\alpha^a = \delta Z^\mu = 0. \qquad (4.5.10)$$

This symmetry also has a chiral counterpart

$$\delta\psi = i\Sigma'\overline{\rho}\psi \qquad\qquad \delta A_\alpha = \epsilon_\alpha{}^\beta\partial_\beta\Sigma'$$

$$\delta\chi_\alpha = -i\Sigma'\overline{\rho}\chi_\alpha \qquad\qquad \delta e_\alpha^a = \delta Z^\mu = 0, \qquad (4.5.11)$$

where $\overline{\rho} = \rho_0\rho_1$ as before, which is also a local symmetry. This pair of local symmetries implies that the self-dual and anti-self-dual components of A_α, *i.e.*, $A_\pm$, act as independent gauge fields for a left-moving and a right-moving local $SO(2)$ group. Altogether, the complete list of local symmetries is sufficient in number to gauge away all components of e_α^a, χ_α, and A_α.

Let us now consider the super-Virasoro algebra of constraints that arise in the quantization of this string theory. We restrict our attention to the bosonic open-string sector. The modes of the Z^μ coordinates are complex oscillators α_m^μ satisfying

$$[\alpha_m^\mu, \overline{\alpha}_n^\nu] = m\delta_{m+n}\eta^{\mu\nu}$$

$$[\alpha_m^\mu, \alpha_n^\nu] = [\overline{\alpha}_m^\mu, \overline{\alpha}_n^\nu] = 0. \qquad (4.5.12)$$

Similarly, for bosonic boundary conditions the Dirac fields ψ_i^μ gives com plex half-integrally moded oscillators b_r^μ with

$$\{b_r^\mu, \overline{b}_s^\nu\} = \delta_{r+s}\eta^{\mu\nu}$$

$$\{b_r^\mu, b_s^\nu\} = \{\overline{b}_r^\mu, \overline{b}_s^\nu\} = 0. \qquad (4.5.13)$$

In terms of these, the Fourier modes of the energy–momentum tensor are

$$L_n = \sum : \alpha_{-m}\overline{\alpha}_{n+m} : + \sum (r + \frac{n}{2}) : b_{-r}\overline{b}_{n+r} :, \qquad (4.5.14)$$

the modes of the supercurrent J are

$$G_r = \sum b_s \cdot \alpha_{r-s} \qquad (4.5.15)$$

$$\overline{G}_r = \sum \overline{b}_s \cdot \overline{\alpha}_{r-s}, \qquad (4.5.16)$$

and the modes of the $SO(2)$ current are

$$T_n = \sum : b_r \cdot \overline{b}_{n-r} : . \qquad (4.5.17)$$

The algebra of the T_m operators is

$$[T_m, T_n] = mD\delta_{m+n}, \qquad (4.5.18)$$

which is an abelian 'Kac–Moody algebra' for the group $SO(2)$. We will encounter Kac–Moody algebras (or affine Lie algebras, as they are often called) in chapter 6, where infinite-dimensional algebras of this type play an important rôle. The $SO(2)$ current has conformal dimension $J = 1$, so

$$[L_m, T_n] = -nT_{m+n}. \tag{4.5.19}$$

The $SO(2)$ charges of the supercurrents are ± 1, and thus

$$[T_m, G_r] = G_{m+r} \tag{4.5.20}$$

$$[T_m, \overline{G}_r] = -\overline{G}_{m+r}. \tag{4.5.21}$$

The Virasoro algebra has twice the anomaly of the $N = 1$ case, since the number of Bose and Fermi coordinates has been doubled,

$$[L_m, L_n] = (m - n)L_{m+n} + \tfrac{1}{4}D(m^3 - m)\delta_{m+n}. \tag{4.5.22}$$

There are no charge-two operators, so

$$\{G_r, G_s\} = \{\overline{G}_r, \overline{G}_s\} = 0. \tag{4.5.23}$$

Perhaps the most interesting bracket in the algebra is

$$\{G_r, \overline{G}_s\} = L_{r+s} + \tfrac{1}{2}(r - s)T_{r+s} + \tfrac{1}{2}D(r^2 - \tfrac{1}{4})\delta_{r+s}. \tag{4.5.24}$$

As in the previous string theories, we require that a physical state satisfy the constraint equations

$$L_n|\phi\rangle = T_n|\phi\rangle = 0 \qquad n > 0 \tag{4.5.25}$$

$$G_r|\phi\rangle = \overline{G}_r|\phi\rangle = 0 \qquad r > 0 \tag{4.5.26}$$

$$(L_0 - a)|\phi\rangle = 0 \tag{4.5.27}$$

$$T_0|\phi\rangle = 0. \tag{4.5.28}$$

The last equation implies that all states with a nonzero $U(1)$ charge are unphysical. This can be interpreted as a sort of $U(1)$ confinement principle. Let us now find the preferred values of a and D by looking for zero-norm physical states.

Consider a state of the form $|\phi\rangle = G_{-1/2}|\tilde{\phi}\rangle$, where

$$(T_0 + 1)|\tilde{\phi}\rangle = 0 \qquad (4.5.29)$$

$$G_r|\tilde{\phi}\rangle = \overline{G}_r|\tilde{\phi}\rangle, \qquad r > 0. \qquad (4.5.30)$$

This state is zero norm and physical if it is annihilated by $\overline{G}_{1/2}$

$$\overline{G}_{1/2}|\phi\rangle = (L_0 - \tfrac{1}{2}T_0)|\tilde{\phi}\rangle = (L_0 + \tfrac{1}{2})|\tilde{\phi}\rangle. \qquad (4.5.31)$$

This vanishes if $L_0|\phi\rangle = 0$. Thus the preferred choice is $a = 0$. This means that the ground state is a massless scalar. Thus the spectrum has no tachyons, even before any possible GSO truncations are considered.

Now let us construct zero-norm states that determine the critical dimension. Consider

$$|\phi\rangle = (T_{-1} + \lambda_1 L_{-1} + \lambda_2 G_{-1/2}\overline{G}_{-1/2})|\tilde{\phi}\rangle, \qquad (4.5.32)$$

where

$$T_0|\tilde{\phi}\rangle = (L_0 + 1)|\tilde{\phi}\rangle = 0 \qquad (4.5.33)$$

$$G_r|\tilde{\phi}\rangle = \overline{G}_r|\tilde{\phi}\rangle = 0 \qquad r > 0. \qquad (4.5.34)$$

Using the extended super-Virasoro algebra it is straightforward to derive that

$$G_{1/2}|\phi\rangle = (-1 + \lambda_1 + \lambda_2)G_{-1/2}|\tilde{\phi}\rangle \qquad (4.5.35)$$

$$\overline{G}_{1/2}|\phi\rangle = (1 + \lambda_1)\overline{G}_{-1/2}|\tilde{\phi}\rangle \qquad (4.5.36)$$

$$T_1|\phi\rangle = (D - \lambda_2)|\tilde{\phi}\rangle. \qquad (4.5.37)$$

The conclusion therefore is that for $\lambda_1 = -1$ and $\lambda_2 = 2$ the state $|\phi\rangle$ is zero norm and physical for $D = 2$. Thus this is the critical dimension of the theory. It is easy to show that there are ghost states for $D > 2$.

The conclusion that $D = 2$ is the critical dimension means that in the light-cone gauge there are no transverse oscillators at all. Therefore it would appear that the massless scalar ground state is the only propagating degree of freedom in the theory (at least for this sector). However, subtleties in the quantization of the theory have been pointed out recently, and this statement may require revision. If it is true that the theory has just a single scalar mode, this mode must be governed by some local quantum field theory, but which theory this is has not yet been determined. It has also not been determined which Fermi fields, if any, should accompany this 'scalar'.

In summary, the $N = 2$ extension of the superstring construction gives a highly symmetrical two-dimensional theory and an interesting generalization of the super-Virasoro algebra. It seemingly cannot be given the usual interpretation of a string theory since the critical dimension is $D = 2$ and there are no transverse excitations of a string in this dimension. Perhaps it enters physics in some other and as yet unknown way.

4.5.2 The $N = 4$ Theory

The $N = 4$ string theory has an even richer mathematical structure, and an even poorer physical outcome. The local $SO(2)$ is replaced by a local $SU(2)$, and there are four local supersymmetries and superconformal symmetries. The physical coordinates consist of four X^μ type variables and four ψ^μ Majorana spinors. The former are $SU(2)$ singlets and the latter are a pair of $SU(2)$ doublets.

The action principles and symmetry transformations can be written down as before. However, we jump over all that and go straight to the gauge algebra. The Virasoro algebra anomaly undergoes another doubling

$$[L_m, L_n] = (m - n)L_{m+n} + \tfrac{1}{2}D(m^3 - m)\delta_{m+n}. \qquad (4.5.38)$$

The local $SU(2)$ algebra results in a Kac–Moody algebra

$$[T_m^a, T_n^b] = i\epsilon_{abc}T_{m+n}^c + 2Dm\delta_{m+n}\delta_{ab}. \qquad (4.5.39)$$

There are four G_r^α operators, which transforms as a pair of $SU(2)$ doublets. Thus

$$[T_m^a, G_r^\alpha] = \tfrac{1}{2}\lambda_{\alpha\beta}^a G_{m+r}^\beta, \qquad (4.5.40)$$

where, in terms of 2×2 Pauli matrices τ^a,

$$\lambda^a = \begin{pmatrix} \tau^a & 0 \\ 0 & \tau^a \end{pmatrix}. \qquad (4.5.41)$$

The most interesting bracket is

$$\{G_r^\alpha, G_s^\beta\} = 2\delta^{\alpha\beta}L_{r+s} - 2(r-s)\lambda_{\alpha\beta}^a T_{r+s}^a + 2D(r^2 - 1/4)\delta^{\alpha\beta}\delta_{r+s}. \qquad (4.5.42)$$

The operator identities can be derived from explicit representations in terms of α and b oscillators, in a more or less obvious way, and so all Jacobi identities are certain to be satisfied.

This scheme leads to a negative critical dimension $(D = -2)$, and thus would seem to have no sensible interpretation as a string theory.

4.6   Summary

The string action has been generalized to include Majorana spinor coordinates $\psi^\mu(\sigma, \tau)$ that are the supersymmetry partners of the space-time coordinates $X^\mu(\sigma, \tau)$. By including two-dimensional supergravity fields, as well, a formulation with local superconformal symmetry can be achieved. A covariant gauge choice gives rise to constraint conditions that correspond to the vanishing of the two-dimensional energy–momentum tensor and supersymmetry current. The constraint operators satisfy an infinite graded algebra, known as the super-Virasoro algebra. These constraints leave a transverse ghost-free spectrum for ten-dimensional space-time. Only in this dimension does light-cone gauge quantization give a Lorentz-invariant quantum theory.

For the fermionic coordinates $\psi^\mu(\sigma, \tau)$ there are two possible choices of boundary conditions. One choice results in integer-frequency modes and describes strings that are space-time fermions. The other choice gives half-integer-frequency modes and describes strings that are space-time bosons. If a projection of the spectrum onto states of even world-sheet fermion number is made, the spectrum becomes free of tachyons. Moreover, there are then an equal number of bosons and fermions at every mass level, suggesting that the theory has space-time supersymmetry. That property is rather obscure in the formulation described in this chapter. It is proved in the next chapter by using a different formulation.

Appendix 4.A   Super Yang–Mills Theories

The open superstring theory can be approximated at low energy by a supersymmetric Yang–Mills theory. Such theories in D dimensions are described by an action of the form

$$S = \int d^D x (-\frac{1}{4}F^2 + \frac{i}{2}\overline{\psi}\Gamma \cdot D\psi), \tag{4.A.1}$$

where $F_{\mu\nu}$ is the nonabelian field strength formed from a vector potential A_μ

$$F^a_{\mu\nu} = \partial_\mu A^a_\nu - \partial_\nu A^a_\mu + gf^a{}_{bc}A^b_\mu A^c_\nu. \tag{4.A.2}$$

The fields A^a_μ and ψ^a are both in the adjoint representation of a semisimple

Lie group. The symbol D is the Yang–Mills covariant derivative

$$(D_\mu \psi)^a = \partial_\mu \psi^a + g f^a{}_{bc} A_\mu^b \psi^c. \tag{4.A.3}$$

The number of physical fermionic modes described by the spinor field ψ is a power of two that depends on the space-time dimension and the type of spinor (Dirac, Majorana, Weyl, etc.). In fact, the spinor representation of $SO(n)$ has a dimension that is always a power of two, as we will see in the appendix 5.A. The vector A_μ describes $D-2$ physical modes corresponding to the various possible transverse polarizations. Supersymmetry requires that the number of physical boson and fermion modes should be equal. Thus, supersymmetry of the minimal Lagrangian (4.A.1) without adding any other fields requires that $D-2$ should be a power of two. The first cases in which $D-2$ is a power of two are $D = 3, 4, 6$, and 10, and these prove to be the interesting cases.

A Majorana spinor in $D = 3$ has one physical mode, a Majorana spinor in $D = 4$ has two, a Weyl spinor in $D = 6$ has four and a Majorana–Weyl spinor in $D = 10$ has eight. These numbers coincide in each case with $D-2$, and these are the four cases in which the minimal Lagrangian might be supersymmetric. Above $D = 10$ the number of components of any spinor greatly exceeds that of the vector, and supersymmetric Yang-Mills theories do not exist.

The supersymmetry transformations that leave (4.A.1) invariant are

$$\delta A_\mu^a = \frac{i}{2} \bar{\epsilon} \, \Gamma_\mu \psi^a$$
$$\delta \psi^a = -\frac{1}{4} F_{\mu\nu}^a \Gamma^{\mu\nu} \epsilon. \tag{4.A.4}$$

If (4.A.4) is inserted in (4.A.1), the terms proportional to $\epsilon \psi$ cancel in any dimension. More delicate are the terms proportional to $\epsilon \psi^3$. The variation of the A field in the covariant derivative gives a term of the form

$$f_{abc} \bar{\epsilon} \, \Gamma_\mu \psi^a \overline{\psi}^b \Gamma^\mu \psi^c. \tag{4.A.5}$$

Since this is the only term with three ψ fields that arises in the variation of the action, it must vanish if supersymmetry is to hold. Remarkably, using the total antisymmetry of f_{abc} it is possible to prove that it does in fact vanish for the four types of spinors (in $D = 3, 4, 6$ and 10) listed above.

We prove this for Majorana–Weyl spinors in $D = 10$, which is the case of paramount interest to us. Removing the spinors from (4.A.5), and using the antisymmetry of f_{abc}, we need to show that

$$(\Gamma^0\Gamma^\mu)_{mn}(\Gamma^0\Gamma_\mu)_{pq} + (\Gamma^0\Gamma^\mu)_{mp}(\Gamma^0\Gamma_\mu)_{qn} + (\Gamma^0\Gamma^\mu)_{mq}(\Gamma^0\Gamma_\mu)_{np} \quad (4.A.6)$$

vanishes. Here we may assume that the spinor indices m, n, p and q are all projected onto, say, positive chirality, though we do not write the projection operator explicitly, because the spinors in (4.A.5) are all Weyl. We also note that $\Gamma^0\Gamma^\mu_{mn}$ is a symmetric matrix, and that the second and third terms get interchanged under $m \leftrightarrow n$, so that the entire expression has $m \leftrightarrow n$ symmetry.

To establish that (4.A.6) vanishes we regard the expression as a matrix labeled by indices m and n, treating p and q as irrelevant additional labels. Alternatively, we can attach anticommuting spinors ψ_1^p and ψ_2^q and obtain

$$(\Gamma^\mu)_{mn}\overline{\psi}_1\Gamma_\mu\psi_2 + (\Gamma^\mu\psi_1)_m(\overline{\psi}_2\Gamma_\mu)_n - (\Gamma^\mu\psi_2)_m(\overline{\psi}_1\Gamma_\mu)_n. \quad (4.A.7)$$

An arbitrary matrix M_{mn} can be expanded in the complete basis set $(\Gamma_{\mu_1...\mu_k})_{mn}$, where $k = 0, 1, \ldots, 10$. Therefore an effective strategy for showing that (4.A.7) vanishes is to show that each type of term vanishes. First of all, we note that the terms with k even vanish because of the Weyl projections. Also, the identity

$$\Gamma_{\mu_1...\mu_k} = \pm\frac{1}{(10-k)!}\epsilon_{\mu_1...\mu_{10}}\Gamma^{\mu_{k+1}...\mu_{10}}\Gamma_{11} \quad (4.A.8)$$

and the fact that Γ_{11} can be dropped for Weyl spinors implies that only terms with $k \leq 5$ need to be considered and that the tensor $\Gamma_{\mu_1...\mu_5}$ can be decomposed into a self-dual and an anti-self-dual piece only one of which contributes. Moreover, $\Gamma^0\Gamma_\mu$ and $\Gamma^0\Gamma_{\mu_1...\mu_5}$ are symmetric, whereas $\Gamma^0\Gamma_{\mu_1\mu_2\mu_3}$ is antisymmetric. Therefore we only need to consider the $k = 1$ and $k = 5$ terms. As a check on this counting we note that a symmetric 16×16 matrix has

$$\frac{16 \cdot 17}{2} = 10 + \frac{1}{2} \cdot \frac{10 \cdot 9 \cdot 8 \cdot 7 \cdot 6}{5!} = 136 \quad (4.A.9)$$

components.

Multiplying (4.A.7) by $(\Gamma_\rho)_{nm}$ and contracting indices gives

$$\mathrm{tr}(\Gamma^\mu\Gamma_\rho)\overline{\psi}_1\Gamma_\mu\psi_2 - \overline{\psi}_2\Gamma_\mu\Gamma_\rho\Gamma^\mu\psi_1 + \overline{\psi}_1\Gamma_\mu\Gamma_\rho\Gamma^\mu\psi_2$$
$$= -16\overline{\psi}_1\Gamma_\rho\psi_2 - 8\overline{\psi}_2\Gamma_\rho\psi_1 + 8\overline{\psi}_1\Gamma_\rho\psi_2 = 0, \quad (4.A.10)$$

which establishes that it has no Γ_ρ piece. Repeating for $(\Gamma_{\rho_1...\rho_5})_{nm}$ we obtain

$$-\overline{\psi}_2\Gamma_\mu\Gamma_{\rho_1\ldots\rho_5}\Gamma^\mu\psi_1 + \overline{\psi}_1\Gamma_\mu\Gamma_{\rho_1\ldots\rho_5}\Gamma^\mu\psi_2 = 2\overline{\psi}_1\Gamma_\mu\Gamma_{\rho_1\ldots\rho_5}\Gamma^\mu\psi_2. \quad (4.\text{A}.11)$$

However, in D dimensions,

$$\Gamma_\mu\Gamma_{\rho_1\ldots\rho_k}\Gamma^\mu = (-1)^{k+1}(D-2k)\Gamma_{\rho_1\ldots\rho_k}. \quad (4.\text{A}.12)$$

Taking $D = 10$ and $k = 5$ we see that (4.A.11) vanishes and the proof is complete.

5. Space-Time Supersymmetry in String Theory

The description of superstrings given in chapter 4 suffers from one striking drawback. It is extremely difficult to understand the origins of space-time supersymmetry. Bosonic strings are described by choosing one set of boundary conditions and fermionic ones by choosing another set. Then, lo and behold, there is a symmetry relating the two sets of states. It is certainly necessary to have this symmetry, as we have already argued, to have consistent interactions of the gravitino field contained in the massless closed-string multiplet. We also showed that when the GSO conditions are imposed there are an equal number of bosons and fermions at every mass level, as is necessary for a linear realization of the supersymmetry.

In this chapter we describe a formalism that leads to the same theory in a way that makes supersymmetry manifest. We begin by describing a covariant world-sheet action with space-time supersymmetry. While it seems to be difficult to quantize this action covariantly, it can be quantized in light-cone gauge. Though the resulting formalism is not manifestly Lorentz invariant, it can be shown to be Lorentz invariant in $D = 10$. Supersymmetry is obvious; the GSO conditions are automatically built in from the outset without having to make any truncations; bosonic and fermionic strings are unified in a single Fock space.

5.1 The Classical Theory

We begin with the classical theory of the manifestly supersymmetric superstring action. Its symmetries are subtle and need to be understood before attempting quantization. An important ingredient is a new type of local fermionic symmetry on the world sheet. It is not ordinary supersymmetry, although it is certainly related to it. In order to illustrate how it works we first give a brief description of the superparticle, which involves fewer complications than the string and is therefore easier to analyze in detail.

5.1.1 The Superparticle

Recall that in §1.3.1 and §2.1.1 we described a relativistic particle of mass

m by the world-line action

$$S = \tfrac{1}{2} \int (e^{-1}\dot{x}^2 - em^2)d\tau. \tag{5.1.1}$$

The auxiliary coordinate e can be identified as the square root of a one-dimensional metric. An important virtue of this formula is the existence of a smooth limit $m \to 0$. In generalizing (5.1.1) to the superparticle we set $m = 0$, since the mass term is not relevant for the subsequent generalization to the superstring.

In addition to the local reparametrization symmetry $\tau \to f(\tau)$, the action (5.1.1) is also invariant under global space-time Poincaré transformations generated by

$$\delta x^\mu = a^\mu + b^\mu{}_\nu x^\nu \tag{5.1.2}$$

$$\delta e = 0, \tag{5.1.3}$$

where $b_{\mu\nu}$ is antisymmetric.

In chapter 4 we established manifest world-sheet supersymmetry by adding fermionic coordinates to the two-dimensional world-sheet coordinates σ, τ. Here, likewise, we achieve space-time supersymmetry by generalizing Minkowski space, with its 'bosonic' coordinates x^μ, to a superspace with fermionic coordinates as well as bosonic coordinates. If there are to be N supersymmetries, we introduce N anticommuting spinor coordinates $\theta^{Aa}(\tau)$, $A = 1, 2, \ldots, N$. The index a is that of a space-time spinor appropriate to D dimensions. For a general Dirac spinor $a = 1, \ldots, 2^{D/2}$. For the most part, however, we shall be interested in spinors satisfying Majorana and Weyl restrictions as reviewed in §4.3.3.

Supersymmetry is realized in superspace in the usual way. Introducing infinitesimal Grassmann parameters ϵ^A, spinors of the same type as the corresponding θ^A coordinates, the transformation formulas are

$$\begin{aligned} \delta\theta^A &= \epsilon^A & \delta x^\mu &= i\bar{\epsilon}^A \Gamma^\mu \theta^A \\ \delta\bar{\theta}^A &= \bar{\epsilon}^A & \delta e &= 0. \end{aligned} \tag{5.1.4}$$

Here ϵ^A is a constant (τ-independent) spinor. These and subsequent formulas are written in a form appropriate for Majorana spinors.

We now generalize the bosonic point particle, which propagates in Minkowski space, to a supersymmetric point particle propagating in superspace. Many supersymmetric actions can be written, since both $\dot{x}^\mu - i\bar{\theta}^A \Gamma^\mu \dot{\theta}^A$ and $\dot{\theta}^{Aa}$ are invariant under supersymmetry. Many Lorentz-invariant Lagrangians can be built from out of them. The simplest and most straightforward generalization of (5.1.1) utilizes the first invariant giving

$$S = \tfrac{1}{2} \int e^{-1} (\dot{x}^\mu - i\overline{\theta}^A \Gamma^\mu \dot{\theta}^A)^2 d\tau. \qquad (5.1.5)$$

This obviously is Lorentz invariant and supersymmetric and so has the full super-Poincaré symmetry. It gives the equations of motion

$$p^2 = 0, \qquad \dot{p}^\mu = 0, \qquad \Gamma \cdot p \, \dot{\theta} = 0, \qquad (5.1.6)$$

where we define

$$p^\mu = \dot{x}^\mu - i\overline{\theta}^A \Gamma^\mu \dot{\theta}^A. \qquad (5.1.7)$$

Since $(\Gamma \cdot p)^2 = -p^2 = 0$, the matrix $\Gamma \cdot p$ has half the maximum possible rank. Furthermore, θ always appears multiplied by $\Gamma \cdot p$. As a result, half of its components are actually decoupled from the theory! This is a consequence of a far from obvious additional symmetry of (5.1.5). The new symmetry is a local fermionic symmetry, which we now describe.

Let $\kappa^{Aa}(\tau)$ denote N infinitesimal Grassmann spinor parameters. (The labeling is the same as for ϵ^{Aa}, except that the κ^{Aa} are allowed to depend on τ.)

Consider the transformation

$$\delta\theta^A = i\Gamma \cdot p\kappa^A \qquad (5.1.8)$$

$$\delta x^\mu = i\overline{\theta}^A \Gamma^\mu \delta\theta^A \qquad (5.1.9)$$

$$\delta e = 4e\dot{\overline{\theta}}^A \kappa^A. \qquad (5.1.10)$$

Note that the relation between δx and $\delta\theta$ has the opposite sign from that for an ϵ transformation. The above transformations are actually a symmetry of (5.1.5). The proof that S is invariant starts with

$$\delta p^\mu = 2i\dot{\overline{\theta}}^A \Gamma^\mu \delta\theta^A. \qquad (5.1.11)$$

Thus

$$\delta p^2 = 4i\dot{\overline{\theta}}^A \Gamma \cdot p\delta\theta^A = 4p^2\dot{\overline{\theta}}^A \kappa^A. \qquad (5.1.12)$$

It follows that $e^{-1}p^2$ is invariant for

$$\delta e^{-1} = -4e^{-1}\dot{\overline{\theta}}^A \kappa^A, \qquad (5.1.13)$$

which is equivalent to (5.1.10).

The κ transformation is not ordinary supersymmetry either on the world-line or in space-time. In fact, the action contains no world-line spinors at all. To see what it is, let us consider the algebra obtained by commuting two κ transformations.

Let δ_1 and δ_2 represent κ variations with parameters κ_1 and κ_2, respectively. Then

$$
\begin{aligned}
[\delta_1, \delta_2]\theta^A &= i\Gamma^\mu \kappa_2^A \delta_1 p_\mu - (1 \leftrightarrow 2) \\
&= -2i\Gamma_\mu \kappa_2^A \overline{\dot{\theta}}^B \Gamma^\mu \Gamma \cdot p\kappa_1^B - (1 \leftrightarrow 2) \qquad (5.1.14) \\
&= (2i\Gamma_\mu \kappa_2^A \overline{\dot{\theta}}^B \Gamma \cdot p\Gamma^\mu \kappa_1^B + 4i\Gamma \cdot p\kappa_2^A \overline{\dot{\theta}}^B \kappa_1^B) - (1 \leftrightarrow 2).
\end{aligned}
$$

We are dealing here with a symmetry for which the action is lacking the auxiliary coordinates required for off-shell closure of the algebra. Therefore we must use equations of motion to demonstrate closure of the algebra. The equation $\Gamma \cdot p\dot{\theta} = 0$ eliminates the first term in the last expression in (5.1.14). This leaves

$$
[\delta_1, \delta_2]\theta^A = i\Gamma \cdot p\kappa^A + \text{eq. of motion term} \qquad (5.1.15)
$$

for the choice

$$
\kappa^A = 4\kappa_2^A \overline{\dot{\theta}}^B \kappa_1^B - (1 \leftrightarrow 2). \qquad (5.1.16)
$$

Thus the commutator of two κ transformations is a κ transformation! This rather bizarre result is only possible because there is no on-shell conserved charge associated with κ. The conserved quantities that one might attempt to derive from κ invariance all vanish by the equations of motion. The structure 'constants' are actually not constant, but coordinate dependent, just as in supergravity theories.

There is still one more local symmetry of (5.1.5). This one is bosonic and involves a scalar parameter $\lambda(\tau)$,

$$
\begin{aligned}
\delta\theta^A &= \lambda\dot{\theta}^A \\
\delta x^\mu &= i\overline{\theta}^A \Gamma^\mu \delta\theta^A \qquad (5.1.17) \\
\delta e &= 0.
\end{aligned}
$$

This is identical to the ξ reparametrization transformation formula for θ, but different for the other coordinates. It has no additional implications for the on-shell theory beyond those that follow from the ξ and κ symmetries.

The quantization of the superparticle action is nontrivial because of the phase-space constraint

$$\pi_\theta^A = i\Gamma \cdot \pi_x \theta^A, \tag{5.1.18}$$

where π_x and π_θ^A are the momenta conjugate to the x and θ coordinates. The Dirac bracket prescription leads to complicated expressions that are apparently impossible to disentangle without breaking the manifest Lorentz invariance of the equations. The same is true for the superstring case to which we turn next. It is for this reason that we will work in light-cone gauge.

5.1.2 The Supersymmetric String Action

Given the bosonic string action of chapter 2, namely

$$S_{bos} = -\frac{1}{2\pi} \int d^2\sigma \sqrt{h} \, h^{\alpha\beta} \partial_\alpha X \cdot \partial_\beta X \tag{5.1.19}$$

and the superparticle action of the preceding section, there is an obvious guess for a supersymmetric superstring action,

$$S_1 = -\frac{1}{2\pi} \int d^2\sigma \sqrt{h} \, h^{\alpha\beta} \Pi_\alpha \cdot \Pi_\beta, \tag{5.1.20}$$

where

$$\Pi_\alpha^\mu = \partial_\alpha X^\mu - i\bar{\theta}^A \Gamma^\mu \partial_\alpha \theta^A, \tag{5.1.21}$$

This obviously possesses local reparametrization invariance and N global supersymmetries. It is not the action we want, however. The local κ symmetry of the superparticle action is lost in this generalization. As a result θ describes twice as many degrees of freedom as it should. Also, the equations of motion constitute a complicated nonlinear system that is quite intractable. Fortunately it is possible to add a second term S_2 ($S = S_1 + S_2$), so that the resulting action S does have local κ symmetry. As a result, half the components of θ are again decoupled, and the equations of motion can be completely solved, at least in a particular gauge.

The construction that restores the local κ symmetry does not work for arbitrary N, in contrast to the superparticle case. We need to take $N \leq 2$, so that there are at most two supersymmetries. We will present formulas in terms of the coordinates θ^1 and θ^2 as appropriate to the $N = 2$ case. The $N = 1$ or $N = 0$ cases can then be obtained by setting one or both of the θ's equal to zero. The extra term in the action that completes the description of the supersymmetric superstring action is

$$
\begin{aligned}
S_2 = \frac{1}{\pi} \int d^2\sigma \{ &-i\epsilon^{\alpha\beta} \partial_\alpha X^\mu (\bar{\theta}^1 \Gamma_\mu \partial_\beta \theta^1 - \bar{\theta}^2 \Gamma_\mu \partial_\beta \theta^2) \\
&+ \epsilon^{\alpha\beta} \bar{\theta}^1 \Gamma^\mu \partial_\alpha \theta^1 \bar{\theta}^2 \Gamma_\mu \partial_\beta \theta^2 \}.
\end{aligned}
\tag{5.1.22}
$$

The alternating symbol $\epsilon^{\alpha\beta}$ is a tensor density, which explains why there is no factor of $\sqrt{h}$. In fact, the term S_2 is completely independent of $h^{\alpha\beta}$. Therefore it does not contribute to the energy–momentum tensor $T_{\alpha\beta}$.

The far from obvious term S_2 clearly has local reparametrization symmetry and global Lorentz symmetry. A little work is required to verify that it also has global $N = 2$ supersymmetry. To check this we consider transformations $\delta\theta^A = \epsilon^A$ and $\delta X^\mu = i\bar{\epsilon}^A \Gamma^\mu \theta^A$, where ϵ^{Aa} is independent of σ and τ. Substituting in (5.1.22) certain terms trivially cancel, with the relative strength of the two terms as indicated. Terms such as

$$
\epsilon^{\alpha\beta} \partial_\alpha X^\mu \bar{\epsilon}^1 \Gamma_\mu \partial_\beta \theta^1
\tag{5.1.23}
$$

are total derivatives and can be dropped. This leaves only a term proportional to

$$
A = \epsilon^{\alpha\beta} \bar{\epsilon}^1 \Gamma^\mu \partial_\alpha \theta^1 \bar{\theta}^1 \Gamma_\mu \partial_\beta \theta^1
\tag{5.1.24}
$$

and a similar expression with 2's instead of 1's. To analyze this term we write out components (and drop the superscripts)

$$
A = \bar{\epsilon} \Gamma^\mu \dot{\theta} \bar{\theta} \Gamma_\mu \theta' - \bar{\epsilon} \Gamma^\mu \theta' \bar{\theta} \Gamma_\mu \dot{\theta} = A_1 + A_2,
\tag{5.1.25}
$$

where

$$
A_1 = \frac{2}{3} [\bar{\epsilon} \Gamma^\mu \dot{\theta} \bar{\theta} \Gamma_\mu \theta' + \bar{\epsilon} \Gamma^\mu \theta' \bar{\dot{\theta}} \Gamma_\mu \theta + \bar{\epsilon} \Gamma^\mu \theta \bar{\theta'} \Gamma_\mu \dot{\theta}]
\tag{5.1.26}
$$

$$
\begin{aligned}
A_2 &= \frac{1}{3} [\bar{\epsilon} \Gamma^\mu \dot{\theta} \bar{\theta} \Gamma_\mu \theta' + \bar{\epsilon} \Gamma^\mu \theta' \bar{\dot{\theta}} \Gamma_\mu \theta - 2\bar{\epsilon} \Gamma^\mu \theta \bar{\theta'} \Gamma_\mu \dot{\theta}] \\
&= \frac{1}{3} \frac{\partial}{\partial \tau} [\bar{\epsilon} \Gamma^\mu \theta \bar{\theta} \Gamma_\mu \theta'] - \frac{1}{3} \frac{\partial}{\partial \sigma} [\bar{\epsilon} \Gamma^\mu \theta \bar{\theta} \Gamma_\mu \dot{\theta}].
\end{aligned}
\tag{5.1.27}
$$

A_2 is a total derivative and can be dropped. This leaves A_1, which we rewrite in the form

$$
A_1 = 2\bar{\epsilon} \Gamma_\mu \psi_{[1} \bar{\psi}_2 \Gamma^\mu \psi_{3]}.
\tag{5.1.28}
$$

The bracket implies that the spinors $(\psi_1, \psi_2, \psi_3) = (\theta, \theta', \dot{\theta})$ are antisymmetrized. This expression has exactly the same structure as one in

Appendix 4.A, which arose in studying the supersymmetry of super Yang–Mills theories.

We stated there that there are four circumstances under which the expression vanishes, and the same is true in the present case. Thus S_2 is supersymmetric only in the following four cases:

   (*i*) $D = 3$ and θ is Majorana;
   (*ii*) $D = 4$ and θ is Majorana or Weyl;
   (*iii*) $D = 6$ and θ is Weyl;
   (*iv*) $D = 10$ and θ is Majorana–Weyl.

Thus (in this formalism) the *classical* superstring theory exists only in these four cases.[*] This is the counterpart of the statement that the classical bosonic string theory exists for any dimension. This result should not be a surprise, since supersymmetry is well-known to restrict the possible values of D even at the classical level. Quantum considerations will single out the $D = 10$ case as special, of course.

5.1.3 The Local Fermionic Symmetry

We have shown that for suitable spinor types in $D = 3$, 4, 6 or 10 the action S_2 preserves the global super-Poincaré and local reparametrization symmetries of S_1. We now wish to show that the sum $S_1 + S_2$ has a local fermionic symmetry that is not possessed by either term separately. In the superparticle case this symmetry involved $\delta\theta^A = i\Gamma \cdot p\kappa^A$. In the string case the analog of p^μ is the expression Π_α^μ in (5.1.21). Therefore in order to have a world-sheet covariant formula it becomes necessary for the parameter κ to carry a world-sheet vector index as well. What we need is something quite unusual in the study of supersymmetry theory. From the two-dimensional world-sheet point of view we have a theory without any spinors at all that has local fermionic symmetries that transform as vectors. The various fermionic quantities that appear are spinors in the D-dimensional sense, of course.

The infinitesimal parameters κ now carry three indices $\kappa^{A\alpha a}$. The index $A = 1, 2$ corresponds to the label on θ^A. It can be truncated to a single value to describe the $N = 1$ case. The index $\alpha = 0, 1$ is a world-sheet vector index and a is a D-dimensional space-time index corresponding to one of the four possible types of spinors. The spinor index a is suppressed in most formulas. In two dimensions a vector representation is reducible (because the Lorentz group is abelian; its irreducible representations are

---

[*] When θ is Weyl but not Majorana, $\bar{\theta}\Gamma^\mu\partial_\alpha\theta$ must be explicitly replaced by $(\bar{\theta}\Gamma^\mu\partial_\alpha\theta - \partial_\alpha\bar{\theta}\Gamma^\mu\theta)/2$.

one dimensional). The decomposition of a vector into what might be called self-dual and anti-self-dual pieces is conveniently achieved using the projection tensors

$$P_\pm^{\alpha\beta} = \tfrac{1}{2}(h^{\alpha\beta} \pm \epsilon^{\alpha\beta}/\sqrt{h}), \tag{5.1.29}$$

which satisfy the projection conditions

$$P_\pm^{\alpha\beta} h_{\beta\gamma} P_\pm^{\gamma\delta} = P_\pm^{\alpha\delta} \tag{5.1.30}$$

$$P_\pm^{\alpha\beta} h_{\beta\gamma} P_\mp^{\gamma\delta} = 0. \tag{5.1.31}$$

The κ^A parameters are restricted to be anti-self-dual for $A = 1$ and self-dual for $A = 2$. Thus

$$\kappa^{1\alpha} = P_-^{\alpha\beta} \kappa_\beta^1 \tag{5.1.32}$$

$$\kappa^{2\alpha} = P_+^{\alpha\beta} \kappa_\beta^2. \tag{5.1.33}$$

It will turn out that $A = 1$ describes right-moving modes and symmetries whereas $A = 2$ describes left-moving modes and symmetries.

Let us now suppose, in analogy with the superparticle case, that

$$\delta\theta^A = 2i\Gamma \cdot \Pi_\alpha \kappa^{A\alpha} \tag{5.1.34}$$

$$\delta X^\mu = i\bar{\theta}^A \Gamma^\mu \delta\theta^A, \tag{5.1.35}$$

with $\delta h_{\alpha\beta}$ still to be determined. Then

$$\delta L_1 = -\sqrt{h}\, h^{\alpha\beta} \Pi_\alpha \cdot \delta\Pi_\beta - \tfrac{1}{2}\delta(\sqrt{h}\, h^{\alpha\beta})\Pi_\alpha \cdot \Pi_\beta, \tag{5.1.36}$$

where

$$\delta\Pi_\alpha^\mu = 2i\partial_\alpha\bar{\theta}^A \Gamma^\mu \delta\theta^A. \tag{5.1.37}$$

The Lagrangian L_2 can be rewritten in the form

$$L_2 = -i\epsilon^{\alpha\beta}\Pi_\alpha^\mu(\bar{\theta}^1\Gamma_\mu\partial_\beta\theta^1 - \bar{\theta}^2\Gamma_\mu\partial_\beta\theta^2) + \theta^4 \text{ terms.} \tag{5.1.38}$$

Varying the explicit θ's in the $\Pi\theta^2$ piece of L_2 and combining with the first term in the variation of L_1 gives a term that can be precisely canceled by the second term in δL_1 for the choice

$$\delta_\kappa(\sqrt{h}\,h^{\alpha\beta}) = -16\sqrt{h}(P_-^{\alpha\gamma}\overline{\kappa}^{1\beta}\partial_\gamma\theta^1 + P_+^{\alpha\gamma}\overline{\kappa}^{2\beta}\partial_\gamma\theta^2). \qquad (5.1.39)$$

Since $\sqrt{h}\,h^{\alpha\beta}$ is unimodular and symmetric, it is important that the right-hand side of (5.1.39) be symmetric and traceless. The self-duality properties of κ^1 and κ^2 as well as identities such as

$$P_+^{\alpha\gamma}P_+^{\beta\delta} = P_+^{\beta\gamma}P_+^{\alpha\delta} \qquad (5.1.40)$$

ensure that this is the case. Note that this construction requires that L_1 and L_2 have exactly the relative coefficient given. (One can change the sign of L_2, since this corresponds to the interchanging of θ^1 and θ^2.)

To complete the proof of the local κ symmetry it is still necessary to consider the variation of Π_α^μ and the θ^4 terms in (5.1.38). Doing this one finds that many terms cancel leaving

$$2\overline{\theta}^1\Gamma_\mu\theta^{1\prime}\overline{\dot{\theta}}^1\Gamma^\mu\delta\theta^1 - 2\overline{\theta}^1\Gamma_\mu\dot{\theta}^1\overline{\theta}^{1\prime}\Gamma^\mu\delta\theta^1$$
$$- 2\overline{\theta}^1\Gamma_\mu\delta\theta^1\overline{\dot{\theta}}^1\Gamma^\mu\theta^{1\prime} + \text{similar terms with } \theta^2. \qquad (5.1.41)$$

This is exactly the combination that cancels for the four special types of spinors listed above. Thus the local κ symmetry can be implemented in just those cases when the global ϵ symmetry can.

In addition to the local reparametrization and fermionic symmetries there is a further local bosonic symmetry of action $S_1 + S_2$. One way of discovering it is by considering the algebra of κ transformations. The closure requires further local bosonic transformations with parameter λ_α

$$\delta\theta^1 = \sqrt{h}P_-^{\alpha\beta}\partial_\beta\theta^1\lambda_\alpha$$
$$\delta\theta^2 = \sqrt{h}P_+^{\alpha\beta}\partial_\beta\theta^2\lambda_\alpha$$
$$\delta X^\mu = i\overline{\theta}^A\Gamma^\mu\delta\theta^A \qquad (5.1.42)$$
$$\delta(\sqrt{h}h^{\alpha\beta}) = 0.$$

The proof that $S_1 + S_2$ is invariant under these transformations requires manipulations similar to those used to verify the local κ symmetry.

The equations of motion for the supersymmetric superstring action are

$$\Pi_\alpha \cdot \Pi_\beta = \tfrac{1}{2}h_{\alpha\beta}h^{\gamma\delta}\Pi_\gamma \cdot \Pi_\delta$$

$$\Gamma \cdot \Pi_\alpha P_-^{\alpha\beta}\partial_\beta\theta^1 = 0$$

$$\Gamma \cdot \Pi_\alpha P_+^{\alpha\beta}\partial_\beta\theta^2 = 0 \qquad (5.1.43)$$

$$\partial_\alpha[\sqrt{h}(h^{\alpha\beta}\partial_\beta X^\mu - 2iP_-^{\alpha\beta}\overline{\theta}^1\Gamma^\mu\partial_\beta\theta^1 - 2iP_+^{\alpha\beta}\overline{\theta}^2\Gamma^\mu\partial_\beta\theta^2)] = 0.$$

The first of these corresponds to $T_{\alpha\beta} = 0$. These are complicated nonlinear equations, but we will show later that they collapse to simple free theory equations in the light-cone gauge.

In the point-particle case, the supersymmetric action with N θ coordinates possesses a manifest global $SO(N)$ symmetry corresponding to a rotation of these variables into one another. In the superstring case we were only able to emulate the superparticle construction for $N = 0$, 1 or 2. However, there is no global rotation symmetry in any of these cases. In the $N = 2$ case the $SO(2)$ symmetry is explicitly broken by S_2. One implication of this fact is that a global $U(1)$ symmetry of the type IIB supergravity theory in $D = 10$ (described in §13.5.2) is not preserved by its superstring extension.

5.1.4 Type I and Type II Superstrings

In ten dimensions – the critical value for superstring theory – we have seen that the θ coordinates in the superstring action must be chosen to be Majorana–Weyl spinors. This means, in particular, that θ^1 and θ^2 must each be assigned a definite handedness. The overall meaning of left and right is a matter of convention, but there are two physically distinct possibilities. Either θ^1 or θ^2 are chosen to have the same handedness or to have the opposite handedness. In the case of closed strings the only boundary conditions that are imposed are periodicity in σ. This is possible in either case, since it does not relate θ^1 and θ^2. For open strings, on the other hand, we will show in §5.2.1 that θ^1 and θ^2 must be equated at the ends of the strings. Since a left-handed spinor cannot equal a right-handed one, this is only possible in the case when θ^1 and θ^2 have the same handedness.

A superstring theory based on open superstrings is called a type I superstring theory. It will be explained in §5.2.1 that the open-string boundary conditions reduce the space-time supersymmetry to $N = 1$ only, which is part of the motivation for the name 'type I'. A theory of open strings can have Yang–Mills group-theory quantum numbers corresponding to any classical group introduced by attaching charges at the ends of the strings. This method, known as the Chan–Paton method, will be explained in detail in §6.1.1. Any group choice is consistent for a classical (tree-level) theory of interacting open strings. However, at the quantum level consistency conditions explained in chapter 10 require the unique choice $SO(32)$. It will be explained in §6.1.1 that when the group is orthogonal or symplectic the strings are unoriented. Thus the unique type I superstring theory that appears to be quantum mechanically consistent is one based on the group $SO(32)$. It consists of interacting unoriented open and closed strings. Closed strings are required in the quantum theory, since as will be explained in chapter 8, open strings must be allowed to join ends giving rise to closed strings. Since the closed strings have no free ends, they are necessarily singlets of the Yang–Mills group.

Let us now consider theories based on closed superstrings only. If θ^1 and θ^2 have opposite handedness the resulting theory necessarily involves oriented strings, since θ^1 describes modes that propagate one way around the string while θ^2 describes modes that propagate in the opposite direction. This theory has two conserved $D = 10$ supersymmetries of opposite handedness. (The equation $\delta\theta^A = \epsilon^A$ shows that the chirality of the supercharge is directly controlled by the chirality of the corresponding θ coordinate.) The theory with two conserved supercharges of opposite chirality is called type IIA. This theory turns out, as we will see, to be left-right symmetric (nonchiral). It has no freedom to introduce a Yang-Mills group.

The remaining possibility is to base a theory of closed superstrings on two θ coordinates of the same handedness. In this case one has the option of symmetrizing the left- and right-moving modes to define a theory of unoriented closed strings or to not do so leaving a theory of oriented closed strings. In the former case one is back to the closed-string sector of the $SO(32)$ theory. As has already been mentioned, consistency requires including $SO(32)$ open strings at the same time, giving the type I superstring theory. If no restrictions are made, so that one has a theory of oriented closed strings, there are two space-time supersymmetries of the same handedness. This is the type IIB superstring theory. It is obviously a left-right asymmetric (chiral) theory. In chapter 13 we will present evidence for the quantum mechanical consistency of type IIB superstring theory. It also has no freedom to introduce a Yang-Mills group.

There is yet another possibility for constructing a consistent supersymmetric string theory. It is based on using only one θ coordinate rather than two. In this case, one is led to the heterotic string theories, which are introduced in the next chapter.

5.2 Quantization

The structure of phase-space constraints in the supersymmetric superstring action makes covariant quantization very difficult, perhaps even impossible. The problems that arise will be discussed §5.4. Fortunately, quantization works very nicely in the light-cone gauge, so we concentrate on its description. This leads to a formalism that allows many explicit amplitude calculations to be done much more easily than is possible by other methods. A number of such calculations will be described in subsequent chapters.

5.2.1 Light-Cone Gauge

In order to be specific, the following discussion assumes that $D = 10$ and that the spinors θ^1 and θ^2 are Majorana–Weyl. The analysis can be repeated with only minor changes for $D = 3$, 4 or 6 using the appropriate spinor types for each case. However, for those dimensions the Lorentz algebra commutator $[J^{i-}, J^{j-}]$ fails to vanish. In the $D = 10$ case with two θ coordinates there are two physically distinct alternatives explained in §5.1.4. Either θ^1 and θ^2 have the same chirality (handedness) or they have opposite chirality.

The local reparametrization and Weyl invariances and the local κ fermionic symmetries allow for a number of gauge choices to be made. Specifically, if no anomalies of the quantum theory invalidate the choice, the former allows us to set $h_{\alpha\beta} = \eta_{\alpha\beta}$, as usual. (We ignore here global questions that arise in loop diagrams.) Unlike the world-sheet actions of previous chapters this does not yet linearize the equations of motion. To achieve this we must also use the residual reparametrization invariance and κ symmetry to make light-cone gauge choices as follows. Firstly we can use the κ symmetry to enforce the condition

$$\Gamma^+ \theta^1 = \Gamma^+ \theta^2 = 0. \qquad (5.2.1)$$

These conditions utilize a light-cone component of the ten-dimensional Dirac matrices

$$\Gamma^\pm = \frac{1}{\sqrt{2}}(\Gamma^0 \pm \Gamma^9). \qquad (5.2.2)$$

The matrices Γ^+ and Γ^- are nilpotent, *i.e.*,

$$\left(\Gamma^+\right)^2 = \left(\Gamma^-\right)^2 = 0, \qquad (5.2.3)$$

but their sum is nonsingular. It follows that exactly half the eigenvalues of each must be zero. Therefore the gauge choice (5.2.1) amounts to setting half the components of θ equal to zero. This is the number of components that are decoupled as a consequence of the local κ symmetries. For reasons to be explained soon the equations for X^+ and X^i obtained from (5.1.43) are simply free wave equations. Then, as for the bosonic theory, the residual conformal invariance can be used to impose the condition

$$X^+(\sigma, \tau) = x^+ + p^+\tau. \tag{5.2.4}$$

This amounts to setting all the α_n^+ modes with $n \neq 0$ equal to zero, just as in §2.3.3. As discussed in chapter 4, a generic spinor in ten dimensions has 32 components. The Majorana condition makes them real and the Weyl condition sets half equal to zero, leaving 16 real components. The light-cone gauge condition reduces the count by another factor of two leaving just eight real components. The only manifest symmetry in the light-cone gauge is the rotational invariance of the eight transverse dimensions. Thus the eight surviving components of each θ can be regarded as forming an eight-dimensional spinor representation of the transverse $SO(8)$ group, or more precisely its 'covering group' spin(8). The Dynkin diagram of the rank-four Lie algebra $SO(8) = D_4$ possesses a symmetry, referred to as triality, described by a six element automorphism group. The automorphisms permute inequivalent irreducible representations having the same dimensionality. For example, there are three eight-dimensional irreducible representations. One is the fundamental vector representation 8_v, which is obviously real, and the other two are spinor representations, 8_s and 8_c. They are also real; indeed, this follows from triality. We use the letters i, j, k for 8_v labels (*e.g.*, X^i), a, b, c for 8_s labels, and $\dot{a}$, $\dot{b}$, $\dot{c}$ for 8_c labels. (Upper and lower indices are equivalent in each case.) For more details about properties of $SO(2n)$ in general and $SO(8)$ in particular see appendix 5.A.

Using the symbol S for the eight surviving components of θ in the light-cone gauge, we have

$$\sqrt{p^+}\theta^1 \rightarrow S^{1a} \text{ or } S^{1\dot{a}} \tag{5.2.5}$$

$$\sqrt{p^+}\theta^2 \rightarrow S^{2a} \text{ or } S^{2\dot{a}}. \tag{5.2.6}$$

The determination of whether the 8_s or 8_c representation occurs in each case is controlled by the chirality of the corresponding θ in ten dimensions. We may choose S^1 to belong to 8_s as a convention. Then S^2 belongs to 8_s, as well, for type I and IIB theories, but to 8_c for the type IIA theory. In writing explicit formulas we use the symbols S^{1a} and S^{2a} as appropriate to the former case, but the latter possibility should be borne in mind.

The equations of motion (5.1.43) collapse dramatically in the light-cone gauge. The important point to realize is that $\Gamma^+\theta = 0$ implies that $\bar{\theta}\Gamma^\mu \partial_\alpha \theta$ vanishes unless $\mu = -$. This is obvious for $\mu = +$, while for $\mu = i$ one can insert $1 = (\Gamma^+\Gamma^- + \Gamma^-\Gamma^+)/2$ and note that both terms vanish due to Γ^+ multiplying θ either to the left or the right. As a result one is left with

$$\left(\frac{\partial^2}{\partial\sigma^2} - \frac{\partial^2}{\partial\tau^2}\right)X^i = 0 \tag{5.2.7}$$

$$\left(\frac{\partial}{\partial\tau} + \frac{\partial}{\partial\sigma}\right)S^{1a} = 0 \tag{5.2.8}$$

$$\left(\frac{\partial}{\partial\tau} - \frac{\partial}{\partial\sigma}\right)S^{2a} = 0. \tag{5.2.9}$$

These equations are identical to those satisfied by X^i, ψ^i_- and ψ^i_+ in the light-cone gauge version of the formalism of chapter 4. The only difference is that the Fermi coordinates now belong to a spinor representation of spin(8) rather than a vector representation. Because of triality this makes many equations essentially isomorphic, but there are also profound differences.

The equations of motion for the light-cone gauge string coordinates given above can be obtained from the action (with the string tension explicitly included)

$$S_{l.c.} = -\frac{1}{2}\int d^2\sigma(T\partial_\alpha X^i\partial^\alpha X^i - \frac{i}{\pi}\overline{S}^a\rho^\alpha\partial_\alpha S^a), \tag{5.2.10}$$

where S^{1a} and S^{2a} have been combined into a two-component Majorana world-sheet spinor S^a. S^{1a} and S^{2a} can be regarded separately as one-component Majorana–Weyl world-sheet spinors describing right- and left-moving degrees of freedom, respectively. Something rather remarkable has just happened. In the covariant action of §5.1.2, the variables θ^{Aa} transform as world-sheet scalars. Yet, by the time the light-cone gauge has been fixed, the remaining nonzero components metamorphose into world-sheet spinors!

The quantization of the X^i coordinate determined by (5.2.10) is the same as in the preceding chapters. The S^{Aa} coordinates have canonical anticommutation relations

$$\{S^{Aa}(\sigma,\tau), S^{Bb}(\sigma',\tau)\} = \pi\delta^{ab}\delta^{AB}\delta(\sigma - \sigma'). \tag{5.2.11}$$

In order to give explicit mode expansions of the S coordinates, we still need to specify boundary conditions. For open strings this requires relating S^1 and S^2 at the ends of the strings, just as we did for the ψ

coordinates in §4.1. If we insist that at least one of the supersymmetries of the action is not broken by the boundary conditions then it is important that there is a zero mode of the S coordinates. This means that we do not have the freedom we had in the discussion of §4.1 concerning relative sign conventions at the two ends of the string; space-time supersymmetry is only valid if we make the same choice at each end. The appropriate choices are

$$S^{1a}(0,\tau) = S^{2a}(0,\tau) \tag{5.2.12}$$

$$S^{1a}(\pi,\tau) = S^{2a}(\pi,\tau). \tag{5.2.13}$$

Because of the global supersymmetry transformation formula $\delta\theta^A = \epsilon^A$, we see that this choice requires equating ϵ^1 and ϵ^2, thereby reducing the supersymmetry to $N = 1$. Hence the name 'type I'. We also see that the boundary conditions require that S^1 and S^2 belong to the same spin(8) representation. If one were to introduce a minus sign in (5.2.13), analogous to that used for the bosonic superstring sector in the formalism of chapter 4, this would completely destroy the supersymmetry. Under special circumstances this can also lead to interesting theories.[*]

The open-string mode expansions that follow from (5.2.10) are

$$S^{1a}(\sigma,\tau) = \frac{1}{\sqrt{2}} \sum_{-\infty}^{\infty} S_n^a e^{-in(\tau-\sigma)} \tag{5.2.14}$$

$$S^{2a}(\sigma,\tau) = \frac{1}{\sqrt{2}} \sum_{-\infty}^{\infty} S_n^a e^{-in(\tau+\sigma)}. \tag{5.2.15}$$

Reality of these coordinates implies that

$$S_{-m}^a = \left(S_m^a\right)^\dagger. \tag{5.2.16}$$

The canonical anticommutation relations, expressed in terms of the expansion coefficients, become

$$\{S_m^a, S_n^b\} = \delta^{ab}\delta_{m+n}. \tag{5.2.17}$$

As always, the only boundary condition for closed strings is periodicity

$$S^{Aa}(\sigma,\tau) = S^{Aa}(\sigma+\pi,\tau), \tag{5.2.18}$$

so that the mode expansion becomes

$$S^{1a}(\sigma,\tau) = \sum S_n^a e^{-2in(\tau-\sigma)} \tag{5.2.19}$$

---

[*] In chapter 9 we shall discuss a theory of this type – the $SO(16) \times SO(16)$ theory – which has no space-time supersymmetry.

$$S^{2a}(\sigma, \tau) = \sum \tilde{S}^a_n e^{-2in(\tau+\sigma)}, \qquad (5.2.20)$$

with independent sets of modes for right and left movers. If S^1 and S^2 belong to different representations ($\mathbf{8_s}$ and $\mathbf{8_c}$), this describes type IIA superstrings, which are necessarily oriented. Otherwise it gives type IIB superstrings if the strings are oriented or type I closed strings if they are not.

The equivalence of the formulation of superstrings given in this chapter and the one in chapter 4 can be understood as follows. The light-cone gauge action in (5.2.10) can be related to the light-cone gauge action (with $T = 1/\pi$ again)

$$S'_{l.c.} = -\frac{1}{2\pi} \int d^2\sigma (\partial_\alpha X^i \partial^\alpha X^i - i\overline{\psi}^i \rho^\alpha \partial_\alpha \psi^i) \qquad (5.2.21)$$

of chapter 4. In one case the world-sheet spinor S^a belongs to the $\mathbf{8_s}$ representation of spin(8) and in the other case the world-sheet spinor ψ^i belongs to the $\mathbf{8_v}$ representation of spin(8). This is a meaningful distinction and not just a convention about names, since in both cases X^i belongs to the $\mathbf{8_v}$ representation.

The relationship between the light-cone theories can be described by bosonizing the fermions ψ^i and then re-fermionizing them. This is done by first introducing four real scalars by [†]

$$\frac{1}{\sqrt{\pi}} \epsilon^{\alpha\beta} \partial_\beta \phi_1 = \overline{\psi}^1 \rho^\alpha \psi^2$$

$$\frac{1}{\sqrt{\pi}} \epsilon^{\alpha\beta} \partial_\beta \phi_2 = \overline{\psi}^3 \rho^\alpha \psi^4$$

$$\frac{1}{\sqrt{\pi}} \epsilon^{\alpha\beta} \partial_\beta \phi_3 = \overline{\psi}^5 \rho^\alpha \psi^6 \qquad (5.2.22)$$

$$\frac{1}{\sqrt{\pi}} \epsilon^{\alpha\beta} \partial_\beta \phi_4 = \overline{\psi}^7 \rho^\alpha \psi^8.$$

Next we reshuffle the scalars into the combinations

$$\sigma_1 = \tfrac{1}{2}(\phi_1 + \phi_2 + \phi_3 + \phi_4)$$

$$\sigma_2 = \tfrac{1}{2}(\phi_1 + \phi_2 - \phi_3 - \phi_4)$$

$$\sigma_3 = \tfrac{1}{2}(\phi_1 - \phi_2 + \phi_3 - \phi_4) \qquad (5.2.23)$$

$$\sigma_4 = \tfrac{1}{2}(\phi_1 - \phi_2 - \phi_3 + \phi_4).$$

[†] Bosonization of fermions was described in chapter 3, where it was shown that upon bosonization a fermion current J^α is expressed as the derivative of a scalar field. The formulas were written there in holomorphic or light-cone coordinates; in any coordinate system $J^\alpha \sim \epsilon^{\alpha\beta} \partial_\beta \phi$, which is obviously conserved.

Finally we re-fermionize them, introducing eight new Majorana fermions S^a with

$$\frac{1}{\sqrt{\pi}}\epsilon^{\alpha\beta}\partial_\beta\sigma_1 = \overline{S}^1\rho^\alpha S^2$$

$$\frac{1}{\sqrt{\pi}}\epsilon^{\alpha\beta}\partial_\beta\sigma_2 = \overline{S}^3\rho^\alpha S^4$$

$$\frac{1}{\sqrt{\pi}}\epsilon^{\alpha\beta}\partial_\beta\sigma_3 = \overline{S}^5\rho^\alpha S^6 \qquad\qquad (5.2.24)$$

$$\frac{1}{\sqrt{\pi}}\epsilon^{\alpha\beta}\partial_\beta\sigma_4 = \overline{S}^7\rho^\alpha S^8.$$

The $SO(8)$ or spin(8) symmetry is not particularly manifest in this transformation. The only $SO(8)$ generators that are manifest in the construction are the four corresponding to the currents used in (5.2.22). They commute with each other and are in fact a maximal commuting set of $SO(8)$ generators. A maximal commuting set of generators is known as a Cartan subalgebra. Because of the factors of $1/2$ in (5.2.23), the S^a defined in (5.2.24) have half-integral quantum numbers under the Cartan subalgebra and transform as spinors of $SO(8)$. In fact, they transform in the $\mathbf{8_s}$ representation, while $\mathbf{8_c}$ could be obtained by changing some signs in (5.2.24). This assertion can be proved rigorously by study of the explicit bosonization formulas $\psi \sim e^{i\phi^+}$ discussed in chapter 3, but we do not attempt this and here thus view the above as a heuristic explanation of the relation between the different formalisms. The rationale behind some of the above steps should become clearer in the discussion of triality in appendix 5.A.

There is a very important subtlety here. Bosonization of fermions in infinite volume is comparatively straightforward, but in finite volume (on a string, for instance), there are various complications, some of which we saw in chapter 3. In particular the bosonization and subsequent re-fermionization introduce changes in the fermion boundary conditions, as a result of which the Lagrangians in (5.2.10) and (5.2.21) are almost, but not quite, equivalent. In fact, the distinctions are of fundamental importance. As we have seen by studying the representations of the S^a and the ψ^i algebras, the Lagrangian (5.2.10) automatically describes supersymmetric multiplets of bosons and fermions. The corresponding states are obtained from the Lagrangian (5.2.21) by quantizing with one set of boundary conditions to give the fermions and then with another set to give the bosons. After that one must truncate the spectra in each sector to the subspaces with $(-1)^F = +1$.

5.2.2 Super-Poincaré Algebra

One of the advantages of the present formulation compared to that of chapter 4 is that space-time supersymmetry can be described in a relatively simple way. To simplify the discussion we drop the index A, as appropriate for open strings or one sector (left- or right-moving) of closed strings. To see how supersymmetry works in the light-cone gauge, recall that for the covariant action of §5.1.2, the transformation formulas include $\delta\theta = \epsilon$. This formula does not preserve the gauge choice $\Gamma^+\theta = 0$ unless $\Gamma^+\epsilon = 0$. For such ϵ, $\bar\epsilon\Gamma^i\theta = 0$, as can be shown by the reasoning described above. Thus these eight supersymmetries can be described in spin(8) notation by

$$\delta S^a = \sqrt{2p^+}\eta^a \qquad (5.2.25)$$

$$\delta X^i = 0. \qquad (5.2.26)$$

This is obviously an invariance of the light-cone gauge action (5.2.10). The factor of $\sqrt{2p^+}$ is inserted for later convenience.

The other eight components of the supersymmetry transformations have $\Gamma^+\epsilon \neq 0$. They correspond to an $\mathbf{8_c}$ spinor $\epsilon^{\dot a}$. In order to preserve $\Gamma^+\theta = 0$ it is necessary to combine the ϵ transformation with a κ transformation

$$\delta\theta = \epsilon + 2i\Gamma\cdot\Pi_\alpha\kappa^\alpha, \qquad (5.2.27)$$

where κ is chosen in terms of ϵ in such a way as to ensure that the right-hand side is annihilated by Γ^+. The resulting redefined ϵ transformation, expressed in spin(8) notation, is

$$\delta S^a = -i\rho\cdot\partial X^i\gamma^i_{a\dot a}\epsilon^{\dot a}\sqrt{2p^+} \qquad (5.2.28)$$

$$\delta X^i = 2\gamma^i_{a\dot a}\bar\epsilon^{\dot a}S^a/\sqrt{2p^+}. \qquad (5.2.29)$$

The $\gamma^i_{a\dot a}$ are Clebsch–Gordan coefficients for coupling the three inequivalent eight-dimensional representations. An explicit representation and a discussion of their properties are given in appendix 5.B. As a check of these formulas, the reader should verify that they leave the action (5.2.10) invariant. The transformations in (5.2.29) look like supersymmetry transformations of two-dimensional field theory. The $\epsilon^{\dot a}$ would, in that case, be viewed as the parameters of eight two-dimensional supercharges.

The anticommutation of two supersymmetry transformations gives a space-time translation, which is interpreted in the light-cone gauge as the combination of a translation in the world sheet and a translation in the transverse direction. This can be seen by considering the action of two successive transformations (with parameters $\eta^{(1)a}$, $\epsilon^{(1)\dot{a}}$ and $\eta^{(2)a}$, $\epsilon^{(2)\dot{a}}$) on the coordinates. The result is

$$[\delta_1, \delta_2]X^i = \xi^\alpha \partial_\alpha X^i + a^i, \qquad (5.2.30)$$

$$[\delta_1, \delta_2]S^a = \xi^\alpha \partial_\alpha S^a, \qquad (5.2.31)$$

where the equation of motion $\rho \cdot \partial S^a = 0$ has been used in evaluating these expressions. The ξ^α are associated with the anticommutator of two ϵ transformations and are translations in σ and τ given by

$$\xi^\alpha = -2i\epsilon^{(1)}\rho^\alpha\epsilon^{(2)}. \qquad (5.2.32)$$

The a^i is a translation of the transverse coordinates, given by

$$a^i = \sqrt{2}\eta^{(2)}\gamma^i\epsilon^{(1)} - \sqrt{2}\eta^{(1)}\gamma^i\epsilon^{(2)}. \qquad (5.2.33)$$

The conserved η^a and $\epsilon^{\dot{a}}$ supersymmetry charges can be deduced by the Noether method by an extension of the procedure used for obtaining the Poincaré charges in chapter 2. A short-cut to the result is to notice from the expressions (5.2.25) and (5.2.26) that the η^a transformation is simply generated by

$$Q^a = (2p^+)^{1/2}S_0^a, \qquad (5.2.34)$$

whereas the $\epsilon^{\dot{a}}$ transformations in (5.2.28) and (5.2.29) are generated by

$$Q^{\dot{a}} = (p^+)^{-1/2}\gamma^i_{\dot{a}a}\sum_{-\infty}^{\infty} S^a_{-n}\alpha^i_n. \qquad (5.2.35)$$

These 16 charges are the components of a covariant Majorana–Weyl spinor, which satisfies a supersymmetry algebra $\{Q, Q\} \sim (1 \pm \Gamma_{11})\Gamma \cdot p$. In spin(8) notation this equation splits up into three pieces

$$\{Q^a, Q^b\} = 2p^+\delta^{ab} \qquad (5.2.36)$$

$$\{Q^a, Q^{\dot{a}}\} = \sqrt{2}\gamma^i_{a\dot{a}}p^i \qquad (5.2.37)$$

$$\{Q^{\dot{a}}, Q^{\dot{b}}\} = 2H\delta^{\dot{a}\dot{b}}, \qquad (5.2.38)$$

where

$$H = \frac{1}{2p^+} \left((p^i)^2 + 2N \right),$$

(5.2.39)

is the light-cone Hamiltonian in which

$$N = \sum_{m=1}^{\infty} \left(\alpha^i_{-m} \alpha^i_m + m S^a_{-m} S^a_m \right).$$

(5.2.40)

Normal ordering is trivial in this case (just as in the R sector of the RNS model) because of a cancellation of zero-point energies of the α modes and the S modes. The mass-shell condition is simply $H = p^-$. Note that the Q^a are square roots of p^+ and $Q^{\dot{a}}$ are square roots of H.

Let us now consider the Lorentz generators. The covariant Lorentz rotation of a spinor θ generated by $b_{\mu\nu} J^{\mu\nu}$ is proportional to $b_{\mu\nu} \gamma^{\mu\nu} \theta$. Such a transformation preserves $\gamma^+ \theta = 0$ for J^{+i}, J^{ij} and J^{+-}, but not for J^{i-}. Thus the correct transformation in the latter case requires compensating ξ and κ transformations to restore the gauge condition. As a result, all the generators except for J^{i-} are the obvious expressions that follow from inserting the gauge choices in the Noether charges of the covariant action. The J^{i-} are more complicated, as they were in the cases previously considered. The results are similar to those of the previous chapters

$$J^{\mu\nu} = l^{\mu\nu} + E^{\mu\nu} + K^{\mu\nu},$$

(5.2.41)

where

$$l^{\mu\nu} = x^\mu p^\nu - x^\nu p^\mu$$

(5.2.42)

and

$$E^{\mu\nu} = -i \sum_{n=1}^{\infty} \frac{1}{n} (\alpha^\mu_{-n} \alpha^\nu_n - \alpha^\nu_{-n} \alpha^\mu_n).$$

(5.2.43)

In evaluating $E^{\mu\nu}$ it is to be understood that $\alpha^+_n = 0$ (and hence $E^{\mu+} = 0$) and, by solving the Virasoro conditions for the coordinate $X^-(\sigma, \tau)$ as in chapter 2,

$$\alpha^-_n = \frac{1}{2p^+} \sum_{-\infty}^{\infty} (\alpha^i_{n-m} \alpha^i_m + (m - \frac{n}{2}) S^a_{n-m} S^a_m).$$

(5.2.44)

In particular, $\alpha_0^- = H$. The $K^{\mu\nu}$ are given by

$$K^{\mu+} = 0$$

$$K^{ij} = K_0^{ij}$$

$$K^{i-} = \frac{1}{p^+} \sum_{-\infty}^{\infty} K_{-n}^{ij} \alpha_n^j \qquad (5.2.45)$$

$$K_n^{ij} = -\frac{i}{4} \sum_{-\infty}^{\infty} S_{n-m}^a \gamma_{ab}^{ij} S_m^b.$$

The algebra of the α_n^- and K_n^{ij} is exactly the same as it was in chapter 4, where the formulas (in the fermionic sector) involved d_n^i oscillators instead of the S_n^a ones. The reason for this isomorphism can be understood in terms of the triality symmetry described in appendix 5.A. It therefore follows, as a consequence of the algebra in chapter 4, that $[J^{i-}, J^{j-}] = 0$, proving Lorentz invariance for $D = 10$. Now, however, we are in a position to describe the full super-Poincaré algebra. The anticommutator of two supersymmetries has already been described. All that remains is to examine the Lorentz transformations of the supercharges to prove that they transform into one another as the components of a $D = 10$ spinor. The algebra is reasonably straightforward. One finds, for example, that

$$[J^{i-}, Q^a] = -i\sqrt{2p^+} \sum_{n \neq 0} \frac{1}{n} \alpha_{-n}^i [\alpha_n^-, S_0^a] + \sqrt{\frac{2}{p^+}} \sum \alpha_{-n}^j [K_n^{ij}, S_0^a]$$

$$= \frac{i}{\sqrt{2p^+}} \sum_{-\infty}^{\infty} (\gamma^i \gamma^j S_{-n})^a \alpha_n^j = \frac{i}{\sqrt{2}} \gamma_{a\dot{a}}^i Q^{\dot{a}}.$$

$$(5.2.46)$$

For type II theories there are two sets of supercharges. In functional form one has (in the IIB case)

$$p^i = \int_0^\pi d\sigma P_\tau^i(\sigma, \tau) \qquad (5.2.47)$$

$$H = \frac{1}{2\pi p^+} \int_0^\pi d\sigma [\pi^2 (P_\tau^i)^2 + (X^{i\prime})^2 - iS^1 S^{1\prime} + iS^2 S^{2\prime}] \quad (5.2.48)$$

$$Q_A^a = \frac{1}{\pi} (2p^+)^{1/2} \int_0^\pi d\sigma S_A^a \qquad (5.2.49)$$

$$Q_1^{\dot{a}} = \frac{1}{\pi}(p^+)^{-1/2} \int_0^\pi d\sigma (\gamma^i S_1)^{\dot{a}} (\pi P_\tau^i - X'^i) \qquad (5.2.50)$$

$$Q_2^{\dot{a}} = \frac{1}{\pi}(p^+)^{-1/2} \int_0^\pi d\sigma (\gamma^i S_2)^{\dot{a}} (\pi P_\tau^i + X'^i), \qquad (5.2.51)$$

where

$$P_\tau^i(\sigma, \tau) = \frac{1}{\pi}\dot{X}^i(\sigma, \tau) = -i\frac{\delta}{\delta X^i(\sigma, \tau)}. \qquad (5.2.52)$$

There are analogous expressions for the $J^{\mu\nu}$.

5.3   Analysis of the Spectrum

Let us now examine the spectrum of physical states arising in the various theories that we have formulated. As in §2.3.4, we use the light-cone formalism so that we can count physical degrees of freedom directly without analyzing constraints. Even in this context there are a number of different routes we can follow. They include the RNS formulation of §4.3.2, and the spin(8) supersymmetric formulation described here. We mention both in this section, but concentrate mainly on the spin(8) description. In chapter 11 we will describe a light-cone formulation that uses coordinates belonging to representations of a $SU(4) \times U(1)$ subgroup of spin(8); it is useful for functional integral calculations.

5.3.1 Open Superstrings

Open-string boundary conditions lead to a single set of bosonic and fermionic modes, corresponding to standing waves on the string. They also restrict the possible supersymmetry to $N = 1$. Internal symmetry changes can be attached to the ends of open strings by a procedure that was introduced in chapter 1 and will be further described in chapter 6. It will be shown there that this results in oriented strings for $U(n)$ groups and unoriented strings for $SO(n)$ or $USp(n)$ groups. The massless states necessarily belong to the adjoint representation in each case. We ignore these symmetry-group quantum numbers for the discussion in this section.

Let us begin by analyzing the massless sector, which is the ground state of the spectrum since there are no tachyons. As shown in §4.3.2, the formulation with world-sheet supersymmetry, analyzed in the light-cone gauge, gives massless bosons that form a transverse eight-vector $b^i_{-1/2}|0\rangle$.

A 16-component Majorana–Weyl $u(p)$ describes the massless fermions required to represent the Dirac algebra. However, only half of these fermion degrees of freedom, corresponding to an 8_c representation of spin(8), say, propagate once the Dirac equation $\gamma \cdot p\, u = 0$ is imposed. Thus there are eight Bose and eight Fermi modes, corresponding precisely to the particle content of the super Yang–Mills theory in $D = 10$.

Now let us re-examine the massless spectrum in the supersymmetric formulation. In the spin(8) description, the ground state must represent the algebra $\{S_0^a, S_0^b\} = \delta^{ab}$. Because of triality, which is explained in appendix 5.A, this can be achieved in exactly the same way the Clifford algebra $\{\gamma^i, \gamma^j\} = 2\delta^{ij}$ is represented in appendix 5.B

$$S_0^a \sim \begin{pmatrix} 0 & \gamma_{i\dot a}^a \\ \gamma_{\dot a i}^a & 0 \end{pmatrix}. \tag{5.3.1}$$

The representation space in this case is $8_v + 8_c$, which is the complete supermultiplet obtained by two separate constructions in the RNS formulation. Let us represent this 16-dimensional multiplet of massless ground states in Fock-space notation by $|\phi_0\rangle$.

These massless ground states described by $|\phi_0\rangle$ consist of eight Bose states in the 8_v representation of spin(8), denoted by $|i\rangle$, and eight Fermi states in the 8_c representation, denoted by $|\dot a\rangle$. The states are normalized so that

$$\langle i|j\rangle = \delta_{ij}, \qquad \langle \dot a|\dot b\rangle = \delta_{\dot a\dot b}, \tag{5.3.2}$$

and the identity operator in the space of S_0^a is given by

$$I = |i\rangle\langle i| + |\dot a\rangle\langle \dot a|. \tag{5.3.3}$$

In order to study properties of the S_0 operators, it is useful to note the 'Fierz' identity

$$\begin{aligned} S_0^a S_0^b &= \frac{1}{2}\{S_0^a, S_0^b\} + \frac{1}{2}[S_0^a, S_0^b] \\ &= \frac{1}{2}\delta^{ab} + \frac{1}{16}S_0^c \gamma_{cd}^{ij} S_0^d \gamma_{ab}^{ij}, \end{aligned} \tag{5.3.4}$$

which can be checked by, for example, multiplying by γ_{ba}^{kl} and using the properties of the matrices $\gamma_{a\dot a}$ given in appendix 5.B. Notice that the only independent tensor that can be made out of two S_0's, other than δ^{ij}, is

R_0^{ij}, defined by

$$R_0^{ij} = \frac{1}{4} S_0^a \gamma_{ab}^{ij} S_0^b. \tag{5.3.5}$$

This is the zero-mode piece of iK_0^{ij}, defined in (5.2.45) and therefore satisfies the commutation relations

$$\left[R_0^{ij}, R_0^{kl} \right] = \delta^{il} R_0^{jk} - \delta^{ik} R_0^{jl} + \delta^{jk} R_0^{il} - \delta^{jl} R_0^{ik}. \tag{5.3.6}$$

Comparing this with the Lorentz algebra satisfied by K_0^{ij} in §4.3.1 we see that $-iR_0^{ij}$ is the operator that rotates the spin of a state but does not change its mode number. In particular, the massless vector state should transform as follows

$$R_0^{ij} |k\rangle = \delta^{jk} |i\rangle - \delta^{ik} |j\rangle, \tag{5.3.7}$$

and the massless spinor state should transform as

$$R_0^{ij} |\dot{a}\rangle = -\frac{1}{2} \gamma_{\dot{a}\dot{b}}^{ij} |\dot{b}\rangle. \tag{5.3.8}$$

The fact that the operator S_0^a maps the states $|i\rangle$ and $|\dot{a}\rangle$ into each other means that

$$\begin{aligned}
S_0^a |\dot{a}\rangle &= \frac{1}{\sqrt{2}} \gamma_{a\dot{a}}^i |i\rangle \\
S_0^a |i\rangle &= \frac{1}{\sqrt{2}} \gamma_{a\dot{a}}^i |\dot{a}\rangle .
\end{aligned} \tag{5.3.9}$$

The normalizations in these expressions are determined by applying S_0^b to both sides of the equations and using (5.3.4) together with (5.3.7) and (5.3.8).

It is convenient to attach physical wave functions to these states. In light-cone coordinates the sixteen real components of the Majorana-Weyl spinor are represented by $(u^a, u^{\dot{a}})$. As explained in appendix 5.B, the Dirac equation in this basis (in momentum space) takes the form

$$\begin{aligned}
k^+ u^a + \gamma_{a\dot{a}}^i k^i u^{\dot{a}} &= 0 \\
k^- u^{\dot{a}} + \gamma_{\dot{a}a}^i k^i u^a &= 0.
\end{aligned} \tag{5.3.10}$$

Since $k^+ = i\partial/\partial x^-$ is not a time derivative in the light-cone system of coordinates, the first equation expresses a constraint that determines u^a

in terms of $u^{\dot{a}}$

$$u^a = -\frac{1}{k^+}\gamma^i_{a\dot{a}}k^i u^{\dot{a}}. \tag{5.3.11}$$

The components $u^{\dot{a}}$ are therefore the eight physical degrees of freedom satisfying the equation that results from eliminating u^a, which is simply the Klein–Gordon equation $k^2 = 0$. The state $|u\rangle$ is defined by

$$|u\rangle = |\dot{a}\rangle u^{\dot{a}}(k)/\sqrt{k^+}. \tag{5.3.12}$$

The wave function for the vector state is the polarization vector $\zeta^\mu(k)$, which satisfies the physical conditions $k^2 = 0$ and

$$\zeta^\mu(k)k_\mu = 0. \tag{5.3.13}$$

Also, in the light-cone gauge

$$\zeta^+ = 0. \tag{5.3.14}$$

Equation (5.3.13) is a constraint equation that determines ζ^- in terms of the eight unconstrained components ζ^i

$$\zeta^- = \zeta^i(k)k^i/k^+. \tag{5.3.15}$$

This corresponds to the statement that only the transverse degrees of freedom of the Yang–Mills vector potential are independent degrees of freedom. The vector state $|\zeta\rangle$ is defined by

$$|\zeta\rangle = |i\rangle\,\zeta^i(k). \tag{5.3.16}$$

We know from appendix 4.A how the massless fields of supersymmetric Yang–Mills theory transform under supersymmetry transformations. The free-field (or linearized) limit of these transformations corresponds to the supersymmetry transformations of the physical wave functions. In considering the transformations in the light-cone gauge it is necessary to include compensating gauge transformations in order to preserve the condition $A^+ = 0$ in the transformed system. From appendix 4.A we learn that the linearized Yang-Mills supersymmetry transformations, including these compensating gauge transformations with gauge parameter $\Lambda(x)$, are

$$\delta A^\mu = \frac{i}{2}\bar{\epsilon}\Gamma^\mu\psi + \partial^\mu\Lambda \tag{5.3.17}$$

$$\delta\psi = -\frac{1}{4}F_{\mu\nu}\Gamma^{\mu\nu}\epsilon. \tag{5.3.18}$$

The condition $\delta A^+ = 0$ determines that

$$\Lambda = -\frac{1}{2p^+}\bar{\epsilon}\Gamma^+\psi. \tag{5.3.19}$$

As a result the supersymmetry transformations of the transverse components A^i are given by

$$\delta A^i = \frac{i}{2}\bar{\epsilon}\Gamma^i\psi - i\frac{p^i}{2p^+}\bar{\epsilon}\Gamma^+\psi, \tag{5.3.20}$$

while the transformation of ψ is unaltered at the linearized level. To translate these statements into the eight-component $SO(8)$ notation we have to split the 16 components of the Majorana-Weyl parameter ϵ into its $SO(8)$ pieces η^a and $\epsilon^{\dot{a}}$ corresponding to the projections

$$\frac{1}{2}\Gamma^+\Gamma^-\epsilon \sim \eta^a, \qquad \frac{1}{2}\Gamma^-\Gamma^+\epsilon \sim \epsilon^{\dot{a}}, \tag{5.3.21}$$

where $\sim$ means that the absolute normalization is arbitrary. Similarly, the spinor ψ decomposes into the $SO(8)$ pieces ψ^a and $\psi^{\dot{a}}$. Using the fact that

$$\bar{\epsilon}\Gamma^+ = \epsilon\Gamma^-\Gamma^+/\sqrt{2}, \qquad \bar{\epsilon}\Gamma^- = \epsilon\Gamma^+\Gamma^-/\sqrt{2}, \tag{5.3.22}$$

it is easy to pick out the different $SO(8)$ pieces of (5.3.20) and (5.3.18). The transformations of ζ and u are given by the identifications $\zeta \sim A$ and $u \sim \psi$ (making allowance for the fact that the spinor wave function u is made out of ordinary numbers while ψ consists of Grassmann numbers).

These transformations can also be deduced by considering how supersymmetry is realized on the states of the massless multiplet $|u\rangle$ and $|\zeta\rangle$. This entails using the formulas for the zero modes of the supercharges derived earlier

$$Q^a = (2p^+)^{1/2}S_0^a \tag{5.3.23}$$

$$Q^{\dot{a}} = (p^+)^{-1/2}\gamma_{a\dot{a}}^i p^i S_0^a. \tag{5.3.24}$$

One finds for the η transformations, defined in (5.2.25) and (5.2.26),

$$\eta^a Q^a \,|u\rangle = \eta^a(2k^+)^{1/2}S_0^a|\dot{a}\rangle u^{\dot{a}}(k)/\sqrt{k^+}$$

$$= \eta^a\gamma_{a\dot{a}}^i \,|i\rangle \, u^{\dot{a}}(k) = |\tilde{\zeta}\rangle, \tag{5.3.25}$$

and

$$\eta^a Q^a \, |\zeta\rangle = \eta^a (2k^+)^{1/2} S_0^a |i\rangle \zeta^i(k)$$

$$= \eta^a (k^+)^{1/2} \gamma^i_{a\dot{a}} |\dot{a}\rangle \zeta^i(k) = |\tilde{u}\rangle. \qquad (5.3.26)$$

In these expressions $\tilde{\zeta}$ and $\tilde{u}$ are the wave functions that result from η^a supersymmetry transformations on $u^{\dot{a}}$ and ζ^i, respectively. They are given by

$$\tilde{\zeta}^i = \gamma^i_{a\dot{a}} \eta^a u^{\dot{a}}(k) \qquad (5.3.27)$$

and

$$\tilde{u}^{\dot{a}} = \eta^a k^+ \gamma^i_{a\dot{a}} \zeta^i(k). \qquad (5.3.28)$$

These formulas are the same as those obtained above from the Yang–Mills field transformations. Similarly, one finds for the ϵ supersymmetries generated by $Q^{\dot{a}}$ that

$$\epsilon^{\dot{a}} Q^{\dot{a}} \, |\zeta\rangle = |\tilde{\tilde{u}}\rangle, \qquad (5.3.29)$$

and

$$\epsilon^{\dot{a}} Q^{\dot{a}} |u\rangle = |\tilde{\tilde{\zeta}}\rangle, \qquad (5.3.30)$$

where the transformed wave functions in this case are given by

$$\tilde{\tilde{u}}^{\dot{a}} = \frac{1}{\sqrt{2}} (\epsilon \gamma_{ij})^{\dot{a}} k^i \zeta^j + \frac{1}{\sqrt{2}} \epsilon^{\dot{a}} \zeta^i k^i \qquad (5.3.31)$$

and

$$\tilde{\tilde{\zeta}}^j = \frac{1}{\sqrt{2}} \epsilon^{\dot{a}} \gamma^j_{a\dot{a}} u^a + \frac{\sqrt{2}}{k^+} \epsilon^{\dot{a}} u^{\dot{a}} k^j. \qquad (5.3.32)$$

Again these agree with those obtained from Yang–Mills theory.

Now let us examine the massive superstring spectrum. The excited (massive) open-string states are most easily obtained by applying α^i_{-n} and S^a_{-n} excitations to the ground state $|\phi_0\rangle$. The number of fermions at every level made in this way is the same as obtained in the RNS formalism. Half of the states at each level are fermions, so their number is obviously the same as the number one gets by applying α^i_{-n} and d^i_{-n} excitations to an eight-component ground state. The number of bosons is manifestly the same as the number of fermions and therefore also agrees with the GSO projected spectrum. Now, however, we are guaranteed to have complete supersymmetry multiplets, since the construction of §5.2.2 provides an explicit realization of the algebra on the physical spectrum.

At the first excited level the physical states are

$$\alpha_{-1}^i |\phi_0\rangle, \qquad S_{-1}^a |\phi_0\rangle, \qquad (5.3.33)$$

describing a total of 128 bosonic and 128 fermionic modes. It is not hard to demonstrate these fit into spin(9) multiplets, as must happen for massive states. The spin(9) representations that arise are **44 + 84**, corresponding to Young tableaux $\square\square$ + $\begin{array}{c}\square\\\square\end{array}$ for the bosons, and a **128** 'spin 3/2' multiplet for the fermions. This is the same particle content as occurs in 11-dimensional supergravity. This is not too surprising since the representation theory of massive multiplets in D dimensions is the same as that of massless ones in $D + 1$ dimensions.

At the second excited level $(\alpha' M^2 = 2)$ the spectrum contains

$$\alpha_{-2}^i |\phi_0\rangle, \qquad S_{-2}^a |\phi_0\rangle, \qquad \alpha_{-1}^i \alpha_{-1}^j |\phi_0\rangle$$

$$S_{-1}^a S_{-1}^b |\phi_0\rangle, \qquad \alpha_{-1}^i S_{-1}^a |\phi_0\rangle, \qquad (5.3.34)$$

describing a total of 2304 modes. The content can be summarized most succinctly in terms of spin(9) multiplets in the form

$$9 \otimes (44 + 84 + 128). \qquad (5.3.35)$$

This is an irreducible massive $N = 1$ supermultiplet with spins ranging up to three (since **9** contains 'spin 1' and **44** contains 'spin 2'). Irreducible supersymmetry multiplets can be described quite generally as the tensor product of a fundamental supersymmetry multiplet with any irreducible spin representation. In the case at hand, $(44 + 84 + 128)$ is the fundamental massive supermultiplet. At the third excited level there are $15,360$ states forming two supermultiplets given by

$$(44 + 16) \otimes (44 + 84 + 128). \qquad (5.3.36)$$

Asymptotically, the total number of supermultiplets of mass M is an exponentially increasing function of M. This is evident since the total degeneracy increases exponentially, whereas the size of the largest supermultiplet occurring at the nth level only increases polynomially.

The asymptotic density of the states can be calculated by the same techniques employed in §2.3.5. In the present case the degeneracies d_n are given by

$$\sum_{n=0}^{\infty} d_n w^n = \mathrm{tr}\, w^N = 16 \prod_{n=1}^{\infty} \left(\frac{1 + w^n}{1 - w^n} \right)^8, \qquad (5.3.37)$$

where N is the sum of the number operators for the α_n^i modes and the S_n^a modes, (5.2.40), and the trace is over both sets of oscillators. (This is to be distinguished from the 'supertrace' in which there is an extra factor of $(-1)^F$ that gives a minus sign to the contributions of the fermions. The supertrace obviously vanishes in a supersymmetric theory.) The factor of $\prod(1-w^n)^{-8}$ comes from the bosonic trace in just the same way as for the bosonic string. For $n > 0$ the fermionic states that form the basis for S_n^a are either occupied or unoccupied giving a contribution of the form $(1 + w^n)$ for each mode. This results in the infinite product in the numerator. The factor of 16 is the degeneracy of the ground state. The asymptotic behavior of the density of states can be deduced by the same kind of technique that was used in §2.3.5. In this case the asymptotic behavior of (5.3.37) is determined by expanding $\exp\left(\sum \ln[(1-w^n)/(1+w^n)]\right)$ for $w \to 1$, which gives

$$16 \prod_{n=1}^{\infty} \left(\frac{1+w^n}{1-w^n}\right)^8 \sim \exp\left(\frac{2\pi^2}{1-w}\right). \tag{5.3.38}$$

A more precise estimate including the prefactor can be obtained from the generalization of the Hardy–Ramanujan formula,

$$\prod_{n=1}^{\infty} \left(\frac{1+w^n}{1-w^n}\right)^{-1} = \theta_4(0|w) = \left(-\frac{\ln w}{\pi}\right)^{-1/2} \theta_2(0|e^{\pi/\ln w}), \tag{5.3.39}$$

where θ_4 and θ_2 are Jacobi theta functions. The definitions of these functions and their transformations under $w \to \exp(\pi/\ln w)$ are explained in the context of loop-diagram calculations in appendix 8.A. By using the same contour argument as in §2.3.5 we conclude that

$$d_n \sim n^{-11/4} \exp\left(\pi\sqrt{8n}\right) \tag{5.3.40}$$

for $n \to \infty$, or

$$\rho(m) \sim m^{-9/2} \exp(m/m_0) \tag{5.3.41}$$

for $m \to \infty$ with

$$m_0 = \left(\pi\sqrt{8\alpha'}\right)^{-1}. \tag{5.3.42}$$

5.3.2 Closed Superstrings

The description of closed strings requires two sets of modes, one for right movers and one for left movers. In particular, the massless states are described by direct product states $|\phi_0\rangle \times |\tilde{\phi}_0\rangle$. There are two cases to be distinguished according to whether the original two Majorana–Weyl spinors have the same chirality or the opposite chirality. In the case where they are distinct no symmetrization of the the two factors is possible and the massless multiplet necessarily contains $16 \times 16 = 256$ modes. Its spin(8) content is given by the tensor product of two super Yang–Mills multiplets of opposite chirality

$$
\begin{aligned}
(\mathbf{8_v} + \mathbf{8_c}) \otimes (\mathbf{8_v} + \mathbf{8_s}) = {} & (1 + 28 + \mathbf{35_v} + \mathbf{8_v} + \mathbf{56_v})_B \\
& + (\mathbf{8_s} + \mathbf{8_c} + \mathbf{56_s} + \mathbf{56_c})_F,
\end{aligned}
\tag{5.3.43}
$$

where B labels bosonic states and F labels fermionic ones. This is the particle content of type IIA supergravity in $D = 10$. It is the same nonchiral multiplet as one obtains by a trivial dimensional truncation of $D = 11$ supergravity.

If both spinors belong to the the same chirality representation, and no other restrictions are imposed, there are again 256 modes. However, the spin(8) content of the massless multiplet is now given by the tensor product of two super Yang–Mills multiplets of the same chirality

$$
\begin{aligned}
(\mathbf{8_v} + \mathbf{8_c}) \otimes (\mathbf{8_v} + \mathbf{8_c}) = {} & (1 + 28 + \mathbf{35_v} + 1 + 28 + \mathbf{35_c})_B \\
& + (\mathbf{8_s} + \mathbf{8_s} + \mathbf{56_s} + \mathbf{56_s})_F.
\end{aligned}
\tag{5.3.44}
$$

This is the particle content of the chiral type IIB supergravity theory in $D = 10$. It cannot be obtained by dimensional reduction of a theory in a higher dimension. The $\mathbf{35_v}$ represents the graviton, whereas the $\mathbf{35_c}$ corresponds to a fourth-rank antisymmetric self-dual tensor.

When the two spinors are the same type, as in the last case discussed, it is possible to impose a symmetrization restriction keeping only terms that are invariant under interchange of $|\phi_0\rangle$ and $|\tilde{\phi}_0\rangle$. This corresponds to a graded symmetrization in the tensor product of super Yang–Mills multiplets

$$
\begin{aligned}
[(\mathbf{8_v} + \mathbf{8_c}) &\times (\mathbf{8_v} + \mathbf{8_c})]_{\text{graded sym}} \\
&= (\mathbf{8_v} \times \mathbf{8_v})_{\text{sym}} + (\mathbf{8_v} \times \mathbf{8_c}) + (\mathbf{8_c} \times \mathbf{8_c})_{\text{antisym}} \\
&= (1 + 28 + \mathbf{35_v})_B + (\mathbf{8_s} + \mathbf{56_s})_F.
\end{aligned}
\tag{5.3.45}
$$

This is the particle content of chiral type I supergravity in $D = 10$.

In constructing massive closed-string states it is necessary to have equal left-moving and right-moving excitations ($L_0 = \tilde{L}_0$). This is necessary in order that the functional field be independent of the choice of an origin for the σ parameter. It follows that the rule we found for massless levels is also true for massive ones. Namely, the closed-string states at the nth massive level are given by the tensor product of the open-string states at the nth massive level with themselves. This is true even for the type II theories, which do not admit open strings. (The reason they do not is that they have two gravitinos, which would have to couple to two supercurrents to make a consistent theory, but the open-string multiplets only have $N = 1$ supersymmetry.) Note that the massive open-string multiplets are nonchiral, and therefore the massive multiplets of the type IIA and type IIB theories are completely identical. The only difference in the spectrum of the two theories is in the massless sector. Thus the first excited level of either theory has $(256)^2$ states given by

$$(44 + 84 + 128) \otimes (44 + 84 + 128). \tag{5.3.46}$$

The component factors are $N = 1$ massive multiplets, and, as a result, the tensor product is an $N = 2$ massive multiplet. The two supersymmetries act on the two factors separately. In other words, one supercharge acts only on left-moving modes and the other acts only on right-moving modes. For type I closed strings a graded symmetrization is required, leaving $\frac{1}{2}(256)^2$ states. Such multiplets only have one supersymmetry, which is generated by the sum of the two supercharges of the $N = 2$ theory.

The asymptotic density of closed-string states can be inferred from that of the open-string ones. The density for type I and type II closed strings only differ by a factor of two, which is a finer distinction than we are keeping track of. If there are d_n open-string states with $\alpha' m^2 = n$, then there are $(d_n)^2$ closed-string states with $\alpha' m^2 = 4n$. Thus

$$d_n^{cl} = (d_n^{op})^2 \sim n^{-11/2} \exp\left(4\pi\sqrt{2n}\right) \tag{5.3.47}$$

for $n \to \infty$, which implies that

$$\rho^{cl}(m) \sim m^{-10} \exp(m/m_0) \tag{5.3.48}$$

for $m \to \infty$ with

$$m_0 = \left(\pi\sqrt{8\alpha'}\right)^{-1}. \tag{5.3.49}$$

For type I superstrings, the open- and closed-string sectors have the same critical temperature m_0. The open-string states are more numerous asymptotically, however, because of the power of m that multiplies the exponential.

5.4   Remarks Concerning Covariant Quantization

Covariant quantization of the supersymmetric superstring action encounters obstacles that did not occur in the analysis of chapters 2 and 4. The action gives rise to a complicated system of phase-space constraints relating the canonical momenta and the coordinates. However, there is a well-defined calculus for quantizing constrained Hamiltonian systems that was developed by Dirac that can be applied.

To apply Dirac's procedure it is necessary to distinguish two classes of constraints, called 'first class' and 'second class'. Constraints that form a closed algebra are called first-class constraints. The Virasoro and super-Virasoro constraints in the preceding chapters were of this type. We described in those chapters two ways of dealing with the constraints while preserving manifest covariance. In the old method, they are imposed as weak operator conditions on the string Fock space. In the modern approach, Faddeev–Popov ghost coordinates are introduced and a nilpotent BRST charge is identified. The techniques used in those constructions are generally valid for handling first-class constraints. Constraints that do not form a closed algebra are called second-class constraints. When such constraints arise, Dirac's procedure calls for replacing the Poisson brackets (or commutators) by modified expressions called Dirac brackets.

In the case of the superstring action with space-time supersymmetry there are second-class as well as first-class constraints. The occurrence of second-class constraints arises from the fact that the Grassmann momenta, P_θ^A, conjugate to the θ^A coordinates in (5.1.20) and (5.1.22) are not independent phase-space variables. This can be seen from the defining equations

$$P_\theta^A = \frac{\delta S}{\delta \partial_\tau \theta^A}, \tag{5.4.1}$$

which express P_θ^A as functions of X^μ, P^μ, θ^A and their σ derivatives. It can be shown, in fact, that half the fermionic constraints are first class, and, together with $T_{\alpha\beta}$, form an extension of the Virasoro algebra . The other half of the fermionic constraints are second class. If this fact is ignored, and as a formal procedure, one attempts to construct Dirac brackets, treating all the fermionic constraints as if they were second class, the resulting expressions are singular. They involve denominators that vanish when the equations of motion are satisfied. An interesting algebraic structure can be developed by ignoring the problem and proceeding anyway. However, it is unclear what meaning, if any, can be attributed to the formulas that result. It is probably preferable to do a careful separation of the first- and second-class constraints and define Dirac brackets for the latter ones by the standard formulas. The difficulty then is that it is impossible to achieve the desired separation of fermionic constraints in a manifestly covariant way. The combined constraint conditions have 16 components, corresponding to a Majorana–Weyl spinor. Since no $D = 10$ representation is eight-dimensional, it is necessary to reduce the manifest symmetry to the transverse spin(8) group, for example. Pursuing the analysis in this way does enable one to derive the light-cone-gauge of §5.2.1 in a reasonably rigorous manner. It does not help to achieve the objective of covariant quantization, however.

What should one conclude from all this? There are several different possible points of view, each of which seems to be preferred by some of the workers in this field. One is that one should accept the light-cone gauge as perfectly satisfactory and natural and not insist on manifest covariance. The loss of manifest covariance, however, does not seem promising for efforts to understand the conceptual underpinnings of string theory in a deeper way. A second position is that one should use the RNS formulation of superstrings, since it can be quantized covariantly. The trouble with this is that space-time supersymmetry is very difficult to describe in this approach, and it is just as fundamental a part of the super-Poincaré group as Lorentz invariance. There has been much recent progress in developing the RNS formalism, however.

The most ambitious hope would be that there is a clever modification or reinterpretation of the supersymmetric superstring action that allows quantization to be achieved while maintaining the full super-Poincaré group as a manifest symmetry. This might even provide a foundation for formulating a field theory of superstrings that has full superspace general coordinate invariance (as well as perhaps the infinite extension appropriate to a string theory). It might actually be easier to discover such an off-shell superfield formulation for superstrings than for $D = 10$ super Yang–Mills theory for which one is also lacking. At the very least, the super Yang–Mills theory probably requires an infinite number of auxiliary fields to achieve manifest supersymmetry (off-shell closure of the algebra). Such a prospect does not sound unnatural in the string context. Some proposals have been made, but it is still too early to assess how fruitful they are likely to be.

5.5 Summary

A covariant superstring action that has manifest space-time supersymmetry has been presented. This action has a peculiar local fermionic symmetry in addition to reparametrization invariance. Covariant gauge fixing gives a complicated system of first-class and second-class constraints that cannot be disentangled while preserving manifest covariance. Therefore, unless some suitable modification of the formulas can be found, covariant quantization appears to be impossible in this formulation.

In the light-cone gauge the formulas for the Fermi coordinates collapse to those of a free two-dimensional spinor field. They belong to a spinor representation 8_s of the transverse rotation group spin(8), in contrast to the case in chapter 4 where the Fermi coordinates in the light-cone gauge belonged to the vector representation 8_V of spin(8). The equivalence of the two formulations was argued by a bosonization and re-fermionization. The correspondence is not trivial inasmuch as the algebra of the 8_s spinors automatically requires $D = 10$ supersymmetry multiplets. To obtain the same spectrum from the 8_V spinors it was necessary to use different boundary conditions for bosons and fermions and to restrict the spectrum in each sector to states of even world-sheet fermion number.

The unification of the bosonic and fermionic sectors and the automatic incorporation of the GSO conditions achieved in the approach of this chapter is very helpful for simplifying many calculations. It will be exploited extensively in the subsequent chapters.

Appendix 5.A   Properties of $SO(2n)$ Groups

This chapter depends heavily on a number of group-theoretic facts about $SO(8)$. Subsequent chapters require properties of $SO(16)$ and $SO(32)$, as well. Therefore, in this appendix we assemble a number of relevant facts about $SO(2n)$ groups, in general. When speaking about these groups it is often desirable to make a distinction between the group of rotations in $2n$ dimensions, which is $SO(2n)$, and various possible 'covering groups'.

Recall that $SO(3)$ has a covering group, $SU(2)$, which has representations (half-integral spin) in which a rotation by 2π is represented by -1. This means that $SO(3)$ can be identified with the quotient space $SU(2)/Z_2$ in which the elements of the center of $SU(2)$, namely the 2×2 matrices 1 and -1, are identified. Topologically, $SU(2)$ is simply connected and $SO(3)$ is not. Also, the representations of $SU(2)$ fall into two conjugacy classes (integer spin and half-integer spin). Only integer spin representations are representations of $SO(3)$.

For the $SO(2n)$ groups the story is similar but not identical. In this case, the simply connected covering group, spin($2n$), has a four element center, and (as we will show) its representations fall into four conjugacy classes. These groups are also called D_n in the Cartan classification. Of course, the distinction between the various covering groups is a global topological distinction. Any of the versions is the same in the neighborhood of the identity element, so there is no distinction at the level of the Lie algebras.

The group $SO(2n)$, or its various covering groups, have rank n, meaning that they have a maximal subset of n commuting generators. If we denote as J_{kl} the $SO(2n)$ generator whose nonzero elements are $(J_{kl})_{mp} = \delta_{km}\delta_{lp} - \delta_{kp}\delta_{lm}$, then a natural choice to make for the n commuting generators is given by $W_k = J_{2k-1,2k}$, $k = 1, \ldots, n$. W_k carries out a rotation in the $(2k-1, 2k)$ plane. A maximal set of commuting generators is known as a Cartan subalgebra, so the W_k form a Cartan subalgebra of $SO(2n)$. Since these commuting operators can be simultaneously diagonalized, it is natural in any irreducible representation to choose a basis of states that are simultaneously eigenstates of all the W_k. If we think of the W_k as defining directions in an n-dimensional vector space called the weight space, then the eigenvalues of the W_k for a given state form a vector in this space called a weight vector. A convenient way to characterize a representation is often to describe the weights that arise in that representation. This is familiar to particle physicists in the case of $SU(3)$ where the third component of isotopic spin I_3 and the hypercharge Y are customarily chosen to label weight vectors.

If we denote an n component vector $(0, \ldots, 0, 1, 0, \ldots, 0)$ with the 1 in the ith position as u_i, then the $2n$ weights of the fundamental representation are given by $\pm u_i$, with $i = 1, 2, \ldots, n$, as the reader should be able to easily verify. Actually, the u_i give a basis for the weight space, so the weights of any $SO(2n)$ representation can be expanded in terms of the u_i. For example, the adjoint representation has $n(2n-1)$ dimensions. The $2n(n-1)$ generators that are not in the Cartan subalgebra are represented by the weights $\pm u_i \pm u_j$ $(i \neq j)$, as the reader should also verify. Weight vectors of the adjoint representation are called root vectors. For future reference, note that the weights $\pm u_i$ of the fundamental representation are vectors of length squared one, while the nonzero weights of the adjoint representation are vectors $\pm u_i \pm u_j$ of length squared two. Among the roots one can select n 'simple, positive' roots that form a basis in which the others can be expressed as linear combinations with integer coefficients, which are all non-negative or all non-positive. In the case of $SO(2N)$ the simple, positive roots are $e_1 = u_1 - u_2$, $e_2 = u_2 - u_3$, $\ldots, e_{n-1} = u_{n-1} - u_n$, and $e_n = u_{n-1} + u_n$.

The $SO(2n)$ groups are examples of groups that are 'simply laced'. This means that all of the root vectors have the same length. In such cases it is convenient to normalize things as we have done, so that the length squared of a root vector is 2. One then defines the Cartan matrix, which

is the $n \times n$ matrix

$$A_{ij} = \frac{2e_i \cdot e_j}{e_i \cdot e_i} = e_i \cdot e_j, \qquad (5.\text{A}.1)$$

where the e_i are the simple, positive roots. The diagonal elements of the Cartan matrix are 2. In the case of simply laced Lie algebras, the off-diagonal elements are all 0 or -1.

Figure 5.1. The Dynkin diagram for $SO(2n)$ is sketched in (*a*). That of $SO(8)$, sketched in (*b*), is symmetric under permutations of the three legs.

One then defines a 'Dynkin diagram', which consists of n dots representing the simple roots with connections that represent the angles between them. The rules for constructing the Dynkin diagram are particularly simple in the simply laced case with which we will deal. Dots in the Dynkin diagram that represent orthogonal roots e_i and e_j are not connected. Dots that represent roots e_i and e_j with $e_i \cdot e_j = -1$ are connected by a single line. In the case of $SO(2n)$, a choice of simple positive roots was described earlier, and is easily seen to give the Dynkin diagram shown in fig. 5.1*a*. The Dynkin diagram of $SO(2n)$ has for any n a two-fold symmetry under reflections through the horizontal plane in fig. 5.1*a*, but in the case of $SO(8)$ the symmetry is larger. The $SO(8)$ Dynkin diagram in fig. 5.1*b* is invariant under a six-element group consisting of arbitrary permutations of the three legs. It is a general fact in the theory of simple Lie groups that a Lie algebra can by uniquely reconstructed from its Dynkin diagram, and consequently automorphisms (symmetries) of the Dynkin diagram always give rise to automorphisms of the Lie algebra.[*] In the case of $SO(2n)$ the two-fold symmetry of the Dynkin diagram corresponds to 'parity', a reflection $x_k \to -x_k$ of any one element of the fundamental $2n$-dimensional representation of $SO(2n)$. Among other things, this symmetry exchanges the positive- and negative-chirality representations of $SO(2n)$, which we will be describing shortly. The larger symmetry of the $SO(8)$ Dynkin diagram means that in this case there are

---

[*] They are, in fact, the so-called outer automorphisms, which do not correspond to conjugation by an element of the Lie algebra.

extra automorphisms that exchange representations that would not be related by any symmetries for other $SO(2n)$ groups. In fact, it turns out that the extra symmetry, which is known as 'triality', relates the spinor representations of $SO(8)$ to the vector representations. We will try to give a self-contained (but not complete or rigorous) account of this that does not require knowledge of Dynkin diagrams.

Let us first give a direct construction of the spinor representation of $SO(2n)$. We begin with a system whose symmetry is $U(n)$ only. We introduce n fermion creation operators $b_i^*, i = 1, \ldots, n$ and the corresponding annihilation operators $b^j, j = 1, \ldots, n$. The anticommutation relations are the usual ones,

$$\{b^j, b^k\} = \{b_j^*, b_k^*\} = 0; \qquad \{b^i, b_j^*\} = \delta_j^i. \tag{5.A.2}$$

Of course, this system can be represented in a Hilbert space of dimension 2^n. There is a 'Fock vacuum' $|\Omega\rangle$ annihilated by the annihilation operators; the other states are $|\Omega_j\rangle = b_j^* |\Omega\rangle$, $|\Omega_{j_1 j_2}\rangle = b_{j_1}^* b_{j_2}^* |\Omega\rangle$, etc. In general we denote by $|\Omega_{j_1 j_2 \ldots j_k}\rangle$ the state obtained by acting on the Fock vacuum by the product of k creation operators $b_{j_1}^* \ldots b_{j_k}^*$. This system manifestly possesses a $U(n)$ symmetry under which (say) b^j transforms as n and b_k^* as $\bar{n}$. The $U(n)$ generators are $[b^i, b_j^*]/2$.

Less obvious is the fact that this system actually possesses a natural $SO(2n)$ symmetry (or actually spin($2n$), since the $SO(2n)$ quantum numbers turn out to be half integral). In fact, define the 'gamma matrices' $\gamma_k, \ k = 1, \ldots, 2n$ by

$$\begin{aligned} \gamma_k &= (b^k + b_k^*), & k &= 1, \ldots, n; \\ \gamma_k &= (b^{k-n} - b_{k-n}^*)/i & k &= n+1, \ldots, 2n. \end{aligned} \tag{5.A.3}$$

They are readily seen to obey the Dirac anticommutation relations

$$\{\gamma_k, \gamma_l\} = 2\delta_{kl} \tag{5.A.4}$$

and as a result the expressions

$$\sigma_{kl} = [\gamma_k, \gamma_l]/4 \tag{5.A.5}$$

obey the commutation relations

$$[\sigma_{kl}, \sigma_{jm}] = \delta_{lj}\sigma_{km} \pm \text{permutations} \tag{5.A.6}$$

of $SO(2n)$. Thus we have constructed a 2^n-dimensional representation of $SO(2n)$, known as the spinor representation. It is not quite irreducible, however. The operator

$$\overline{\gamma} = i^{n(2n-1)}\gamma_1\gamma_2\cdots\gamma_{2n} \tag{5.A.7}$$

commutes with all of the $SO(2n)$ generators; here the phase factor $i^{n(2n-1)}$ has been included so that $\overline{\gamma}^2 = 1$. Spinor states of $\overline{\gamma} = +1$ or -1 are known as spinors of positive or negative chirality, respectively. Simple manipulations with fermion creation and annihilation operators show that these representations are irreducible, so that $SO(2n)$ has two irreducible spinor representations $S_\pm$ each of dimension 2^{n-1}, namely the spinors of positive and negative chirality. Moreover, the fermion creation operators b_k^* anticommute with $\overline{\gamma}$ and so reverse chirality. If thus $|\Omega\rangle$ has positive chirality then the positive chirality spinors of $SO(2n)$ are states made by acting on $|\Omega\rangle$ with an even number of creation operators

$$S_+ = \oplus_{k \; even} |\Omega_{j_i\ldots j_k}\rangle \tag{5.A.8}$$

while the negative chirality spinors of $SO(2n)$ are those obtained by acting on the Fock vacuum with an odd number of chirality operators.

Now let us determine the weights of the spinor representation. The Cartan generators can be taken as the n commuting operators

$$W_k = [b^k, b_k^*]/2 \quad k = 1,\ldots,n, \tag{5.A.9}$$

The eigenvalues of each W_k are evidently $\pm 1/2$. The 2^n-dimensional vector space in which we constructed the spinor representation has weights $\frac{1}{2}(\pm 1, \pm 1, \ldots, \pm 1)$ corresponding to independent choices of whether the kth fermion level is filled or empty. The description of the chirality operator makes it clear that the weights of one irreducible spinor representation of definite chirality are $\frac{1}{2}(\pm 1, \pm 1, \ldots, \pm 1)$ with an even number of $+$ signs while the other has weights with an odd number of $+$ signs. The length squared of any of these weight vectors is $n/4$. At this point we notice something very special about the case $n = 4$, corresponding to spin(8). Precisely in this case, the spinor weights have length squared one, just like the vector weights $\pm u_i$, so it is only in this case that there might be a symmetry between the vector and spinor representations of spin$(2n)$. This is indeed the case in which the Dynkin diagram has an enlarged symmetry corresponding to a symmetry between vectors and spinors. The transformation of weight space that turns vectors into spinors can be written explicitly; it is

$$u_1 \rightarrow (u_1 + u_2 + u_3 + u_4)/2$$
$$u_2 \rightarrow (u_1 + u_2 - u_3 - u_4)/2$$
$$u_3 \rightarrow (u_1 - u_2 + u_3 - u_4)/2 \qquad (5.A.10)$$
$$u_4 \rightarrow (u_1 - u_2 - u_3 + u_4)/2.$$

This transformation of the weights is indeed an automorphism of the $SO(8)$ Lie algebra, because it is an orthogonal rotation of the weight space that preserves lengths and angles, and it permutes the 24 nonzero weights $\pm u_i \pm u_j$ of the adjoint representation among themselves. Moreover, it is essentially the same transformation that we used in §5.2.1 in bosonizing and re-fermionizing in order to indicate the relation between space-time vector and space-time spinor fermions. Together with 'parity,' which is an automorphism of any $SO(2n)$, the transformation (5.A.10) generates the six-element 'triality' symmetry.

Let us now very briefly explain what conjugacy classes of representations are and why spin($2n$) has four of them. We have seen that the states in an irreducible representation can be described by their weights, which are just a set of points in an n-dimensional weight space. Consider the set of all possible weights of all possible representations. They form a lattice, which is called the weight lattice and denoted Λ_W. Now consider the lattice consisting of integral linear combinations of the weights of the adjoint representation. This lattice, known as the root lattice Λ_R, is clearly a sublattice of the weight lattice Λ_W.

Two representations are said to belong to the same conjugacy class if and only if their weight vectors differ by a vector in the root lattice. In the case of spin($2n$) the four representations we have described (fundamental, adjoint, and two spinor ones) belong to different conjugacy classes. This is easily demonstrated by examining the weight vectors given above. To show that there are no other conjugacy classes requires a somewhat more sophisticated analysis that we can only sketch here. For this purpose we invoke a theorem from the theory of Lie algebras. The theorem states that when the simple roots (of a simply laced algebra) are normalized to satisfy $e_i \cdot e_i = 2$, as we have done, the weight lattice is the dual of the root lattice. This means that it is generated by a set of basis vectors e_i^*, $i = 1, \ldots, n$, that satisfy $e_i^* \cdot e_j = \delta_{ij}$. The number of conjugacy classes must then be given by the ratio of the volume of a unit cell of the root lattice to one of the weight lattice, since two weights inside the same cell of the root lattice obviously cannot differ by a root vector. Now all we need to do is to calculate the two volumes. It is easy to see that the determinant of the Cartan matrix $A_{ij} = e_i \cdot e_j$ gives the square of the volume of a unit cell of the root lattice Λ_R. Similarly, its inverse gives the square of the volume of a unit cell of $\Lambda_R^* = \Lambda_W$. Thus $\det A_{ij}$ gives the ratio of the two volumes and hence the total number of conjugacy classes. Using the explicit expression for the simple roots of $SO(2n)$ given above it is not difficult to show that $\det A = 4$ for $SO(2n)$. Therefore the four representations we have described are representatives of all four of the conjugacy classes. Let us simply remark without further explanation that the only other simply laced groups are $SU(n)$, which has n conjugacy classes, and the exceptional groups E_n, which have $9 - n$ conjugacy classes.

Appendix 5.B   The spin(8) Clifford Algebra

The description of Dirac matrices for spin(8) requires a Clifford algebra with eight anticommuting matrices. Because of their importance, and the fact that they are useful building blocks for ten-dimensional Dirac matrices, we describe them explicitly in this appendix.

The Dirac algebra of spin(8) requires 16-dimensional matrices corresponding to the reducible $\mathbf{8_s} + \mathbf{8_c}$ representation of spin(8). These matrices can be written in the block form

$$\gamma^i = \begin{pmatrix} 0 & \gamma^i_{a\dot{a}} \\ \gamma^i_{\dot{b}b} & 0 \end{pmatrix}, \tag{5.B.1}$$

where $\gamma^i_{\dot{a}a}$ is the transpose of $\gamma^i_{a\dot{a}}$. The equations $\{\gamma^i, \gamma^j\} = 2\delta^{ij}$ are satisfied if

$$\gamma^i_{a\dot{a}}\gamma^j_{\dot{a}b} + \gamma^j_{a\dot{a}}\gamma^i_{\dot{a}b} = 2\delta^{ij}\delta_{ab}, \quad i,j = 1\ldots 8, \tag{5.B.2}$$

and similarly with dotted and undotted indices interchanged. A specific set of matrices $\gamma^i_{a\dot{a}}$ that satisfy these equations, expressed as direct products of 2×2 blocks, are

$$
\begin{array}{ll}
\gamma^1 = \epsilon \times \epsilon \times \epsilon & \gamma^2 = 1 \times \tau_1 \times \epsilon \\
\gamma^3 = 1 \times \tau_3 \times \epsilon & \gamma^4 = \tau_1 \times \epsilon \times 1 \\
\gamma^5 = \tau_3 \times \epsilon \times 1 & \gamma^6 = \epsilon \times 1 \times \tau_1 \\
\gamma^7 = \epsilon \times 1 \times \tau_3 & \gamma^8 = 1 \times 1 \times 1,
\end{array}
\tag{5.B.3}
$$

where $\epsilon = i\tau_2$, and τ_i are the Pauli matrices. We define

$$\gamma^{ij}_{ab} = \tfrac{1}{2}(\gamma^i_{a\dot{a}}\gamma^j_{\dot{a}b} - \gamma^j_{a\dot{a}}\gamma^i_{\dot{a}b}) \tag{5.B.4}$$

and similarly for $\gamma^{ij}_{\dot{a}\dot{b}}$.

The ten-dimensional Majorana–Weyl spinors described in chapter 4 can also be understood in terms of the matrices described here. Recall that Majorana spinors in $D = 10$ have 32 real components (in a Majorana representation), but that the Weyl condition $\Gamma_{11}\lambda = \lambda$ eliminates half of them. In terms of the transverse $SO(8)$ subgroup of $SO(9,1)$, the surviving 16 components are given by $\mathbf{8_s} + \mathbf{8_c}$. Therefore the Dirac matrices can be re-expressed in this basis in terms of the γ^i of (5.B.1)– at least the eight transverse ones can. A ninth one that anticommutes with these eight is given by $\gamma^9 = \gamma^1\gamma^2 \dots \gamma^8 = \begin{pmatrix} 1 & 0 \\ 0 & -1 \end{pmatrix}$. There is no tenth anticommuting 16×16 matrix, however. This makes sense because in the subspace $\Gamma_{11} = 1$ that we are considering the product of the ten Dirac matrices should be one. The upshot of this is that the Dirac equation of a Majorana–Weyl spinor $(\lambda_s^a, \lambda_c^{\dot{a}})$ decomposes into the pair of eight-component equations

$$
\begin{aligned}
(\partial_0 + \partial_9)\lambda_s^a + \gamma_{a\dot{a}}^i \partial_i \lambda_c^{\dot{a}} = 0 \\
(\partial_0 - \partial_9)\lambda_c^{\dot{a}} + \gamma_{\dot{a}a}^i \partial_i \lambda_s^a = 0.
\end{aligned}
\tag{5.B.5}
$$

For a Majorana–Weyl spinor of the opposite chirality the $\partial_0 \pm \partial_9$ would be interchanged.

6. Nonabelian Gauge Symmetry

If superstrings are to describe nature they must account not only for general coordinate invariance and local supersymmetry, but also for the local gauge symmetries that underly the other forces. Indeed, nonabelian gauge symmetry is more obviously needed than local supersymmetry! One possibility is that the gauge symmetries are not present at all in the ten-dimensional world, but arise only upon reduction to four dimensions. This idea, which seems to be forced upon us if we try to describe nature with type II superstrings, proves to have enormous difficulties. We will discuss these issues to some extent in chapter 14.

A more promising possibility is that gauge symmetries are present already in the ten-dimensional world. In this case, compactification from ten to four dimensions may play a role as part of an initial stage of symmetry breaking. In this chapter we investigate how gauge symmetries can be introduced in $D = 10$, and defer the study of compactification and symmetry breaking to later chapters.

Two entirely different procedures for incorporating gauge interactions are known. In the first one, which was already briefly described in §1.5.3, internal symmetry charges are placed at the ends of open strings. In the second the charges are distributed on closed strings. The second procedure leads to heterotic strings, which seem to have many advantages. Indeed, if one of the presently known superstring theories turns out to be the correct theory, it is almost bound to be the $E_8 \times E_8$ heterotic theory. We discuss in this chapter both approaches to incorporating gauge interactions, but with an emphasis on the approach that leads to heterotic strings.

6.1   Open Strings

We have already discussed in §1.5.3 how nonabelian symmetries can be introduced by attaching charges to the ends of open-string states. This issue was discussed there because of its historic role in the hadronic interpretation of the theory. We now return to the subject and clarify some points that were left open in §1.5.3.

Attaching charges to open strings is of course only relevant in string theories that do have such open strings. Of the supersymmetric string theories, open strings are present only in the type I theory. The type II theory, since it has $N = 2$ supersymmetry, cannot be coupled to open strings, which can at most have $N = 1$ supersymmetry. Open strings are not possible in the heterotic theory – which is introduced in this chapter – for a different although related reason. Even in type I theory, including gauge quantum numbers by attaching charges at the end of open strings means that the closed strings are neutral.

6.1.1 The Chan–Paton Method

Figure 6.1. An open string with a 'quark' at one end and an 'antiquark', transforming oppositely under the symmetry group, at the other end. If the quarks lie in a complex representation, we draw as in (a) an arrow from the 'quark' to the 'antiquark', depicting the intrinsic orientation of the string; if the representation is real, there is no arrow as in (b).

A technique for introducing $U(2)$ or $U(3)$ gauge symmetry in the open-string sector of the bosonic string theory was proposed by Chan and Paton in 1969. They were merely trying to incorporate global symmetries in the model, but in fact the occurrence in the open-string spectrum of massless gauge mesons (something not yet properly understood at that time) turned out to mean that in consistently formulated string theories the quantum numbers that they introduced are actually gauge quantum numbers. A generalization of the Chan-Paton work to $U(n)$ symmetry was immediately apparent; other classical groups – $SO(n)$ and $USp(n)$ – are possible if one considers unoriented open strings rather than the oriented ones that were emphasized for phenomenological reasons in the early days of dual models. For reasons that were explained in chapter 4, type I superstrings *must* be unoriented, so the generalization is quite relevant.

We begin by considering any semisimple gauge group G and an arbitrary representation R, which may or may not be complex ($R \neq \overline{R}$ or $R = \overline{R}$). Let n be the dimension of the representation R. As in fig. 6.1, we suppose that a 'quark' transforming in the R representation resides at one end of the open string, and an 'antiquark' transforming as $\overline{R}$ resides at the other

end.[*] The quark and antiquark each have n possible states, so there are
a total of n^2 possibilities. If R is complex this construction introduces an
intrinsic orientation as depicted in fig. 6.1a, whereas if R is real the string
can be unoriented as in fig. 6.1b.

Let us discuss the quantum numbers of the various mass levels of an
open string that has such charges attached at the ends. We have seen in
chapter 2 that $X^i(\sigma, \tau)$ has a mode expansion

$$X^i(\sigma, \tau) = x^i + p^i \tau + i \sum_{n \neq 0} \frac{1}{n} \alpha_n^i e^{-in\tau} \cos n\sigma. \tag{6.1.1}$$

The parameter σ runs from $\sigma = 0$ to $\sigma = \pi$ along the length of the string.
Suppose we wish to interchange the names of the two ends and run the
parametrization in the opposite direction. Then we would consider instead

$$X^i(\pi - \sigma, \tau) = x^i + p^i \tau + i \sum_{n \neq 0} \frac{1}{n} (-1)^n \alpha_n^i e^{-in\tau} \cos n\sigma. \tag{6.1.2}$$

Therefore the reversal of the parametrization corresponds to the replace-
ment $\alpha_n^i \to (-1)^n \alpha_n^i$. (Similar statements can be made for the Fermi
coordinates of superstrings in either the formulation of chapter 4 or that
of chapter 5.) In the absence of Chan–Paton factors a string state $|\Lambda\rangle$
is simply a state in the oscillator Hilbert space. Now suppose there is a
quark a at one end of the string and an antiquark $\bar{b}$ at the other end,
where a and $\bar{b}$ are n-valued labels associated with the representations R
and $\bar{R}$ of G. Then the state $|\Lambda\rangle$ should also carry these labels, so let us
write $|\Lambda; a, \bar{b}\rangle$, the first label referring to the $\sigma = 0$ end of the string and
the second one to the $\sigma = \pi$ end of the string.

The question now arises whether the spectrum should be allowed to
contain all n^2 states $|\Lambda; a, \bar{b}\rangle$ or whether some restriction should be im-
posed. One handle on this question is provided by the massless vector
particles. Massless vector mesons in consistent interacting theories always transform in the adjoint representation of the gauge group, which
must be a compact semisimple Lie group plus possible additional $U(1)$
factors. In fact, in the next chapter we will see that the three-vector couplings of the open-string theories are just those of Yang–Mills theories.
(This statement is exact for type I superstrings but only true to $O(\alpha')$ in

---

[*] By 'quarks' and 'antiquarks' we only mean labels representing certain quantum
numbers. To attach mass or spin at the ends of open strings would ruin the
analysis described in the previous chapters.

the case of the bosonic string theory, as we will see.) Thus we conclude that when $|\Lambda; a, \bar{b}\rangle$ describes a massless vector, the quantum numbers $a\bar{b}$ must be restricted to run over the adjoint representation only. In case a and $\bar{b}$ run over the n and $\bar{n}$ of $U(n)$, this is automatically obeyed, since in $U(n)$, $n \times \bar{n}$ is the adjoint representation. However, for other cases of interest, the product $R \times \bar{R}$ contains more than the adjoint representation of G, and we may wish to impose some restriction. Only one suitable restriction is known. In case the representations R and $\bar{R}$ are equivalent, there is no real difference between the two ends of the string, and it makes sense to require that the quantum wave function of the string is invariant under the reversal $\sigma \to \pi - \sigma$. This corresponds to saying that

$$|\Lambda(\alpha_n^i); b, a\rangle = \epsilon|\Lambda((-1)^n \alpha_n^i); a, b\rangle, \qquad (6.1.3)$$

where we may pick the phase ϵ to be ± 1, and we have used the information gained above that under $\sigma \to \pi - \sigma$, the transformation of the oscillators is $\alpha_n^i \to (-1)^n \alpha_n^i$. In (6.1.3), we write ab rather than $a\bar{b}$ for the labels at the ends of the strings, since the twisting symmetry postulated there only makes sense if the charges at the two ends of the string are indeed equivalent. Strings subject to the condition (6.1.3) are known as unoriented strings. Recalling the expression for the number operator N in terms of oscillators, (6.1.3) can be written alternatively as

$$|\Lambda; ab\rangle = \pm(-1)^{N-N_0} |\Lambda; ba\rangle, \qquad (6.1.4)$$

where N_0 is the eigenvalue of N for massless states. We are using here the fact that the oscillator α_n^i has $N = n$.

Consider as an example $SO(n)$ with R and $\bar{R}$ both equal to the fundamental n-dimensional representation. In this case the adjoint corresponds to antisymmetric matrices, and so the massless vector must satisfy (6.1.3) with the minus sign. For bosonic strings, $(-1)^N$ is -1 for the massless vector, so we must pick $\epsilon = 1$; for superstrings $(-1)^N$ is $+1$ for the massless vector and we must pick $\epsilon = -1$. Either way, if we refer to states in which $\alpha' M^2$ is even or odd as even or odd levels, respectively, then for the even levels the string states transform under $SO(N)$ like an antisymmetric $n \times n$ matrix, and for the odd levels they transform like a symmetric $n \times n$ matrix. Another interesting case is that in which the gauge group is $USp(n)$ and R and $\bar{R}$ are the fundamental representation of $USp(n)$. In this case, the adjoint representation is the *symmetric* part of $R \times R$, so we must pick $\epsilon = -1$ for bosonic strings and $\epsilon = +1$ in the supersymmetric case. The quantum numbers of even and odd mass levels are now, respectively, the symmetric and antisymmetric part of $R \times R$.

The examples that have just been sketched seem to be the only ones that make sense, for the following reason. As we have already said, in order to obtain a consistent gauge theory it is necessary that the massless vector particles belong to the adjoint representation of G only. Let us denote antihermitian $n \times n$ matrices representing the algebra by $\lambda^i_{a\bar{b}}$, where $i = 1, 2, \ldots, \dim G$. In general this is a subset of all possible antihermitian matrices. All of the $n \times n$ antihermitian matrices taken together would represent $U(n)$ in the complex case or $SO(n)$ in the real case. If we wish the gauge group to be, say, a proper subgroup of $U(n)$, then the λ^i do not range over all $n \times n$ antihermitian matrices. So the question arises whether additional restrictions can be consistently imposed. There is a potential clash here with unitarity. In arbitrary M-particle tree amplitudes, massless poles appear; we must ask whether these massless poles appear only for states that transform in the adjoint representation of G. In ordinary Yang–Mills theories this is always true provided only that the commutator $[\lambda^i, \lambda^j]$ is again one of the λ^k. This is always true regardless of the choice of G or R simply because the λ's represent the algebra. In the case of type I string theories there is an additional condition that provides restrictions on both G and R.

In order to explain how restrictions arise, let us recall some general features of M-particle open-string tree amplitudes from chapter 1. Let $A(1, 2, \ldots, M)$ be the M-particle scattering diagram for open strings that do *not* carry gauge quantum numbers. The A amplitudes have cyclic symmetry in the M lines, with possible minus signs in certain cases. If the group quantum numbers of the particles are described by matrices $\lambda_1, \lambda_2, \ldots, \lambda_M$, then the complete tree amplitude is given by

$$T(1, 2, \ldots, M) = \sum \operatorname{tr}(\lambda_1 \lambda_2 \ldots \lambda_M) A(1, 2, \ldots, M), \qquad (6.1.5)$$

where the sum runs over all $(M-1)!$ cyclically inequivalent permutations of the external lines. We do not claim that the structure (6.1.5) should be taken on faith, it is simply the only more or less consistent scheme that anyone has been able to find.

The primitive amplitude $A(1, \ldots, M)$ has poles in channels composed of cyclically consecutive lines and in no other channels. The residues of these poles have certain simple factorization properties that can be symbolically represented by saying that in the limit as $s \to m^2$

$$A(1, \ldots, M) \sim \frac{1}{m^2 - s} \sum_X A(1, 2, \ldots, P, X) A(X, P+1, \ldots, M),$$

$$(6.1.6)$$

where $s = -(k_1 + k_2 + \cdots + k_P)^2$, and X runs over all states in the spectrum

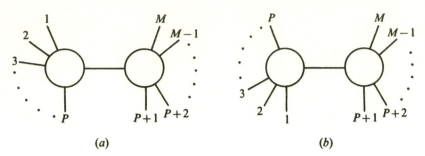

Figure 6.2. The primitive amplitude $A(1,\ldots,M)$ has a pole in the invariant energy $s = -(k_1 + k_2 + \cdots k_P)^2$, whose residue factorizes into a product of primitive amplitudes as shown in (a). Another primitive amplitude that has a pole with the same residue is depicted in (b).

with mass m. This is shown in fig. 6.2a. We have not yet proved such behavior; it is a problem that we will turn to in the next chapter. But the factorization of (6.1.6) is a requirement of unitarity, so it must hold if the theory is to make sense. To show that unitarity (if obeyed in the absence of group theory factors) is still present when those factors are included, we must show that the factorization as in (6.1.6) still holds with the λ matrices taken into account. This gives restrictions on the λ matrices.

There are additional terms in $T(1, 2, \ldots, M)$ that give additional contributions to the residue of the particular pole X of the form given in (6.1.6). Such a term arises from the primitive amplitude $A(P, P - 1, \ldots, 1, P + 1, \ldots, M)$ as depicted in fig. 6.2b. It is also possible to reverse the order of all M lines. Using (6.1.5), we see that the sum of the residues of the X pole in T is given by

$$\mathrm{tr}(\lambda_1 \ldots \lambda_P \lambda_{P+1} \ldots \lambda_M) A(1, \ldots, P, X) A(X, P + 1, \ldots, M)$$
$$+\mathrm{tr}(\lambda_P \ldots \lambda_1 \lambda_{P+1} \ldots \lambda_M) A(P, \ldots, 1, X) A(X, P + 1, \ldots, M)$$
$$+\mathrm{tr}(\lambda_1 \ldots \lambda_P \lambda_M \ldots \lambda_{P+1}) A(1, \ldots, P, X) A(X, M, \ldots, P + 1)$$
$$+\mathrm{tr}(\lambda_P \ldots \lambda_1 \lambda_M \ldots \lambda_{P+1}) A(P, \ldots, 1, X) A(X, M, \ldots, P + 1). \quad (6.1.7)$$

In order to relate these four terms and deduce restrictions on the λ's we need to relate the various A factors that appear. Specifically, we require a rule relating an expression of the form $A(1, 2, \ldots, N)$ to one in the reverse order $A(N, \ldots, 2, 1)$. They correspond to the same world-sheet diagrams just looked at from opposite sides, so the resulting amplitudes are the same up to an adjustable sign that depends on the choice of ϵ in (6.1.4).

So

$$A(1,\ldots,N) = \prod_{i=1}^{N}(\epsilon(-1)^{N_i})A(N,\ldots,1), \qquad (6.1.8)$$

where N_i is the mass squared of the ith strings. In particular, for N massless vector particles the phase is $(-1)^N$.

Let us now suppose that all M external lines are massless vector particles and the internal X pole is such a state, as well. Then since

$$A(P,\ldots,1,X) = (-1)^{P+1}A(1,\ldots,P,X), \qquad (6.1.9)$$

and so forth, the terms in (6.1.7) can be combined as follows

$$\text{tr}[(\lambda_1\ldots\lambda_P - (-1)^P\lambda_P\ldots\lambda_1)(\lambda_{P+1}\ldots\lambda_M - (-1)^{M-P}\lambda_M\ldots\lambda_{P+1})]$$
$$\times A(1,\ldots,P,X)A(X,P+1,\ldots,M).$$

$$(6.1.10)$$

Now the trace in this expression needs to be factorized into a pair of terms appropriate to accompany each of the A factors. Specifically we want to write

$$\text{tr}[(\lambda_1\ldots\lambda_P - (-1)^P\lambda_P\ldots\lambda_1)(\lambda_{P+1}\ldots\lambda_M - (-1)^{M-P}\lambda_M\ldots\lambda_{P+1})]$$

$$= \sum_\alpha \text{tr}[(\lambda_1\ldots\lambda_P - (-1)^P\lambda_P\ldots\lambda_1)\lambda_\alpha]$$

$$\times \text{tr}[\lambda_\alpha^T(\lambda_{P+1}\ldots\lambda_M - (-1)^{M-P}\lambda_M\ldots\lambda_{P+1})].$$

$$(6.1.11)$$

This is valid if the λ_α are a complete set of $n \times n$ matrices normalized so that $\text{tr}(\lambda_\alpha\lambda_\beta^T) = \delta_{\alpha\beta}$. However, if matrices of the form $\lambda = \lambda_1\ldots\lambda_P - (-1)^P\lambda_P\ldots\lambda_1$ span a subspace of the space of all $n \times n$ matrices, then the validity of (6.1.11) is more delicate. For example, if $\lambda_1,\ldots,\lambda_P$ are arbitrary real antisymmetric matrices (corresponding to generators of $SO(n)$), then the combinations λ are themselves antisymmetric and it suffices to restrict the λ_α to be antisymmetric, as well.

The crucial consistency condition required by unitarity is that quantum numbers of the massless vector particles that occur as poles in T in (6.1.5) should be restricted to the adjoint representation of the same group as we started with for the external particles. More generally, the quantum numbers of *any* pole of T should correspond to particles of the spectrum that we began with. Also, the two factors in the residue should describe the T amplitudes for the two halves of the cut diagram.

The consistency condition is satisfied provided all matrices of the form

$$\lambda = \lambda_1 \ldots \lambda_P - (-1)^P \lambda_P \ldots \lambda_1 \qquad (6.1.12)$$

are also matrices belonging to the algebra of the λ_i. For example, the case $P = 2$ says that $[\lambda_1, \lambda_2]$ is in the algebra. This is certainly true, simply because the λ's form a representation of the symmetry group. The conditions with $P > 2$ are a stronger requirement, however, that restrict the possible groups and the possible representations to which the λ matrices can belong.

6.1.2 Allowed Gauge Groups and Representations

The general solution of the conditions described above is not difficult to obtain. We have a linear space L_a of antihermitian matrices λ_i that is closed under commutation, a fact we represent symbolically by $[\lambda, \lambda] = \lambda$. We can also obtain hermitian matrices μ as anticommutators $\{\lambda, \lambda\} = \mu$. These matrices describe the possible quantum numbers that can arise at the odd mass levels. More generally, the same reasoning as above implies that the set of all possible μ matrices is given by expressions of the form

$$\mu = \lambda_1 \lambda_2 \ldots \lambda_P + (-1)^P \lambda_P \ldots \lambda_2 \lambda_1. \qquad (6.1.13)$$

The set of all possible hermitian μ matrices form a linear space L_h closed under anticommutation, $\{\mu, \mu\} = \mu$.

Let us now consider the space L consisting of all matrices ρ formed as real linear combinations of elements of L_a and L_h – in other words all matrices of the form $\rho = a\mu + b\lambda$, where a and b are real. The conditions given above are satisfied if and only if L is closed under multiplication, in other words $\rho_1 \rho_2 \in L$ for all $\rho_1, \rho_2 \in L$. Thus not only must L_a be a Lie algebra, but L must also be an algebra. Given L it is easy to reconstruct L_h and L_a as its hermitian and antihermitian parts. The question therefore is what L's are allowed and what Lie algebras do they give rise to. The general solution to this problem is provided by a theorem of Wedderburn. Before giving the result, let us first state the theorem for complex coefficients. In this case the relevant theorem states that the only irreducible complex algebras (closed under multiplication) are the matrix algebras $GL(n, C)$. This result is relevant if the original algebra contains $i = \sqrt{-1}$. In this case the antihermitian λ matrices generate the Lie algebra $U(n)$. It is important to include all of $U(n)$, including the $U(1)$ factor, and not just $SU(n)$ to satisfy the theorem.

If the original algebra does not contain i then it must be a real form of $GL(n, C)$. We learn from Wedderburn (and from an analysis given for string theory by Marcus and Sagnotti) that there are only two real forms that give Lie algebras for L_h. One of these real forms is $GL(n, R)$, whose antihermitian part gives $SO(n)$ as the algebra L_a. The other possible real form is $U^*(2n)$ (a real version of $U(2n)$ that happens to be noncompact), whose anithermitian part gives $USp(2n)$ (the algebra of $n \times n$ matrices with quaternionic components) as the algebra L_a. Altogether, the only possible solutions consist of $n \times n$ matrices with elements that are either real, complex, or quaternionic. It is obvious that each of these types of matrices defines an algebra closed under multiplication.

This analysis shows that only classical Lie algebras can be incorporated by the Chan–Paton method. The exceptional algebras, such as E_8, cannot be accommodated. Moreover, the fundamental representation must be used in each case. This representation is real for $SO(n)$, pseudoreal for $USp(2n)$, and complex for $U(n)$. Therefore the strings are unoriented in the first two cases and oriented in the latter one.

There is a simple argument that for superstrings the $U(n)$ case must lead to one-loop inconsistencies; it will be substantiated by our later analysis. Certain nonplanar one-loop diagrams describe transitions in which open strings join ends to turn into closed strings. This was sketched in §1.5.6. Since a $U(n)$ open string is oriented the resulting closed strings must also be oriented. But oriented closed superstrings have $N = 2$ supersymmetry, whereas unoriented ones have $N = 1$ supersymmetry, as explained in §5.3.2. This includes the supergravity multiplet, in particular. However, the open-string states only contain $N = 1$ supermultiplets. Since it is not possible to couple $N = 2$ supergravity to $N = 1$ matter consistently, the $U(n)$ case must be anomalous. This reasoning allows any $SO(n)$ or $USp(2n)$ choice, even though it will turn out that they too are unacceptable for more subtle reasons except for the case of $SO(32)$.

There are a few cases of theories having the same gauge group that are nonetheless distinct theories. The $SO(3)$ and $USp(2)$ theories have the same states at the even mass levels, but at the odd levels the hermitian $GL(3, R)$ matrices describe the representations $\mathbf{5} + \mathbf{1}$, whereas the hermitian $GL(1, H)$ matrices contain a singlet only. Similar distinctions can be made between the pairs $SO(2)$ and $U(1)$, $SO(4)$ and $USp(2) \times USp(2)$, $SO(5)$ and $USp(4)$.

In the following sections we discuss the other general approach to including gauge interactions; we will find another way to get $SO(32)$, and, more importantly, we will learn that $E_8 \times E_8$ is possible.

6.2   Current Algebra on the String World Sheet

We have seen that it is possible to place charges (which we called 'quarks') at the ends of open strings. Is there some place else that the charges might be placed? The endpoints of an open string are the only preferred points on the string, and on a closed string there are no preferred points at all. So if we are not to place the charges at the ends of an open string, we must be 'democratic', and contemplate charges that are not localized in any particular way but can be distributed throughout the string.

How can we introduce such a continuous charge distribution? Let us return to the bosonic string action, which we write here in the conformally gauge-fixed form:

$$S = -\frac{1}{2\pi} \int d^2\sigma \partial_\alpha X_\mu \partial^\alpha X^\mu. \tag{6.2.1}$$

In chapter 4, searching to generalize this action, we added fermion degrees of freedom ψ^μ propagating on the string. We took the ψ^μ to transform in the vector representation of $SO(1,9)$, but to carry no other quantum numbers. As an alternative, if we wish to introduce internal symmetry groups, we could introduce fermions that are Lorentz singlets but carry some internal quantum numbers. For example, introducing n real Majorana fermions λ^A, $A = 1, \ldots, n$, we can write

$$S = -\frac{1}{2\pi} \int d^2\sigma \left(\partial_\alpha X_\mu \partial^\alpha X^\mu - i\overline{\lambda}^A \rho^\alpha \partial_\alpha \lambda^A \right). \tag{6.2.2}$$

Equation (6.2.2) has a manifest $SO(n)$ symmetry acting on the A index. Actually, the global symmetry in (6.2.2) is much larger. Writing out (6.2.2) explicitly in terms of left- and right-moving modes $\lambda_\pm^A$, (6.2.2) becomes

$$S = -\frac{1}{2\pi} \int d^2\sigma \left(\partial_\alpha X_\mu \partial^\alpha X^\mu - 2i\lambda_-^A \partial_+ \lambda_-^A - 2i\lambda_+^A \partial_- \lambda_+^A \right). \tag{6.2.3}$$

Equation (6.2.3) is evidently invariant under separate $SO(n)$ rotations for left or right movers. There is thus an $SO(n)_L \times SO(n)_R$ global symmetry. We could, if we wish, simply delete the left or right movers from (6.2.3), giving a system with (say) n left-moving Majorana fermions only and $SO(n)_L$ symmetry.

One might ask if (6.2.3) makes any sense in string theory. The consistency of string theory is so delicate that at first sight this might seem unlikely. However, although some of the details of the consistency of (6.2.3) are actually rather subtle, it is not too difficult to see that at a rough level most of the consistency problems in (6.2.3) can be over-come. Indeed, this follows to a large extent from the equivalence between bosons and fermions in two dimensions that we have uncovered in §3.2.4. Bosonization of fermions in two dimensions means that at least in the infinite volume limit, two Majorana fermions in $1+1$ dimensions are equiv-alent to one real boson. In string theory, we usually are concerned with $(1 + 1)$-dimensional quantum field theory not in the infinite volume limit but on a finite interval or circle, but we have seen in chapter 3 that, under suitable conditions, bosonization of fermions remains valid in this situa-tion. This means that under suitable conditions, (6.2.3) is equivalent to a theory of bosons only. Thus, let n be even, say $n = 2d$. Then the $2d$ Ma-jorana fermions in (6.2.3) are equivalent to d real bosons ϕ^i, $i = 1, \ldots, d$. We could thus rewrite (6.2.3) as

$$\tilde{S} = -\frac{1}{2\pi} \int d^2\sigma \left(\partial_\alpha X_\mu \partial^\alpha X^\mu + \partial_\alpha \phi^i \partial^\alpha \phi^i \right). \qquad (6.2.4)$$

Here there are $D + d$ free bosons, just as in the conformal gauge-fixed Veneziano model. We therefore know that consistency is more or less attainable (depending on whether the requirements to make (6.2.4) com-patible with (6.2.3) are really compatible with other requirements of string theory) provided that $D + d = 26$, the critical dimension of the Veneziano model. For instance, if for some reason one wants $D = 10$, then one should take $n = 32$; one might reasonably aim to get from (6.2.4) a ten-dimensional string theory with $SO(32) \times SO(32)$ internal symmetry.

One may, in fact, wonder what is the difference between the X^μ and the ϕ^i in (6.2.4). Why does (6.2.4) not have the full $SO(D+d-1,1)$ Lorentz symmetry rotating between the X^μ and ϕ^i? How can (6.2.4) have instead a Lorentz symmetry that is only $SO(D-1,1)$ and an internal symmetry $SO(2d) \times SO(2d)$? In fact, how can such a large internal symmetry as $SO(2d) \times SO(2d)$ emerge in (6.2.4)? With the d free bosons ϕ^i, an $SO(d)$ global symmetry is certainly possible, but how can this be promoted to $SO(2d) \times SO(2d)$? The answer to these questions is that, remarkably enough, d free bosons can give an $SO(2d) \times SO(2d)$ symmetry; doing this requires very special conditions on the zero modes of the bosons of which we got a first taste in §3.2.4; and those special conditions prevent any enlargement of the Lorentz group beyond $SO(D-1,1)$. These points should become clearer in the next section.

Before plunging into the study of (6.2.3) and (6.2.4), it is useful to describe what these represent in a slightly more abstract way. Considering for instance (6.2.3), we can naturally define the fermion currents such as

Figure 6.3. The c-number anomaly in the fermion current commutator in $1+1$ dimensions comes from the diagram sketched here.

the left-moving current (now using σ in place of σ^+)

$$J_+^a(\sigma) = \frac{1}{2\pi} T_{AB}^a \lambda_+^A(\sigma) \lambda_+^B(\sigma). \tag{6.2.5}$$

The $n \times n$ matrix representation of the generators is now called T^a instead of λ^a to avoid confusion with the fermion fields. They satisfy the algebra

$$[T^a, T^b] = i f^{abc} T^c, \tag{6.2.6}$$

so that the currents obey the commutation relations

$$[J_+^a(\sigma), J_+^b(\sigma')] = i f^{abc} J_+^c(\sigma) \delta(\sigma - \sigma') + \frac{ik}{4\pi} \delta^{ab} \delta'(\sigma - \sigma'). \tag{6.2.7}$$

Here the first term would arise by canonical manipulations, and the second term, the Schwinger term, is a c-number anomaly that can be evaluated from the diagram of fig. 6.3. The required computation is not at all difficult; it can be done just like the computation of the Virasoro anomaly in §3.2.2. *

The Lie algebra (6.2.7) is called the affine Lie algebra $\widehat{SO}(n)$ with central extension, the central extension being the Schwinger term. Notice that for any finite-dimensional Lie algebra G we can define, as in (6.2.7), the corresponding affine Lie algebra $\hat{G}$. Evaluation of the diagram in fig. 6.3 shows that if the currents in (6.2.5) are constructed as we have described from a single n-plet of fermions in the fundamental representation of $SO(n)$, then $k = 1$.

More generally, we could introduce free fermions transforming not in the fundamental representation of $SO(n)$ but in some other real representation R.[†] The currents defined in (6.2.5) still obey the affine Lie algebra

---

  * In particular, there is no need to do any integrals to evaluate fig. 6.3; the evaluation is best done in coordinate space, where the diagram is just given by the product of the two fermion propagators.

  † We may as well consider only real representations since hermiticity (and $SO(n)$ invariance) of the fermion action would force us to extend a complex or pseudoreal representation to a real one by adding the complex conjugate representation.

of (6.2.7), but with a modified value of k. Indeed, in computing k from the diagram of fig. 6.3 with two currents attached to a single fermion loop, the group theory factor that arises is $\mathrm{tr}_R T^a T^b$, the trace of the product of generators in the representation R. Hence, k is proportional to this trace. To make a precise statement, let T be any generator of $SO(n)$, and let tr_R and tr_F denote traces in the R representation and the fundamental representation of $SO(n)$, respectively. Denote the value of k in the fundamental representation F (which is the n-component vector of $SO(n)$) as k_F, and denote the value in a general representation R as k_R. Then

$$k_R/k_F = \mathrm{tr}_R T^2 / \mathrm{tr}_F T^2. \tag{6.2.8}$$

We have normalized (6.2.7) so that $k = 1$ for the fundamental representation. Depending on the choice of the not necessarily irreducible representation R, k can be an arbitrary positive integer.[‡] For instance, $k = n$ can be obtained by taking R to consist of n 'flavors', that is, n copies of the fundamental representation of $SO(n)$.

It is interesting that regardless of the choice of representation in (6.2.5), k is always an integer. This fact has a deep significance in the theory of affine Lie algebras. The algebra (6.2.7) has 'good' representations[§] only if k is properly quantized. This statement holds for not just for $\widehat{SO}(n)$ but for the affine Lie algebra $\hat{G}$ based on an arbitrary finite-dimensional Lie algebra G. For any $\hat{G}$, the coefficient k in (6.2.7) is quantized; only the allowed quantum depends on G. Unfortunately, to explain why k must be quantized for 'good' representations would take us too far afield here.

For $SO(n)$, the general theory shows that k may be an arbitrary nonnegative integer, and in particular the n free fermions that give $k = 1$ are in some sense the minimal or fundamental model. Suppose, however, that we wish to consider another group such as an exceptional group. It is in fact the exceptional group E_8 that will interest us most. Some of its properties are described in appendix 6.A and others will emerge in sections below. For the time being, let us note that the representations of E_8 are all real, and the smallest dimension of a nontrivial E_8 representation

---

[‡] That the ratio in (6.2.8) is always an integer is a group-theoretic fact of life, which is easily checked in simple examples. There is a special exception; for $n = 3$ a half-integer value is possible by taking two copies of the spinor of $SO(3)$ (two copies since a single spinor is pseudoreal, not real).

[§] Technically, the 'good' representations are the so-called integrable highest weight representations. They are roughly unitary representations in a Hilbert space in which the energy associated with the Virasoro generators defined later is bounded below.

is 248. Since the 248-dimensional representation E_8 (which in fact is the adjoint representation) is real, the existence of this representation means that E_8 can be taken to act on 248 free Majorana fermions in $1 + 1$ dimensions. This gives the smallest value of k that can be straightforwardly achieved by realizing E_8 in terms of free fermions.[¶]

In contrast to the case of $\widehat{SO}(n)$, realizing $\hat{E}_8$ on free fermions does not give the minimal possible value of k. With a normalization in which the general theory permits k to be an arbitrary integer, the straightforward realization of E_8 on 248 free fermions gives $k = 30$. How can we find a minimal theory in which $k = 1$?

We wish to answer this question for practical reasons, not just for cultural ones. First of all, we are interested in the E_8 current algebra (6.2.7) because, by analogy with our initial comments about $SO(n)$, any system that realizes this current algebra would give a possible way of incorporating E_8 in string theory. The 248 free fermions on which we could straightforwardly realize E_8 would be equivalent (in infinite volume, and even in finite volume if we are careful with boundary conditions) to 124 real bosons. As 124 exceeds the critical dimension of the Veneziano model, we cannot trade in a 26-dimensional Veneziano model for a theory in less than 26 dimensions that has E_8 symmetry by using 124 real bosons or 248 Majorana fermions to represent E_8. We must find how to realize E_8 on a smaller number of degrees of freedom.

A clue to this is provided by our introductory discussion of $SO(n)$ for even n, i.e., $n = 2d$. While the minimal $k = 1$ theory has in that case a fairly obvious realization (6.2.3) in terms of n free fermions, it also has a less obvious realization (6.2.4) (whose subtleties we have not yet examined) in terms of $d = n/2$ free bosons. While (6.2.3) is more obvious than (6.2.4) if our interest is $SO(n)$, life is rather different if we wish to incorporate other groups such as exceptional groups in string theory. It turns out that for some of the exceptional groups (namely E_6, E_7, and E_8, but not G_2 and F_4) (6.2.4) generalizes much more straightforwardly than (6.2.3) to give a realization of the minimal current algebra structure of $k = 1$. One of our tasks is therefore to investigate the intricacies of (6.2.4) for various Lie groups, the interesting application being to $E_8 \times E_8$. We will also see how to describe E_8 (with $k = 1$) in terms of fermions, but this too requires a fairly elaborate discussion.

[¶] We say 'straightforwardly', because we will soon learn that there is a nonstraightforward way to realize E_8 with only 16 fermions, giving a much smaller value of k.

There is another way to understand the fact that putting current algebra on the string world sheet is a promising way to obtain internal symmetries in gauge theory. Suppose that we find *any* representation of the affine Lie algebra (6.2.7). Then we can define (for example, for closed strings)

$$T_{++}(\sigma) = \frac{\pi^2}{\beta} \sum_a : J_+^a(\sigma) J_+^a(\sigma) : = 2 \sum L_n e^{-2in\sigma}, \qquad (6.2.9)$$

where $\beta = k + \frac{1}{2}c_2$ and c_2 is the quadratic Casimir operator in the adjoint representation given by

$$f^{acd} f^{bcd} = c_2 \delta^{ab}. \qquad (6.2.10)$$

These operators can be shown, by virtue of the affine Lie algebra (6.2.7), to obey a Virasoro algebra

$$[L_m, L_n] = (m - n)L_{m+n} + \frac{c}{12}(m^3 - m)\delta_{m+n}, \qquad (6.2.11)$$

where

$$c = \frac{2k \dim G}{c_2 + 2k}. \qquad (6.2.12)$$

While it is relatively easy to see from (6.2.9) and from the affine Lie algebra (6.2.7) that the commutator (6.2.11) has the indicated structure, extracting the precise numerical coefficients requires great care in regularizing the operator product in (6.2.9). The reader may wish to start by checking the abelian case in which c_2 vanishes.

Equation (6.2.11) is promising because the Virasoro algebra is certainly an essential part of string theory. Actually, quantization of k means that the values of c that can appear in (6.2.11) are quantized. For instance, in the case of E_8 (for which $c_2 = 60$), the minimal choice $k = 1$ corresponds to $c = 8$, a realization of E_8 current algebra with the same Virasoro anomaly as eight free bosons. This might suggest that just as $SO(2d)$ can be realized with d free bosons, E_8 could be realized with eight of them. Whether or not this suggestion seems plausible at this point, it proves to be correct, as we will see.

6.3   Heterotic Strings

Before plunging into a search for more general realizations of current algebra than those given in the preceding section, let us discuss how these are applied in string theory. One possibility is already indicated in (6.2.3) and (6.2.4): one can replace n dimensions of the Veneziano model with a left- and right-moving current algebra of $c = n$. The difficulty with this scheme is that (although it is indeed a way to incorporate nonabelian gauge symmetry in the Veneziano model) the resulting theory has the two basic drawbacks of the Veneziano model: a tachyon, and no fermions. Indeed, (6.2.4) *is* the Veneziano model (or more precisely the Shapiro–Virasoro closed-string sector of the Veneziano model), except that there are special conditions on the ϕ^i that we have not yet discussed.

Yet another possibility is to begin with one of the superstring theories described in previous chapters, and replace some of the space-time coordinates with a left- and right-moving current algebra. It turns out that in this way one can incorporate nonabelian global symmetries, but not nonabelian gauge symmetries.

Yet another possibility, which was introduced by Gross, Harvey, Martinec and Rohm, uses the fact that in closed-string theories, left- and right-moving modes are decoupled. It is therefore possible to envisage a closed-string theory in which the left-moving modes are of one type, and the right-moving modes are of another type altogether. To incorporate space-time supersymmetry (which at once ensures the presence of fermions and the absence of tachyons) one takes the right-moving modes to be superstring modes of the type discussed in previous chapters. To incorporate gauge degrees of freedom, one takes left-moving modes to include a suitable current algebra. This hybridization of two different kinds of modes has been referred to as 'heterosis'. The following action describes an example of a construction of this type:

$$S = -\frac{1}{2\pi} \int d^2\sigma \Big(\sum_{\mu=0}^{9} (\partial_\alpha X^\mu \partial^\alpha X_\mu - 2i\psi_-^\mu \partial_+ \psi_{\mu-}) - 2i \sum_{A=1}^{n} \lambda_+^A \partial_- \lambda_+^A \Big).$$

$$(6.3.1)$$

Here ψ^μ, $\mu = 0, \ldots, 9$ transform as the vector representation of the Lorentz group; λ^A, $A = 1, \ldots, n$ are Lorentz singlets, but carry some internal quantum numbers. Both ψ^μ and λ^A are Majorana–Weyl fermions.

The right-moving modes are ψ_-^μ and the right-moving part of X^μ. These are the same as the right-moving modes of one of the type II models, so the critical dimension is ten, and this is why we have set $D = 10$ in (6.3.1). There is a supersymmetry between X^μ and ψ_-^μ, just as in chapter 4. The precise formula is

$$\delta X^\mu = i\epsilon\psi_-^\mu, \qquad \delta\psi_-^\mu = \epsilon\partial_- X^\mu, \qquad (6.3.2)$$

which is the same formula as in chapter 4, except that the supersymmetry generator ϵ has only a positive-chirality component. Because of this symmetry, the quantization of the theory involves introducing commuting ghosts. Only half as many of them arise, however; one obtains only the modes $v_{3/2}$ and $u_{-1/2}$, which are related to the positive-chirality supersymmetry generator. With the ghosts included, the critical dimension ten for the right-moving can be computed, just as in §4.4.3. Of course, instead of talking about the ghosts, we could use any of a variety of methods described in chapter 4 for determining the critical dimension of the right-moving modes; or we could replace the ψ^μ of (6.3.1) by an equivalent set of light-cone gauge variables described in chapter 5. There is space-time supersymmetry if we make the GSO projection on the right-moving modes.

On the other hand, the left-moving modes are the left-moving part of the X^μ and the λ^A. As there is no left-moving supersymmetry, the only left-moving ghosts are the reparametrization ghosts, which are enough to cancel the contributions of 26 bosons. As we only have ten X^μ in (6.3.1), the rest of the Virasoro anomaly must be canceled by the λ^A. Since (as we know from bosonization of fermions) two Majorana fermions or one Dirac fermion make up the Virasoro anomaly of a boson, we need 32 λ^A in (6.3.1). If, therefore, the λ^A all obey the same boundary conditions, they carry an $SO(32)$ symmetry; further analysis below will reveal the occurrence of massless gauge mesons of $SO(32)$, so that $SO(32)$ is in fact a gauge symmetry. There is a more subtle possibility in which the λ^A do not all obey the same boundary conditions, and one gets $E_8 \times E_8$ symmetry instead of $SO(32)$. This also will be discussed below.

Instead of introducing the 32 λ^A in (6.3.1), we could introduce any current algebra with $c = 16$. We thus have further motivation for our search for more general realizations of current algebra.

6.3.1 The SO(32) Theory

We begin with a detailed description of the $SO(32)$ theory, in order to temporarily postpone some subtleties that arise in formulating $E_8 \times E_8$ current algebra, even though it is the $E_8 \times E_8$ theory that is far more interesting. We will describe the low-lying states that arise in the quantization of (6.3.1) in the case in which all 32 of the λ^A obey the same boundary conditions, so that the symmetry is $SO(32)$. It is possible to be relatively brief because separate quantization of the left- or right-moving modes has already been carried out in previous chapters.

Just as in the RNS model, the fermionic coordinate λ^A can obey either periodic or antiperiodic boundary conditions. Thus, there are two distinct Fock spaces that we can try to construct; we will explore the properties of both of them. Of course, if we are willing to break $SO(32)$ symmetry, there are many more possibilities; we could assign periodic boundary conditions to some of the fermions and antiperiodic boundary conditions to others. We assume $SO(32)$ symmetry here, so that we are limited to the $SO(32)$ conserving sectors with periodic or antiperiodic boundary conditions.

The periodic sector, denoted P, is the analog of the Ramond sector explored in chapter 4. It has

$$\lambda^A(\sigma) = \sum_{-\infty}^{\infty} \lambda_n^A e^{-2in\sigma}, \qquad (6.3.3)$$

with $n \in Z$ and canonical anticommutation relations

$$\{\lambda_m^A, \lambda_n^B\} = \delta^{AB}\delta_{m+n}. \qquad (6.3.4)$$

The antiperiodic sector, denoted A, is the analog of the NS sector of superstrings. It is described in similar fashion using modes λ_r^A with $r \in Z + 1/2$ and the anticommutation relations

$$\{\lambda_r^A, \lambda_s^B\} = \delta^{AB}\delta_{r+s}. \qquad (6.3.5)$$

In our previous studies of closed-string theories we have seen that it was necessary to construct separate Virasoro operators L_m and $\tilde{L}_m$ from right and left-moving modes. A physical state $|\Omega\rangle$ was required to obey $L_m|\Omega\rangle = \tilde{L}_m|\Omega\rangle = 0$, for $m > 0$, while for $m = 0$ the requirement was

$$(L_0 - a)|\Omega\rangle = (\tilde{L}_0 - \tilde{a})|\Omega\rangle = 0. \qquad (6.3.6)$$

Here a (and $\tilde{a}$) is a normal-ordering constant, which enters in the process of closing the Lorentz algebra. L_0 (or $\tilde{L}_0$) is of the form $p^2/8 + N$ (or $p^2/8 + \tilde{N}$), where N and $\tilde{N}$ are constructed from oscillator coordinates. In the present context we describe the right-moving modes using the transverse spinor degrees of freedom described in chapter 5. Thus (in terms of transverse modes)

$$N = \sum_1^{\infty}(\alpha_{-n} \cdot \alpha_n + nS_{-n}^a S_n^a). \qquad (6.3.7)$$

(The equivalent formula in terms of ψ^μ modes is described in chapter 4.)

For the left-moving modes on the other hand

$$\tilde{N} = \sum_{1}^{\infty}(\tilde{\alpha}_{-n} \cdot \tilde{\alpha}_n + n\lambda_{-n}^A\lambda_n^A) \tag{6.3.8}$$

for the P sector, and similarly

$$\tilde{N} = \sum_{1}^{\infty}\tilde{\alpha}_{-n} \cdot \tilde{\alpha}_n + \sum_{1/2}^{\infty}r\lambda_{-r}^A\lambda_r^A \tag{6.3.9}$$

for the A sector.

Now we must determine the constants a and $\tilde{a}$ in (6.3.6). The first is easy; we have $a = 0$, just as in chapter 5, because of supersymmetry. As for $\tilde{a}$, we can borrow the results of chapters 2, 3, and 4. In chapter 2 we learned that the contribution to the normal ordering constant from a physical Bose coordinate is 1/24, while at the end of §3.2.4 we learned that the normal ordering constant from a half integrally moded Fermi coordinate is 1/48, and from an integrally moded Fermi coordinate is $-1/24$. The latter results were also obtained in another way in chapter 4. We can combine these results to determine the value of $\tilde{a}$ for the two sectors of the theory under discussion:

$$\tilde{a}_A = \frac{8}{24} + \frac{32}{48} = 1$$
$$\tilde{a}_P = \frac{8}{24} - \frac{32}{24} = -1. \tag{6.3.10}$$

Since $a = 0$, the first equation in (6.3.6) is a mass-shell condition

$$p^2 = -8N, \tag{6.3.11}$$

which (since N is a nonnegative operator) immediately tells us that there are no tachyons. This was to be expected, because of supersymmetry. Upon combining the two equations in (6.3.6), we learn that in the A sector

$$\tfrac{1}{4}(\text{mass})^2 = N + \tilde{N} - 1, \tag{6.3.12}$$

where $N = \tilde{N} - 1$, and in the P sector

$$\tfrac{1}{4}(\text{mass})^2 = N + \tilde{N} + 1, \tag{6.3.13}$$

where $N = \tilde{N} + 1$. Since according to (6.3.11) massless states have $N = 0$, we see that massless states in the A sector must have $\tilde{N} = 1$, while massless states in the P sector would have to have $\tilde{N} = -1$. Since $\tilde{N}$ is a positive semi-definite operator, there are no massless states in the P sector, and we can concentrate on the A sector.

The space of massless particles is just the tensor product of the space of right-moving modes with $N = 0$ and the space of left-moving modes with $\tilde{N} = 1$ (in the A sector). For the right-moving modes, the space of states of $N = 0$ is just the $D = 10$ super Yang–Mills multiplet, just as in the open superstring of chapter 5. These massless modes are explicitly of the form

$$|i\rangle_R \quad \text{and} \quad |\dot{a}\rangle_R, \qquad (6.3.14)$$

representing the bosonic and fermionic modes belonging to the 8_v and 8_c representations of spin(8), respectively. For the left-moving modes with $\tilde{N} = 1$ there are two possibilities. We have first

$$\tilde{\alpha}^i_{-1}|0\rangle_L . \qquad (6.3.15)$$

These states are $SO(32)$ singlets, of course. There are eight transverse physical states in (6.3.15); they transform as a vector under the transverse rotation group. The other states of $\tilde{N} = 1$ are

$$\lambda^A_{-1/2}\lambda^B_{-1/2}|0\rangle_L . \qquad (6.3.16)$$

These states are Lorentz singlets and transform in the adjoint representation (the antisymmetric second-rank-tensor representation) of $SO(32)$. Taking the tensor product of the supersymmetric Yang–Mills multiplet (6.3.14) with the eight states (6.3.15) gives the 16×8 states of the $N = 1$ supergravity multiplet. Specifically, in the Bose sector

$$|i\rangle_R \otimes \tilde{\alpha}^j_{-1}|0\rangle_L \qquad (6.3.17)$$

gives 64 states consisting of a symmetric traceless tensor (graviton), an antisymmetric tensor, and a scalar. Similarly, in the Fermi sector

$$|\dot{a}\rangle_R \otimes \tilde{\alpha}^i_{-1}|0\rangle_L \qquad (6.3.18)$$

gives 64 states that decompose into a gravitino $\chi^{\dot{a}}_i$, which satisfies

$$\gamma^i_{a\dot{a}}\chi^{\dot{a}}_i = 0 \qquad (6.3.19)$$

and therefore has 56 independent components, and an eight-component spinor λ_a corresponding to the contraction $\gamma^i_{a\dot{a}}|\dot{a}\rangle_R \otimes \tilde{\alpha}^i_{-1}|0\rangle_L$. The tensor product of (6.3.14) with the $32 \cdot 31/2 = 496$ states (6.3.16) gives the

$16 \cdot 496$ states of the super Yang–Mills multiplet for the group $SO(32)$. Notice that the massless spin-one modes automatically appear in the adjoint representation of the group, even though this was not obviously guaranteed by the construction *a priori*. The adjoint representation is the one that must arise if the theory is to make sense at the interacting level as we know from the limiting low-energy Yang-Mills theory.

In the above we have described both boundary conditions on the fermions that are compatible with $SO(32)$ symmetry, namely the P and A sectors. The P sector did not give rise to any massless particles, so we might be tempted to discard it as being uninteresting. However, when we study loop diagrams in chapter 9, we will find that unitarity requires inclusion of the P as well as the A sector; also, in §6.4 below, we give an alternative construction of the theory in which the P and A sectors automatically appear together. Anticipating these points, we consider both the P and A sectors in our subsequent presentation of the model.

Thinking about the A sector, which gives the massless particles, the condition $N = \tilde{N} - 1$ has the following interesting consequence. The operator N defined in (6.3.7) above only has integer eigenvalues, while $\tilde{N}$ has both integer and half-integer eigenvalues. The half-integer eigenvalues of $\tilde{N}$ are obtained by acting with an odd number of λ^A oscillators on the ground state. The half-integer eigenvalues of $\tilde{N}$ do not contribute to the physical spectrum, since there is no choice of right-moving state that would enable one to obey $N = \tilde{N} - 1$. Therefore, the condition $N = \tilde{N} - 1$ forces us to discard states containing an odd number of λ^A oscillators. This is strikingly reminiscent of the GSO projection, which yields space-time supersymmetry. It is natural to suspect that we should carry out the GSO projection in the P sector also. We will learn in chapter 9 that this is indeed necessitated by unitarity at the one loop level; we also will find that the GSO-like projection automatically arises (in both P and A sectors) in the alternative construction in §6.4 below. In the P sector, the GSO-like projection means that physical states have eigenvalue $+1$ of the operator $(-1)^F$ that anticommutes with all of the λ^A. This operator is

$$(-1)^F = \overline{\lambda}_0 (-1)^{\sum_1^\infty \lambda^A_{-n} \lambda^A_n}, \qquad (6.3.20)$$

where

$$\overline{\lambda}_0 = \lambda_0^1 \lambda_0^2 \cdots \lambda_0^{32} \qquad (6.3.21)$$

is the product of the zero-mode fermions.

Let us now examine the spectrum at the first excited level. The states at this level can be formed in either of two ways. If we use the A sector for

the left-moving modes we must take $N = 1$ and $\tilde{N} = 2$, whereas if we use
the P sector for the left-moving modes $N = 1$ and $\tilde{N} = 0$ are required.
In either case the right-moving sector is contributing the $N = 1$ multiplet
of the open superstring, which we saw in chapter 5 corresponds to the
$D = 10$ supermultiplet of spin(9) representations $(44 + 84)_B + 128_F$.
This is to be tensored with the states made from the left-moving sector.
The possible A sector states with $\tilde{N} = 2$ are

$$\tilde{\alpha}^i_{-2}|0\rangle_L, \quad \tilde{\alpha}^i_{-1}\tilde{\alpha}^j_{-1}|0\rangle_L, \quad \tilde{\alpha}^i_{-1}\lambda^A_{-1/2}\lambda^B_{-1/2}|0\rangle_L,$$

$$\lambda^A_{-1/2}\lambda^B_{-3/2}|0\rangle_L, \quad \lambda^A_{-1/2}\lambda^B_{-1/2}\lambda^C_{-1/2}\lambda^D_{-1/2}|0\rangle_L. \tag{6.3.22}$$

For the P sector the only possibility with $\tilde{N} = 0$ is the spin(32) spinor $|a\rangle$
satisfying the GSO-like projection condition (analogous to a Weyl con-
dition in the space-time context) $\overline{\lambda}_0|a\rangle = |a\rangle$. The dimension of this
representation is 2^{15}, a rather large number! Altogether the A and P
sectors together contribute 73 764 modes. Tensoring with the 256 right-
moving modes gives a total of 18 883 584 states at the first excited level of
the heterotic string. The construction ensures that they form a supersymmetry multiplet, because the supersymmetry transformations act only on
the right-moving factor. Similarly, they form spin(32) multiplets, because
those transformations act on the multiplets of the left-moving factor.

We explained in appendix 5.A that the representations of spin($2n$) fall
into four conjugacy classes. In the construction described above we see
that A sector representations consist entirely of tensor representations
of even rank because there are always an even number of λ excitations.
All such representations belong to the same conjugacy class as the adjoint
representation. In the P sector we have found one of the two fundamental
spinor representations occurring at the first excited level. The fact that
all allowed states can be obtained from this one with an even number of
λ excitations (to ensure $(-1)^F = 1$) means that all other states occurring
in this sector at higher mass levels belong to the same conjugacy class.
As a result the complete spectrum of the theory only involves two of the
four conjugacy classes of spin(32). This is what is being alluded to when
one describes the group symmetry as being spin(32)/Z_2, rather than, say,
spin(32) or $SO(32)$.

6.3.2 The $E_8 \times E_8$ Theory

In the foregoing we have described the basic structure of heterotic strings
with $SO(32)$ gauge symmetry and $D = 10$ spacetime supersymmetry.

The physical space is a tensor product of right-moving modes that include supersymmetric degrees of freedom and left-moving modes that are responsible for gauge symmetries. Tachyons are absent because of supersymmetry. Having described the basic structure, our next task is to explain how this structure can be implemented in a way that gives a more interesting and unusual gauge symmetry group – not $SO(32)$ but the exceptional group $E_8 \times E_8$.

In constructing the spin$(32)/Z_2$ theory we assigned to all 32 components of λ^A the same boundary conditions – A or P – in order to be consistent with the symmetry. If we are willing to break $SO(32)$ symmetry, we could contemplate many more possibilities, with one boundary conditions for some of the λ^A and another boundary condition for others. At first sight it may seem that there are a bewildering array of possibilities. Moreover, they all seem to involve some loss of symmetry. We will now see, however, that most of these choices run afoul of the $N = \tilde{N} - 1$ constraint, but that there is one option that is rather attractive. It does not really involve a loss of symmetry: while some of the spin$(32)/Z_2$ symmetry is lost, additional gauge symmetry appears in an unexpected way. In fact, we will arrive at a symmetry group, namely $E_8 \times E_8$, which in some sense is just as large as $SO(32)$ (they both have 496 generators) but far more interesting.

Consider attempting to construct a theory with a symmetry smaller than spin(32), perhaps spin$(n) \times$ spin$(32 - n)$. To do so, we split the 32 left-moving fermions up into a group of n and a group of $32 - n$. There is no reason that the two groups of fermions must obey the same boundary conditions, if the gauge symmetry is to be only $SO(n) \times SO(32-n)$. Thus, we can assign the P sector and A sector boundary conditions to the two sets of Fermi oscillators separately. There are then four possible sectors, which we denote AA, AP, PA and PP, where the first label refers to the boundary conditions obeyed by the first n components of λ and the second label to those obeyed by the remaining $32 - n$ components. As a result, there are four distinct possibilities for the normal ordering constant $\tilde{a}$ in the relation $N = \tilde{N} - \tilde{a}$. These constants can be calculated by the rules described in §6.3.1. Recalling that the normal-ordering constant of a boson is $+1/24$, while that of a periodic fermion is $-1/24$, and that of an antiperiodic fermion is $+1/48$, we get

$$\tilde{a}_{AA} = \frac{8}{24} + \frac{n}{48} + \frac{32-n}{48} = 1 \qquad (6.3.23)$$

$$\tilde{a}_{AP} = \frac{8}{24} + \frac{n}{48} - \frac{32-n}{24} = \frac{n}{16} - 1 \qquad (6.3.24)$$

$$\tilde{a}_{PA} = \frac{8}{24} - \frac{n}{24} + \frac{32-n}{48} = 1 - \frac{n}{16} \qquad (6.3.25)$$

$$\tilde{a}_{PP} = \frac{8}{24} - \frac{n}{24} - \frac{32-n}{24} = -1. \tag{6.3.26}$$

To see the implication of the above formulas we recall that N only has integer eigenvalues. Similarly, $\tilde{N}$ takes integer eigenvalues in the P sector. In the A sector $\tilde{N}$ has integer eigenvalues for states of even world-sheet fermion number and half-integer eigenvalues for states of odd world-sheet fermion number. For values of n not divisible by eight, there are no physical states in the AP and PA sectors. Discarding those sectors, and keeping only the AA and PP sectors in which all 32 λ^A obey the same boundary conditions, we are back to the spin$(32)/Z_2$ theory considered earlier. For n divisible by eight there are essentially three possibilities: (i) $n = 32$ or 0, (ii) $n = 16$, (iii) $n = 8$ or 24. Cases (i) and (ii) give integer values for all four $\tilde{a}$'s, whereas case (iii) gives half-integer values for $\tilde{a}_{AP}$ and $\tilde{a}_{PA}$. Case (i) is again precisely the spin$(32)/Z_2$ theory considered in the previous section. Case (ii) appears to give a spin$(16) \times$ spin(16) theory, whereas case (iii) is a candidate for a spin$(24) \times$ spin(8) theory. Let us consider case (ii), as it turns out to be the one of most interest. Case (iii) is a theory that suffers from various one-loop anomalies, and will not be considered further.

In case (ii) we are assigning independent A or P boundary conditions to each set of 16 λ coordinates separately. As we have already said, the current algebra for this case looks like that of a spin$(16) \times$ spin(16) theory, and that is all the symmetry one might reasonably expect. The spin$(16) \times$ spin(16) gauge mesons arise in the AA sector in the above construction. Since $\tilde{a} = 1$ in the AA sector, one makes massless left-moving states in this sector (just as in the spin$(32)/Z_2$ theory) by acting on the ground state with two $\lambda^A_{-1/2}$ oscillators, each contributing $+1/2$ to the eigenvalue of $\tilde{N}$. The resulting states are of the form

$$\lambda^A_{-1/2}\lambda^B_{-1/2} |0\rangle_L, \tag{6.3.27}$$

They transform as follows under spin$(16) \times$ spin(16):

$$\begin{aligned}
&(\mathbf{120}, \mathbf{1}) && \text{if} && A, B = 1, \dots, 16; \\
&(\mathbf{1}, \mathbf{120}) && \text{if} && A, B = 17, \dots, 32; \\
&(\mathbf{16}, \mathbf{16}) && \text{if} && A = 1, \dots, 16, \ B = 17, \dots, 32.
\end{aligned} \tag{6.3.28}$$

Here $\mathbf{16}$ and $\mathbf{120}$ denote, respectively, the vector and adjoint representations of $SO(16)$. Taken together, these 496 states with the indicated spin$(16) \times$ spin(16) content would fill out the adjoint representation of spin(32), so one may wonder whether we are somehow discovering a new theory with spin(32) gauge group. But now we find a surprise. Since the normal-ordering constants are $\tilde{a} = 0$ in the AP and PA sectors, additional massless states come from those sectors. Indeed, massless left-moving states in the AP or PA sector are states of $\tilde{N} = 0$. Quantization of fermion zero modes causes the states in the PA (or AP) sector to transform as spinors of the first (or second) spin(16) group, just as in the Ramond sector discussed in chapter 4. So, if we denote the two spinor representations of spin(16) as **128** and **128′**, respectively, the massless left-moving states in the PA and AP sectors are

$$PA : (\mathbf{128}, \mathbf{1}) \oplus (\mathbf{128'}, \mathbf{1})$$
$$AP : (\mathbf{1}, \mathbf{128}) \oplus (\mathbf{1}, \mathbf{128'}), \tag{6.3.29}$$

When tensored with right-moving massless states, the left-moving massless states in (6.3.28) and (6.3.29) give rise to supersymmetric Yang–Mills multiplets, so if the theory is to make sense they must form a Lie algebra. However, there simply is no Lie algebra with the spin(16) × spin(16) content of (6.3.28) and (6.3.29), so we must find a projection that eliminates some of the states.

The required projection is a variant of the GSO projection. Upon separating the 32 world-sheet fermions into two groups of 16, as we have done in the foregoing, there are two candidates for the operator $(-1)^F$ that is used in the GSO projection. We can define an operator $(-1)^{F_1}$ that anticommutes with the first group of 16 λ^A and commutes with the second group, and a second operator $(-1)^{F_2}$ that commutes with the first group and anticommutes with the second group. The GSO-like projection that extracts a Lie algebra from (6.3.28) and (6.3.29) consists of keeping precisely the states that are invariant under both $(-1)^{F_1}$ and $(-1)^{F_2}$. If we postulate that in the AA sector the ground state $|0\rangle_L$ is even under both $(-1)^{F_1}$ and $(-1)^{F_2}$, then it is easy to see which states in (6.3.28) obey the GSO-like projection. The states in question must be created by acting on $|0\rangle_L$ with two fermions that are both from the first group of 16 λ^A or both from the second group, so the surviving states transform as

$$(\mathbf{120}, \mathbf{1}) \oplus (\mathbf{1}, \mathbf{120}). \tag{6.3.30}$$

As for (6.3.29), in both the AP and PA sectors, the two multiplets in (6.3.29) have opposite eigenvalues of $(-1)^F$, so one multiplet survives from each pair; which one survives is immaterial, since the difference between the **128** and **128′** of spin(16) is a matter of convention. Picking a convention, we may say that the surviving states from (6.3.29) transform as

$$(\mathbf{128}, \mathbf{1}) \oplus (\mathbf{1}, \mathbf{128}). \tag{6.3.31}$$

As we explain in appendix 6.A, the $\mathbf{120} \oplus \mathbf{128}$ of spin(16) make up the exceptional Lie algebra E_8, so (6.3.30) and (6.3.31) together correspond to $E_8 \times E_8$. We will see in chapter 9 that the GSO-like projection we have made is required by unitarity, but for the time being we are satisfied to note that (short of discarding (6.3.29) altogether) it is essentially the only projection of (6.3.28) and (6.3.29) that makes a Lie algebra.

The theory as we have presented it manifestly has a spin(16) $\times$ spin(16) current algebra, but that there is an $E_8 \times E_8$ current algebra is not apparent. This enlarged current algebra must exist, however, if the theory makes sense, since the massless gauge bosons of $E_8 \times E_8$ mean that there must be an $E_8 \times E_8$ symmetry, and a theory of $E_8 \times E_8$ symmetry that has spin(16) $\times$ spin(16) current algebra must have $E_8 \times E_8$ current algebra. To construct the extra conserved currents of $E_8 \times E_8$ is somewhat difficult. We will see how it can be done at the end of §7.3. Later in this chapter we will give an alternative construction of the theory with an explicit demonstration of the occurrence of $E_8 \times E_8$ current algebra.

At the first excited level there are a large number of possible states, all of which also piece together into $E_8 \times E_8$ multiplets. The 256 $N = 1$ right-moving modes can be tensored with the $\tilde{N} = 0$ PP states $|a;b\rangle_L$, the $\tilde{N} = 1$ AP states

$$\tilde{\alpha}^i_{-1}|0;a\rangle_L, \qquad \lambda^A_{-1/2}\lambda^B_{-1/2}|0;a\rangle_L, \qquad \lambda'^A_{-1}|0;\dot{a}\rangle_L \tag{6.3.32}$$

or the PA state

$$\tilde{\alpha}^i_{-1}|a;0\rangle_L, \qquad \lambda'^A_{-1/2}\lambda'^B_{-1/2}|a;0\rangle_L, \qquad \lambda^A_{-1}|\dot{a};0\rangle_L. \tag{6.3.33}$$

A dotted spinor index refers to the spinor of the opposite chirality. In other words $\bar{\gamma}|a\rangle = |a\rangle$, whereas $\bar{\gamma}|\dot{a}\rangle = -|\dot{a}\rangle$. Finally, one must also include the $\tilde{N} = 2$ AA states

$$\tilde{\alpha}^i_{-2}|0\rangle_L, \quad \tilde{\alpha}^i_{-1}\tilde{\alpha}^j_{-1}|0\rangle_L, \quad \tilde{\alpha}^i_{-1}\lambda^A_{-1/2}\lambda^B_{-1/2}|0\rangle_L, \quad \tilde{\alpha}^i_{-1}\lambda'^A_{-1/2}\lambda'^B_{-1/2}|0\rangle_L$$

$$\lambda^A_{-1/2}\lambda^B_{-1/2}\lambda^C_{-1/2}\lambda^D_{-1/2}|0\rangle_L, \quad \lambda'^A_{-1/2}\lambda'^B_{-1/2}\lambda'^C_{-1/2}\lambda'^D_{-1/2}|0\rangle_L,$$

$$\lambda^A_{-1/2}\lambda^B_{-3/2}|0\rangle_L, \quad \lambda'^A_{-1/2}\lambda'^B_{-3/2}|0\rangle_L, \quad \lambda^A_{-1/2}\lambda^B_{-1/2}\lambda'^C_{-1/2}\lambda'^D_{-1/2}|0\rangle_L.$$
$$\tag{6.3.34}$$

Altogether, this gives a total of 73 764 modes, just as in the spin(32)/Z_2 theory. Later we will show that there are actually an equal number of states at every mass level of the two heterotic string theories.

6.4 Toroidal Compactification

The fermionic construction of current algebra on which we have concentrated so far is simple to describe, but also has some drawbacks. The construction of an $E_8 \times E_8$ multiplet may seem somewhat artificial, and the proof of $E_8 \times E_8$ symmetry is incomplete (the key steps in completing it will be explained at the end of §7.3). We now begin a discussion of the 'bosonic' realization of current algebra, which requires more preliminary discussion, but gives a simpler way to demonstrate E_8 symmetry.

In (6.3.1), we are really interested in a theory with left-moving current algebra only. However, we begin our discussion of the bosonic realization of current algebra by discussing a theory (in fact, (6.2.4)) in which there is both a left- and a right-moving current algebra. The reason for this is that the required formulas have other applications (to compactification of string theory), and also some of the required structures are more easily understood if one begins with a discussion of the left–right symmetric case. Thus, we consider first the Veneziano model formulated not on M^{26} (26-dimensional Minkowski space) but on $M^{25} \times S^1$, with S^1 being a circle. This is a special case of 'compactification' from 26 to $26 - d$ dimensions. The goal is to treat the d toroidal bosons as the ϕ^i of (6.2.4), and to show that with a proper formulation, an unexpected nonabelian symmetry group emerges.

After this example is behind us we shall turn to the case of primary interest, the bosonized description of the heterotic string. In this construction the 32 left-moving fermionic coordinates are replaced by 16 bosonic coordinates. Together with the ten left-moving bosons of the ten-dimensional string theory, this means that the left-moving modes consist of 26 bosons; one can think of them as the 26 free bosons of the Veneziano model. The right-moving modes on the other hand are supersymmetric modes – 10 bosons and (in the RNS framework) 10 fermions, just as in our discussion in §6.3 above. In this description, therefore, the heterotic string can be viewed as a hybrid of left-moving bosonic string and a right-moving fermionic string. The 16 extra bosons of the left-movers are compactified on particular 16-dimensional tori, leading to $E_8 \times E_8$ or $SO(32)$ symmetry.

This construction makes the emergence of the $E_8 \times E_8$ symmetry easier to understand than the fermionic one. It also gives the first hint of why the particular constructions in the previous section are special. Specifically, we will discover that the two supersymmetric theories correspond to the only two even self-dual lattices in 16 dimensions. While this feature appears only to be a curiosity in the context of the discussion of this chapter, we will discover in chapter 9 that it is actually of fundamental importance for the consistency of one-loop diagrams.

6.4.1. Compactification on a Circle

To illustrate the basic idea of how nonabelian symmetries arise from toroidal compactification, let us begin with the simplest example. Consider bosonic closed strings in $D = 26$, and suppose one of the spatial coordinates is compactified into a circle. For this coordinate one identifies $x \equiv x + 2\pi R n$, where R is the radius of the circle and n is an arbitrary integer. The most general configuration for $X(\sigma, \tau)$ satisfying the two-dimensional wave equation and the closed-string boundary conditions then becomes

$$X(\sigma, \tau) = x + p\tau + 2L\sigma + \frac{i}{2}\sum_{n \neq 0}\frac{1}{n}[\alpha_n e^{-2in(\tau - \sigma)} + \tilde{\alpha}_n e^{-2in(\tau + \sigma)}], \quad (6.4.1)$$

where

$$p = \frac{m}{R}, \qquad L = nR. \quad (6.4.2)$$

The integer m labels the allowed momentum eigenvalues. The restriction to integer m is needed so that the quantum wave function $e^{ip \cdot x}$ is single-valued (invariant under $x \to x + 2\pi R$). The integer n is the number of times that the string wraps around the circle. When this 'winding number' is nonzero, $X(\sigma, \tau)$ describes a soliton string state. Such states have no counterpart in the uncompactified case, since their energy diverges as $R \to \infty$. Clearly, such states exist whenever the spatial manifold contains noncontractible closed curves, or in other words whenever its fundamental group π_1 is not zero.

The mode expansion in (6.4.1) can be decomposed into a sum of right-moving and left-moving pieces

$$X(\sigma, \tau) = X_R(\tau - \sigma) + X_L(\tau + \sigma) \quad (6.4.3)$$

$$X_R(\tau - \sigma) = x_R + \left(\frac{p}{2} - L\right)(\tau - \sigma) + \frac{i}{2}\sum_{n \neq 0}\frac{1}{n}\alpha_n e^{-2in(\tau - \sigma)} \quad (6.4.4)$$

$$X_L(\tau + \sigma) = x_L + \left(\frac{p}{2} + L\right)(\tau + \sigma) + \frac{i}{2}\sum_{n \neq 0}\frac{1}{n}\tilde{\alpha}_n e^{-2in(\tau + \sigma)}. \quad (6.4.5)$$

The zero-frequency parts of the Virasoro conditions $T_{++} = T_{--} = 0$ in this case give $L_0 = \tilde{L}_0 = 1$, where

$$L_0 = \frac{1}{2}(\frac{1}{2}p - L)^2 + N + (p_\mu)^2/8$$

$$\tilde{L}_0 = \frac{1}{2}(\frac{1}{2}p + L)^2 + \tilde{N} + (p_\mu)^2/8. \quad (6.4.6)$$

This implies that

$$\tfrac{1}{4}(\text{mass})^2 = N + \tilde{N} - 2 + \frac{m^2}{4R^2} + n^2 R^2 \qquad (6.4.7)$$

and

$$N - \tilde{N} = pL = mn, \qquad (6.4.8)$$

where

$$N = \sum_{n=1}^{\infty}(\alpha^{\mu}_{-n}\alpha_{n\mu} + \alpha_{-n}\alpha_n) \qquad (6.4.9)$$

$$\tilde{N} = \sum_{n=1}^{\infty}(\tilde{\alpha}^{\mu}_{-n}\tilde{\alpha}_{n\mu} + \tilde{\alpha}_{-n}\tilde{\alpha}_n). \qquad (6.4.10)$$

The last two terms in the mass formula (6.4.7) give the contributions of the internal momentum and winding energy to the 25-dimensional mass. In the formulas for N and $\tilde{N}$, α^{μ}_n and $\tilde{\alpha}^{\mu}_n$ refer to the first 25 dimensions, while α_n and $\tilde{\alpha}_n$ refer to the compactified 26th dimension. Since a shift of σ by an amount σ_0 is generated by the unitary operator

$$U(\sigma_0) = \exp[2i(N - \tilde{N} - pL)\sigma_0], \qquad (6.4.11)$$

the $N - \tilde{N}$ equation ensures that the physics does not depend on the choice of origin for the σ coordinate.

Let $|m, n\rangle$ denote a ground state of the Fock space having internal momentum m/R and winding eigenvalue n. The 25-vector momentum p^{μ} is not shown explicitly. Let us now construct massless vector states. Two of them are given by

$$(\alpha^{\mu}_{-1}\tilde{\alpha}_{-1} \pm \alpha_{-1}\tilde{\alpha}^{\mu}_{-1})\,|0;0\rangle\,, \qquad (6.4.12)$$

since these states have $N = \tilde{N} = 1$ and $p = L = 0$. If the string is assumed to be oriented, the massless $D = 26$ spectrum contains a graviton $g_{\mu\nu}$, an antisymmetric tensor $B_{\mu\nu}$, and a scalar. The two vectors can be understood as arising from the decomposition of $g_{\mu\nu}$ and $B_{\mu\nu}$, with respect to the $D = 25$ Lorentz subgroup. If the string is unoriented, $B_{\mu\nu}$ is not present in the spectrum, and only the symmetric combination in (6.4.12) is physical.

Another possibility for making massless vectors is to have states with $p, L \neq 0$. For mass 0, $L_0 = \tilde{L}_0 = 1$ gives

$$2N - 2 + \left(\frac{p}{2} - L\right)^2 = 0 \qquad (6.4.13)$$

$$2\tilde{N} - 2 + \left(\frac{p}{2} + L\right)^2 = 0. \qquad (6.4.14)$$

These can be solved by $N = 1$ and $\tilde{N} = 0$ for $L = p/2$ and $p^2 = 2$. Since $pL = mn = 1$, this requires $m = n = \pm 1$ but this is only possible for $R^2 = 1/2 = \alpha'$. Thus, for this particular value of the radius, there are four more massless vectors given by

$$\alpha^\mu_{-1}|1,1\rangle, \quad \alpha^\mu_{-1}|-1,-1\rangle, \quad \tilde{\alpha}^\mu_{-1}|1,-1\rangle, \quad \tilde{\alpha}^\mu_{-1}|-1,1\rangle. \qquad (6.4.15)$$

On physical grounds, massless vectors must couple to conserved currents for consistency, so at the value of the radius at which the massless vectors just noted appear, there must be an enlarged gauge symmetry. The enlarged gauge symmetry has as a subgroup $U(1)_L \times U(1)_R$ generated by the massless gauge bosons that are present at any radius. The $U(1)_L \times U(1)_R$ quantum numbers are m and n. The counting of states and the $U(1)_L \times U(1)_R$ quantum numbers show that the enlarged gauge symmetry that appears at a special radius is $SU(2)_L \times SU(2)_R$.

It is not difficult to see how some of the low-lying states fit into $SU(2) \times SU(2)$ multiplets. For example, the massless scalars include the nine states

$$\alpha_{-1}\tilde{\alpha}_{-1}|0,0\rangle, \quad |\pm 2, 0\rangle, \quad |0, \pm 2\rangle,$$

$$\alpha_{-1}|1,1\rangle, \quad \alpha_{-1}|-1,-1\rangle, \quad \tilde{\alpha}_{-1}|1,-1\rangle, \quad \tilde{\alpha}_{-1}|-1,1\rangle. \qquad (6.4.16)$$

which form a $(\mathbf{3}, \mathbf{3})$ representation of $SU(2) \times SU(2)$.

For unoriented closed strings, the symmetry is not $SU(2)_L \times SU(2)_R$ but only a diagonal $SU(2)$ symmetry. Of the nine scalars noted in the last paragraph, only six states $(\mathbf{1} + \mathbf{5})$ of the diagonal $SU(2)$ survive.

6.4.2 Fermionization

The appearance of an enlarged gauge symmetry for a special radius is rather striking. It is not the first time that we have seen that phenomenon occur when a scalar field is an angular variable with the correct periodicity. In our discussion of bosonization of fermions in chapter 3, we saw that to bosonize fermions correctly in finite volume, one must introduce a boson that is a scalar field with a definite radius.

It is possible to make a precise connection between these two phenomena and thereby to give, perhaps, a simple explanation of the appearance of an enlarged symmetry when the scalar has the correct radius. Let us concentrate on, say, the left-moving sector of $SU(2)_L \times SU(2)_R$. Suppose that we wish to obtain not $SU(2)_L$ symmetry but $SO(4)_L$. There is a relation, since $SO(4)_L$ is isomorphic to the product $SU(2)_L \times SU(2)'_L$ of two left-moving $SU(2)$ groups. As in the preceding section, we can realize $SO(4)_L$ with four left-moving fermions:

$$S = \frac{i}{\pi} \int d^2\sigma \sum_{i=1}^{4} \psi_+^i \partial_- \psi_+^i. \tag{6.4.17}$$

On the other hand, we can bosonize these fermions, and describe them by two left-moving scalar fields ϕ_1^+ and ϕ_2^+ that obey

$$\partial_+ \phi_1^+ = \psi_1^+ \psi_2^+ \qquad \partial_+ \phi_2^+ = \psi_3^+ \psi_4^+. \tag{6.4.18}$$

In view of the discussion in chapter 3, ϕ_1^+ and ϕ_2^+ must be angular variables of definite periodicity.

Now, the currents $\psi_1^+ \psi_2^+$ and $\psi_3^+ \psi_4^+$ of (6.4.18) are linear combinations of $SU(2)_L$ and $SU(2)'_L$ currents. Specifically, the decomposition of $SO(4)_L$ into $SU(2)_L \times SU(2)'_L$ corresponds to a decomposition of the fermion currents $J_{ij}^+ = \psi_i^+ \psi_j^+$ into self-dual and anti-self-dual pieces $J_{ij}^+ \pm \frac{1}{2}\epsilon_{ijkl}J_{kl}^+$. The linear combinations $\psi_1^+ \psi_2^+ \pm \psi_3^+ \psi_4^+$ are self-dual and anti-self-dual, respectively, and so are generators, respectively, of $SU(2)_L$ and $SU(2)'_L$. If we introduce new boson fields σ_1^+ and σ_2^+ by

$$\sigma_1^+ = (\phi_1^+ + \phi_2^+)/2 \qquad \sigma_2^+ = (\phi_1^+ - \phi_2^+)/2 \tag{6.4.19}$$

then the σ_i^+ correspond in the bosonization formulas to generators, respectively, of $SU(2)_L$ and $SU(2)'_L$:

$$\partial_+ \sigma_1^+ = \tfrac{1}{2}(\psi_1^+ \psi_2^+ + \psi_3^+ \psi_4^+)$$
$$\partial_+ \sigma_2^+ = \tfrac{1}{2}(\psi_1^+ \psi_2^+ - \psi_3^+ \psi_4^+). \tag{6.4.20}$$

Equation (6.4.20) shows that the derivative of σ_1^+ is an $SU(2)_L$ current and so commutes with $SU(2)'_L$; in fact, σ_1^+, and not only its derivative, has this property. Thus σ_1^+ commutes with $SU(2)'_L$, and describes $SU(2)_L$ degrees of freedom only. Likewise, σ_2^+ describes $SU(2)'_L$ degrees of freedom only. Thus, we have shown that a single left-moving boson σ_1^+ can describe $SU(2)_L$ symmetry, provided that it is a periodic variable with the correct periodicity. This is in agreement with the preceding analysis.

The transformation from (6.4.18) to (6.4.19) is strikingly reminiscent of the transformation used in chapter 5 to relate the two different light-cone formulations of superstrings. Indeed, the general understanding of this is as follows. A Lie group G of rank d has d commuting generators $T_1, T_2, \ldots, T_d$. They generate a $U(1)^d$ subgroup of G called its maximal torus. In certain cases, G symmetry can be realized by introducing d left-moving scalars that propagate on the maximal torus of G. (Or $G \times G$ symmetry can be realized by introducing d scalars with both left- and right-moving pieces that propagate on this maximal torus.) For example, $SU(2)$ has rank one; its maximal torus is a circle. $SU(2)$ symmetry can be achieved with a single scalar field. $SO(4)$ and $SO(8)$ have rank two and rank four, respectively, and require the introduction of two or four scalar fields. Since the periodic scalar fields that are required in order to get $SO(4)$ or $SO(8)$ symmetry are variables that 'live' on the maximal torus of $SO(4)$ or $SO(8)$, any natural transformation of the respective maximal tori is a symmetry of these scalar fields. The transformation from (6.4.18) to (6.4.19) is the transformation of the maximal torus of $SO(4)$ that decomposes $SO(4)$ into $SU(2) \times SU(2)$, while the transformation considered in chapter 5 is (as sketched in appendix 5.A) the transformation of the maximal torus that corresponds to triality symmetry.

The groups that can be realized in terms of scalar fields propagating on the maximal torus are the so-called simply laced groups (groups in which all root vectors are of equal length), namely $SU(N)$, $SO(2N)$, and E_6, E_7 and E_8. Since we can understand $SO(2N)$ in terms of free fermions, it is the other cases and especially the exceptional groups about which we can get some insight by working with bosons on the maximal torus.

6.4.3 Bosonized Description of the Heterotic String

We now generalize the previous discussion with a view to getting not $SU(2)_L \times SU(2)_R$ symmetry but G_L, where G is either of the two symmetry groups spin$(32)/Z_2$ or $E_8 \times E_8$ that appear in the supersymmetric heterotic string theories. The way we achieve this is to use 16 left-moving bosonic coordinates, in place of the 32 λ coordinates considered previously, and to show that by compactifying them on a torus of suitable size and shape one can reproduce the results found previously. Finding the correct tori means finding the boundary conditions for the bosonic coordinates that correctly describe the bosonization of the Fermi coordinates with specified boundary conditions. Instead of using a purely deductive approach, it is much easier to simply construct the particular torus that correctly reproduces the desired symmetry group and particle spectrum.

Let us suppose that 16 of the 26 left-moving bosonic coordinates are compactified and write their mode expansions in the form

$$X_L^I = x_L^I + p_L^I(\tau + \sigma) + \frac{i}{2} \sum_{n \neq 0} \frac{1}{n} \tilde{\alpha}_n^I e^{-2in(\tau + \sigma)}, \qquad (6.4.21)$$

where $I = 1, 2, \ldots, 16$. These supplement the usual superstring coordinates for the right-moving modes α_n^i and S_n^a (in the light-cone gauge) where $i = 1, \ldots, 8$ labels the 8_v representation of the transverse space-time symmetry and $a = 1, \ldots, 8$ labels the 8_s representation of the same group. Also present, of course, are the left-moving space-time modes $\tilde{\alpha}_n^i$.

In quantizing the modes of and $X_L^I(\tau + \sigma)$ in the absence of corresponding $X_R^I(\tau - \sigma)$ modes there is one slightly subtle detail. These variables have phase-space constraints, since the conjugate momentum $\dot{X}$ is proportional to X'. Quantization requires the use of Dirac brackets. A short cut to the correct answer is as follows. If both left and right moving modes are present, the standard commutation relation is

$$[x^I, p^J] = i\delta^{IJ}. \qquad (6.4.22)$$

If we write $x^I = x_L^I + x_R^I$, $p^I = p_L^I + p_R^I$, where left-moving modes x_L^I, p_L^I are to commute with right-moving modes x_R^I, p_R^I, then (6.4.22) implies

$$[x_L^I, p_L^J] = \frac{i}{2}\delta^{IJ} = [x_R^I, p_R^J]. \qquad (6.4.23)$$

The factor of $1/2$ here is important. As a result, the operator

$$V_0(K, \tau + \sigma) =: e^{2iK \cdot X_L(\tau + \sigma)} : \qquad (6.4.24)$$

satisfies

$$[p_L^I, V_0(K)] = K^I V_0(K). \qquad (6.4.25)$$

The vector p_L^I corresponds to the quantity $\frac{1}{2}p + L$ in the discussion of §6.4.1. In that example, there was a left-moving mode with $p_L = \frac{1}{2}p + L$ and a right-moving mode with $p_R = \frac{1}{2}p - L$. Now we are supposing that there are no right-moving modes that correspond to the 16 compactified coordinates. This means that the winding numbers and the momenta must be identified by the requirement that $\frac{1}{2}p^I - L^I = 0$.

The compactification of the 16 left-moving coordinates into a torus T^{16} means that coordinates that differ by period vectors are to be identified. Such a torus can be obtained as follows. One begins with 16-dimensional

Euclidean space R^{16}. In this space one introduces 16 independent vectors e_i^I, $i = 1, \ldots, 16$, and one defines a lattice Γ consisting of points of the form $\sum n_i e_i^I$, the n_i being integers. One then forms the torus $T^{16} = R^{16}/\pi\Gamma$. (The factor of π will simplify our formulas.) A point on this torus is a point x^I in R^{16} subject to the equivalence relation

$$x^I \approx x^I + \pi \sum_{i=1}^{16} n_i e_i^I = x^I + 2\pi L^I. \qquad (6.4.26)$$

We introduce the metric tensor

$$g_{ij} = \sum_{I=1}^{16} e_i^I e_j^I \qquad (6.4.27)$$

on R^{16}. We will see that the most interesting case is that in which this matrix has integer entries with diagonal elements $g_{ii} = 2$.

The commutation relation (6.4.23) implies that $p_L^I \sim -(i/2)\partial/\partial x_L^I$. Thus the Kaluza–Klein quantization condition for the allowed momenta K^I is given by the requirement that $\exp(2iK \cdot x)$ be single valued. The requirement on K^I is that $K^I e_i^I$ must be an integer for each $i = 1, \ldots, 16$. The K^I that obey this condition form a lattice $\tilde{\Gamma}$ called the dual lattice of Γ. It is always possible to find 16 vectors e_i^{*I} obeying

$$\sum_{I=1}^{16} e_i^{*I} e_j^I = \delta_{ij}. \qquad (6.4.28)$$

The e_i^{*I} are uniquely determined by (6.4.28). Integer linear combinations of the e_i^{*I} form a lattice $\tilde{\Gamma}$ called the dual lattice of Γ. Allowed momenta K^I must be of the form

$$K^I = \sum_{i=1}^{16} m_i e_i^{*I}. \qquad (6.4.29)$$

The condition $K^I = 2L^I$, which follows from $K = \frac{1}{2}p + L$ and $\frac{1}{2}p - L = 0$ (the conditions asserting that X^I has left-moving components only), gives

$$K^I = \sum_{i=1}^{16} m_i e_i^{*I} = \sum_{i=1}^{16} n_i e_i^I. \qquad (6.4.30)$$

Until now, we have imposed no restriction on the 16 vectors e_i^I (apart from linear independence); there is likewise no restriction on the matrix

g_{ij}. For a generical choice of the e_i^I, (6.4.30) is very restrictive. Indeed, (6.4.30) says that the same vector K^I must belong to both the lattice Γ generated by the e_i^I and the dual lattice $\tilde{\Gamma}$ generated by the e_i^{*I}. Both Γ and $\tilde{\Gamma}$ consist of discrete points in R^{16}, and for generical choices of the e_i^I, these two lattices have no point in common except the origin. Thus, for generical choices of the e_i^I, a momentum K^I obeying (6.4.30) must be zero.

It is unphysical for the zero modes of the string motion to be absent, so we can anticipate from this that the theory with a generical choice of e_i^I is inconsistent. The precise nature of the inconsistency will emerge in chapter 9 when we study modular invariance. For the time being, let us go on and look for instances in which (6.4.30) is not too restrictive.

Under what conditions are there many points that lie in both the lattice Γ and its dual lattice $\tilde{\Gamma}$? Such a phenomenon arises if the matrix elements g_{ij} of the metric g defined in (6.4.27) above are all integers. The lattice Γ is then said to be an *integral* lattice. In this case, $(e_i, e_j) = e_i^I e_j^I$ is an integer for all $1 \leq i, j \leq 16$, so the vectors e_i^I themselves lie in the dual lattice $\tilde{\Gamma}$. Indeed, in this situation the entire lattice Γ generated by the e_i lies in $\tilde{\Gamma}$, so (6.4.30) permits K^I to be an arbitrary element of Γ.

The requirement that Γ should be an integral lattice amounts to a very special restriction on the radius of the torus T^{16} on which the extra 16 left-moving bosons are compactified. We actually wish to impose yet further restrictions. In general, the requirement that Γ be integral does not imply that Γ is identical to $\tilde{\Gamma}$; it may be a sub-lattice thereof. If they actually are identical, then Γ is said to be a self-dual lattice.

Another important notion is the following. If the diagonal elements g_{ii} of g are even and Γ is an integral lattice, then Γ is said to be an *even* lattice. Consider an arbitrary lattice vector $v^I = \sum n_i e_i^I$ (the n_i are integers). Its length squared is $(v, v) = \sum_{ij} n_i n_j g_{ij} = \sum n_i^2 g_{ii} + 2 \sum_{i<j} n_i n_j g_{ij}$, and is even if the g_{ii} are all even; thus an even lattice may be described by saying that the length squared of any lattice vector is even.

The left- and right-moving sectors are related, as usual, by the constraint $L_0 = \tilde{L}_0 - 1$, which ensures invariance under rigid shifts of the σ parameter. Expressed in terms of modes this becomes

$$N = \tilde{N} - 1 + \tfrac{1}{2} \sum_I (p_L^I)^2, \qquad (6.4.31)$$

where N is the superstring number operator and

$$\tilde{N} = \sum_{n=1}^{\infty} \left(\sum_{i=1}^{8} \tilde{\alpha}_{-n}^i \tilde{\alpha}_n^i + \sum_{I=1}^{16} \tilde{\alpha}_{-n}^I \tilde{\alpha}_n^I \right). \qquad (6.4.32)$$

The mass operator is given by

$$\tfrac{1}{4}(\text{mass})^2 = 2N. \tag{6.4.33}$$

Since N and $\tilde{N}$ have integer eigenvalues, we see already that the eigenvalues K^I of the operator p_L^I must have a length-squared that is an even integer. If the lattice Γ is even as well as integral, this is no additional requirement. What happens if Γ is not even? If Γ is an integer lattice that is not even, the points $\sum n_i e_i^I \in \Gamma$ whose length squared is even form a sublattice Γ'. Since (6.4.33) forces the momenta to lie in Γ' (and not just in Γ as we learned from (6.4.30)), we may as well carry out the whole construction using Γ' from the beginning. Thus, we henceforth assume that the lattice Γ is even.

As already noted for the fermionic formulation, (6.4.33) ensures that the spectrum is tachyon-free because N is nonnegative. This is remarkable, because the bosonic string theory, which is being used to describe the left movers, has a tachyon.

Let us now discuss the gauge symmetry of the theory. Massless vector particles can be made by tensoring the $N = 0$ right-moving $SO(1,9)$ vectors with left-moving Lorentz scalars of $\tilde{N} + \tfrac{1}{2}(p_L)^2 = 1$. Sixteen left-moving scalars are made for any choice of the radii by the states $\tilde{\alpha}_{-1}^I |0\rangle$. These states just correspond to the Kaluza–Klein vector gauge fields associated with the $[U(1)]^{16}$ isometry of the torus T. They are no surprise. However, just as we saw in §6.4.1 that $U(1) \times U(1)$ could be enlarged to $SU(2) \times SU(2)$ for a suitably chosen radius, so we can obtain an enlarged gauge symmetry in the present construction if the lattice Γ has points of length squared two. If there are lattice sites with $K^2 = 2$, then the associated state $|K\rangle$ describes left-moving scalars. The additional massless vectors are in one-to-one correspondence with such lattice sites. The restriction to even integral lattices Γ that we have tentatively advocated above means that such enlarged gauge invariance is likely; for instance if the basis vectors e_i^I have length squared two (the smallest possible value in an even integer lattice), then the choice $K^I = e_i^I$ (for any i) is suitable for giving a massless state $|K^I\rangle$.

We have already encountered even integral lattices once previously in this book. In appendix 5.A we described the root and weight lattices associated with $SO(2n)$ groups. We saw that the roots – the nonzero weights of the adjoint representation – were of length-squared equal to two, so that the root lattice was in fact an even integral lattice. The root lattice of $SO(2n)$ is not self-dual, however; its dual lattice includes weight vectors of other representations as well as the adjoint representation.

More generally, on any even lattice the points of length squared two (collectively denoted Λ_2) are always the nonzero roots of some Lie algebra $G.^*$ The massless vector mesons found in the above construction for every point $|K^I\rangle$ of length squared two in Γ give rise in the string theory to gauge fields of a gauge symmetry group isomorphic to G. Since there are 16 left-moving bosons at our disposal in the above construction, we can obtain an arbitrary simply-laced Lie group of rank 16.[†] At first sight we face a plethora of possibilities.

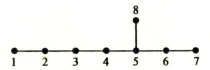

Figure 6.4. The Dynkin diagram of E_8.

What drastically reduces this plethora is the condition that the lattice Γ should be self-dual. In the above, we have seen that the restriction to self-dual Γ is rather natural for left-moving bosons, which should have the same winding numbers and momenta. (For a self-dual Γ (6.4.30) is no restriction.) We have not been able to prove the necessity of a self-dual lattice, and indeed we will not be able to prove this requirement until chapter 9, after developing the concept of modular invariance. Let us however here anticipate the fact that Γ must be an even self-dual lattice and explore the consequences. Such lattices have been thoroughly studied by mathematicians because of their role in number theory. Even self-dual lattices (of Euclidean signature) only exist in dimensions that are a multiple of eight. In eight dimensions there is just one, Γ_8, the root lattice of the exceptional group E_8. In this case the matrix $g_{ij} = e_i \cdot e_j$ is the Cartan matrix of E_8, which is given in a basis of simple positive roots

---

[*] In fact, we prove this in §6.4.4 below by constructing a closed Lie algebra of vertex operators corresponding to points of length squared two on any integer lattice.

[†] A simply-laced Lie group is simply one in which all nonzero roots have the same length squared; the finite dimensional Lie algebras with this property are $SO(2N)$, $SU(N)$, E_6, E_7, and E_8. The common length squared of the nonzero roots of a simply-laced Lie algebra is customarily taken to be two.

by

$$
g_{ij} = \begin{pmatrix}
2 & -1 & 0 & 0 & 0 & 0 & 0 & 0 \\
-1 & 2 & -1 & 0 & 0 & 0 & 0 & 0 \\
0 & -1 & 2 & -1 & 0 & 0 & 0 & 0 \\
0 & 0 & -1 & 2 & -1 & 0 & 0 & 0 \\
0 & 0 & 0 & -1 & 2 & -1 & 0 & -1 \\
0 & 0 & 0 & 0 & -1 & 2 & -1 & 0 \\
0 & 0 & 0 & 0 & 0 & -1 & 2 & 0 \\
0 & 0 & 0 & 0 & -1 & 0 & 0 & 2
\end{pmatrix}. \tag{6.4.34}
$$

This corresponds to the Dynkin diagram shown in fig. 6.4. (Dynkin diagrams were explained briefly in appendix 5.A.)

In 16 dimensions there are two even self-dual lattices, namely the product $\Gamma_8 \times \Gamma_8$ of two copies of Γ_8 and a second one called Γ_{16}, which contains the root lattice of $SO(32)$ as a sublattice. The former choice gives rise in string theory to the gauge group $E_8 \times E_8$, while in the second case the gauge group is $SO(32)$. The $E_8 \times E_8$ and $SO(32)$ theories constructed using these two self-dual lattices coincide with those that we constructed in the fermionic formulation in §6.3. For instance, the lattice Γ_{16} contains the root lattice of $SO(32)$ as well as the weights associated with one of the two spinor conjugacy classes (described in appendix 5.A). Therefore the representations that are included in Γ_{16} correspond to the group spin(32)/Z_2, in agreement with the analysis of §6.3.1. We will say more about the lattices Γ_8 and Γ_{16} later in this section.

6.4.4 Vertex Operator Representations

Having set the stage, we now finally construct the bosonic representation of current algebra. This also permits us to give a full account of the gauge symmetry in the $E_8 \times E_8$ theory.

In the fermionic construction of heterotic strings we were able to prove $SO(32)$ symmetry or $SO(16) \times SO(16)$ symmetry, but it remained somewhat of a mystery how $E_8 \times E_8$ works. In the bosonic formulation we have found that left-moving coordinates are compactified on the maximal torus of $E_8 \times E_8$ with the result that the conjugate momenta are restricted to the lattice $\Gamma_8 \times \Gamma_8$. It is reasonable to expect that this is a framework in which the symmetry can be established, and that is what we intend to explain in this section.[*]

The plan of attack is to describe an explicit construction of the group generators in terms of vertex operators and to verify that they satisfy the correct commutation relations. This constitutes a proof of the symmetry, provided only that they commute with the mass-squared operator, which turns out to be the case. The reason, of course, is that then the space of string states provides a representation space for the group at each mass level.

Let us first make general comments about the form of a simply-laced Lie algebra. If the rank is d, we first pick a maximal commuting subgroup $[U(1)]^d$, called the Cartan subalgebra. Let us denote these commuting generators by H_I

$$[H_I, H_J] = 0. \tag{6.4.35}$$

The eigenvalues of the H_I (for any state in any representation) form a point in a d-dimensional 'weight space' R^d. The other $n-d$ generators (n is the dimension of the group) correspond to root vectors $K \in R^d$. For a simply laced group they all have the same length squared, which we define to be two. They generate then an even integral lattice Γ. Let us denote them by E_K. These operators are 'charged', meaning that they shift the eigenvalues of H_I by K_I. Thus

$$[H_I, E_K] = K_I E_K. \tag{6.4.36}$$

To complete the specification of the Lie algebra in this basis we must give the rule for commuting E_K with $E_{K'}$. The result, if nonzero, must have total charge $K + K'$. Since $K + K'$ is in the even lattice Γ, the possible values of $(K + K')^2$ are $0, 2, 4, 6, 8$. However, the values 4, 6, 8 (which correspond to $K \cdot K' \geq 0$) do not correspond to charges carried by generators, since we are dealing with a simply-laced Lie algebra in which all roots have length squared two. Therefore we must have

$$[E_K, E_{K'}] = 0 \qquad \text{if} \quad K \cdot K' \geq 0. \tag{6.4.37}$$

The commutator $[E_K, E_{K'}]$ must vanish unless there is a root $E_{K+K'}$ in

---

[*] The basic construction was invented by Frenkel, Kac, and Segal, following previous work by Lepowsky and Wilson. It was elucidated for physicists by Goddard and Olive. Early formulations of special cases of this construction were given by Halpern (who was specifically motivated by possible applications to string theory), and by Banks, Horn, and Neuberger, and others.

the Lie algebra with 'charge' $K + K'$.[†] If there is such a root then we must have

$$[E_K, E_{K'}] = \epsilon(K, K') E_{K+K'} \qquad \text{if} \quad K \cdot K' = -1. \qquad (6.4.38)$$

By properly normalizing the E_K, the coefficients $\epsilon(K, K')$ can be chosen to be all ± 1. An explicit construction of $\epsilon(K, K')$ is given in §6.4.5. for cases of interest. The final case is $K + K' = 0$, in which case the charge carried by the commutator is zero, so that the most general possible outcome is a linear combination of the generators of the Cartan subalgebra. For suitable normalizations, one has

$$[E_K, E_{K'}] = \sum_I K^I H_I \qquad \text{if} \quad K + K' = 0. \qquad (6.4.39)$$

This will also emerge in the construction that follows.

Given an even integer lattice Γ, we would now like to construct operators that obey the above commutation relations, with the E_K corresponding to the points of length squared two in Γ. Our success in doing so will indeed constitute a proof of some of the above assertions. To start with, it is clear that the correct pattern of root vectors is achieved if the operators E_K contain the factor $\exp[2iK \cdot x_L]$ and if

$$H^I = p_L^I. \qquad (6.4.40)$$

These two conditions, by themselves, are sufficient to ensure that (6.4.35) and (6.4.36) are satisfied. They are also consistent with the rules for $[E_K, E_{K'}]$. What is still needed, and is certainly not trivial, is to complete the specification of E_K.

Given that E_K contains $\exp(2iK \cdot x_L)$, it is natural to consider the string vertex operator

$$V_0(K, z) =: \exp[2iK \cdot X_L(z)] :, \qquad (6.4.41)$$

where

$$z = \exp[2i(\tau + \sigma)]. \qquad (6.4.42)$$

Indeed, the factor $\exp(2iK \cdot x_L)$ is rather artificial in string theory unless extended as in (6.4.41). Written out explicitly in terms of modes, the

---

[†] Such a root, if it exists, is always unique; it is a general fact about Lie algebras that there are never two generators not in the Cartan subalgebra with the same charge.

vertex operator is

$$V_0(K,z) = \exp(2iK \cdot x_L + K \cdot p_L \ln z)$$
$$\times \exp\left(\sum_1^\infty \frac{1}{n} K \cdot \tilde{\alpha}_{-n} z^n\right) \exp\left(-\sum_1^\infty \frac{1}{n} K \cdot \tilde{\alpha}_n z^{-n}\right).$$

$$(6.4.43)$$

This operator has conformal dimension $\frac{1}{2}K^2$, and thus $J = 1$ for $K \in \Lambda_2$, which is the case of interest. The operators $V_0(K,z)$ are not yet the solution to our problem, because they do not commute with $\tilde{L}_n$ and they have an undesired z dependence. The obvious solution to both of these problems is to extract the zero-frequency part of V_0

$$A_K = \frac{1}{\pi} \int_0^\pi V_0(K,z) d\tau = \oint \frac{dz}{2\pi i z} V_0(K,z). \qquad (6.4.44)$$

In writing this formula we have used the fact that as τ goes from 0 to π, the coordinate z in (6.4.42) goes around the unit circle in the complex plane in a counterclockwise sense. Of course, the radius of the circle can be increased or decreased so long as no singularities are crossed. As it stands there are no apparent singularities. However, the expression is an operator and singularities can arise when it multiplies another operator. This feature in fact plays an important role in the following.

For the integrals in (6.4.44) to be well-defined, it is important that $V_0(K,z)$ not have a branchpoint at the origin. Then for a circle of sufficiently small radius so that no other branchpoints that may arise from singular operator products are enclosed, the integrand is single valued. Fortunately, this is the case because the eigenvalues of p_L all lie in Γ, and for every $p_L \in \Gamma$, $p_L \cdot K$ is an integer. Therefore, for all allowed choices of p_L, $V_0(K,z)$ does not have a branch point at $z = 0$, and the integrand is single valued along a contour of sufficiently small radius. The fact that $V_0(K,z)$ has $J = 1$ for $K \in \Lambda_2$ implies that

$$[\tilde{L}_n, V_0(K,z)] = z \frac{d}{dz}(z^n V_0(K,z)). \qquad (6.4.45)$$

Therefore

$$[\tilde{L}_n, A_K] = \oint \frac{dz}{2\pi i} \frac{d}{dz}(z^n V_0(K,z)) = 0. \qquad (6.4.46)$$

This implies that the operators A_K map physical states into physical states.

The operators A_K are a good guess for the generators E_K. They turn out not to be quite right, but only a relatively simple additional factor is required. To see how this works, the first step is to study the algebra of the A_K operators. The multiplication of two such operators involves the product of the two vertices $V_0(K,z)$ and $V_0(K',w)$ that appear in their integrands. This operator product is singular, of course, at $z = w$. To see this explicitly, it is convenient to re-express the product in normal-ordered form. This can be achieved by using the well-known identity $\exp A \cdot \exp B = \exp B \cdot \exp A \cdot \exp[A,B]$, which is valid when $[A,B]$ is a c number. The $[A,B]$ term consists of

$$-K \cdot K' \sum_1^\infty \frac{1}{n} \left(\frac{w}{z}\right)^n = K \cdot K' \ln(1 - w/z) \qquad (6.4.47)$$

from the $\tilde{\alpha}$ oscillator terms and

$$-\tfrac{1}{2} K \cdot K' \ln(w/z) \qquad (6.4.48)$$

from the zero modes. Putting these facts together we see that

$$\begin{aligned} V_0(K,z)V_0(K',w) =&(wz)^{-K \cdot K'/2}(z-w)^{K \cdot K'} \\ & \times\; : \exp[2iK \cdot X(z) + 2iK' \cdot X(w)] : . \end{aligned} \qquad (6.4.49)$$

However, this formula is only valid for $|w| < |z|$, because otherwise the infinite series that gives the power of $z - w$ is divergent.

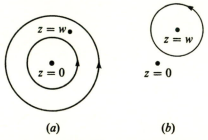

$$(a) \qquad\qquad\qquad (b)$$

Figure 6.5. The z integration contours corresponding to the terms $A_K A_{K'}$ and $A_{K'} A_K$ are inside and outside of $|w|$, respectively, as shown in (a). Therefore the difference corresponds to a contour enclosing the only potentially singular point, namely $z = w$, as in (b).

In applying (6.4.49) to the multiplication $A_K A_{K'}$, it is necessary to take the radius of the integration contour of A_K ($|z|$) to be larger than that for the one $A_{K'}$ ($|w|$) in order that the product be well-defined. When the multiplication is done in the other order, the formula differs in two respects. First of all, the contours must now be taken with $|w| > |z|$, since A_K is to the right of $A_{K'}$. Second, the factor $(z - w)^{K \cdot K'}$ is replaced by $(w - z)^{K \cdot K'}$. This distinction does not involve any subtle choices of branches, only a factor of $(-1)^{K \cdot K'}$, since $K \cdot K'$ only takes integer values. This being the case, the combination of operators for which the integration contours combine nicely is

$$A_K A_{K'} - (-1)^{K \cdot K'} A_{K'} A_K$$

$$= \oint \frac{dw}{2\pi i w} \oint_w \frac{dz}{2\pi i z} (z - w)^{K \cdot K'} (wz)^{-K \cdot K'/2} \qquad (6.4.50)$$

$$\times \, : \exp[2iK \cdot X(z) + 2iK' \cdot X(w)] : \, .$$

In the first term in (6.4.50) the contours are taken with $|z| > |w|$, whereas in the second term $|z| < |w|$. These contours are depicted in fig. 6.5a. Therefore, in taking the difference the z contours combine to give one that only encircles the singularities that are enclosed between these two circles. Since the only potentially singular point between the two contours is $z = w$, the two terms combine to give a z contour enclosing the point w as indicated in (6.4.50) and fig. 6.5b.

As we saw earlier, the only possible values of $K \cdot K'$ are integers ranging from -2 to $+2$. Thus the z contour integration in (6.4.50) immediately gives a vanishing result unless $K \cdot K' = -1$ or -2. (Otherwise the point $z = w$ is nonsingular.) These are easily evaluated using Cauchy's theorem. For $K \cdot K' = -1$, the sum $K + K' \in \Lambda_2$. The pole is simple in this case so one immediately obtains

$$A_K A_{K'} - (-1)^{K \cdot K'} A_{K'} A_K = A_{K+K'} \qquad \text{for} \quad K \cdot K' = -1. \quad (6.4.51)$$

Note that in evaluating the residue we have used

$$\frac{1}{z} (wz)^{-K \cdot K'/2} \to 1. \qquad (6.4.52)$$

The case $K \cdot K' = -2$ is also easy because it corresponds to $K + K' = 0$. In this case the residue requires evaluating

$$\frac{\partial}{\partial z} \left\{ \frac{1}{z} (wz)^{-K \cdot K'/2} \, : \exp[2iK \cdot X(z) + 2iK' X(w)] : \right\} \qquad (6.4.53)$$

at $z = w$. Using $K \cdot K' = -2$ and $K + K' = 0$, this is simply

$$w \frac{\partial}{\partial w} (2iK \cdot X(w)). \qquad (6.4.54)$$

Then the w contour integral picks off the zero-frequency part of this, which is just

$$w\frac{\partial}{\partial w}(K \cdot p_L \ln w) = K \cdot p_L. \qquad (6.4.55)$$

Substituting, $p_L^I = H^I$, we have

$$A_K A_{K'} - (-1)^{K \cdot K'} A_{K'} A_K = \sum K^I H_I \quad \text{for} \quad K \cdot K' = -2. \quad (6.4.56)$$

Thus, altogether, the A_K's have the desired algebra *except* that (6.4.51) involves anticommutators instead of commutators when $K \cdot K'$ is odd. One might think at first that we have stumbled onto a graded Lie algebra (as in supersymmetry), but this is not so. The rule according to which commutators or anticommutators appear in the above formulas does not distinguish two classes of operators (even and odd) but rather depends on their relative labels in a rather peculiar way. To turn this system of commutators and anticommutators into the desired Lie algebra it is necessary to multiply A_K by a 'cocycle' to form the generator E_K.

Let us define the generators of the Lie algebra of G associated with the roots $K \in \Lambda_2$ by

$$E_K = A_K c_K, \qquad (6.4.57)$$

where c_K is a correction factor that we will choose to turn the relations (6.4.51) into commutators. The factor c_K (for each K) will depend on the momentum operator p_L but not on the other oscillators. Thus, for each K, c_K is the operation of multiplying by a certain function of the momenta. Using the algebra of the A_K given in (6.4.51), it is evident that if we want to convert the relations obeyed by the A_K into commutators, we need the $c_K(p)$ factors to satisfy the equation

$$c_K(p - K')c_{K'}(p) = (-1)^{K \cdot K'} c_{K'}(p - K)c_K(p). \qquad (6.4.58)$$

This is enough to convert the relations in (6.4.51) into commutators, but to get a *closed* set of commutators (the operators on the right-hand side of (6.4.51) should be of the same form as those on the left), the product of two c_K's must be another c_K. In fact, we impose

$$c_K(p - K')c_{K'}(p) = \epsilon(K, K')c_{K+K'}(p), \qquad (6.4.59)$$

where the $\epsilon(K, K')$ is ± 1; in fact, from (6.4.51) they can evidently be identified with the factors appearing in (6.4.38), so that our success in finding factors $c_K(p)$ with the stated properties will also establish the general form given in (6.4.38) (and previous equations) for a simply-laced

Lie algebra. By evaluating the product $(c_K c_{K'}) \cdot c_{K''} = c_K \cdot (c_{K'} c_{K''})$ in two different ways using (6.4.59), we can deduce a restriction on the $\epsilon(K, K')$. Specifically,

$$
\begin{aligned}
c_K(p - K' - K'') &\left[c_{K'}(p - K'') c_{K''}(p) \right] \\
&= \epsilon(K', K'') c_K(p - K' - K'') c_{K'+K''}(p) \\
&= \epsilon(K', K'') \epsilon(K, K' + K'') c_{K+K'+K''}(p),
\end{aligned}
\tag{6.4.60}
$$

whereas

$$
\begin{aligned}
\left[c_K(p - K' - K'') c_{K'}(p - K'') \right] & c_{K''}(p) \\
&= \epsilon(K, K') c_{K+K'}(p - K'') c_{K''}(p) \\
&= \epsilon(K, K') \epsilon(K + K', K'') c_{K+K'+K''}(p).
\end{aligned}
\tag{6.4.61}
$$

Therefore we must require that ϵ satisfy the condition

$$
\epsilon(K, K') \epsilon(K + K', K'') = \epsilon(K', K'') \epsilon(K, K' + K''),
\tag{6.4.62}
$$

which is usually described by saying that ϵ is a Z_2-valued two-cocycle.

It only remains to give an explicit construction of c and ϵ in order to complete the proof of the symmetry. So we now turn to that.

6.4.5 Formulas for the Cocycles

We have just seen how to construct generators representing the symmetry group G. The internal momenta $H^I = p_L^I$ represent the Cartan subalgebra and operators $E_K = A_K c_K$ are associated with the roots $K \in \Lambda_2$. The factors $c_K(p)$ depend on the momentum operator p_L^I, written as p for simplicity, but not on the oscillators associated with nonzero frequencies. The proof of G symmetry is incomplete until $c_K(p)$ operators satisfying the fundamental equation

$$
c_K(p - K') c_{K'}(p) = \epsilon(K, K') c_{K+K'}(p)
\tag{6.4.63}
$$

are proved to exist. In this section we present an explicit construction.

In the heterotic string construction a left-moving momentum, associated with the 16 compactified dimensions, takes values corresponding to sites of one of the self-dual even lattices $\Gamma_8 \times \Gamma_8$ or Γ_{16}. Let e_i^I represent the basis vectors of one of these lattices. Then an internal momentum can be written in the form

$$
K^I = \sum_{i=1}^{16} m_i e_i^I,
\tag{6.4.64}
$$

where the m_i are arbitrary integers.

The construction that follows depends on choosing a particular ordering of the 16 basis vectors. We define an ordered $*$ product by

$$K * K' = \sum_{i>j} m_i m'_j e_i \cdot e_j. \tag{6.4.65}$$

This product is always an integer. We then define

$$c_K(p) = (-1)^{p*K}. \tag{6.4.66}$$

It is easy to see that this satisfies (6.4.63) for

$$\epsilon(K, K') = (-1)^{K*K'}. \tag{6.4.67}$$

This automatically satisfies the two-cocycle condition (6.4.62) as well.

This specific construction implies the particular relations

$$\epsilon(K, K')\epsilon(K', K) = (-1)^{K \cdot K'}$$

$$\epsilon(K, 0) = \epsilon(0, K) = 1 \tag{6.4.68}$$

$$c_K(p)c_{K'}(p) = c_{K+K'}(p).$$

There is considerable nonuniqueness in the construction, but any solution of the fundamental equations is equivalent.

6.4.6 The Full Current Algebra

In §6.4.4 and 6.4.5 we have given an explicit construction of the generators of the global $E_8 \times E_8$ symmetry of the heterotic string theory. These charges are distributed along the length of the string. Thus one can also define charge densities associated with each of them. These densities form the Kac–Moody algebra known as affine $E_8 \times E_8$. This algebra was already formulated abstractly in §6.2, where we saw that the left-moving currents $J_+^a(\sigma)$ should form an algebra of the form

$$[J_+^a(\sigma), J_+^b(\sigma')] = if^{abc}J_+^c(\sigma)\delta(\sigma - \sigma') + \frac{k}{4\pi}\delta^{ab}\delta'(\sigma - \sigma'). \tag{6.4.69}$$

In §6.2 we described fermionic representations of affine $SO(2d)$ with $k = 1$. In the case of $E_8 \times E_8$ we were only able to represent the subalgebra $SO(16) \times SO(16)$ explicitly, using 32 fermionic coordinates, but we saw that if there were an extension to the full affine $E_8 \times E_8$, it would give the correct anomaly term ($c = 16$) in the Virasoro algebra.

In the construction of the $E_8 \times E_8$ generators given in the preceding subsections we have used a representation based on 16 left-moving bosonic coordinates. This contributes the same anomaly as 32 fermionic coordinates, namely $c = 16$, in the Virasoro algebra. Therefore by the uniqueness of the representation, this must be an equivalent description of the theory. However, now we have achieved a complete description of the full algebra, not just the $SO(16) \times SO(16)$ subalgebra. Moreover, the currents are easily read off from the previous formulas. Associated with the generators H^I of the Cartan subalgebra are the currents

$$H^I(\sigma) = \frac{1}{\pi} \dot{X}_L^I(\sigma) = \frac{1}{\pi}(p_L^I + \sum_{n \neq 0} \tilde{\alpha}_n^I e^{-2in\sigma}). \tag{6.4.70}$$

Similarly, associated with the generators $E_K = A_K c_K$ one has the currents

$$E_K(\sigma) = \frac{1}{\pi} V(K, z) c_K, \tag{6.4.71}$$

where $z = \exp(2i\sigma)$. The construction ensure that these currents satisfy the affine $E_8 \times E_8$ algebra with $k = 1$.

6.4.7 The E_8 and $\mathrm{spin}(32)/Z_2$ Lattices

In view of their fundamental importance it is worthwhile to describe explicitly the lattices Γ_8 and Γ_{16}.

Let us begin with the group E_8. It is a simply-laced group of rank eight and dimension 248. Therefore there are 240 roots of length squared two in the root lattice. In terms of eight orthonormal unit vectors u_i they are given by

$$\pm u_i \pm u_j \qquad i \neq j \qquad i, j = 1, 2, \ldots, 8 \tag{6.4.72}$$

$$\tfrac{1}{2}(\pm u_1 \pm u_2 \cdots \pm u_8) \quad \text{even \# of + signs.} \tag{6.4.73}$$

The 112 roots in (6.4.72) complete the root system of the spin(16) subalgebra of E_8, and the 128 roots in (6.4.73) describe the weights of one of the spinor representations of spin(16). Together they describe E_8, since its decomposition with respect to a maximal spin(16) subalgebra is given by $\mathbf{248} = \mathbf{120} + \mathbf{128}$ (see appendix 6.A). The lattice Γ_8 is generated by this root system. It is easily seen to be an even integral lattice. To show that it is self-dual, we must show that if (v, e) is an integer for each vector e in Γ_8, then v must be an integral linear combination of vectors in Γ_8. Write $v = \sum v_i u_i$, where v_i are real numbers to be determined. Picking

$e = u_i \pm u_j$, the requirement that $(v, e) = v_i \pm v_j$ be an integer for all i and j (and both choices of the $\pm$ sign) tells us that the v_i are all integers or all half-integers. Consideration of (v, e_0) for $e_0 = (1/2, 1/2, \ldots, 1/2)$ shows that $\sum v_i$ must be even. If the v_i are integers whose sum is even, then $v = \sum v_i u_i$ is an integral linear combination of root vectors $u_i \pm u_j$. If the v_i are half-integers whose sum is even, then $v_i - e_0$ are integers whose sum is even, so v is the sum of the root e_0 and roots of the form $u_i \pm u_j$. All in all, we have established that a vector whose inner product with all roots of E_8 is an integer must be a linear combination of roots of E_8, or in other words we have shown that the root lattice Γ_8 of E_8 is self-dual.

The lattice Γ_{16} is described by a similar construction. Consider the weights

$$\pm u_i \pm u_j \qquad i, j = 1, 2, \ldots, 16 \tag{6.4.74}$$

$$\tfrac{1}{2}(\pm u_1 \pm u_2 \cdots \pm u_{16}) \quad \text{even \# of + signs.} \tag{6.4.75}$$

The 480 weights in (6.4.74) complete the root system of spin(32), and the 2^{15} weights in (6.4.75) are weights of a spinor representation of spin(32). The lattice generated by these representations contains all the points of two conjugacy classes, and can be seen to be self dual just as in the discussion above. Since spin($2n$) has four conjugacy classes (adjoint, vector, spinor, spinor'), it is necessary to choose two of them in order that the volume of a unit cell of the lattice and the dual lattice be the same. Self duality can, in fact, be achieved by taking the adjoint representation conjugacy class (*i.e.*, the root lattice) and one of the spinor conjugacy classes for spin($16n$), with $n = 1, 2, \ldots$. The $n = 1$ case gives Γ_8 and the $n = 2$ case is Γ_{16}. Only in the $n = 1$ case do the additional spinor weights lead to an enlarged Lie algebra.

6.4.8 The Heterotic String Spectrum

The spectrum of low-lying states in the heterotic string theories has already been examined in the fermionic formulation. In this section the spectrum at the first excited level is worked out in detail in the bosonic description. Also the number of states at each excited level will be calculated and shown to agree with the corresponding result in the fermionic description.

A state with compactified left-moving momentum K contributes an amount $\tfrac{1}{2}K^2$ to $\tfrac{1}{4}$(mass)^2. Therefore to compute the degeneracies, we need to know how many points of any given length exist on the lattice. This information is conveniently encoded in the theta function of the lattice Γ, which is defined by

$$\Theta_\Gamma(\tau) = \sum_{w \in \Gamma} e^{\pi i \tau |w|^2}, \tag{6.4.76}$$

where the sum runs over all sites (or weights) w of the lattice. For an even lattice this series can be rewritten in the form

$$\Theta_\Gamma(\tau) = \sum_{n=0}^{\infty} d_n e^{2\pi i n \tau}, \tag{6.4.77}$$

where d_n is the number of lattice sites w with $w \cdot w = 2n$. Thus the degeneracies are just the coefficients of the various frequencies in the expansion (6.4.76).

In §3.2.4 we introduced the modular group of transformations

$$\tau \to \tau' = \frac{a\tau + b}{c\tau + d}, \tag{6.4.78}$$

where $\begin{pmatrix} a & b \\ c & d \end{pmatrix} \in SL(2, Z)$. The entries a, b, c, d are integers and $ad - bc = 1$. A function $G(\tau)$ is called a modular form of weight $2k$ if

$$G(\tau') = (c\tau + d)^{2k} G(\tau) \tag{6.4.79}$$

for all modular group transformations. The theta function of an even self-dual lattice in d dimensions is a modular form of weight $d/2$. (Modular forms are discussed in appendix 6.B, and the proof that the theta function of a self-dual lattice is a modular form of weight $d/2$ is given in appendix 9.B.)

By the uniqueness theorem quoted in appendix 6.B there is just one modular form of weight four (up to normalization) and therefore the theta function for the E_8 lattice is

$$\begin{aligned}
\Theta_{\Gamma_8} &= 1 + 240 \sum_{m=1}^{\infty} \sigma_3(m) e^{2\pi i m \tau}, \\
&= 1 + 240 e^{2\pi i \tau} + 9 \cdot 240 e^{4\pi i \tau} + \cdots,
\end{aligned} \tag{6.4.80}$$

which is a modular form of weight four. The coefficients $\sigma_\alpha(m)$ are defined in appendix 6.B. Since there is a unique modular form of weight eight (see appendix 6.B), the $\Gamma_8 \times \Gamma_8$ and Γ_{16} lattice must have the same theta functions

$$\Theta_{\Gamma_{16}} = \Theta_{\Gamma_8 \times \Gamma_8} = (\Theta_{\Gamma_8})^2 = 1 + 480 \sum_{m=1}^{\infty} \sigma_7(m) e^{2\pi i m \tau} \tag{6.4.81}$$

$$= 1 + 480 e^{2\pi i \tau} + 129 \cdot 480 e^{4\pi i \tau} + \cdots$$

The fact that the two lattices have the same theta functions implies that the two heterotic string theories have exactly the same number of states at every mass level. Let us now work out what those degeneracies are.

The number of right-moving superstring modes at the Nth mass level was seen in §5.3.1 to be given by $d_R(N)$, where

$$\sum_{N=1}^{\infty} d_R(N) x^N = 16 \prod_{n=1}^{\infty} \left(\frac{1+x^n}{1-x^n} \right)^8. \tag{6.4.82}$$

The degeneracy of bosonic left-moving modes at the Nth mass level compactified on $\Gamma_8 \times \Gamma_8$ or Γ_{16} with

$$\tilde{N} + \tfrac{1}{2} p_L^2 - 1 = N \tag{6.4.83}$$

is given by

$$\sum_{N=-1}^{\infty} d_L(N) x^N = \frac{1}{x}(1 + 480 \sum_{1}^{\infty} \sigma_7(m) x^m) \prod_{1}^{\infty} (1 - x^n)^{-24}. \tag{6.4.84}$$

The factor $1/x$ in (6.4.84) corresponds to the -1 term in (6.4.83), the second factor is Θ_Γ and the last one is the partition function for 24 bosonic dimensions. The number of heterotic string states at the Nth mass level is given by the product

$$d(N) = d_R(N) d_L(N), \tag{6.4.85}$$

that arises from tensoring right-moving modes with left-moving ones. For example

$$d(0) = 16 \cdot [480 + 24] \tag{6.4.86}$$

counts the $16 \cdot 8$ states of the supergravity multiplet and the $16 \cdot 496$ states of the super Yang–Mills multiplet in the adjoint of $E_8 \times E_8$ or spin(32)/Z_2.

The asymptotic density of states can be deduced by the methods described in §5.3. One finds that

$$d(n) \sim n^{-11/2} \exp[(\sqrt{2} + 2)2\pi\sqrt{n}], \qquad (6.4.87)$$

which implies a mass density

$$\rho(m) \sim m^{-10} \exp(m/m_0) \qquad (6.4.88)$$

with

$$m_0 = \left[(2 + \sqrt{2})\pi\sqrt{\alpha'}\right]^{-1}. \qquad (6.4.89)$$

The equivalence of the bosonic and fermionic formulations of the heterotic string, established abstractly from the uniqueness theorem for current algebras, can be checked by comparing the number of states at each level in the two different formulations of the model. The contribution to the partition function of eight extra dimensions compactified on an E_8 torus is

$$\text{ch}(\hat{E}_8) = [1 + 240 \sum_1^\infty \sigma_3(n)x^n] \prod_1^\infty (1 - x^m)^{-8}. \qquad (6.4.90)$$

where the first term, the theta function of the lattice, arises from the zero modes, and the infinite product arises from the oscillators associated with the nonzero modes. This function is known as the character function of $\hat{E}_8$ in the basic representation. In the fermionic formulation, including the restriction to even fermion number and adding the contributions of the A and P sectors, the corresponding formula is

$$\text{ch}(\hat{E}_8) = \tfrac{1}{2}[\prod_1^\infty (1 - x^{n-1/2})^{16} + \prod_1^\infty (1 + x^{n-1/2})^{16}] + 128x \prod_1^\infty (1 + x^n)^{16}. \qquad (6.4.91)$$

The fact that these two expressions are equal is analogous to the identity of Jacobi cited in §4.3.3 as evidence of supersymmetry. The analogous identity

$$\text{ch}\left(\widehat{SO}(32)\right) = (1 + 480 \sum_1^\infty \sigma_7(n)x^n) \prod_1^\infty (1 - x^m)^{-16} \qquad (6.4.92)$$

$$= \tfrac{1}{2}[\prod_1^\infty (1 - x^{n-1/2})^{32} + \prod_1^\infty (1 + x^{n-1/2})^{32}] + 2^{15}x^2 \prod_1^\infty (1 + x^n)^{32}$$

also holds. In fact

$$\mathrm{ch}\big(\widehat{SO}(32)\big) = \mathrm{ch}(\hat{E}_8 \times \hat{E}_8) = [\mathrm{ch}(\hat{E}_8)]^2. \qquad (6.4.93)$$

The proof of this statement is based on a theorem in the theory of modular forms cited in appendix 6.B.

We have already described the massless states in the bosonized version of the heterotic string theories in the last subsection. Therefore let us now examine the first excited level, $N = 1$. The 256 right-moving modes are

$$(44, 1) \oplus (84, 1) \oplus (128, 1). \qquad (6.4.94)$$

The first label refers to spin(9) (the 'spin') and the second label to $E_8 \times E_8$ or spin(32)/Z_2. This multiplet was described in §5.3.1. There are 73,764 left-moving modes with $N = 1$ given by

$$(44, 1) \oplus (9, 496) \oplus (1, 69256). \qquad (6.4.95)$$

These are constructed as follows: The massive tensor $(44, 1)$ is given by

$$\tilde{\alpha}^i_{-1} \tilde{\alpha}^j_{-1} |0\rangle, \qquad \tilde{\alpha}^i_{-2} |0\rangle. \qquad (6.4.96)$$

The massive vectors $(9, 496)$ decompose under $SO(8)$ into $(8, 496) \oplus (1, 496)$. The $(8, 496)$ states are given by

$$\tilde{\alpha}^i_{-1} \tilde{\alpha}^I_{-1} |0\rangle, \qquad \tilde{\alpha}^i_{-1} |K\rangle, \qquad K^2 = 2 \qquad (6.4.97)$$

where $K^2 = 2$. The remaining $(1, 496)$ are given by $\tilde{\alpha}^I_{-2} |0\rangle$ and 480 of the states $|K\rangle$ with $K^2 = 4$. Since there are altogether 61,920 lattice sites with $K^2 = 4$ this leaves 61,440 of them, which combine with the 136 states

$$\tilde{\alpha}^I_{-1} \tilde{\alpha}^J_{-1} |0\rangle \qquad (6.4.98)$$

and the 7680 states

$$\tilde{\alpha}^I_{-1} |K\rangle, \qquad K^2 = 2 \qquad (6.4.99)$$

to make the $(1, 69256)$.

In the case of $E_8 \times E_8$

$$69,256 = (248, 248) \oplus 2(1, 1) \oplus (3875, 1) \oplus (1, 3875), \qquad (6.4.100)$$

whereas in the case of spin(32)

$$69,256 = 2^{15} \oplus 35,960 \oplus 527 \oplus 1. \qquad (6.4.101)$$

The **35,960** is a fourth-rank antisymmetric tensor and the **527** is a second-rank symmetric traceless tensor. Tensoring the left- and right-moving modes gives

$$256 \times 73,764 = 18,883,584 \qquad (6.4.102)$$

states for the first excited level of the heterotic string! The spinor representation of spin(32)/Z_2 at this level should be absolutely stable, which might be an interesting prediction for that theory.

6.5   Summary

This chapter has described two methods of introducing nonabelian gauge symmetry in string theories. The Chan-Paton method, which dates back to the early days of dual theory, involves attaching 'charges' at the ends of open strings. This procedure can be used for any classical groups by associating to emitted strings matrices in the fundamental representation. It is not applicable to exceptional groups. In the context of type I superstrings, we will show in chapter 10 that only the choice $SO(32)$ leads to an anomaly-free theory.

The second method of introducing gauge symmetry distributes the charge along the length of a closed string. The charge density satisfies a Kac-Moody algebra, which can also be viewed as a two-dimensional current algebra. Symmetries of this type can be described in either a fermionic or a bosonic form. In the bosonic version, one introduces extra dimensions that are compactified t form the maximal torus of the corresponding Lie algebra. The group generators, or the full current algebra, has a vertex operator representation of the type introduced by Frenkel and Kac and Segal. In the heterotic string theory the momenta conjugate to the torus are required to form a self-dual 16-dimensional torus. This singles out $E_8 \times E_8$ and spin(32)/Z_2 as the unique possibilities.

Appendix 6.A   Elements of E_8

The purpose of this appendix is to give an elementary exposition of some selected properties of the exceptional group E_8 and some of its subgroups. E_8 is the largest and in many ways the most interesting of the exceptional finite dimensional Lie algebras; the others are all subalgebras of E_8. By constructing the foregoing vertex operator representation of E_8 we have already given an explicit description of this algebra and a proof that it exists. But it seems appropriate to give a direct and perhaps more elementary presentation.

We now construct the Lie algebra of E_8 beginning with an $SO(16)$ subalgebra. The generators of $SO(16)$ are operators J_{ij} (with $J_{ij} = -J_{ji}$, so that there are $16 \cdot 15/2 = 120$ of them) that obey the $SO(16)$ Lie algebra

$$[J_{ij}, J_{kl}] = J_{il}\delta_{jk} - J_{jl}\delta_{ik} - J_{ik}\delta_{jl} + J_{jk}\delta_{il}. \tag{6.A.1}$$

To these we adjoin operators Q_α transforming in the positive chirality spinor representation of $SO(16)$. As we know from appendix 5.A, this representation has dimension $2^7 = 128$, so the resulting Lie algebra, if we can construct it, has dimension $120 + 128 = 248$. The statement that the Q_α transform as spinors of $SO(16)$ means that[*]

$$[J_{ij}, Q_\alpha] = (\sigma_{ij})_{\alpha\beta}Q_\beta. \tag{6.A.2}$$

To complete the specification of the 248-element Lie algebra that we are trying to describe, we must define the remaining commutator $[Q_\alpha, Q_\beta]$. (We are trying to define a Lie algebra, not a superalgebra, so this is a commutator, not an anticommutator!) $SO(16)$ group theory uniquely determines this up to normalization to be

$$[Q_\alpha, Q_\beta] = (\sigma_{ij})_{\alpha\beta}J_{ij}. \tag{6.A.3}$$

While $SO(16)$ group theory would permit us to multiply the right-hand side of (6.A.3) by a constant, such a factor can be absorbed in rescaling

---

[*] As in appendix 5.A, $SO(16)$ gamma matrices are denoted as γ^i, $i = 1, \ldots, 16$ and $SO(16)$ generators in the spinor representation as $\sigma_{ij} = [\gamma_i, \gamma_j]/4$. The chirality operator is $\overline{\gamma} = \gamma_1\gamma_2\ldots\gamma_{16}$. Also, we define the antisymmetrized product of n gamma matrices $\gamma_{i_1 i_2 \ldots i_n} = (\gamma_1\gamma_2\ldots\gamma_n \pm \text{permutations})/n!$; thus $\gamma_{i_1 i_2} = 2\sigma_{i_1 i_2}$. Like the $SO(8)$ gamma matrices explicitly constructed in the preceding chapter, $SO(16)$ gamma matrices can be chosen to be real; indeed, they can be constructed similarly to the construction of $SO(8)$ gamma matrices. The positive chirality spinor Q_α can likewise be chosen to be real; thus we need not distinguish between Q_α and $\overline{Q}_\alpha$. Otherwise indeed we would have to add the $\overline{Q}_\alpha$ to make a real Lie algebra.

the Q_α, since none of the previous formulas determine the normalization of Q_α. Thus, if a Lie algebra with the assumed $SO(16)$ content exists, it is given by (6.A.1), (6.A.2), and (6.A.3). There is nothing that can be adjusted.

To determine whether these formulas define a Lie algebra, we must determine whether the Jacobi identity is obeyed. Many of the Jacobi identities are obviously obeyed. The identity involving three J's merely asserts that $SO(16)$ is a Lie algebra; the JJQ identity asserts that the spinors indeed form a representation of $SO(16)$; the QQJ identity can be reduced to the QQQ identity using (6.A.3). Therefore, the really crucial case to check is the Jacobi identity $[[Q_\alpha, Q_\beta], Q_\gamma] + [[Q_\beta, Q_\gamma], Q_\alpha] + [[Q_\gamma, Q_\alpha], Q_\beta] = 0$. Writing this out explicitly with the aid of (6.A.3), the identity that must hold if the E_8 Lie algebra is to exist with the $SO(16)$ content we have assumed is

$$(\sigma_{ij})_{\alpha\beta}(\sigma_{ij})_{\gamma\delta} + (\sigma_{ij})_{\beta\gamma}(\sigma_{ij})_{\alpha\delta} + (\sigma_{ij})_{\gamma\alpha}(\sigma_{ij})_{\beta\delta} = 0. \qquad (6.A.4)$$

Of course, we must establish this identity only in the case that α, β and γ are all spinor indices of the same chirality. This identity has a striking resemblance to the identity that was needed in appendix 4.A in order for supersymmetric Yang–Mills theory to exist; indeed, apart from explaining useful facts about E_8, one purpose of the present discussion is to underscore the resemblance of E_8 to a supersymmetric system.

To establish (6.A.4), note that since the product of two spinors can be expanded in a complete set of Dirac matrices, (6.A.4) will hold if it holds when contracted with $(\gamma_{k_1 k_2 \ldots k_n})_{\alpha\beta}$ for all n and all $k_1, k_2, \ldots, k_n$. Because α and β are indices of the same chirality, we only have to worry about even n. Because (6.A.3) is antisymmetric in α and β, we only need to consider values of n for which $(\gamma_{k_1 k_2 \ldots k_n})_{\alpha\beta}$ is antisymmetric in α and β; using elementary properties of γ matrices, this means $n = 2, 6, 10$, or 14. Actually, because of the identity $\gamma_{i_1 \ldots i_k} = \epsilon_{i_1 \ldots i_{16}} \gamma_{i_{k+1} \ldots i_{16}} \cdot \overline{\gamma}/(16 - k)!$ and the fact that the chirality operator $\overline{\gamma}$ can be dropped when acting on positive chirality spinor indices α or β, it is enough to check the cases $n = 2$ and $n = 6$. (To see in another way why only $n = 2$ and $n = 6$ enter, note that the antisymmetric combination of Q_α and Q_β has

$$\frac{128 \cdot 127}{2} = \frac{16 \cdot 15}{2} + \frac{16 \cdot 15 \cdot 14 \cdot 13 \cdot 12 \cdot 11}{6!} \qquad (6.A.5)$$

independent terms. The two numbers on the right-hand side of (6.A.5) are the number of independent components of $\gamma_{i_1 i_2 \ldots i_k}$ for $k = 2$ and $k = 6$, respectively.)

Contracting (6.A.3) with $\gamma_{i_1 i_2}$ or equivalently with $\sigma_{i_1 i_2}$, we get

$$(\text{tr}_+\sigma_{kl}\sigma_{ij}) \cdot (\sigma_{ij})_{\gamma\delta} - 2(\sigma_{ij}\sigma_{kl}\sigma_{ij})_{\gamma\delta}, \qquad (6.A.6)$$

where tr_+ denotes a trace in the positive-chirality spinor representation. This vanishes (for $SO(16)$) with use of standard Dirac identities. (The two terms in (6.A.6) equal, respectively, $\mp 256(\sigma_{kl})_{\gamma\delta}$.) Contracting with $\gamma_{i_1...i_6}$ gives

$$-2(\sigma_{jk}\gamma_{i_1...i_6}\sigma_{jk})_{\gamma\delta}, \qquad (6.A.7)$$

which again vanishes for $SO(16)$ with the aid of elementary gamma matrix algebra. This completes our elementary construction of the Lie algebra known as E_8.

Just as supersymmetric Yang–Mills theory with minimal field content exists in a few dimensions (3, 4, 6 and 10) of which 10 is only the largest, there are a few cases in which one can build a Lie algebra by adjoining a real spinor to the adjoint representation of $SO(N)$. Apart from the case $N = 16$, which leads to E_8, other cases that work are $N = 9$ and $N = 8$. By adding the 16-component spinor of $SO(9)$ to the thirty-six-dimensional adjoint representation, one can in precise analogy to the above build a Lie algebra known as the exceptional Lie algebra F_4. By adding an eight component positive – or negative – chirality spinor to the 28-dimensional adjoint representation of $SO(8)$, one can build the Lie algebra of $SO(9)$ – in a basis that differs by an $SO(8)$ triality rotation from more standard (and easier) constructions of that algebra.

We now wish to describe some subgroups of E_8. We already have exhibited one maximal subgroup, namely $SO(16)$. $SO(16)$ contains, of course, a subgroup $SO(10) \times SO(6)$. The adjoint representation of $SO(16)$ is easily seen to decompose under $SO(10) \times SO(6)$ as

$$(\mathbf{45}, \mathbf{1}) \oplus (\mathbf{1}, \mathbf{15}) \oplus (\mathbf{10}, \mathbf{6}). \qquad (6.A.8)$$

The first two terms are the adjoint representations of $SO(10)$ and $SO(6)$, while the last is the product of the respective vector representations. How does the spinor of $SO(16)$ transform under $SO(10) \times SO(6)$? The spinor of $SO(16)$ is formed, as we discussed in appendix 5.A, by introducing 16 gamma matrices $\gamma_1 \ldots \gamma_{16}$. The first ten can be regarded as gamma matrices of $SO(10)$, while the last six can be regarded as gamma matrices of $SO(6)$. Thus, the spinor of $SO(16)$ transforms as the product of a spinor of $SO(10)$ and a spinor of $SO(6)$. What about chirality? The $SO(16)$ chirality operator is $\overline{\gamma} = \gamma_1 \ldots \gamma_{16}$ while the $SO(10)$ chirality operator is

$\gamma^{(10)} = \gamma_1 \dots \gamma_{10}$ and the $SO(6)$ chirality operator is $\gamma^{(6)} = \gamma_{11} \dots \gamma_{16}$. Evidently,

$$\overline{\gamma} = \gamma^{(10)} \cdot \gamma^{(6)}. \tag{6.A.9}$$

Thus, the positive–chirality spinor Q_α of $SO(16)$ decomposes into pieces of $\gamma^{(10)} = \gamma^{(6)} = +1$ or $\gamma^{(10)} = \gamma^{(6)} = -1$. We also know from appendix 5.A that the spinors of $SO(10)$ or $S0(6)$ of definite chirality have dimension 16 and 4, respectively. If we denote the positive- and negative-chirality spinors of $SO(10)$ (or $SO(6)$) as $\mathbf{16}$ and $\overline{\mathbf{16}}$, respectively (or as $\mathbf{4}$ and $\overline{\mathbf{4}}$, respectively), then we have established the $SO(10) \times SO(6)$ decomposition of the $\mathbf{128}$ of $SO(16)$:

$$\mathbf{128} = (\mathbf{16}, \mathbf{4}) \oplus (\overline{\mathbf{16}}, \overline{\mathbf{4}}). \tag{6.A.10}$$

(6.A.8) and (6.A.10) together give the $SO(10) \times SO(6)$ decomposition of the adjoint representation of E_8.

We would now like to describe another exceptional group, namely E_6, as a subgroup of E_8. As a preliminary, note that in the $\mathbf{4}$ of $SO(6)$ the $SO(6)$ generators are hermitian 4×4 complex matrices, which must be traceless since $SO(6)$ is a simple Lie algebra; hence they are $SU(4)$ generators. Hence, the $SO(6)$ Lie algebra is a subalgebra of the $SU(4)$ Lie algebra. Noting that these Lie algebras both have 15 generators, we conclude that the $SO(6)$ Lie algebra cannot be a proper subalgebra of the $SU(4)$ Lie algebra; they must be the same algebra.[*] We also learn in this way that the fundamental $\mathbf{4}$ and $\overline{\mathbf{4}}$ of $SU(4)$ are the positive- and negative-chirality spinors of $SO(6)$. Conversely, the fundamental $\mathbf{6}$ of $SO(6)$ is the antisymmetric second-rank tensor of $SU(4)$, which would have $4 \cdot 3/2 = 6$ components.

In the above, instead of speaking of an $SO(10) \times SO(6)$ subalgebra of E_8, we could speak of $SO(10) \times SU(4)$. Now, $SU(4)$ has an obvious $U(1) \times SU(3)$ subalgebra. If we use a superscript to denote the $U(1)$ charge, the $\mathbf{4}$ of $SU(4)$ decomposes as $\mathbf{1}^3 \oplus \mathbf{3}^{-1}$ under $U(1) \times SU(3)$. The $\mathbf{6}$ of $SU(4)$ (being as we just noted the antisymmetric product of two $\mathbf{4}$'s) transforms as $\mathbf{3}^2 \oplus \overline{\mathbf{3}}^{-2}$ under $U(1) \times SU(3)$, and the adjoint representation of $SU(4)$ (being $\mathbf{4} \otimes \overline{\mathbf{4}}$ with a singlet removed) transforms as $\mathbf{8}^0 \oplus \mathbf{3}^{-4} \oplus \overline{\mathbf{3}}^4 \oplus \mathbf{1}^0$ under $SU(3)$ (the $\mathbf{8}$ is the adjoint of $SU(3)$). Combining these facts with (6.A.8) and (6.A.10), we get the $SO(10) \times U(1) \times SU(3)$

---

[*] This can also be established by constructing the $SU(N)$ and $SO(N)$ Dynkin diagrams and then noting that $SU(4)$ and $SO(6)$ have the same Dynkin diagram.

content of the adjoint of E_8:

$$248 = ((\mathbf{45},\mathbf{1})^0 \oplus (\mathbf{1},\mathbf{1})^0 \oplus (\mathbf{16},\mathbf{1})^3 \oplus (\overline{\mathbf{16}},\mathbf{1})^{-3})$$
$$\oplus ((\mathbf{16},\mathbf{3})^{-1} \oplus (\mathbf{10},\mathbf{3})^2 \oplus (\mathbf{1},\mathbf{3})^{-4}) \qquad (6.A.11)$$
$$\oplus ((\overline{\mathbf{16}},\overline{\mathbf{3}})^1 \oplus (\mathbf{10},\overline{\mathbf{3}})^{-2} \oplus (\mathbf{1},\overline{\mathbf{3}})^4) \oplus (\mathbf{1},\mathbf{8})^0.$$

Of note here is that the adjoint representation $\mathbf{248}$ contains 78 generators that are $SU(3)$ singlets. As the commutator of two $SU(3)$ singlets must be an $SU(3)$ singlet, we conclude that these 78 generators form a closed subalgebra of E_8; it is known as the exceptional Lie algebra E_6. Evidently, E_6 has a maximal subalgebra $SO(10) \times U(1)$ with the adjoint representation of E_6 decomposing under that subalgebra as

$$\mathbf{78} = \mathbf{45}^0 \oplus \mathbf{16}^3 \oplus \overline{\mathbf{16}}^{-3} \oplus \mathbf{1}^0. \qquad (6.A.12)$$

What is more, in (6.A.11) there are 27 copies of the $\mathbf{3}$ of $SU(3)$. These must be mapped into themselves by E_6 transformations, so E_6 must have a 27-dimensional representation with $SO(10) \times U(1)$ content

$$\mathbf{27} = \mathbf{16}^{-1} \oplus \mathbf{10}^2 \oplus \mathbf{1}^{-4}. \qquad (6.A.13)$$

As a check, note that as $16 \cdot (-1) + 10 \cdot 2 + 1 \cdot (-4) = 0$, the trace of the $U(1)$ generator in the $\mathbf{27}$ of E_6 is zero. This is in agreement with the fact that the trace of any generator of a simple Lie algebra vanishes in any representation. (This also proves that the $\mathbf{27}$ is irreducible, since the $U(1)$ trace would not vanish if we drop some pieces from (6.A.13).) The complex conjugate of the $\mathbf{27}$ is an E_6 representation

$$\overline{\mathbf{27}} = \overline{\mathbf{16}}^1 \oplus \mathbf{10}^{-2} \oplus \mathbf{1}^4. \qquad (6.A.14)$$

Equations (6.A.13) and (6.A.14) are obviously not isomorphic to one another, so the $\mathbf{27}$ and $\overline{\mathbf{27}}$ of E_6 are complex representations, inequivalent to their complex conjugates. E_6 is indeed the only exceptional Lie algebra that does have complex representations. Study of (6.A.11) shows further that the adjoint representation of E_8 transforms as

$$\mathbf{248} = (\mathbf{78},\mathbf{1}) \oplus (\mathbf{1},\mathbf{8}) \oplus (\mathbf{27},\mathbf{3}) \oplus (\overline{\mathbf{27}},\overline{\mathbf{3}}) \qquad (6.A.15)$$

under $E_6 \times SU(3)$.

Use of an $SU(2)$ subgroup of $SU(3)$ would enable us to discover in a similar way a new exceptional algebra, E_7, with 133 generators. We will not enter into the details. Apart from E_6, E_7 and E_8, the other exceptional Lie algebras are F_4 and G_2. We have already noted above that F_4 can be constructed starting from $SO(9)$ just as we constructed E_8 starting from $SO(16)$, by adding a spinor. The $SO(9)$ construction can obviously be embedded in the $SO(16)$ construction, and this gives a natural embedding of F_4 in E_8. The subgroup that commutes with F_4 then turns out to be G_2, rather as in the above construction we found E_6 as the subgroup of E_8 that commutes with $SU(3)$. Thus, E_8 contains $G_2 \times F_4$ as a subgroup. (G_2 has many other and perhaps simpler characterizations. It is the symmetry group of the octonion multiplication table; it is the subgroup of $SO(7)$ that leaves fixed an element of the spinor representation.)

We conclude by briefly mentioning some other maximal subgroups of E_8 and E_6. E_8 has the maximal subgroup $SU(5) \times SU(5)$, with decomposition

$$\mathbf{248} = (\mathbf{24}, \mathbf{1}) \oplus (\mathbf{1}, \mathbf{24}) \oplus (\mathbf{5}, \overline{\mathbf{10}}) \oplus (\mathbf{10}, \mathbf{5}) \oplus (\overline{\mathbf{5}}, \mathbf{10}) \oplus (\overline{\mathbf{10}}, \overline{\mathbf{5}}). \quad (6.A.16)$$

E_6 has a maximal subgroup $SU(6) \times SU(2)$, with the decompositions

$$\mathbf{78} = (\mathbf{35}, \mathbf{1}) \oplus (\mathbf{1}, \mathbf{3}) \oplus (\mathbf{20}, \mathbf{2})$$
$$\mathbf{27} = (\overline{\mathbf{15}}, \mathbf{1}) \oplus (\mathbf{6}, \mathbf{2}). \quad (6.A.17)$$

Here $\mathbf{2}$ is the fundamental representation of $SU(2)$ and $\mathbf{3}$ the adjoint representation. $\mathbf{6}$ is the fundamental representation of $SU(6)$, $\overline{\mathbf{15}}$ the antisymmetric product of two $\overline{\mathbf{6}}$'s, $\mathbf{35}$ the adjoint representation, and $\mathbf{20}$ the third rank self-dual antisymmetric tensor. E_6 has a maximal subgroup $SU(3) \times SU(3) \times SU(3)$, which will be of considerable significance in chapter 16. It has the decomposition

$$\mathbf{78} = (\mathbf{8}, \mathbf{1}, \mathbf{1}) \oplus (\mathbf{1}, \mathbf{8}, \mathbf{1}) \oplus (\mathbf{1}, \mathbf{1}, \mathbf{8}) \oplus (\mathbf{3}, \mathbf{3}, \mathbf{3}) \oplus (\overline{\mathbf{3}}, \overline{\mathbf{3}}, \overline{\mathbf{3}})$$
$$\mathbf{27} = (\mathbf{3}, \overline{\mathbf{3}}, \mathbf{1}) \oplus (\mathbf{1}, \mathbf{3}, \overline{\mathbf{3}}) \oplus (\overline{\mathbf{3}}, \mathbf{1}, \mathbf{3}). \quad (6.A.18)$$

These subgroups and decompositions can be discovered (with some patience) along lines similar to the above. For instance, to obtain (6.A.18), one can pick a maximal $SU(3) \times U(1) \times SU(2) \times SU(2)$ subgroup[*] of $SO(10) \in E_6$ (the $\mathbf{10}$ of $SO(10)$ decomposing as $(\mathbf{3}, \mathbf{1}, \mathbf{1})^1 \oplus (\overline{\mathbf{3}}, \mathbf{1}, \mathbf{1})^{-1} \oplus (\mathbf{1}, \mathbf{2}, \mathbf{2})^0$). Using (6.A.12) to decompose the $\mathbf{78}$ of E_6 under this, one finds 16 generators that commute with $SU(3)$. From their $U(1) \times SU(2) \times SU(2)$ content these 16 operators are seen to generate $SU(3) \times SU(3)$, and with some effort one arrives at (6.A.18).

---

[*] This subgroup is used in grand unified $SO(10)$ models.

Appendix 6.B   Modular Forms

In §3.3 we introduced the modular transformations

$$\tau \to \tau' = \frac{a\tau + b}{c\tau + d}, \tag{6.B.1}$$

where a, b, c, and d are integers and $ad - bc = 1$. This set of transformations forms a group called the modular group, isomorphic to $SL(2, Z)$. A function $G(\tau)$ is called a modular form of weight $2k$ if

$$G(\tau') = (c\tau + d)^{2k} G(\tau) \tag{6.B.2}$$

for all modular group transformations. Modular forms are of fundamental importance in number theory, but to explain this would unfortunately take us too far afield. Examples of modular forms of weight $2k$ are the Eisenstein series

$$G_{2k}(\tau) = \sum_{(m,n)\neq(0,0)} (m + n\tau)^{-2k}, \tag{6.B.3}$$

which converges for $k = 2, 3, \ldots$. That (6.B.3) defines a modular form can be proved without too much difficulty. The intuitive idea is that the sum in (6.B.3) runs over all lattice points of a lattice in the complex plane whose lattice structure is determined (in a manner described in §3.3) in terms of the complex number τ. Modular transformations map this lattice into itself (this was the motivation for introducing them in §3.3) so (6.B.3) transforms simply under modular transformations. A basic theorem in the theory of modular forms states than an arbitrary holomorphic modular form of weight $2k$ can be expressed as a polynomial in G_4 and G_6.[†] Since the weights of modular forms are additive under multiplication, the only modular form of weight eight is G_4^2. The lowest weight for which there is more than one independent modular form is weight 12, for which there are two independent modular forms, namely G_4^3 and G_6^2.

The functions G_{2k} can be rewritten in the alternative form

$$G_{2k}(\tau) = 2\zeta(2k)[1 + c_{2k} \sum_{m=1}^{\infty} \sigma_{2k-1}(m)e^{2\pi i m\tau}], \tag{6.B.4}$$

[†] The proof of this theorem is not difficult. For a readable account see J.P. Serre, 'A course in arithmetic' (Springer, 1973).

where ζ is the Riemann zeta function,

$$\zeta(z) = \sum_{n=1}^{\infty} n^{-z}, \tag{6.B.5}$$

$$c_{2k} = \frac{(2\pi i)^{2k}}{(2k-1)!\,\zeta(2k)}, \tag{6.B.6}$$

and

$$\sigma_{\alpha}(k) = \sum_{d|k} d^{\alpha}. \tag{6.B.7}$$

The sum in (6.B.7) runs over all positive integer divisors of k. Thus $\sigma_{\alpha}(1) = 1$, $\sigma_{\alpha}(2) = 1 + 2^{\alpha}$, and so forth. The first few values of c_{2k} are $c_4 = 240$, $c_6 = -504$, and $c_8 = 480$.

It can be shown that the theta function of an even self-dual lattice in d dimensions is a modular form of weight $d/2$. This will be shown in appendix 9.B. By the uniqueness theorem for modular forms, the theta function for the E_8 lattice must therefore be proportional to G_4. The zero-frequency term is 1, which fixes the normalization. Thus

$$\Theta_{\Gamma_8} = 1 + 240 \sum_{m=1}^{\infty} \sigma_3(m) e^{2\pi i m \tau}, \tag{6.B.8}$$

which is a modular form of weight four. Since there is a unique modular form of weight eight, given by G_4^2 or G_8, the $\Gamma_8 \times \Gamma_8$ and Γ_{16} lattices must have the same theta functions

$$\Theta_{\Gamma_{16}} = \Theta_{\Gamma_8 \times \Gamma_8} = \left(\Theta_{\Gamma_8}\right)^2 = 1 + 480 \sum_{m=1}^{\infty} \sigma_7(m) e^{2\pi i m \tau}$$

$$= \left(1 + 240 \sum_{m=1}^{\infty} \sigma_3(m) e^{2\pi i m \tau}\right)^2. \tag{6.B.9}$$

7. Tree Amplitudes

A brief description of string interactions was presented in the introductory chapter. Feynman diagrams of string theory were shown to correspond to world sheets that describe the space-time history of strings joining and splitting. After continuation to a Euclidean metric, the diagrams can be classified by their topology. The order of a particular diagram in the perturbation expansion is determined by the number of handles (as well as the number of windows in the case of theories with open strings) and the number of external lines. The leading terms for a given process correspond to the tree approximation. Once a tree approximation (corresponding to a consistent classical field theory) is known, the full quantum theory is in principle determined by unitarity (modulo the possible appearance of parameters such as θ angles that do not show up perturbatively). The quantum theory itself is consistent if finite S matrix elements satisfying unitarity, causality, and the other usual requirements can be obtained. In practice one ordinarily investigates these matters in the context of perturbation theory, checking, in particular, for nonrenormalizable divergences or anomalies that would imply a breakdown of gauge invariance.

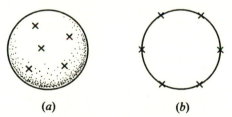

(a) $\qquad\qquad\qquad$ (b)

Figure 7.1. World-sheet configurations for tree diagrams of (a) closed and (b) open strings.

The amplitude for the scattering of M on-mass-shell closed-string states is described in tree approximation by a world sheet that is topologically a sphere with the M external particles attached at M specific points z_i on the surface, as shown in fig. 7.1a. In principle, one integrates over all

353

geometries of this topology and all values of the points z_i, up to conformal equivalence. At tree level, as was discussed in §1.4.3, an arbitrary metric on the sphere is conformally equivalent to the standard round metric on the sphere (which, if we remove a point, can be projected on the plane), and the sphere itself is invariant under an $SL(2,C)$ subgroup of the conformal group. Therefore there is no geometric integral to be done and three of the z_i can be fixed at three arbitrarily chosen points on the sphere. In this diagram there obviously is no natural ordering of the particles. Being the unique diagram at this order it must give total symmetry among identical bosons or total antisymmetry among identical fermions.

In the case of open strings the world sheet has boundaries, and emitted open-string states are attached to boundaries. The cyclic ordering among sets of particles attached to common boundaries is meaningful. As we discussed in chapter 1 and in more detail in §6.1, Yang–Mills symmetry can be incorporated by weighting inequivalent cyclic orderings by suitable group-theoretic factors. The tree approximation corresponds to the disk diagram, which has one boundary, as shown in fig. 7.1b. In this case the cyclic ordering of all the open-string states must be specified. Closed-string states can be attached to the interior of the surface at the same time to describe a tree amplitude for a 'mixed' process.

The purpose of this chapter is to describe the explicit construction of tree amplitudes in considerable detail. For the most part we use 'old-fashioned' operator methods, which are actually quite effective. Compared to the prescription described in chapter 1, the operator methods make certain properties, like unitarity, more obvious at the cost of making other properties, like crossing symmetry, less apparent. In any case the various methods are closely linked; the operator methods provide, as we will discuss, a concrete recipe for building up a world sheet out of components. The latter statement is in fact central in yet another way of thinking about interactions, which will be presented in chapter 11.

7.1   Bosonic Open Strings

The understanding of string theory has not yet been developed to the point where one can write down a Lagrangian and follow a standard prescription to deduce rules for constructing Feynman diagrams that provide the loop expansion of the full quantum theory.[*] In chapter 1 we described

---

[*] There has been recent progress in this direction, but we will not attempt to survey these developments in this book.

a certain plausible prescription for string-theory amplitudes as sums over world sheets of all possible intrinsic geometries, defined up to conformal equivalence, and showed that it gave well-defined tree amplitudes. Here we wish to describe a different approach that gives the same answers. It dates back to the very early days of dual string theory. The basic idea is to postulate a certain reasonable set of rules for constructing diagrams in terms of vertices and propagators and to demonstrate that they give rise to scattering amplitudes that incorporate all the desirable features that are required for a sensible quantum theory. These rules are roughly of the form that one might expect to deduce from a Lagrangian, even though none is known that gives precisely these rules. The prescription turns out to give satisfactory results for on-mass-shell S matrix elements only, a fact that makes it extremely difficult to work backwards to deduce a string field theory action from which they can be derived. Also, the extension to loop amplitudes has some nontrivial elements. The operator methods that we employ here make unitarity manifest, especially at tree and one-loop order, where the methods can be implemented with particular efficiency. In this section we shall consider tree amplitudes for the scattering of open strings.

7.1.1 The Structure of Tree Amplitudes

The basic ingredients in constructing Feynman diagrams are propagators and interaction vertices, so we will discuss these in turn. For an ordinary scalar boson field ϕ with mass m^2, which obeys the Klein–Gordon equation $(\Box + m^2)\phi = 0$, the standard Feynman propagator is simply $(\Box + m^2)^{-1}$, the inverse of the Klein–Gordon operator. For free open strings the closest analog of the Klein–Gordon equation is the mass-shell condition $(L_0 - 1)|\phi\rangle = 0$, which can regarded as an infinite-component generalization of the Klein–Gordon equation. A plausible guess for a propagator (suppressing the usual $-i\epsilon$) would therefore be

$$\Delta = (L_0 - 1)^{-1} = \int\limits_0^1 z^{L_0-2}dz. \qquad (7.1.1)$$

The other basic ingredient in a Feynman diagram is the interaction vertex. At tree level or one-loop level, any string diagram can be represented in a form in which every interaction vertex is coupled to at least

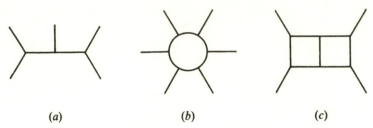

(a) (b) (c)

Figure 7.2. Construction of on-shell tree or one-loop diagrams requires, as in (a) and (b), an interaction vertex for emission of an external particle from an internal line. At two-loop level, vertices for interaction of three internal lines (which are much more difficult to deal with in the case of string theory) are encountered for the first time, as in (c).

one external particle , as in figs. 7.2a and 7.2b.[†] These external particles, in contrast to internal lines, correspond to physical on-shell string states.

We thus must find an interaction vertex or 'vertex operator' for emission (or absorption) of an external state by an internal line. We discussed heuristically in chapter 1 some of the properties of these 'vertex operators', and in particular why they can be expected to be local operators in the quantum field theory that governs the propagation of the string. We discussed in §2.2.3 some of the mathematical properties of vertex operators in a form suitable for the formalism used here. For every physical state Λ of momentum k^{μ} in the open-string spectrum we associate a vertex operator

$$V_{\Lambda}(k, \tau) = e^{i\tau L_0} V_{\Lambda}(k, 0) e^{-i\tau L_0} \qquad (7.1.2)$$

that describes emission of the Λ state from the $\sigma = 0$ end of an open string at proper time τ. It is convenient to define $z = e^{i\tau}$. The vertex operator of (7.1.2) is required to have conformal dimension $J = 1$, in the sense that

$$[L_m, V_{\Lambda}(k, z)] = \left(z^{m+1} \frac{d}{dz} + m z^m \right) V_{\Lambda}(k, z). \qquad (7.1.3)$$

Vertex operators are constructed from normal-ordered expressions based on

$$X^{\mu}(z) = x^{\mu} - i p^{\mu} \ln z + i \sum_{n \neq 0} \frac{1}{n} \alpha_n^{\mu} z^{-n} \qquad (7.1.4)$$

[†] A diagram with more than one loop always contains at least one vertex coupling three internal lines, as shown in fig. 7.2c.

and its derivatives. (We set the open-string Regge slope $\alpha' = 1/2$ throughout this chapter.) For example, the tachyon vertex is given by

$$V_0(k, z) =: e^{ik \cdot X(z)} : \tag{7.1.5}$$

with $k^2 = 2$. It is convenient to write explicitly $V_0 = Z_0 W_0$ with Z_0 being the zero mode operator

$$Z_0 = \exp(ik \cdot x + k \cdot p \ln z) = e^{ik \cdot x} z^{k \cdot p + 1} = z^{k \cdot p - 1} e^{ik \cdot x}, \tag{7.1.6}$$

and W_0 being the remaining factor

$$W_0 = \exp(k \cdot \sum_{n=1}^{\infty} \frac{1}{n} \alpha_{-n} z^n) \exp(-k \cdot \sum_{n=1}^{\infty} \frac{1}{n} \alpha_n z^{-n}) \tag{7.1.7}$$

(which is defined to be normal ordered). Similarly, the vertex operator for a massless vector particle of momentum k^μ and polarization $\zeta^\mu(k)$ is given by

$$V(\zeta, k, z) = \zeta \cdot \dot{X}(z) e^{ik \cdot X(z)}. \tag{7.1.8}$$

In this case we must set $k^2 = \zeta \cdot k = 0$, and then normal ordering is immaterial. The dot denotes differentiation with respect to $\tau = -i \ln z$.

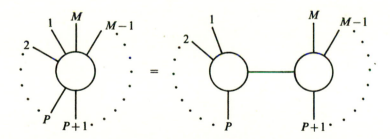

Figure 7.3. Tree-level unitarity requires that in an M-particle amplitude for particles $1\,2\,3 \ldots M$, the residue in the $1\,2\,3 \ldots P \rightarrow P+1 \ldots M$ channel should be the product of the $1\,2\,3 \ldots P\,X$ and $X\,P+1 \ldots M$ tree amplitudes.

The structure of perturbation theory in field theories based on point particles suggests that an M-particle tree amplitude for open-string states is given by a sequence of vertex operators and propagators according to the simple formula

$$A_M = g^{M-2} \langle \phi_1 | V_2(k_2) \Delta V_3(k_3) \cdots \Delta V_{M-1}(k_{M-1}) | \phi_M \rangle. \qquad (7.1.9)$$

This is an attempt at a string theoretic analog of the tree-level Feynman diagram of fig. 7.2a, with the vertex operator playing the role of an on-mass-shell interaction vertex. Equation (7.1.9) has the virtue that at least some aspects of tree-level unitarity are manifest. Equation (7.1.9) has poles that arise from poles of the propagators. Since the propagator is the inverse of $(L_0 - 1)$, its poles are states of $L_0 - 1 = 0$ or, in other words, on-mass-shell states. Unitarity also requires the following. Given a tree level process for M particles 1 2 3 ... M, the residue of a pole in a subprocess such as 1 2 3 ... $P \to (P+1) \ldots M$ should factorize as the product of tree amplitudes for subprocesses 1 2 3 ... P X and X $(P+1) \ldots M$, as depicted in fig. 7.3. A more delicate requirement of unitarity is that 'timelike' on-mass-shell states (which have negative norm) should not appear in (7.1.9). This will be shown to follow from the basic property (7.1.3) of vertex operators. Although it is not evident from (7.1.9), there are also poles in the other subprocesses, such as 2 3 ... $P \to (P + 1) \ldots M$ 1. The fact that the expression is actually symmetric under cyclic permutations of the external particles is, of course, the property of duality. It is part of what makes string theory so interesting. Duality is discussed later in this chapter.

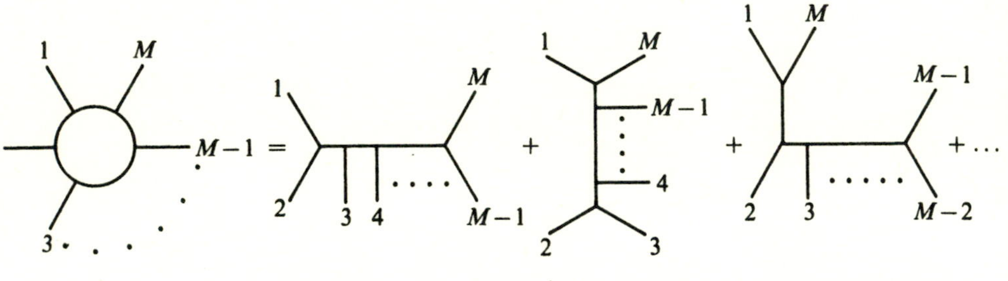

Figure 7.4. In field theory the tree amplitude for M particles 1 2 3 ... M is written as a sum of many Feynman diagrams corresponding to poles in different subchannels.

While (7.1.9) is written as an ansatz for a string-theoretic generalization of the one particular tree-level diagram of fig. 7.2a, in field theory there would be many different diagrams corresponding to poles in different subchannels, as in fig. 7.4. We therefore might at first expect to have to generalize all of those diagrams to string theory, but here we encounter a surprise that is one of the characteristic features of string theory. While written as a string-theoretic generalization of just one Feynman diagram in fig. 7.2a, (7.1.9) actually generalizes all of them. This one formula is the entire string theoretic tree amplitude. As was described in chapter 1, string theory actually began historically with the search for formulas that would have this duality property.

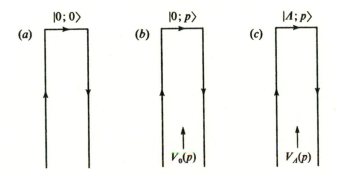

Figure 7.5. The Feynman path integral represents the ground-state wave function of an open string (the zero-momentum tachyon $|0;0\rangle$) in terms of a path integral on a semi-infinite strip, as in (a). If one wishes instead the state $|0;k\rangle$, which is the lowest state of momentum k, one must insert in the far past an operator that carries momentum k, the natural choice in this case being the tachyon vertex operator, $V_0(k)$, as in (b). More, generally, the wave function of any physical state $|\Lambda;k\rangle$ can be computed by an insertion on a semi-infinite strip of $V_\Lambda(k)$ in the far past, as in (c).

To shed some light on how duality can arise, and to make contact with the formulation of tree amplitudes that was given in chapter 1, requires the following digression. Let us for definiteness take the initial and final states in (7.1.9) to be tachyons of momentum k_1 and k_M, respectively. At first sight it might appear that to evaluate (7.1.9) requires using some explicit form of the initial and final state wave functions that appear in (7.1.9). While it is certainly possible to work that way (and in some of our computations we indeed use the Fock-space description of the wave functions), the Feynman path integral gives a well-known alternative way to describe the low-lying states of a quantum system. For example, the ground-state wave function $|0;0\rangle$ (recall that $|0;k\rangle$ is a tachyon state of momentum k) can be represented by a path integral on a semi-infinite strip, as shown in fig. 7.5a. As is often convenient with Feynman path integrals, we do a world-sheet Wick rotation, so as to interpret the strip in the figure as a piece of Euclidean world sheet of positive signature. If one wishes to compute in an analogous way not the ground-state wave function but the wave function for some other state, it is necessary to insert in the far past an operator with the quantum numbers of the desired state. For instance, if one wishes to obtain the tachyon state $|0; k\rangle$ one inserts in the far past, as in fig. 7.5b, a minimal operator of momentum k^μ; this is the tachyon vertex operator $V_0(k) =: e^{ik \cdot X}:$. More generally, as in fig. 7.5c, any physical state $|\Lambda; k\rangle$ can be obtained by inserting in the far past a suitable operator – namely its vertex operator $V_\Lambda(k)$. Thus,

$$|\Lambda; k\rangle = \lim_{\tau \to +i\infty} e^{-i\tau} V_\Lambda(k, \tau)|0; 0\rangle. \qquad (7.1.10)$$

Here, we write $\tau \to +i\infty$ to underscore the fact that the path integral in (7.1.10) is on a Euclidean world sheet of imaginary τ. Likewise, representing a final state as a path integral on a semi-infinite strip that extends to $\tau = -i\infty$, we have a formula analogous to (7.1.10) for final states:

$$\langle \Lambda; k| = \lim_{\tau \to -i\infty} e^{i\tau} \langle 0; 0|V_\Lambda(k, \tau). \qquad (7.1.11)$$

These formulas will be useful. The factors of $z^{-1} = e^{-i\tau}$ and $z = e^{i\tau}$ in (7.1.10) and (7.1.11) are required to compensate for the factors arising from

$$Z_0 |0; 0\rangle = e^{ik \cdot x} z^{k \cdot p + 1} |0; 0\rangle = z |0; k\rangle \qquad (7.1.12)$$

and

$$\langle 0; 0| Z_0 = \langle 0; 0| z^{k \cdot p - 1} e^{ik \cdot x} = \frac{1}{z} \langle 0; k| . \qquad (7.1.13)$$

The reader is invited to verify (7.1.10) and (7.1.11) for tachyons using the explicit form of (7.1.7) to get (7.1.10) for $z \to 0$ or (7.1.11) for $z \to \infty$. We have written the factor of i explicitly in (7.1.10) and (7.1.11) for emphasis, but henceforth we work with $\tau' = -i\tau$.

In (7.1.9), apart from initial and final states and vertex operators, we also have propagators

$$\Delta = (L_0 - 1)^{-1} = \int_0^\infty d\tau e^{-\tau(L_0 - 1)}. \qquad (7.1.14)$$

As $L_0 - 1$ is the 'Hamiltonian' that governs propagation of a string, $e^{-\tau(L_0-1)}$ is the operator that propagates an open string through imaginary time τ and thereby creates a strip of width π and length τ, as in fig. 7.6a. Putting the pieces together, we see in fig. 7.6b that (7.1.9) actually generates an integral on an infinite strip of width π, and the following

(a)

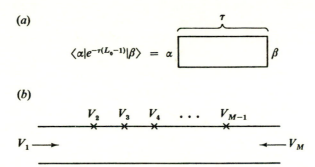

(b)

Figure 7.6. The propagator $(L_0 - 1)^{-1} = \int d\tau e^{-\tau(L_0-1)}$ propagates a string through imaginary time τ and thereby generates a strip of width π and length τ, as in (a). The tree $\langle 1|V_2(L_0-1)^{-1}V_3 \ldots V_{M-1}|M\rangle$ therefore can be understood in terms of an integral on an infinite strip, with semi-infinite extensions to the left and right corresponding to initial and final states $|M\rangle$ and $\langle 1|$; this is sketched in (b).

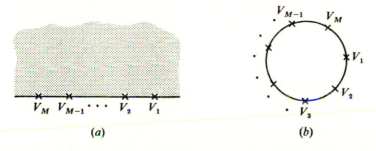

(a)   (b)

Figure 7.7. The infinite strip of fig. 7.6 can be conformally mapped onto the upper half plane, as in (a), or onto the unit disk, as in (b).

additional structure. There are M vertex operators inserted on the strip, one each in the far past and the far future for initial and final states, and $M - 2$ inserted at finite times $0, \tau_1, \tau_2 \ldots, \tau_{M-1}$; and there is an instruction to integrate over the τ_k. This can be converted into a form in which the 'initial' and 'final' states $|1\rangle$ and $|M\rangle$ are not singled out by a simple conformal mapping. Indeed, a simple conformal map from $u = \sigma + i\tau$ to $v = e^{iu}$ maps the strip in fig. 7.6b onto the upper-half v plane; a further conformal map from v to $w = (v - i)/(v + i)$ maps the upper-half v plane onto the unit disc in the w plane. These conformal mappings are symmetries because of the conformal invariance of the theory. (Of course, there may be anomalies in the conformal invariance, so that it must be verified to be a symmetry by explicit calculation, as is explained below.) These mappings are sketched in fig. 7.7. In the case of mapping onto the upper-half plane, the vertex operators show up, as in fig. 7.7a, on the real axis; in mapping onto the unit disc, as in fig. 7.7b, the vertex operators show up on the boundary of the disc. The figures sketched in fig. 7.7 are the ones we have discussed in chapter 1. The virtue of making contact in this way between the previous picture and (7.1.9) is that we now know that the manifestly cyclically symmetric expression of fig. 7.7b is equivalent to the expression (7.1.9) in which the correct poles have been exhibited at least in one channel.

7.1.2 Decoupling of Ghosts

Now let us show that only physical positive norm states, and not ghosts or other unphysical states, appear as poles in (7.1.9), which we repeat here for convenience:

$$A_M = g^{M-2}\langle\phi_1|V_2(k_2)\Delta V_3(k_3)\cdots\Delta V_{M-1}(k_{M-1})|\phi_M\rangle. \qquad (7.1.15)$$

The external states, described by $\langle\phi_1|, |\phi_M\rangle$, and the $M-2$ vertex operators V_i, are all required to satisfy the mass-shell condition $(L_0-1)|\phi\rangle = 0$ and the Virasoro subsidiary conditions $L_n|\phi\rangle = 0$ for $n > 0$, since A_M is only meant to describe amplitudes for physical states. We wish to show that the physical states that appear as poles in (7.1.15) have the same property. Consider a particular propagator Δ in (7.1.15). Since $L_0 = \frac{1}{2}p^2 + N$, where N has eigenvalue $0, 1, 2, \ldots$, poles occur at values of p^2 corresponding to physical mass values. The residue of Δ at one of these mass values is the unit operator in this particular mass sector, which can be expanded in the form $\sum_i |\psi_i\rangle\langle\psi_i|$, where $|\psi_i\rangle$ are orthonormal states (finite in number) that span this sector. The essential requirement is that only those states $|\psi_i\rangle$ that satisfy the physical state conditions $L_n|\phi\rangle = 0$ for $n > 0$ contribute to the residue. We do not have to worry about the L_0 condition, since the contribution to the pole certainly comes from states that obey this!

We do not quite have to show that

$$|\alpha\rangle = V_N\Delta V_{N+1}\cdots\Delta V_{M-1}|\phi_M\rangle \qquad (7.1.16)$$

obeys the conditions $L_m|\alpha\rangle = 0, m > 0$. Only the part of $|\alpha\rangle$ that is on mass shell (annihilated by $L_0 - 1$) contributes to the residue of the pole in (7.1.15) and must be annihilated by $L_m, m > 0$. Thus if P_k is the operator that projects on states of $L_0 = k$, then what we must prove is that the state $|\beta\rangle$ defined by

$$|\beta\rangle = P_1|\alpha\rangle \tag{7.1.17}$$

obeys $L_m|\beta\rangle = 0, m > 0$. Since $[L_0, L_m] = -mL_m$, we have $L_m P_1 = P_{1-m}L_m$, and hence what we wish to prove is that

$$P_{1-m}L_m|\alpha\rangle = 0, \quad m > 0. \tag{7.1.18}$$

Since $P_{1-m}(-L_0 - m + 1) = 0$, (7.1.18) is equivalent to

$$P_{1-m}(L_m - L_0 - m + 1)|\alpha\rangle = 0, \quad m > 0, \tag{7.1.19}$$

and (substituting the definition of α) this will follow if we can show that

$$(L_m - L_0 - m + 1)V_N\Delta\cdots V_{M-1}|\phi_M\rangle = 0 \quad \text{for } m > 0. \tag{7.1.20}$$

This proves to be a convenient way to formulate the problem.

Since the vertex operators have conformal dimension $J = 1$, it follows from the definition given in §2.2.3 that they satisfy (7.1.3) and hence (subtracting the $m = 0$ equation and evaluating at $z = 1$)

$$(L_m - L_0 - m + 1)V(1) = V(1)(L_m - L_0 + 1). \tag{7.1.21}$$

Using the Virasoro algebra it is also easy to show that

$$(L_m - L_0 + 1)\frac{1}{L_0 - 1} = \frac{1}{L_0 + m - 1}(L_m - L_0 - m + 1). \tag{7.1.22}$$

Together, (7.1.21) and (7.1.22) imply that

$$(L_m - L_0 - m + 1)V\frac{1}{L_0 - 1} = V\frac{1}{L_0 + m - 1}(L_m - L_0 - m + 1). \tag{7.1.23}$$

Therefore the factor $L_m - L_0 - m + 1$ in (7.1.20) can be pushed step by step all the way to the right until one encounters $(L_m - L_0 + 1)|\phi_M\rangle$, which vanishes because $|\phi_M\rangle$ is physical. This establishes (7.1.20).

Another way to express the fact that only physical positive-norm states are created in scattering processes is to say that if $|\phi_1\rangle$ is orthogonal to all of the physical states then (7.1.15) vanishes. In §2.2.2 we defined a state that is orthogonal to all physical states as a spurious state. We showed there that an arbitrary spurious state can be written as

$$|\psi\rangle = \sum_{m=1}^{\infty} L_{-m}|\chi_m\rangle, \qquad (7.1.24)$$

where

$$(L_0 - 1 + m)|\chi_m\rangle = 0, \qquad (7.1.25)$$

so that $(L_0 - 1)|\psi\rangle = 0$. Since the physical states are defined to be the subspace of the Fock space that satisfies the Virasoro conditions they are annihilated by operators of the form $\sum_1^{\infty} X_n L_n$. Therefore, if ϕ_1 is of the form (7.1.24) then (7.1.15) vanishes.

7.1.3 Cyclic Symmetry

Equation (7.1.15) has a property, known as duality, that has no counterpart in field theory for point particles. This is possible because (7.1.15) actually has poles in channels other than those in which propagators are displayed explicitly. These poles arise as divergences in the infinite sums that are implicit in the operator multiplications. These extra poles make possible the remarkable fact that A_M is invariant under a cyclic transformation of the M external particles, $(1\,2\ldots M) \to (M\,1\ldots M-1)$. This is necessary in order that it correspond to a world sheet that is conformally equivalent to a disk with the M particles attached to the boundary in a given cyclic order. The proof of the cyclic symmetry is our next objective. Once it is established, it follows that the unitarity property applies to all channels, not just the ones we checked above. The fact that a string theory diagram can describe poles in several different crossed channels makes the correspondence with point-particle theories that describe a low-energy limit somewhat nontrivial. It also accounts for the fact that the counting of diagrams is quite different from that of point-particle theories. This becomes particularly dramatic in theories of oriented closed strings, where there is a unique diagram at each order of the perturbation expansion, as was discussed in chapter 1.

We have already discussed cyclic symmetry and duality heuristically, but now let us discuss these matters more precisely with the aim of showing that the integration measure that arises actually has the necessary properties. In order to demonstrate cyclic symmetry and better understand the structure of tree amplitudes, let us try to evaluate (7.1.15). Substituting (7.1.1) and (7.1.2) and using $(L_0 - 1)|\phi_1\rangle = (L_0 - 1)|\phi_M\rangle = 0$, we get

$$A_M = g^{M-2} \int\limits_0^1 \frac{dz_3 \cdots dz_{M-1}}{z_3 \cdots z_{M-1}} \langle \phi_1 | V(k_2, 1) V(k_3, z_3) \cdots$$

$$V(k_{M-1}, z_3 \cdots z_{M-1}) | \phi_M \rangle.$$

$$= g^{M-2} \int\limits_0^1 \left(\prod_{i=3}^{M-1} \theta(y_{i-1} - y_i) \frac{dy_i}{y_i} \right) \langle \phi_1 | V(k_2, y_2) V(k_3, y_3) \cdots$$

$$V(k_{M-1}, y_{M-1}) | \phi_M \rangle,$$

$$(7.1.26)$$

where $y_2 = 1$. Here we have made the change of variables

$$y_i = z_3 z_4 \ldots z_i \qquad i = 3, \ldots, M-1. \qquad (7.1.27)$$

The fact that $0 < z_i < 1$ implies that $y_{i-1} > y_i$, which has been incorporated with step functions $\theta(y_{i-1} - y_i)$. At this point the expression for the amplitude does not treat the M particles symmetrically. The next step is to substitute our previous equations

$$|\phi_M \rangle = \lim_{y_M \to 0} y_M^{-1} V(k_M, y_M) |0; 0\rangle \qquad (7.1.28)$$

and

$$\langle \phi_1 | = \lim_{y_1 \to \infty} \langle 0; 0 | y_1 V(k_1, y_1), \qquad (7.1.29)$$

where $|0; 0\rangle$ represents a zero-momentum Fock-space ground state. This enables us to identify

$$\langle \phi_1 | V(k_2, y_2) \cdots V(k_{M-1}, y_{M-1}) | \phi_M \rangle$$

$$= \frac{y_1}{y_M} \langle 0; 0 | V(k_1, y_1) V(k_2, y_2) \cdots V(k_M, y_M) |0; 0\rangle \qquad (7.1.30)$$

in the limit $y_1 \to \infty$ and $y_M \to 0$.

Let us pause to discuss the properties of the state $|0; 0\rangle$ (the Fock-space ground state with zero momentum), which appears in (7.1.30). In some sense this state is the vacuum or ground state of the two-dimensional quantum field theory that describes the propagation of the bosonic open string. That we have not discussed it more already is a reflection of the unusual physical interpretation of the free string compared to other two-dimensional quantum field theories. The state $|0; 0\rangle$ is not a 'physical

state', since it is not annihilated by $L_0 - 1$ (instead, it is annihilated by L_0).
However, it has another property that in a sense is equally important and
that singles it out uniquely. In the entire Fock space, $|0; 0\rangle$ is the unique
state that is annihilated by the anomaly free $SL(2, R)$ group generated
by L_{-1}, L_0, and L_1. Although just a finite-dimensional subalgebra of the
infinite Virasoro algebra, $SL(2, R)$ plays a special role because it is not
affected by the anomaly or central term in the Virasoro algebra, so it
is a true symmetry, and it annihilates the vacuum. If one thinks of the
free string as a two-dimensional field theory with an enormous symmetry
group, then in some sense $|0; 0\rangle$ is the vacuum and $SL(2, R)$, which is not
anomalous and leaves the vacuum invariant, is the unbroken symmetry
group of the theory. In any case, $SL(2, R)$ is the basic tool for proving all
simple properties of the bosonic tree amplitudes that are not obvious on
more elementary grounds.

Infinitesimally, L_m generates a transformation $\delta y \sim y^{m+1}$. This is
visible, for instance, in (7.1.3), which we repeat for convenience:

$$[L_m, V_\Lambda(k, y)] = \left(y^{m+1} \frac{d}{dy} + m y^m \right) V_\Lambda(k, y). \tag{7.1.31}$$

The $y^{m+1} d/dy$ term in (7.1.31) would be present in the action of L_m
on a field of any conformal dimension; the $m y^m$ term is present because
the vertex operator has nonzero conformal spin $J = 1$. An arbitrary
infinitesimal $SL(2, R)$ transformation can be written as

$$y \to y' = y + \lambda_{-1} + \lambda_0 y + \lambda_1 y^2, \tag{7.1.32}$$

where the λ_m are arbitrary small parameters. Exponentiating the action
of the L_m gives the finite transformations

$$e^{\lambda_1 L_1} : y \to y' = \frac{y}{1 - \lambda_1 y}$$

$$e^{\lambda_0 L_0} : y \to y' = e^{\lambda_0} y \tag{7.1.33}$$

$$e^{\lambda_{-1} L_{-1}} : y \to y' = y + \lambda_{-1},$$

where the λ_m's are now finite. Combining these gives the general form of
an $SL(2, R)$ transformation

$$y \to y' = \frac{ay + b}{cy + d}. \tag{7.1.34}$$

Note that although four parameters appear in (7.1.34), one could be absorbed in a rescaling of a, b, c and d. A convenient convention is to set

$ad - bc = 1$, since then $\begin{pmatrix} a & b \\ c & d \end{pmatrix}$ belongs to the group $SL(2, R)$ of 2×2 real matrices of determinant one.

We pause to explain briefly 'why' 2×2 real matrices have a natural nonlinear action on one real variable y. The 2×2 real matrices of $SL(2, R)$ certainly act naturally on a two-dimensional real vector space,

$$\begin{pmatrix} v_1 \\ v_2 \end{pmatrix} \rightarrow \begin{pmatrix} a & b \\ c & d \end{pmatrix} \begin{pmatrix} v_1 \\ v_2 \end{pmatrix}. \tag{7.1.35}$$

Setting $y = v_1/v_2$, we find that y transforms as in (7.1.34).

The subgroup $SL(2, R)$ of the conformal group is called the 'Möbius group' or the 'projective group'. The reader is urged to check explicitly that the three parameter family of transformations in (7.1.34) is generated infinitesimally by (7.1.33). Assuming that a, b, c and d are real, (7.1.34) obviously maps the real axis to itself. It is somewhat less obvious, but not difficult to check, that if we analytically continue to complex values of y, then $SL(2, R)$ transformations map the upper half y plane to itself. In fact, $SL(2, R)$ can be shown to be the group of all one-to-one mappings of the upper half plane to itself of the form $y \rightarrow \tilde{y}$, with $\tilde{y}$ being an analytic function of y. A transformation $y \rightarrow \tilde{y}$, with $\tilde{y}$ being an analytic and one-to-one function of y (in some region under study) is a conformal mapping; we may say that the mappings of the upper half plane onto itself form a subgroup of the full conformal group, namely the Möbius group $SL(2, R)$.

It was for essentially this reason that $SL(2, R)$ (and $SL(2, C)$, which arises in the same way in the theory of closed strings) appeared in chapter 1. Indeed, there we began with a generally covariant two-dimensional action; we then chose a coordinate system in which the metric took the form $ds^2 = e^{\phi}|dy|^2$. This coordinate system is not quite unique, since a conformal mapping $y \rightarrow \tilde{y}$ transforms the metric to $ds^2 = e^{\phi} \cdot |dy/d\tilde{y}|^2 \cdot |d\tilde{y}|^2$, which is of the same general form. It is really because of this that $SL(2, R)$ emerges as the 'symmetry of the vacuum' upon covariant quantization. We recall that in chapter 1 the conformal mappings of the string world sheet into itself appeared as the remnant of the original general covariance that was left intact after picking the conformal gauge. A residual Faddeev–Popov gauge fixing was needed to fix this remaining symmetry.

Returning to our study of tree amplitudes, further study of (7.1.33) shows that for $n = 1, 0, -1$, the transformation from y to $y' = e^{\lambda L_n} y$ obeys

$$\left(\frac{y'}{y}\right)^{n+1} = (cy + d)^{-2} = (a - cy')^2. \tag{7.1.36}$$

As a verification of this formula we need simply note that each expression gives 1 for $n = -1$, e^λ for $n = 0$, and $(1 - \lambda y)^{-2}$ for $n = 1$. Therefore, since the exponentiated form of (7.1.3) implies that

$$e^{\lambda L_n} y^n V(k, y) e^{-\lambda L_n} = (y')^n V(k, y'), \qquad (7.1.37)$$

it follows (by dividing by y^{n+1}) that for any $SL(2, R)$ transformation

$$\Lambda(T) \frac{V(k, y)}{y} \Lambda^{-1}(T) = (a - cy')^2 \frac{V(k, y')}{y'}. \qquad (7.1.38)$$

Using the $SL(2, R)$ invariance of the vacuum,

$$\Lambda(T) |0; 0\rangle = |0; 0\rangle, \qquad (7.1.39)$$

we deduce that

$$I_M(k, y) = \langle 0; 0| \frac{V(k_1, y_1)}{y_1} \cdots \frac{V(k_M, y_M)}{y_M} |0; 0\rangle \qquad (7.1.40)$$

transforms under Möbius transformations into

$$I_M(k, y) = I_M(k, y') \prod_{i=1}^{M} (a - cy_i')^2 \qquad (7.1.41)$$

when all M y_i are subjected to a common $SL(2, R)$ transformation. Equation (7.1.41) is often expressed by saying that I_M transforms under $SL(2, R)$ with Möbius weight 2.

The M-point tree amplitude in (7.1.26) is of the form

$$A_M = g^{M-2} \int d\mu_M(y) I_M(k, y), \qquad (7.1.42)$$

where $d\mu_M(y)$ is the measure for the y_i integrations that appears in our previous formulas (apart from the powers of y_i absorbed into I_M). We would now like to re-express this measure in a more symmetrical form than in (7.1.26).

Let us first consider the transformation of a single differential dy_i. In view of (7.1.34), this transforms as follows

$$dy_i' = d\left(\frac{ay_i + b}{cy_i + d}\right) = \frac{(ad - bc)dy_i}{(cy_i + d)^2} = \frac{dy_i}{(cy_i + d)^2}, \qquad (7.1.43)$$

or

$$dy_i = \frac{dy_i'}{(a - cy_i')^2}. \qquad (7.1.44)$$

The most obvious guess is that a 'nice' measure would be $d\mu_M(y) = \prod dy_i$. This gives a Möbius-invariant formula, but one that does not coincide with the measure that actually appears in (7.1.42). There are several discrepancies. The first is that the y_i coordinates in (7.1.42) are ordered on the boundary of the world sheet. This is easily accommodated by including a product of step functions $\prod \theta(y_{i-1} - y_i)$ in our naive guess for what a 'good' measure should be. Somewhat more subtle is that, precisely because of the Möbius invariance, $\prod dy_i$ overcounts equivalent configurations by a divergent factor (the volume of the noncompact group $SL(2, R)$). We must therefore remove this infinite factor by choosing just one representative of each conformal equivalence class. This is very similar to the usual procedure of factoring out of an infinite volume factor in the integral over the vector potential in Yang–Mills theory to account for the overcounting of gauge-equivalent configurations.

Since the Möbius group has three real parameters (the three λ's in (7.1.32)), it is possible to find an $SL(2, R)$ mapping that sets any three of the y_i variables – let us call them y_A, y_B and y_C – to arbitrarily chosen fixed values called y_A^0, y_B^0 and y_C^0. The divergent group volume factor can then be removed simply by not integrating over y_A, y_B, and y_C. In doing this we must include the analog of a Faddeev-Popov determinant, which will cancel the factors $\prod (a - cy_i')^2$ of (7.1.44) for the unintegrated coordinates. The required determinant is the Jacobian of the transformation from y_A, y_B, and y_C to λ_{-1}, λ_0, and λ_1. From (7.1.32), this Jacobian is

$$\frac{\partial(y_A, y_B, y_C)}{\partial(\lambda_{-1}, \lambda_0, \lambda_1)} = \begin{vmatrix} 1 & 1 & 1 \\ y_A & y_B & y_C \\ y_A^2 & y_B^2 & y_C^2 \end{vmatrix}. \tag{7.1.45}$$

The determinant is readily evaluated, and gives $(y_A - y_B)(y_A - y_C)(y_B - y_C)$. This factor is easily seen to transform in the required way because

$$y_A - y_B = \frac{y_A' - y_B'}{(a - cy_A')(a - cy_B')}. \tag{7.1.46}$$

Altogether, we conclude that the desired volume element, which transforms the same way as the naive guess $\prod dy_i$, is

$$d\mu_M(y) = \delta(y_A - y_A^0)\delta(y_B - y_B^0)\delta(y_C - y_C^0)$$

$$\times (y_A - y_B)(y_A - y_C)(y_B - y_C) \prod_{i=2}^{M} \theta(y_{i-1} - y_i) \prod_{j=1}^{M} dy_j. \tag{7.1.47}$$

Now we have achieved our goal. Equation (7.1.47) is indeed the correct measure that appears in our previous formulas; for instance, the previous formula in (7.1.26) corresponds to the particular choices $y_1^0 \to \infty$, $y_2^0 \to 1$, $y_M^0 \to 0$. What we have gained is an understanding that (7.1.47) comes from a more symmetrical starting point by something akin to Faddeev-Popov gauge fixing; we are now free to make other choices instead.

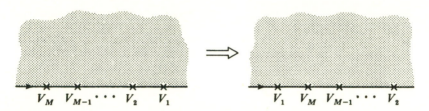

Figure 7.8. An $SL(2, R)$ transformation can turn the 'last' vertex operator in a string of M vertices into the 'first' one.

Now let us return to the question of cyclic symmetry of the amplitudes. An arbitrary $SL(2, R)$ transformation maps the real axis onto itself monotonically, preserving the cyclic ordering of the y_i coordinates. However, as in fig. 7.8, an $SL(2, R)$ transformation can be chosen to map the 'last' vertex operator on the real axis all the way out to $y = +\infty$, whence it returns from $y = -\infty$ to finite values as the 'first' in line. This is potentially what we need to prove the cyclic symmetry required by unitarity and the world-sheet interpretation provided that

$$V(1)V(2) \cdots V(M) = V(M)V(1) \cdots V(M-1). \qquad (7.1.48)$$

Then (7.1.41) is true even for $SL(2, R)$ transformations that induce a cyclic permutation. In order to be quite explicit, let us consider the special case of tachyon emission vertices, even though the result is general. The key identity that is required is

$$V_0(k_1, y_1)V_0(k_2, y_2) = V_0(k_2, y_2)V_0(k_1, y_1) \exp[\pi i k_1 \cdot k_2 \epsilon(y_1 - y_2)], \quad (7.1.49)$$

where $\epsilon(x)$ is $+1$ for $x > 0$ and -1 for $x < 0$. One proof is given in appendix 7.A. An alternative proof can be based on the well-known identity

$$e^A e^B = e^B e^A e^{[A,B]}, \qquad (7.1.50)$$

which is valid when $[A, B]$ is a c number. In the present application the required commutator is

$$[ik_1 \cdot X(y_1), ik_2 \cdot X(y_2)] = \pi i k_1 \cdot k_2 \epsilon(y_1 - y_2). \qquad (7.1.51)$$

The commutator in this equation is a little subtle to define. However, the derivative of both sides with respect to y_1 gives a formula that is unambiguous and easy to verify. The integration constant is determined by the requirement that the commutator be odd under interchange of y_1 and y_2. Equations (7.1.50) and (7.1.51) can easily be used to derive (7.1.49), which holds for other vertex operators as well. Equation (7.1.49) is what we need to derive (7.1.48); indeed, bringing $V(M)$ all the way through for $y_1 > y_2 \cdots > y_M$ gives a phase

$$\exp[\pi i k_M \cdot \sum_1^{M-1} k_i] = \exp(-\pi i k_M^2) = 1, \qquad (7.1.52)$$

using momentum conservation and the mass-shell condition. This completes the proof of cyclic symmetry.

7.1.4 Examples

Let us begin with the simplest examples, namely vertices for three on-shell particles. We will consider each of the combinations of tachyons and massless vector bosons in turn, using (7.1.15). In all cases we do not bother to write the obvious momentum-conservation delta function. For the tachyons we simply have

$$g\langle 0; k_1 | V_0(k_2) | 0; k_3 \rangle = g. \qquad (7.1.53)$$

For two tachyons and one massless vector we have

$$
\begin{aligned}
g\langle 0; k_1 | V(\zeta, k_2) | 0; k_3 \rangle &= g\langle 0; k_1 | \zeta \cdot \dot{X}(1) V_0(k_2) | 0; k_3 \rangle \\
&= g\zeta \cdot k_3 = g\zeta \cdot (k_3 - k_1)/2.
\end{aligned}
\qquad (7.1.54)
$$

The last identity uses the fact that $\zeta \cdot k_2 = 0$ and momentum conservation $k_1 + k_2 + k_3 = 0$. Equation (7.1.54) is the form of the coupling of a massless vector to charged scalars in field theory. In particular, it is gauge invariant (it vanishes, using the mass-shell conditions and momentum conservation, if one sets $\zeta = k_2$). The proof of cyclic symmetry ensures that the same result is given by putting the vector in one of the end states

$$g\langle 0; k_3 | V_0(k_1) \zeta \cdot \alpha_{-1} | 0; k_2 \rangle = -g\zeta \cdot k_1 = g\zeta \cdot k_3. \qquad (7.1.55)$$

For two massless vectors and one tachyon we have (putting the vectors at the ends)

$$g\langle 0; k_1 | \zeta_1 \cdot \alpha_1 V_0(k_2)\zeta_3 \cdot \alpha_{-1} | 0; k_3 \rangle$$

$$= g\langle 0; k_1 | \zeta_1 \cdot \alpha_1 e^{k_2 \cdot \alpha_{-1}} e^{-k_2 \cdot \alpha_1} \zeta_3 \cdot \alpha_{-1} | 0; k_3 \rangle \qquad (7.1.56)$$

$$= g(\zeta_1 \cdot \zeta_3 - \zeta_1 \cdot k_2 \; \zeta_3 \cdot k_2).$$

One can easily check that this is gauge invariant (vanishes if ζ_1 is replaced by k_1 or ζ_3 is replaced by k_3) by using $k_1^2 = k_3^2 = 0$, $k_2^2 = 2$, $\zeta_1 \cdot k_1 = \zeta_3 \cdot k_3 = 0$.

Now let us consider the three massless vector vertex. It is given by

$$g\langle 0; k_1 | \zeta_1 \cdot \alpha_1 V(\zeta_2, k_2)\zeta_3 \cdot \alpha_{-1} | 0; k_3 \rangle$$

$$= g\langle 0; k_1 | \zeta_1 \cdot \alpha_1 \; \zeta_2 \cdot (\alpha_{-1} + p + \alpha_1) e^{k_2 \cdot \alpha_{-1}} e^{-k_2 \cdot \alpha_1} \zeta_3 \cdot \alpha_{-1} | 0; k_3 \rangle$$

$$= g[\zeta_1 \cdot k_2 \; \zeta_2 \cdot \zeta_3 + \zeta_2 \cdot k_3 \; \zeta_3 \cdot \zeta_1$$

$$+ \zeta_3 \cdot k_1 \; \zeta_1 \cdot \zeta_2 + \zeta_1 \cdot k_2 \; \zeta_2 \cdot k_3 \; \zeta_3 \cdot k_1]$$

$$= g \; \zeta_1^\mu \; \zeta_2^\nu \; \zeta_3^\rho \; t_{\mu\nu\rho},$$

$$(7.1.57)$$

where

$$t_{\mu\nu\rho} = k_{2\mu}\eta_{\nu\rho} + k_{3\nu}\eta_{\rho\mu} + k_{1\rho}\eta_{\mu\nu} + k_{2\mu}k_{3\nu}k_{1\rho}. \qquad (7.1.58)$$

Since this vertex has total antisymmetry in the three external lines it would vanish when we sum over cyclic permutations unless it is multiplied by a group theory factor $\text{tr}(\lambda^a\lambda^b\lambda^c) \sim f^{abc}$. Thus it is present for nonabelian theories, but drops out in the abelian case. The first three terms in (7.1.57) correspond to the standard cubic interaction derived from $\text{tr}F^2$ in Yang–Mills theory. The last term is an $O(\alpha')$ correction that corresponds to an additional term of the form $\text{tr}(F_\mu{}^\nu F_\nu{}^\rho F_\rho{}^\mu)$ in the low-energy effective action.

Next let us consider the four-tachyon amplitude given by

$$A_4 = g^2 \langle 0; k_1 | V_0(k_2)\Delta V_0(k_3) | 0; k_4 \rangle$$

$$= g^2 \int_0^1 \frac{dx}{x} \langle 0; k_1 | V_0(k_2, 1)V_0(k_3, x) | 0; k_4 \rangle . \qquad (7.1.59)$$

Using (7.1.6), the zero-mode part of the matrix element in the integrand is just

$$x^{k_3 \cdot k_4 + 1} = x^{-s/2 - 1}, \qquad (7.1.60)$$

where $s = -(k_3 + k_4)^2$. The nonzero mode part can be evaluated using the coherent-state methods described in appendix 7.A as follows

$$\langle 0| \exp(-k_2 \cdot \sum_1^\infty \frac{\alpha_n}{n}) x^{\sum_1^\infty \alpha_{-n} \alpha_n} \exp(k_3 \cdot \sum \frac{\alpha_{-n}}{n}) |0\rangle$$

$$= \prod_{n=1}^\infty \prod_{\mu=0}^{D-1} \langle 0| \exp(-k_{2\mu} \frac{a_n^\mu}{\sqrt{n}}) x^{n a_{n\mu}^\dagger a_n^\mu} \exp(k_3^\mu \frac{a_{n\mu}^\dagger}{\sqrt{n}}) |0\rangle \qquad (7.1.61)$$

$$= \prod_{n\mu} \exp(-k_{2\mu} k_3^\mu \frac{x^n}{n}) = (1-x)^{k_2 \cdot k_3} = (1-x)^{-\frac{1}{2}t - 2}.$$

Thus altogether

$$A_4 = g^2 \int_0^1 x^{-\frac{1}{2}s - 2}(1-x)^{-\frac{1}{2}t - 2} dx \qquad (7.1.62)$$

$$= g^2 B(-\tfrac{1}{2}s - 1, -\tfrac{1}{2}t - 1) = g^2 B(-\alpha(s), -\alpha(t)),$$

where B is the Euler beta function

$$B(a, b) = \int_0^1 x^{a-1}(1-x)^{b-1} dx = \frac{\Gamma(a)\Gamma(b)}{\Gamma(a+b)}, \qquad (7.1.63)$$

and $\alpha(s) = 1 + s/2$. Thus we recover the Veneziano amplitude. The Regge asymptotic behavior of this amplitude was described in §1.1.2.

Let us now consider the generalization of this result to the tree amplitude with M external tachyons. To do this efficiently it is convenient to use the more symmetrical formulation given in (7.1.40), (7.1.42), and (7.1.47). Appendix 7.A gives a detailed derivation of the correlation function corresponding to M external tachyons. The result is

$$\langle 0| V_0(k_1, y_1) \cdots V_0(k_M, y_M) |0\rangle = y_1 y_2 \cdots y_M \prod_{i<j} (y_i - y_j)^{k_i \cdot k_j}, \qquad (7.1.64)$$

a formula that was also derived from path integrals in chapter 1. Therefore

$$I_M(k, y) = \prod_{i<j} (y_i - y_j)^{k_i \cdot k_j}, \qquad (7.1.65)$$

which substituted in (7.1.42) gives the Koba–Nielsen formula

$$A_M = g^{M-2}(y_A^0 - y_B^0)(y_A^0 - y_C^0)(y_B^0 - y_C^0) \int\limits_{-\infty}^{\infty} dy_1 dy_2 \cdots dy_M$$

$$\times \, \delta(y_A - y_A^0)\delta(y_B - y_B^0)\delta(y_C - y_C^0) \prod_{i=1}^{M} \theta(y_{i-1} - y_i) \prod_{i<j}(y_i - y_j)^{k_i \cdot k_j}.$$

$$(7.1.66)$$

As another example let us consider a tree with M external massless vector particles. In this case the vertex operator $V(\zeta, k, z)$ in (7.1.8) contains the extra factor $\zeta \cdot \dot{X}(z)$. A convenient way to deal with this factor is to replace it by $\exp(\zeta \cdot \dot{X})$, with the understanding that the part of the answer linear in ζ is to be extracted at the end of the calculation. Having done this the oscillator algebra again separates into a product of individual factors for each oscillator. This makes the evaluation relatively easy. We now have effective vertex operators of the form $V = \exp(ik \cdot X + \zeta \cdot \dot{X})$, where normal ordering is immaterial because $k^2 = k \cdot \zeta = 0$, and ζ^2 terms are not being kept. The correlation function of vertices of this type is readily calculated using the formulas given in appendix 7.A. One finds that

$$\langle \frac{V(\zeta_1, k_1, y_1)}{y_1} \cdots \frac{V(\zeta_M, k_M, y_M)}{y_M} \rangle$$

$$= \exp \sum_{i<j} \left(k_i \cdot k_j \log(y_i - y_j) + \frac{\zeta_i \cdot \zeta_j}{(y_i - y_j)^2} - \frac{k_i \cdot \zeta_j}{y_i - y_j} + \frac{\zeta_i \cdot k_j}{y_i - y_j} \right).$$

$$(7.1.67)$$

The result is therefore

$$A_M = g^{M-2} \int d\mu_M(y) F_M(\zeta, k, y) I_M(k, y), \qquad (7.1.68)$$

where $d\mu_M(y)$ and I_M are given in (7.1.47) and (7.1.65)and where F_M is the part of

$$\exp \sum_{i \neq j} \left(\frac{1}{2} \frac{\zeta_i \cdot \zeta_j}{(y_i - y_j)^2} - \frac{k_i \cdot \zeta_j}{y_i - y_j} \right), \qquad (7.1.69)$$

which is linear in each of the polarization vectors. In particular, for $M = 3$ this result agrees with the one given in (7.1.57).

7.1.5 Tree-Level Gauge Invariance

We now discuss the analog of gauge invariance in tree-level open-string processes. For massless external states, this was discussed in §1.5.1.

In §7.1.2 we demonstrated that spurious states decouple from a tree diagram whose other states are physical. A subclass of the spurious states of particular interest are the zero-norm states. These states, first discussed in §2.2.2, are spurious states that simultaneously satisfy the physical-state conditions. The pair of conditions – spurious plus physical – was shown in §2.2.2 to imply that such states have zero norm. The decoupling of zero-norm states in string theory is of fundamental importance because it is related to gauge invariance. Indeed, a typical example of a zero-norm state is a massless vector meson of longitudinal polarization $\zeta^\mu = k^\mu$. The decoupling of this zero-norm state corresponds to the gauge invariance that is familiar in Yang–Mills theory, or in other words to invariance of the amplitudes under $\zeta^\mu \to \zeta^\mu + \epsilon k^\mu$.

To understand the decoupling of zero-norm states, note that scattering amplitudes containing zero-norm states can be computed just like scattering amplitudes for any other physical states. In particular, there is a vertex operator of conformal dimension $J = 1$ associated with every state that satisfies the Virasoro conditions, whether or not it has zero norm. However, if any vertex operators corresponding to zero-norm states are used in the construction of A_M, the amplitude must vanish. This is an immediate consequence of what has already been proved if one combines the results of the preceding sections. Cyclic symmetry allows A_M to be re-expressed with the zero-norm state moved to the end of the expression, where $\langle \phi_1 |$ appears in (7.1.15). Then the theorem on the decoupling of spurious states implies that the expression vanishes.

This reasoning is perfectly fine, but it is instructive to give a second demonstration of the decoupling that does not explicitly invoke the cyclic symmetry. Let us consider the longitudinally polarized massless vector meson as a typical example of a zero-norm state. As we discussed in §2.2.2 the operator

$$V(k, \tau) = -i\frac{d}{d\tau}(e^{ik \cdot X(\tau)}) = k \cdot \dot{X} e^{ik \cdot X(\tau)} \tag{7.1.70}$$

describes the emission of such a state. The corresponding Fock-space state, given by the construction in (6.21), is $k \cdot \alpha_{-1} |0; k\rangle = L_{-1} |0; k\rangle$, which is easily seen to have zero norm as a consequence of $k^2 = 0$. We certainly want states of this type to be decoupled from physical processes, since the decoupling of this zero-norm state is the statement of on-shell gauge invariance.

Let us now prove that a longitudinal massless vector decouples in tree amplitudes. We already know this is true when the state is written at the 'end' of the tree, but let us now suppose instead that it is emitted

somewhere in the middle. Without any increased difficulty we can extend
the discussion to a class of emission vertices that includes the longitudinal
massless vector as a special case. Specifically, suppose that $V_Z(k,\tau) = -i\frac{d}{d\tau}W(k,\tau)$, where W is any operator of conformal dimension $J = 0$ so
that V_Z has $J = 1$. In this case, the 'Schrödinger equation' gives

$$V_Z = -i\frac{d}{d\tau}W = [L_0, W] = [L_0 - 1, W]. \qquad (7.1.71)$$

Inserting this as one of the vertices in the formula for a tree amplitude, the
$L_0 - 1$ factors next to W either cancel against an adjacent propagator or
else annihilate against the states $\langle\phi_1|$ or $|\phi_M\rangle$ at the ends of the expression.
Now a term with a canceled propagator is a holomorphic function of the
Mandelstam invariant ($s = -p^2$) of the corresponding channel, since the
propagator itself is the only potential source of singularities. On the other
hand, tree amplitudes have Regge asymptotic behavior, as discussed in
chapter 1. Combining these two facts, it follows that there are regions of
the invariant energy variables in which the term in question is analytic and
vanishes as $|s| \to \infty$. Thus by a standard theorem of complex analysis,
the amplitude must be identically zero in the region described, and zero
everywhere else as well by analyticity.

An alternative argument is to note that a pair of vertex operators has
an operator product expansion of the form

$$V_1(k_1,\tau)V_2(k_2,0) \sim \tau^{k_1 \cdot k_2 - n} f(\tau), \qquad (7.1.72)$$

as $\tau \to 0$ for some integer n, where $f(\tau)$ is analytic at $\tau = 0$. When the
propagator is canceled, two adjacent vertex operators are multiplied at
equal τ. The right-hand side of (7.1.72) then becomes ill-defined. Since it
vanishes at $\tau = 0$ for $k_1 \cdot k_2 > n$, the only sensible interpretation consistent
with analyticity is for it to vanish for all other values of $k_1 \cdot k_2$ as well.

The fact that a term with a canceled propagator vanishes is used frequently in string calculations, and is referred to as 'the canceled propagator argument'. String theories seem to be similar to field theories with
enormous gauge invariances; the decoupling theorem for zero-norm states
is analogous to the statement that the on-shell tree amplitudes respect
those gauge invariances. As in point-particle theories, string theories can
have a breakdown of gauge invariance arising at the one-loop level, due to
effects known as anomalies. This is discussed in detail in chapter 10. The
canceled propagator argument can give erroneous results if used carelessly
in unregulated divergent formulas, such as occur for certain open-string
loop diagrams.

7.1.6 The Twist Operator

In the construction of A_M we have used vertex operators that describe emission from the $\sigma = 0$ edge of the world sheet. There is no reason not to allow emissions from the $\sigma = \pi$ edge as well. This basically entails using a vertex operator built from $X(\sigma = \pi, \tau)$ instead of $X(\sigma = 0, \tau)$. Equivalently, we can introduce a 'twist operator' Ω that maps σ to $\pi - \sigma$ and so has the property

$$V(\sigma = \pi) = \Omega V(\sigma = 0)\Omega^{-1}. \tag{7.1.73}$$

Clearly it is necessary that $\Omega^2 = 1$, so that two twists are equivalent to none at all. Replacing $\sigma = 0$ by $\sigma = \pi$, for open-string boundary conditions, amounts to multiplying each oscillator α_n (as well as S_n in the case of superstrings) by the phase $(-1)^n$. This is achieved by the choice

$$\Omega = \pm(-1)^N, \tag{7.1.74}$$

where N is the usual number operator ($\sum_1^\infty \alpha_{-n} \cdot \alpha_n$ in the case of bosonic strings). The overall phase of Ω is not determined by (7.1.73). The correct choice for bosonic strings is the plus sign, since states with N even are charge-conjugation (C) even and N odd states are C odd. In the case of superstrings (in the supersymmetric formulation) the situation is reversed and the minus sign choice is appropriate.

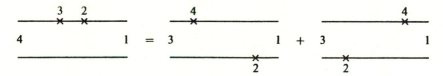

Figure 7.9. Twist identity for a four-particle amplitude showing that when the emissions are from opposite edges of the world sheet both time orderings correspond to the same cyclic ordering of the particles.

As a specific example of the use of the twist operator, let us consider the four-tachyon amplitude. The following identity is suggested by fig. 7.9

$$\langle 1| V(2)\Delta V(3) |4\rangle = \langle 1| V(2)\Delta\Omega V(4) |3\rangle + \langle 1| V(4)\Delta\Omega V(2) |3\rangle. \tag{7.1.75}$$

The two terms on the right-hand side of this equation correspond to the two possible τ orderings of the particle #2 and #4 emissions. Both correspond to a world sheet with the external states attached in the same cyclic order. Thus it is reasonable to expect that the sum reproduces the left-hand side.

The left-hand side of (7.1.75) has already been evaluated and shown to yield

$$
g^2 \int_0^1 y^{-2-s/2}(1-y)^{k_2 \cdot k_3}\, dy = g^2 B\big(-\alpha(s), -\alpha(t)\big), \qquad (7.1.76)
$$

where $s = -(k_1 + k_2)^2$ and $t = -(k_2 + k_3)^2$. The two terms on the right-hand side are easily evaluated by the techniques described in §7.1.4. The effect of the twist is to replace $1 - y$ by $1 + y$. Thus the first term on the right-hand side gives

$$
g^2 \int_0^1 x^{-2-s/2}(1+x)^{k_2 \cdot k_4}\, dx = g^2 \int_0^{1/2} y^{-2-s/2}(1-y)^{k_2 \cdot k_3}\, dy, \qquad (7.1.77)
$$

where we have made a change of variables $y = x/(x+1)$. Similarly, the second term gives the integral from $1/2$ to 1, so that the sum reproduces (7.1.76), as expected.

The twist operator is just a convenience, not a necessity, in calculating tree amplitudes. Even in the case of one-loop amplitudes it can be avoided for oriented world sheets by using $V(0)$ vertices for one boundary and $V(\pi)$ vertices for the other one. At the one loop level for unoriented open strings one meets a world sheet with the topology of the Möbius strip and the use of the twist operator becomes almost unavoidable.

7.2 Bosonic Closed Strings

A closed-string tree diagram has a world sheet that is topologically a sphere. By a conformal transformation it can be chosen to be geometrically a standard round sphere, which in turn can be mapped by stereographic projection onto the entire complex plane. The separation of left- and right-moving modes in closed-string theories, emphasized in previous chapters, implies a close relationship between closed-string amplitudes and pairs of open-string ones. Thus, whereas an open-string tree is generically of the form $\int dy f(y)$, with y denoting a set of real variables, a closed-string tree takes the form $\int d^2 z f_R(z) f_L(\bar{z})$, where $f_R(z)$ and $f_L(\bar{z})$ are expressions, similar to $f(y)$, associated with right- and left-moving modes. The integration is over the entire complex z plane.

7.2.1 Construction of Tree Amplitudes

In our discussion of closed strings in §2.1.3, we learned that the string co-ordinate $X^\mu(\sigma, \tau)$ decomposes into a sum of right-moving and left-moving pieces, $X_R^\mu(\tau - \sigma) + X_L^\mu(\tau + \sigma)$, and that the two pieces have independent mode expansions in terms of two sets of oscillators $\{\alpha_m^\mu\}$ and $\{\tilde{\alpha}_m^\mu\}$. Since each set separately can describe the Fock space of open-string states, the Fock-space description of closed-string states is given by direct products of open-string states. The only restriction is $L_0 = \tilde{L}_0$ (to which we will return), which implies that each sector contributes equally to the mass. Also, a symmetrization between the two sectors is required in the case of unoriented strings.

It is a plausible first guess that the emission vertex for the closed-string ground state is described by $: \exp(ik \cdot X) :$, just as in the case of open strings. We may argue, as in the open-string case, that the closed-string tachyon state carries no quantum numbers except its momentum, which is correctly incorporated in the operator $V =: \exp(ik \cdot X) :$. Since $X = X_L + X_R$, where X_L and X_R are constructed from left- and right-moving modes and commute with one another, our candidate vertex operator factorizes:

$$V(\sigma, \tau) = V_R(\tau - \sigma) V_L(\tau + \sigma). \tag{7.2.1}$$

The individual factors are separately just like open-string emission vertices:

$$V_L = e^{ik \cdot X_L}, \ V_R = e^{ik \cdot X_R}. \tag{7.2.2}$$

It is important, though, to remember from §2.1.3 that for closed strings,

$$\alpha_0^\mu = \tilde{\alpha}_0^\mu = \tfrac{1}{2} p^\mu, \tag{7.2.3}$$

while the usual open-string formula is simply $\alpha_0^\mu = p^\mu$ without the factor of $1/2$. The factor of $1/2$ in (7.2.3) means roughly that while V_L and V_R in (7.2.2) are each constructed as open-string tachyon vertex operators, they carry momentum $\tfrac{1}{2} p^\mu$ instead of p^μ. It is as though half of the momentum of the string were carried by left-moving modes and half by right-moving modes.

Is a factorization as in (7.2.1) special to tachyons, or might it be a general rule? We recall that the closed-string Fock space is constructed in a very simple way as a tensor product of left- and right-moving open-string Fock spaces. It is therefore very plausible that one should build a general closed-string vertex operator V as a product of left- and right-moving vertex operators, as in (7.2.1). Such operators are in one-to-one correspondence with the closed-string Fock space, and are easily seen to possess the necessary physical properties of vertex operators. For example, if V_L and V_R are left- and right-moving vertex operators of conformal dimension one, then the product $V_L V_R$ has total conformal dimension two, which we saw in chapter 1 is the correct value for a closed-string vertex operator.

Unlike the open-string case where the emission must occur from a boundary of the world sheet, closed-string emissions occur from the interior of the world sheet. Physically, the vertex operator

$$V(k, \tau, \sigma) = V_R(\tfrac{1}{2}k, \tau - \sigma) V_L(\tfrac{1}{2}k, \tau + \sigma) \qquad (7.2.4)$$

describes an emission from a point on the world sheet with coordinates τ and σ. However, what we require in the operator formalism is the complete amplitude for emission at time τ, irrespective of the value of σ. Therefore we define a closed-string vertex operator as a superposition over all possible σ values,

$$V^{cl}(k, \tau) = \frac{1}{\pi} \int_0^\pi d\sigma V_R(\tfrac{1}{2}k, \tau - \sigma) V_L(\tfrac{1}{2}k, \tau + \sigma)$$

$$= \frac{1}{\pi} \int_0^\pi d\sigma e^{-2i\sigma(L_0 - \tilde{L}_0)} V_R(\tfrac{1}{2}k, \tau) V_L(\tfrac{1}{2}k, \tau) e^{2i\sigma(L_0 - \tilde{L}_0)}. \qquad (7.2.5)$$

Continuing as in the case of open-string trees, one can introduce a propagator

$$\Delta = \tfrac{1}{2}(L_0 + \tilde{L}_0 - 2)^{-1} = \tfrac{1}{2} \int_0^1 \rho^{L_0 + \tilde{L}_0 - 3} d\rho \qquad (7.2.6)$$

and define a tree amplitude by

$$A_M = \kappa^{M-2} \langle \phi_1 | V_2 \Delta V_3 \cdots \Delta V_{M-1} | \phi_M \rangle$$
$$+ \text{ permutations of vertices.} \qquad (7.2.7)$$

These formulas are quite similar to the open-string formulas and again have the virtue that unitarity (at least in the exhibited channels) is manifest. An important difference from the open-string case is that there is no well-defined ordering for the $M - 2$ emitted particles, and the amplitude includes the sum over all permutations of their vertices. Since the vertices are integrated over all possible values of σ (so that the expression is invariant under 'twisting' any propagator) this sum of terms includes all possible orderings of the M external states. This is related to another significant feature that does not have an analog in the open-string case, namely, that the physical closed-string states $|\phi\rangle$ obey a constraint $(L_0 - \tilde{L}_0)|\phi\rangle = 0$. The propagator defined in (7.2.6) propagates closed-string states that do not necessarily obey this constraint. We can easily modify (7.2.6) to a propagator that propagates states that are necessarily annihilated by $L_0 - \tilde{L}_0$:

$$\Delta = \int_0^1 d\rho \int_0^{2\pi} \frac{d\phi}{4\pi} \rho^{L_0 + \tilde{L}_0 - 3} e^{i\phi(L_0 - \tilde{L}_0)}. \qquad (7.2.8)$$

Since in (7.2.7) the initial and final states are annihilated by $L_0 - \tilde{L}_0$ and the vertices (because of the σ integrals in (7.2.5)) commute with $L_0 - \tilde{L}_0$, the modification (7.2.8) of the propagator does not affect the amplitudes, but it simplifies the rest of the discussion. One is then free, of course, to drop the σ integrations in the vertex operator (7.2.5). It is convenient now to introduce the complex variable

$$z = \rho e^{i\phi}, \qquad (7.2.9)$$

and rewrite the propagator (7.2.8) (defining $d^2 z = \rho d\rho d\phi = dx dy$)

$$\Delta = \frac{1}{4\pi} \int_{|z|\leq 1} \frac{d^2 z}{|z|^2} z^{L_0 - 1} \bar{z}^{\tilde{L}_0 - 1}. \qquad (7.2.10)$$

7.2.2 Examples

Equation (7.2.7) has an especially simple implication for three-point functions. To describe it, let us represent a closed-string state ϕ as a product of a right-moving state ρ and a left-moving state λ. In this case the vertex operator can be represented as a product of a right-moving factor and a left-moving factor

$$V_2(k_2) \equiv V(\phi_2, k_2) = V_R(\rho_2, \tfrac{1}{2}k_2) V_L(\lambda_2, \tfrac{1}{2}k_2) \qquad (7.2.11)$$

evaluated at $\tau = 0$ (or $z = 1$). Since $L_0 = \tilde{L}_0$ for the end states that multiply the vertex, the integral in (7.2.5) can be dropped. We can represent the end states as products of left-moving and right-moving factors, as well,

$$\langle\phi_1| = \langle\rho_1|_R \otimes \langle\lambda_1|_L \,, \quad |\phi_3\rangle = |\rho_3\rangle_R \otimes |\lambda_3\rangle_L \,. \tag{7.2.12}$$

As a result

$$A_3^{cl}(\phi_1, \phi_2, \phi_3) = A_3^{op}(\rho_1, \rho_2, \rho_3) \times A_3^{op}(\lambda_1, \lambda_2, \lambda_3), \tag{7.2.13}$$

which is a simple product of open-string couplings for the left-moving and right-moving modes. In each case, the ρ_i and λ_i states should be though of as carrying momentum $\frac{1}{2}k_i$. For example, the three-graviton vertex can be expressed in terms of open-string three-photon vertices as follows

$$A_3 = \kappa\zeta_1^{\mu\mu'}\zeta_2^{\nu\nu'}\zeta_3^{\rho\rho'}t_{\mu\nu\rho}(k/2)t_{\mu'\nu'\rho'}(k/2), \tag{7.2.14}$$

where $t_{\mu\nu\rho}$ is the tensor given in (7.1.58). The graviton polarization tensors $\zeta^{\mu\nu}(k)$ are symmetric and traceless and satisfy $k_\mu\zeta^{\mu\nu} = 0$. Amplitudes involving massless antisymmetric tensors and dilatons can also be extracted from the same formulas. Since $t_{\mu\nu\rho}$ contains an $O(\alpha')$ correction to the leading term, the three graviton coupling in (7.2.14) contains both $O(\alpha')$ and $O(\alpha'^2)$ corrections to the three-graviton vertex of general relativity. They correspond to specific R^2 and R^3 terms in an effective action.

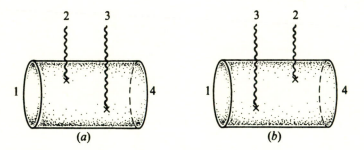

Figure 7.10. The two diagrams that contribute to the four-particle closed-string tree amplitude consist of a contribution with an s-channel propagator shown in (a) and a diagram with a u-channel propagator shown in (b). The sum of these terms gives the fully crossing-symmetric amplitude.

Returning to Euclidean world-sheet metric, let us re-express the integrand of (7.2.5) in terms of the complex variables defined in (7.2.9). The amplitude is the sum of terms corresponding to the different orderings of the emitted particles. In each term the propagators are given by integrating the parameter z in the region $|z| \leq 1$. When these terms are combined the result is that the z's are integrated independently over the whole complex plane. As a simple illustrative example, consider the two tree diagrams that contribute to the four-particle amplitude with external tachyons shown in fig. 7.10. The amplitude has the form

$$A_4 = \kappa^2 \langle k_1 | V_2(k_2) \Delta V_3(k_3) | k_4 \rangle + \kappa^2 \langle k_1 | V_3(k_3) \Delta V_2(k_2) | k_4 \rangle. \qquad (7.2.15)$$

Using the integral representation for Δ in (7.2.10) the factors of $z^{L_0} \bar{z}^{\tilde{L}_0}$ can be moved to the right in the first term using

$$V(k, z, \bar{z}) = z^{L_0} \bar{z}^{\tilde{L}_0} V(k, 1) z^{-L_0} \bar{z}^{-\tilde{L}_0} = V_R(\tfrac{1}{2}k, z) V_L(\tfrac{1}{2}k, \bar{z}). \qquad (7.2.16)$$

In the second term on the right-hand side of (7.2.15) we choose to move these factors to the left. The result is

$$
\begin{aligned}
A_4 = {}& \frac{\kappa^2}{4\pi} \int\limits_{|z| \leq 1} \frac{d^2 z}{|z|^2} \langle k_1 | V_2(k_2, 1, 1) V_3(k_3, z, \bar{z}) | k_4 \rangle \\
& + \frac{\kappa^2}{4\pi} \int\limits_{|z| \geq 1} \frac{d^2 z}{|z|^2} \langle k_1 | V_3(k_3, z, \bar{z}) V_2(k_2, 1, 1) | k_4 \rangle,
\end{aligned}
\qquad (7.2.17)
$$

where we have changed integration variables from z to $1/\bar{z}$ in the second term. The two terms can clearly be combined into a single formula in which z is integrated over the whole complex plane keeping the vertices ordered according to the magnitude of their arguments.

This example generalizes to the M-particle case in which there are $M-3$ variables that are integrated over the whole complex plane. The amplitude can now be rewritten in manifestly symmetric form by the same method as was used in the case of open strings in the last section. This involves writing the end states as infinite τ limits of vertex insertions on the world sheet. The general M-particle tree has the form (analogous to (7.1.42))

$$A_M = 4\pi \left(\frac{\kappa}{4\pi}\right)^{M-2} \int d\mu_M(z) I_M^R(\tfrac{1}{2}k, z) I_M^L(\tfrac{1}{2}k, \bar{z}), \qquad (7.2.18)$$

where (as in (7.1.40))

$$I_M^R(\tfrac{1}{2}k, z) = \langle 0 | T\{V_R(\tfrac{1}{2}k_1, z_1) \cdots V_R(\tfrac{1}{2}k_M, z_M)\} | 0 \rangle / (\Pi z_i) \qquad (7.2.19)$$

$$I_M^L(\tfrac{1}{2}k, \bar{z}) = \langle 0 | T\{V_L(\tfrac{1}{2}k_1, \bar{z}_1) \cdots V_L(\tfrac{1}{2}k_M, \bar{z}_M)\} | 0 \rangle / (\Pi \bar{z}_i). \qquad (7.2.20)$$

The 'T ordering' means that V factors are arranged so that $|z| = \rho$ decreases from one factor to the next, and the different orderings correspond to the sum over permutations in (7.1.73). The boundaries between different $|z|$ orderings have no invariant meaning, and it is only the full combination of terms that makes sense.

To understand duality, we now must discuss the residual symmetry that plays for closed strings the role played by $SL(2, R)$ for open strings. When one quantizes the string in the Hamiltonian picture used in chapter 2, with a world-sheet metric of signature $(-+)$, the vacuum, or zero-momentum ground state, $|0; 0\rangle$ is invariant under the six-parameter group $SL(2, R) \times SL(2, R)$. One $SL(2, R)$ factor is generated by L_1, L_0, L_{-1} and the other by $\tilde{L}_1$, $\tilde{L}_0$, $\tilde{L}_{-1}$. In calculating string scattering amplitudes, it is much more convenient to use a world-sheet metric of signature $(++)$. Indeed, it is such a metric that appears naturally in the above formulas, without being put in by hand. With a Euclidean metric, the symmetry group $SL(2, R) \times SL(2, R)$ is replaced by the group $SL(2, C)$ of 2×2 complex matrices of determinant one. The groups $SL(2, C)$ and $SL(2, R) \times SL(2, R)$ are closely related and both have real dimension six, but they are quite different groups.

Since the closed-string world sheet is a sphere, which can be conformally mapped to the complex plane, the residual symmetry group for closed strings should be the group of conformal transformations of the complex plane (including the point at infinity). To understand why $SL(2, C)$ is the group of such transformations, note that a 2×2 complex matrix can act on a two-dimensional complex vector space:

$$\begin{pmatrix} v_1 \\ v_2 \end{pmatrix} \rightarrow \begin{pmatrix} a & b \\ c & d \end{pmatrix} \begin{pmatrix} v_1 \\ v_2 \end{pmatrix}. \tag{7.2.21}$$

The formula is the same as (7.1.35), but now the variables are all complex, since we are considering $SL(2, C)$ instead of $SL(2, R)$. As in the discussion of (7.1.35), we define $z = v_1/v_2$ and find that z transforms as

$$z \rightarrow \frac{az + b}{cz + d}. \tag{7.2.22}$$

Thus, we learn that $SL(2, C)$ acts as a group of conformal mappings of the complex z plane (including the point at infinity). Since the variables are complex, $SL(2, C)$ has twice as many degrees of freedom as $SL(2, R)$; this corresponds to the doubling of degrees of freedom in going from open strings to closed strings.[*]

---

[*] In chapter 15, we will interpret v_1 and v_2 in these formulas as homogeneous coordinates for CP^1; the formula $z = v_1/v_2$ is the stereographic projection from CP^1 (minus the 'point at infinity', namely $v_2 = 0$) to the z plane.

Even though $SL(2, C)$ does not factorize as a product of a symmetry group of left-moving modes and one of right-moving modes, the two sets of modes are decoupled in many formulas. For example, as a consequence of (7.1.41),

$$I_M^R(\tfrac{1}{2}k, z) = I_M^R(\tfrac{1}{2}k, z') \prod_i (a - c\, z_i')^2$$

$$I_M^L(\tfrac{1}{2}k, \overline{z}) = I_M^L(\tfrac{1}{2}k, \overline{z}') \prod_i (\overline{a} - \overline{c}\, \overline{z}_i')^2.$$

(7.2.23)

This is described by saying that I_M^R and I_M^L have Möbius weight (2,0) and (0,2), so that, altogether, the product $I_M^L I_M^R$ has Möbius weight (2,2) under $SL(2, C)$ transformations.

As in the case of open strings, we now wish to re-express our integration measure in a more symmetrical way. Let us try to guess what would be a measure with the right properties; we will then check our result by showing that with suitable choices it reduces to the form of (7.2.7), which is 'correct' in the sense that it is manifestly unitary (at least in the channel exhibited). As in the open-string discussion, in this way we get a more general and more symmetrical description of the measure than that in (7.2.7).

The desired measure cannot include an integral over the z_i throughout the whole complex plane without any restriction, since this would again include an infinite factor corresponding to the volume of the group $SL(2, C)$. As before, we deal with this by using the $SL(2, C)$ symmetry to fix three of the z_i, denoted z_A, z_B, z_C, to values z_A^0, z_B^0, z_C^0. These are twice as many conditions as in the open-string case since the z are complex, but the group also has twice as many parameters, so the mapping is always possible. The measure must have the transformation law

$$d\mu_M(z) = d\mu_M(z') \prod_{i=1}^{M} |a - c z_i'|^{-4}$$

(7.2.24)

in order that the amplitude be $SL(2, C)$ invariant. By the same algebra as was described in the open-string case, one easily sees that the correct choice is

$$d\mu_M(z) = |z_A - z_B|^2 |z_A - z_C|^2 |z_B - z_C|^2$$
$$\times \delta^2(z_A - z_A^0)\delta^2(z_B - z_B^0)\delta^2(z_C - z_C^0) \prod_{i=1}^{M} d^2 z_i.$$

(7.2.25)

with arbitrary choices of A, B and C and of z_A^0, z_B^0 and z_C^0. The operator formula in (7.2.7) corresponds to the particular choice $z_1^0 = \infty$, $z_2^0 = 1$

and $z_M^0 = 0$. The angular phase part of the z integrations ensures that $L_0 = \tilde{L}_0$ at all poles. This is a direct consequence of the σ integration in (7.2.5), which ensures that there is no preferred point on the string.

As a specific example, let us consider the amplitude for the scattering of M closed-string tachyons. In the units we are using ($\alpha' = 1/2$ for open strings), the leading Regge trajectory for closed strings is given by $\alpha(s) = 2 + s/4$. This describes a massless state corresponding to the tachyon in the theory with $\alpha(s) = 0$ at $k^2 = -s = 8$. Similarly, there is a spin 2 state at $\alpha(s) = 2$ that has $k^2 = 0$ and corresponds to the massless graviton. The amplitude is given by identifying I_M^R and I_M^L with I_M of the open-string theory. The result is thus

$$A_M = 4\pi \left(\frac{\kappa}{4\pi}\right)^{M-2} \int d\mu_M(z) \prod_{i<j} |z_i - z_j|^{k_i \cdot k_j/2}. \tag{7.2.26}$$

In particular, setting $z_1 = \infty$, $z_2 = 1$, $z_3 = z$ and $z_4 = 0$, gives for the case $M = 4$

$$A_4 = \frac{\kappa^2}{4\pi} \int |z|^{k_3 \cdot k_4/2} |1 - z|^{k_2 \cdot k_3/2} d^2 z. \tag{7.2.27}$$

Integrals of this form can be explicitly evaluated by substituting the representation

$$|z|^{-A} = \frac{1}{\Gamma(A/2)} \int_0^\infty t^{A/2-1} \exp(-t|z|^2) dt \tag{7.2.28}$$

and similarly for $|1 - z|^{-B}$. This turns the z integration into a Gaussian that can be explicitly carried out. The remaining t integrals can then be evaluated using the Euler beta function formula (7.1.63). In this way one finds that

$$I = \int |z|^{-A}|1 - z|^{-B}d^2z$$

$$= \frac{1}{\Gamma(A/2)\Gamma(B/2)} \int\limits_0^\infty t^{A/2-1}u^{B/2-1}dtdu$$

$$\times \int\limits_{-\infty}^\infty dxdy \exp\{-t(x^2 + y^2) - u[(1 - x)^2 + y^2]\}$$

$$= \frac{\pi}{\Gamma(A/2)\Gamma(B/2)} \int\limits_0^\infty \frac{dtdu}{t+u}t^{A/2-1}u^{B/2-1}\exp\left(-\frac{tu}{t+u}\right)$$

(7.2.29)

$$= \pi\frac{\Gamma(\frac{1}{2}(A+B)-1)}{\Gamma(A/2)\Gamma(B/2)} \int\limits_0^1 \lambda^{-A/2}(1-\lambda)^{-B/2}d\lambda$$

$$= B(1 - \frac{A}{2}, 1 - \frac{B}{2}, \frac{A+B}{2} - 1),$$

where

$$B(a, b, c) = \pi\frac{\Gamma(a)\Gamma(b)\Gamma(c)}{\Gamma(a+b)\Gamma(b+c)\Gamma(c+a)}.$$ (7.2.30)

Applying this to the evaluation of (7.2.27), using

$$k_3 \cdot k_4 = -\tfrac{1}{2}s - 8, \qquad k_2 \cdot k_3 = -\tfrac{1}{2}t - 8,$$ (7.2.31)

and

$$s + t + u = \sum m_i^2 = -32,$$ (7.2.32)

gives

$$A_4 = \frac{\kappa^2}{4\pi}B\left(-\tfrac{1}{2}\alpha(s), -\tfrac{1}{2}\alpha(t), -\tfrac{1}{2}\alpha(u)\right),$$ (7.2.33)

which is the Virasoro formula. The Regge trajectories are given by $\alpha(s) = \frac{1}{4}s + 2$ and satisfy $\alpha(s) + \alpha(t) + \alpha(u) = -2$.

This formula is manifestly crossing symmetric and it is easy to check that this expression has poles at $\alpha(s), \alpha(t), \alpha(u) = 0, 2, \ldots$ corresponding to the ground-state tachyon (at s, t, or $u = -8$) and the infinite set of excited states. The asymptotic behavior of the amplitude at high energy ($s \to \infty$) and fixed t is easy to evaluate by using Stirling's approximation

(which was discussed in the context of the open-string Veneziano formula in chapter 1)

$$\Gamma(a) \sim \sqrt{2\pi}a^{a-1/2}e^{-a}, \tag{7.2.34}$$

This formula is valid for $|a| \to \infty$ in any direction in the complex a plane apart from the negative real a axis. Applying this to the expression (7.2.33) with $|s| \to \infty$, $u \sim -s$ and fixed t gives the Regge behavior

$$A_4 \sim \frac{\kappa^2}{4}e^{-i\pi\alpha(t)/2}\frac{\Gamma(-\alpha(t)/2)}{\Gamma(1+\alpha(t)/2)}\left(\frac{s}{8}\right)^{\alpha(t)}, \tag{7.2.35}$$

as long as s is kept slightly away from the real axis. As with the Veneziano model, this tree-level result has to be interpreted in an average sense for the physical values of real, positive s, since the amplitude has poles along the real axis. The poles move away from the real axis when loop effects are taken into account.

The question of gauge invariance of closed-string amplitudes is again related to the question of whether zero-norm states decouple from tree diagrams. Consider, for example, a tree diagram with a longitudinally-polarized graviton of momentum k^μ (such as particle #2 in fig. 7.10). An arbitrary longitudinal polarization tensor can be written as a sum of terms of the form $\zeta^{\mu\nu} = k^\mu \zeta^\nu$, where $k^2 = 0$ and $\zeta \cdot k = 0$. The vertex for describing the emitted longitudinally polarized graviton has the form

$$V(k, \tau) = \frac{1}{\pi}\int d\sigma k_\mu \dot{X}_R^\mu(\tau - \sigma)\zeta_\nu \dot{X}_L^\nu(\tau + \sigma)e^{ik\cdot X_R(\tau-\sigma)}e^{ik\cdot X_L(\tau+\sigma)}$$

$$= -\frac{i}{\pi}\int d\sigma \frac{d}{d\tau}\left(e^{ik\cdot X_R(\tau-\sigma)}\right)\zeta_\nu \dot{X}_L^\nu(\tau + \sigma)e^{ik\cdot X_L(\tau+\sigma)}. \tag{7.2.36}$$

When inserted in either of the diagrams of fig. 7.10 the τ derivative causes a propagator to be canceled as in the case of open strings. The result is not zero in either diagram separately in this case because the adjoining vertices, which are at the same value of τ, are not at the same value of σ. The expressions for the vertices involve integrating over all σ values independently. However, in the combination of fig. 7.10 (a) and (b) the σ integrals cancel everywhere except in the vicinity of the singular point where the two vertices coincide on the world-sheet. As we saw in (7.1.72), the product of two vertices at the same point is singular and is consistently interpreted as giving a vanishing result. This argument generalizes to tree diagrams with an arbitrary number of external states.

7.2.3 Relationship to Open-String Trees

The relation between open- and closed-string three-particle vertices

$$A_3^{cl} \sim A_{3R} A_{3L},   \tag{7.2.37}$$

illustrated in (7.2.14) can be generalized to M-particle amplitudes. The most naive guess, consisting of $A_{MR} A_{ML}$ for a particular cyclic ordering, is not correct since it only has cyclic symmetry and not total symmetry in general. It happens that in the $M = 3$ case the total symmetry of A_{3R} and A_{3L} for tachyons (or antisymmetry for photons and so forth) results in total symmetry for A_3^{cl}.

To find the analog of (7.2.37) in the four-tachyon case, we note that

$$B(a, b)B(b, c) = \frac{1}{\pi}\Gamma(b)\Gamma(a + c)B(a, b, c).   \tag{7.2.38}$$

Therefore, using $a + b + c = 1$ and the standard formula $\Gamma(b)\Gamma(1 - b) = \pi/\sin \pi b$, we have

$$B(a, b, c) = \sin \pi b B(a, b)B(b, c).   \tag{7.2.39}$$

In terms of the four-tachyon amplitudes this gives

$$A_4^{cl}(s, t, u) \sim (\sin \pi t/8) A_4^{op}(s/4, t/4) A_4^{op}(t/4, u/4).   \tag{7.2.40}$$

This formula is not an accident, but a special case of a more general relation.

A general four-particle amplitude involving arbitrary excited states differs from (7.2.27) by some additional polynomial dependence on z and $\bar{z}$ in the integrand, with coefficients involving the momenta and polarization tensors. Therefore in the general case one needs to consider terms of the form

$$I(n_1, n_2, n_3, n_4) = \int d^2z\, z^{k_3 \cdot k_4/4 + n_1}(1 - z)^{k_2 \cdot k_3/4 + n_2}$$
$$\times (\bar{z})^{k_3 \cdot k_4/4 + n_3}(1 - \bar{z})^{k_2 \cdot k_3/4 + n_4}.   \tag{7.2.41}$$

By writing $z = x + iy$ and deforming the y integration contour this can be expressed as an integral with z and $\bar{z}$ treated as independent real variables integrated from $-\infty$ to $+\infty$. However, one must be careful to choose the correct branch of the integrand in doing this. Denoting as η and χ the real variables that correspond to z and $\bar{z}$, the contours in each case go over or under the branch points at 0 and 1 depending on whether the other variable is less than zero, between 0 and 1, or greater than 1.

Next one deforms the contour to enclose branch cuts, so that the integrands can be replaced by discontinuities. The analysis requires considerable care, and we omit the details. The result is that

$$I(n_1, n_2, n_3, n_4) = -\sin(\pi k_2 \cdot k_3/4) \int_0^1 d\xi \xi^{k_3 \cdot k_4/4 + n_1} (1 - \xi)^{k_2 \cdot k_3/4 + n_2}$$

$$\times \int_1^\infty d\eta \eta^{k_3 \cdot k_4/4 + n_3} (1 - \eta)^{k_2 \cdot k_3/4 + n_4}.$$

$$(7.2.42)$$

This is precisely $\sin(\pi t/8)$ times an $s - t$ open-string amplitude times a $t - u$ open-string amplitude. In particular, putting all the n equal to zero, it reproduces the four-tachyon result derived above.

Equation (7.2.42) provides a basis for generalizing the result in (7.2.40) to other four-particle processes. For example, the four-graviton amplitude is related to the four-photon one by

$$A_4^{\mu\mu'\nu\nu'\rho\rho'\lambda\lambda'} \sim (\sin \pi t/8) A_4^{\mu\nu\rho\lambda}(s/4, t/4) A_4^{\mu'\nu'\rho'\lambda'}(t/4, u/4). \quad (7.2.43)$$

In the case of M-particle functions the relation takes the form

$$A_M^{cl} \sim \sum_{P,P'} e^{i\pi F(P,P')} A_M^{op}(P) \overline{A}_M^{op}(P'), \quad (7.2.44)$$

where P and P' refer to particular cyclic orderings of the M external lines, A_M^{op} and $\overline{A}_M^{op}$ are M-particle open-string tree amplitudes associated with right- and left-moving modes (which can be different in general) with the usual substitution $k \to k/2$, and $F(P, P')$ are appropriate phase factors. The number of inequivalent terms grows rapidly with M. Whereas there is only one for $M = 4$ and two for $M = 5$, there are already 12 for $M = 6$. These relationships are applicable to all closed-string theories. For example, heterotic string amplitudes can be related to products of trees made from open superstrings and open bosonic strings, a fact that will be exploited later.

7.3   Superstrings in the RNS Formulation

The construction of superstring amplitudes involves a number of new features in addition to those described in the preceding sections for bosonic strings. Some calculations are carried out most easily using the covariant

RNS formulation of §4.2 and others are better understood using the supersymmetric light-cone quantization of §5.2. We begin with a discussion of the covariant RNS formulation.

7.3.1 Open-String Tree Amplitudes in the Bosonic Sector

The analogies between amplitudes in the $D = 26$ bosonic string theory and the $D = 10$ superstring theory are closest for the bosonic sector in the RNS formulation. Even here, there are new issues introduced by the G gauge symmetries implied by world-sheet supersymmetry. Just as in the 26-dimensional case, we can define a tree amplitude by

$$A_M = g^{M-2} \langle \phi_1 | V(2) \Delta V(3) \cdots \Delta V(M-1) | \phi_M \rangle \qquad (7.3.1)$$

and demonstrate that for appropriately chosen definitions of V, Δ and $|\phi\rangle$ it satisfies tree unitarity and cyclic symmetry. In the present case tree unitarity requires that intermediate-state poles satisfy subsidiary constraint conditions associated with superconformal generators G_r as well as those associated with Virasoro generators L_n. Also, the analysis requires a peculiar ingredient that does not appear in the purely bosonic theory. In chapter 4, we formulated the supersymmetric string in Fock space in a very natural seeming way. It turns out, however, that there are other equivalent ways of formulating this theory, known as different 'pictures'. What is more, the proof that the bosonic tree amplitudes (7.3.1) obey cyclic symmetry requires study of two different pictures, called F_1 and F_2, which must be shown to be equivalent. The F_1 picture is more convenient for establishing cyclic symmetry, whereas the F_2 picture is more convenient for proving that all poles correspond to physical states.

The formulation of the bosonic sector in chapter 4 involved what is usually called the F_2 picture. In this formulation, the wave equation is $(L_0 - 1/2) |\phi\rangle = 0$, and the subsidiary conditions for physical states are $G_r |\phi\rangle = 0$ for $r > 0$. (This implies that $L_n |\phi\rangle = 0$ for $n > 0$, as well.) In analogy with the construction in §7.1.1, it is natural to define the propagator in this formulation to be given by

$$\Delta = (L_0 - \tfrac{1}{2})^{-1}. \qquad (7.3.2)$$

If we also require that physical-state vertex operators have conformal dimension $J = 1$, then it follows by exactly the same reasoning as before that intermediate poles satisfy the L_n conditions. However, the G_r conditions are a stronger requirement, and therefore physical vertices have to be restricted further. The appropriate rule was given in §4.2.3, where we introduced physical vertices for the construction of the spectrum generating algebra used in the proof of the no-ghost theorem. The requirement is that there exist an operator W of conformal dimension $J = 1/2$ such that V is expressible in the form

$$V = \{G_r, W\} \tag{7.3.3}$$

for all r. This implies that V has $J = 1$, so the L_n conditions are certainly satisfied. For even G-parity emissions (satisfying the GSO conditions) the operator W is fermionic and V is bosonic in the world-sheet sense. For odd G emissions, on the other hand, W is bosonic and V is fermionic, so that a commutator is required in place of the anticommutator in (7.3.3). Since $G_r^2 = L_{2r}$, (7.3.3) implies that

$$[G_r, V] = [L_{2r}, W] \tag{7.3.4}$$

while $V = [G_r, W]$ would imply that

$$\{G_r, V\} = [L_{2r}, W]. \tag{7.3.5}$$

The proof of tree unitarity requires showing that a spurious state decouples from a physical tree, *i.e.*,

$$\langle \tilde{\phi} | G_r V(2) \Delta V(3) \cdots \Delta V(M - 1) | \phi_M \rangle = 0, \qquad r > 0 \tag{7.3.6}$$

where $\langle \tilde{\phi} |$ is annihilated by $L_0 - 1/2 + r$ and is otherwise unrestricted. The proof involves manipulating G_r to the right. For the commutator term we replace $[L_{2r}, W]$ by

$$(L_{2r} - L_0 - r + \tfrac{1}{2})W - W(L_{2r} - L_0 + \tfrac{1}{2}) = 0, \tag{7.3.7}$$

which is justified by the mass-shell condition for $\langle \tilde{\phi} |$ and the canceled propagator argument. The vanishing of (7.3.7) is a consequence of the requirement that W has $J = 1/2$. The next step is to bring G_r past a propagator using

$$G_r \frac{1}{L_0 - 1/2} = \frac{1}{L_0 + r - 1/2} G_r. \tag{7.3.8}$$

It can then be brought past the subsequent vertices and propagators in the same way until it finally annihilates against $|\phi_M\rangle$.

The trouble with the F_2 picture is that (7.1.28) and (7.1.29), which were essential in the proof of cyclic symmetry, do not hold. Instead we have

$$|\phi_M\rangle = \lim_{y_M \to 0} y_M^{-1/2} W(k_M, y_M) |0; 0\rangle \qquad (7.3.9)$$

and

$$\langle\phi_1| = \lim_{y_1 \to \infty} \langle 0; 0| y_1^{1/2} W(k_1, y_1), \qquad (7.3.10)$$

which leads to a counterpart of (7.1.30) in which $M - 2$ states are described by V operators and two are described by W operators. To prove cyclic symmetry it is much better to describe all M external states by V operators, and this is what the F_1 picture (the first one discovered) achieves.

The fact that the powers $y_1^{1/2}$ and $y_M^{-1/2}$ occur in these formulas rather than y_1 and y_2, as in (7.1.28) and (7.1.29), is a reflection of the fact that the W vertices have conformal dimension $J = 1/2$, whereas the V vertices have conformal dimension $J = 1$. As an example that illustrates this, we note that in the case of the tachyon vertex operator $W =: e^{ik \cdot X} :$, the factor Z_0 defined in (7.1.6) becomes

$$Z_0 = e^{ik \cdot x + k \cdot p \ln z} = e^{ik \cdot x} z^{k \cdot p + 1/2} = z^{k \cdot p - 1/2} e^{ik \cdot x}, \qquad (7.3.11)$$

since now $k^2 = 1$ in place of the value $k^2 = 2$ used earlier. As a result one now has

$$Z_0 |0; 0\rangle = z^{-1/2} |0; k\rangle \qquad (7.3.12)$$

and

$$\langle 0; 0| Z_0 = \langle 0; k| z^{1/2}, \qquad (7.3.13)$$

which are the key to verifying (7.3.9) and (7.3.10) for the tachyon vertex.

7.3.2 The F_1 Picture

To explain the F_1 formalism, we begin by noting that an F_2 physical state satisfies

$$\langle\phi| G_{1/2} G_{-1/2} = \langle\phi|, \qquad (7.3.14)$$

since $G_{1/2} G_{-1/2} = 2L_0 - G_{-1/2} G_{1/2}$ and $\langle\phi| G_{-1/2} = \langle\phi| (L_0 - \frac{1}{2}) = 0$. Therefore the tree amplitude (7.3.1) can be written in the form

$$A_M = g^{M-2} \langle\phi_1| G_{1/2} G_{-1/2} V(2) \Delta \cdots V(M-1) |\phi_M\rangle. \qquad (7.3.15)$$

Next we push the $G_{-1/2}$ factor to the right, using exactly the same reasoning as in the proof of tree unitarity. In this way we obtain the alternative formulation

$$A_M = g^{M-2} \langle \tilde{\phi}_1 | V(2) \tilde{\Delta} V(3) \cdots \tilde{\Delta} V(M-1) | \tilde{\phi}_M \rangle, \qquad (7.3.16)$$

where

$$|\tilde{\phi}\rangle = G_{-1/2} |\phi\rangle \qquad (7.3.17)$$

and

$$\tilde{\Delta} = (L_0 - 1)^{-1}. \qquad (7.3.18)$$

The latter expression is a consequence of (7.3.8). Equation (7.3.16) is the F_1 formulation of the amplitude. The state $|\tilde{\phi}\rangle$ in (7.3.17) is the Fock-space state of the F_1 picture that corresponds to the Fock-space state $|\phi\rangle$ in the F_2 picture. Since $\tilde{\Delta}$ is the propagator in the F_1 picture, the wave equation becomes $(L_0 - 1)|\tilde{\phi}\rangle = 0$, just as in the bosonic string theory.

Note that a physical state $|\tilde{\phi}\rangle$ in the F_1 picture is *not* annihilated by $G_{1/2}$. In fact,

$$G_{1/2}|\tilde{\phi}\rangle = G_{1/2}G_{-1/2}|\phi\rangle = (2L_0 - G_{-1/2}G_{1/2})|\phi\rangle = |\phi\rangle. \qquad (7.3.19)$$

This equation and (7.3.17) show how to go back and forth between F_1 and F_2 states. The G_r with $r \geq 3/2$ do annihilate F_1 states, because

$$G_r|\tilde{\phi}\rangle = G_r G_{-1/2}|\phi\rangle = (2L_{r-1/2} - G_{-1/2}G_r)|\phi\rangle = 0 \qquad (7.3.20)$$

for $r \geq 3/2$.

The relationship between F_1 and F_2 states can also be understood as follows. Each physical state has an associated $J = 1/2$ emission vertex W as well as a $J = 1$ one V given by $V = \{G_r, W\}$ or $V = [G_r, W]$. States in the F_2 picture are made using the the W vertex operator according to (7.3.9) and (7.3.10). States in the F_1 picture, on the other hand, are made by the same construction as ones in the bosonic string theory, namely

$$|\tilde{\phi}_M\rangle = G_{-1/2} \lim_{y_M \to 0} y_M^{-1/2} W(y_M) |0;0\rangle = \lim_{y_M \to 0} y_M^{-1} V(k_M, y_M) |0;0\rangle \qquad (7.3.21)$$

$$\langle \tilde{\phi}_1 | = \lim_{y_1 \to \infty} \langle 0;0| y_1^{1/2} W(y_1) G_{1/2} = \lim_{y_1 \to \infty} \langle 0;0| y_1 V(k_1, y_1), \qquad (7.3.22)$$

where we have used the fact that $G_{1/2}$ and $G_{-1/2}$ annihilate the vacuum. Therefore we end up with the $SL(2, R)$ invariant formula

$$A_M = g^{M-2} \int d\mu_M(y) (\prod y_i)^{-1} \langle 0;0| V(k_1, y_1) \cdots V(k_M, y_M) |0;0\rangle, \qquad (7.3.23)$$

just as in §7.1.3. The proof of cyclic symmetry in this formulation is identical to the one given in that section. Equation (7.3.23) actually has a larger symmetry than $SL(2, R)$. It extends to the superalgebra $OSp(1|2)$, generated by L_0, $L_{\pm 1}$, $G_{\pm 1/2}$, since these are now the operators that annihilate the vacuum $|0; 0\rangle$.

States in the F_1 picture that are annihilated by all the G_r with $r \geq 1/2$ and satisfy the F_1 mass-shell condition $(L_0 - 1)|\tilde{\phi}\rangle = 0$ constitute a new class of spurious states without counterpart in the F_2 picture. (Equation (7.3.19) would give 0 for the corresponding F_2 state.) For example, the F_1 Fock-space ground state exactly corresponds to the tachyon at $k^2 = 2$ in the bosonic string theory. (The tachyon at $k^2 = 1$ is described by $G_{-1/2}|0; k\rangle = k \cdot b_{-1/2}|0; k\rangle$ in the F_1 picture.) Whereas the $k^2 = 2$ tachyon is physical in the bosonic string theory, it is unphysical in the superstring theory. Its decoupling from trees, all of whose other states are physical, is established by the same methods as that of other spurious states. Namely, consider a tree amplitude formulated in the F_1 picture as in (7.3.16). Now suppose that $|\tilde{\phi}_M\rangle$ is a spurious state annihilated by $G_{1/2}$. Such a tree can be shown to vanish by using (7.3.17) to substitute $\langle \tilde{\phi}_1 | = \langle \phi_1 | G_{1/2}$ and then pushing the $G_{1/2}$ to the right until it annihilates against $|\tilde{\phi}_M\rangle$, just as in the previous decoupling proofs.

To recapitulate, we have demonstrated the equivalence between the F_1 and F_2 pictures and shown how to transform between them. The F_2 picture was used to demonstrate that tree amplitudes contain physical-state poles only in the channels corresponding to explicitly displayed propagators and that the residues of these poles are tree amplitudes of the same form, as required by unitarity. The F_1 picture was used to give a manifestly $SL(2, R)$-invariant formula for the tree amplitudes in which cyclic symmetry can be readily established by the same reasoning used for the bosonic string theory. This extends the factorization proof to all channels.

7.3.3 Examples

The vertex operator for tachyon emission is given by $W_T =: e^{ik \cdot X} :$, so

$$V_T(k) = [G_r, : e^{ik \cdot X} :] = k \cdot \psi : e^{ik \cdot X} : . \qquad (7.3.24)$$

The tachyon mass-shell condition, $k^2 = 1$, ensures that W_T has $J = 1/2$. Even though this operator incorporates all the super-Virasoro conditions it is not altogether physical, because it describes emission of an odd G state violating the GSO conditions. A physical boson-emission vertex must be given by a bosonic operator, which V_T is not. The first physical bosonic state in the spectrum with the correct G parity is the massless vector. Its emission vertex, which was utilized in chapter 4 in the construction of the spectrum generating algebra, is

$$V(\zeta, k) = \{G_r, \zeta \cdot \psi e^{ik \cdot X}\} = (\zeta \cdot \dot{X} - \zeta \cdot \psi k \cdot \psi)e^{ik \cdot X}. \qquad (7.3.25)$$

As usual, we must require $k^2 = \zeta \cdot k = 0$.

Let us now construct explicit examples of tree amplitudes. A good warm-up exercise, even though it is involves states that are not present in the form of the theory that we will eventually wish to study, is the M-tachyon amplitude. In the early days of the RNS model, before the importance of the GSO conditions and space-time supersymmetry were appreciated, these were the first amplitudes calculated for this theory. In those days the tachyon was identified as a π meson with a slightly wrong mass.

A useful trick in carrying out the calculation is to exponentiate the factor $k \cdot \psi$, so that once again the dependence on the various oscillators factorizes. Since ψ is fermionic it must be multiplied by a fermionic parameter in the exponential. This leads us to introduce the representation

$$\frac{1}{\sqrt{y}}V_T(y) = \int d\theta \exp\{ik \cdot X(y) + \theta k \cdot \psi(y)/\sqrt{y}\}, \qquad (7.3.26)$$

where θ is a Grassmann number and normal-ordering is understood. The factor $1/\sqrt{y}$ is convenient since

$$\langle 0| \frac{\psi^\mu(y_i)}{\sqrt{y_i}} \frac{\psi^\nu(y_j)}{\sqrt{y_j}} |0\rangle = \frac{1}{y_i - y_j}\eta^{\mu\nu}. \qquad (7.3.27)$$

In fact, for this reason some authors find it natural to include an extra factor of y^{-J} in the definition of an operator of conformal dimension J. In order to be consistent with almost all the literature, we choose not to do that.

The exponent occurring in (7.3.26) is essentially $ik \cdot Y(y, \theta)$, where Y is the superspace field introduced in §4.1.2. There are some differences of detail, however, relating to the fact that ψ and θ in §4.1.2 were two-component Majorana spinors, whereas here they are the one-component expressions that survive after imposing open-string boundary conditions. We could derive the formulas of this section by systematically following a superspace approach. We have chosen to work with component fields instead to keep things as simple as possible. Even so, as (7.3.26) demonstrates, we are naturally led to expressions in the form that the superspace approach would yield.

Combining (7.3.27) and the correlation function

$$\langle 0| X^\mu(y_i) X^\nu(y_j) |0\rangle = -\eta^{\mu\nu} \ln(y_i - y_j)\lambda, \qquad (7.3.28)$$

explained in appendix 7.A, gives the expression for the correlation function of two vertices

$$\langle \frac{V_T(y_i)}{y_i} \frac{V_T(y_j)}{y_j} \rangle = \int d\theta_i d\theta_j \exp \left\{ k_i \cdot k_j \left(\ln(y_i - y_j) - \frac{\theta_i \theta_j}{y_i - y_j} \right) \right\}$$

$$= \int d\theta_i d\theta_j \exp \left\{ k_i \cdot k_j \ln(y_i - y_j - \theta_i \theta_j) \right\}. $$

$$(7.3.29)$$

In the second step we have used the fact that a function of a Grassmann variable can be, at most, linear in that variable. Generalizing to the M-particle correlation functions, making use of the formulas in appendix 7.A, one deduces that the M-tachyon amplitude is

$$A_M = \int d\mu_M(y) I_M(k, y), \qquad (7.3.30)$$

where

$$I_M(k, y) = \langle 0| \frac{V(k_1, y_1)}{y_1} \cdots \frac{V(k_M, y_M)}{y_M} |0\rangle$$

$$= \int (\prod d\theta_i) \prod_{i<j} (y_i - y_j - \theta_i \theta_j)^{k_i \cdot k_j}. $$

$$(7.3.31)$$

Since the integrand contains an even number of θ_i, the amplitude vanishes unless M is even, consistent with the notion of a multiplicatively conserved 'G-parity' quantum number. I_M is invariant under the world-sheet supersymmetry transformation

$$\delta y_i = \theta_i \epsilon, \qquad \delta \theta_i = \epsilon, \qquad (7.3.32)$$

since the combination $y_i - y_j - \theta_i \theta_j$ is invariant (as is the combination $\theta_i - \theta_j$, which does not occur in the formula). This transformation is generated by $G_{-1/2}$. There is also a fermionic transformation

$$\delta y_i = y_i \theta_i \eta, \qquad \delta \theta_i = y_i \eta, \qquad (7.3.33)$$

generated by $G_{1/2}$ under which $d\mu_M \cdot I_M$ is also invariant. This follows from the closure of the $OSp(1|2)$ algebra.

Let us now consider the more important case of M massless vector particles. The massless vector satisfies the GSO condition, so this is a physically relevant process. The mathematics again becomes tractable once we express the vertex operator as an exponential linear in oscillators. This is achieved by writing the vertex $V(\zeta, k)$ of (7.3.25) in the form

$$\frac{1}{y}V(\zeta, k, y) = \int d\phi d\theta \exp(ik \cdot X + \theta\phi\zeta \cdot \dot{X}/y$$
$$+ \theta k \cdot \psi/\sqrt{y} - \phi\zeta \cdot \psi/\sqrt{y}). \tag{7.3.34}$$

As usual, the dot denotes a τ derivative $(d/d\tau = iyd/dy)$. Two auxiliary Grassmann variables have been used in writing (7.3.34). The variable θ is the usual world-sheet superspace coordinate, whereas ϕ has no such interpretation.

The amplitude can now be evaluated using the identities

$$\langle 0|\, X^\mu(y_i)\dot{X}^\nu(y_j)/y_j\, |0\rangle = \frac{i}{y_i - y_j}\eta^{\mu\nu} \tag{7.3.35}$$

$$\langle 0|\, \frac{\dot{X}^\mu(y_i)}{y_i}\, \frac{\dot{X}^\nu(y_j)}{y_j}\, |0\rangle = \frac{1}{(y_i - y_j)^2}\eta^{\mu\nu}, \tag{7.3.36}$$

given in appendix 7.A, in addition to (7.3.27) and (7.3.28). It then follows that the integrand of (7.3.23) is given by

$$\langle 0|\, \frac{V(\zeta_1, k_1, y_1)}{y_1} \cdots \frac{V(\zeta_M, k_M, y_M)}{y_M}\, |0\rangle$$

$$= \int (\prod d\theta_i) \prod_{i<j} (y_i - y_j - \theta_i\theta_j)^{k_i \cdot k_j} F_M(\zeta, k, y, \theta), \tag{7.3.37}$$

where

$$F_M = \int (\prod d\phi_i) \prod_{i<j} \exp\left[\frac{(\theta_i - \theta_j)(\phi_i\zeta_i \cdot k_j + \phi_j\zeta_j \cdot k_i)}{y_i - y_j}\right.$$
$$\left. - \frac{\phi_i\phi_j\zeta_i \cdot \zeta_j}{y_i - y_j} - \frac{\theta_i\theta_j\phi_i\phi_j\zeta_i \cdot \zeta_j}{(y_i - y_j)^2}\right]. \tag{7.3.38}$$

This can be rewritten in the alternative form

$$F_M = \int (\prod d\phi_i) \prod_{\text{all } (ij)} \exp\left[\frac{(\theta_i - \theta_j)\phi_i\zeta_i \cdot k_j - \frac{1}{2}\phi_i\phi_j\zeta_i \cdot \zeta_j}{y_i - y_j - \theta_i\theta_j}\right], \tag{7.3.39}$$

which makes the world-sheet supersymmetry manifest.

The explicit evaluation of the θ and ϕ integrations, while trivial in principle, does give rise to very tedious combinatorics. The resulting formulas exhibit quite a few cancellations, moreover. The reason for this can be understood as a manifestation of the existence of an equivalent F_2 formulation in which two of the V/y factors are replaced by $W/\sqrt{y}$ factors. The simplest example of this is the three-vector interaction for which we find that the $(k \cdot \zeta)^3$ terms of (7.1.57) and (7.1.58) cancel leaving

$$A_3 = gk_2 \cdot \zeta_1 \zeta_2 \cdot \zeta_3 + \text{perms.} \tag{7.3.40}$$

Unlike the formula in the bosonic string theory, this agrees with the coupling of Yang–Mills theory without any $O(\alpha')$ corrections. The $M = 4$ case will be discussed in §7.4.2.

7.3.4 Tree Amplitudes with One Fermion Line

(a) $\qquad\qquad\qquad\qquad (b)$

Figure 7.11. An external fermion emitting bosons is pictured in (a); in (b) we consider the case in which all boson emissions occur on one side.

There is one more class of tree amplitudes that can be studied in the RNS formulation without confronting the complexities of fermion-emission vertices. Amplitudes with two external fermions and M external bosons can be pictured in terms of a diagram in which a fermionic string runs from $+\infty$ to $-\infty$ with boson emissions occurring along the boundaries of the world sheet, as shown in fig. 7.11a. For simplicity we only discuss the case in which all of the emissions occur from one edge of the string, say the $\sigma = 0$ edge as in fig. 7.11b, but there is no problem using vertices for $\sigma = \pi$ to describe emissions from the other edge, as explained in §7.1.6.

The tree amplitude for two fermions and M bosons is given by

$$A_{2,M} = \langle \psi_1 | W_1 S W_2 \ldots S W_M | \psi_2 \rangle, \tag{7.3.41}$$

where $|\psi_1\rangle$ and $|\psi_2\rangle$ are physical fermion states satisfying $F_n |\psi\rangle = 0$ for $n \geq 0$. The W's represent emission of physical bosonic states from a fermionic string. The propagator, S, is the obvious generalization of the Dirac propagator for a spin 1/2 particle. Namely, since the wave equation is $F_0 |\psi\rangle = 0$, we set

$$S = F_0^{-1} = F_0/L_0. \qquad (7.3.42)$$

To establish tree unitarity in the fermionic channels, we must ensure that the intermediate fermionic poles satisfy the F_n or L_n subsidiary conditions. (The two are equivalent for states annihilated by F_0, because $\{F_0, F_n\} = 2L_n$ and $[F_0, L_n] = -(n/2)F_n$.)

It is convenient to define auxiliary vertex operators

$$V = \{F_n, W\}, \qquad (7.3.43)$$

just as in (7.3.3). As in that case, W is required to have conformal dimension $J = 1/2$ and V is required to be independent of n, which implies that it has conformal dimension $J = 1$. We are dealing with more than an analogy here. Bosonic and fermionic strings are distinguished by boundary conditions that require comparing $\psi(\sigma, \tau)$ at $\sigma = 0$ and $\sigma = \pi$. However, the emissions are local at one particular value of σ and cannot depend on global considerations. The similarity can be made more pronounced by using (7.3.42) and (7.3.43) to recast (7.3.41). Start with the propagator between W_1 and W_2 and commute the F_0 in the numerator to the right. This gives a commutator term with W_2 replaced by V_2 (because of (7.3.43)). The commuted term has a factor $F_0 S = 1$ between W_2 and W_3, and therefore vanishes by the canceled propagator argument. The reasoning can be repeated for the subsequent propagators until one ends up with

$$A_{2,M} = \langle \psi_1 | W_1 \Delta V_2 \dots \Delta V_M | \psi_2 \rangle, \qquad (7.3.44)$$

where $\Delta = L_0^{-1}$. Note that even in this form one of the boson emissions is represented by a W rather than a V. If one were to replace this W by V, the expression would vanish, which is certainly not what we want. This can be seen by inserting F_0 between $\langle \psi_1 |$ and W_1. On the one hand this gives zero due to the wave equation for $\langle \psi_1 |$, and on the other it can be pushed to the right so that W_1 also gets replaced by V_1.

The study of cyclic symmetry and tree unitarity in bosonic channels requires facing the problem of fermion-emission vertices. Without them there is no possibility of repeating the arguments used to establish these properties in the previous sections.

7.3.5 Fermion-Emission Vertices

Just as we have constructed operators that represent the emission and absorption of bosons, so it is natural to go on and construct operators that represent the emission and absorption of fermions. Indeed, qualitative arguments presented in chapter 1 show that there must exist a vertex operator for emission and absorption of *any* state. Nonetheless, the construction of a satisfactory fermion-emission vertex in the RNS formalism proves to be a rather difficult task. There are several, related ways to express this difficulty. First of all, the physical fermion states transform as spinors of the Lorentz group $SO(1,9)$, so the fermion vertex operator must be a Lorentz spinor. But in the RNS model the elementary fields $X^\mu(\sigma,\tau)$ and $\psi^\mu(\sigma,\tau)$ transform as Lorentz *vectors*. To construct a spinor operator out of vectors is a challenging and at first sight perhaps a seemingly impossible task. Clearly the vertex operator cannot be simply $e^{ik \cdot X}$ times a polynomial in the elementary fields.

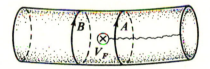

Figure 7.12. Here we sketch a closed-string world sheet on which a fermion vertex operator V_F has been inserted. The world sheet has been cut just before and just after insertion of V_F to reveal two circles A and B. If the RNS fermions ψ^μ are periodic in circumnavigating A, they are antiperiodic in circumnavigating B, and vice versa. This must mean that a 'cut' in the ψ^μ field originates at the V_F insertion; ψ^μ changes sign in crossing the cut.

A related way to express the difficulty is to note that emission of a fermion must turn fermions into bosons and must turn bosons into fermions. But in the RNS model, fermions and bosons correspond respectively to states in which $\psi^\mu(\sigma + \pi) = +\psi^\mu(\sigma)$ and states in which $\psi^\mu(\sigma + \pi) = -\psi^\mu(\sigma)$. The fermion vertex operator V_F, which turns fermions into bosons and vice versa, must therefore change the boundary conditions of the ψ^μ field! This can be better described, in fact, by saying that V_F must create a cut in the field ψ^μ, as indicated in fig. 7.12. For instance, in fig. 7.12 we sketch a situation in which a cut in ψ^μ originates at the point of insertion of V_F and heads into the 'future' (to the right in the diagram); if the only antiperiodicity of ψ^μ arises from the sign change in crossing this cut, then in the figure the state on string A is a fermion (ψ^μ periodic) while the state on B is a boson (ψ^μ antiperiodic). Just

as it is hard to see at first sight how to make space-time spinors from
the vectors X^μ and ψ^μ, so it is at first sight hard to see how to form an
operator that can create or annihilate such a cut.

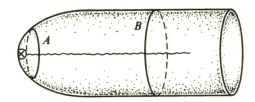

Figure 7.13. The purpose of this figure is to clarify the nature of the state created by
the fermion vertex operator, whose insertion is indicated by the symbol $\otimes$.

We must pause here to attempt to clear up a point that is rather confusing at first sight. Since the fermion fields $\psi^\mu(\sigma, \tau)$ are periodic in the
fermion sector and antiperiodic in the boson sector, one might think that
an operator that creates a cut in ψ^μ would be a *boson*-emission vertex!
The purpose of fig. 7.13 is to elucidate this point. A fermion vertex operator V_F has been inserted at the end of a cigar-like world sheet; a 'cut'
in the ψ^μ field emanates from the point where the vertex operator is inserted. If one parallel transports the ψ^μ field on a small loop, marked
A in the figure, around the vertex operator, it changes sign because of
crossing the cut. However, to compare to the canonical quantization of
the model and determine whether the state created by V_F is a boson or a
fermion, one must 'cut' the world sheet in a region where it is a standard,
flat cylinder of the type considered in the canonical formalism. Thus,
instead of parallel transport around the curve marked A, it is more appropriate to consider parallel transport around the curve marked B. In
circumnavigating the curve marked B in fig. 7.13, the ψ^μ field picks up
a factor of -1 from crossing the cut, but it also picks up a factor of -1
from integrating the spin connection around the curve B. That the spin
connection gives for parallel transport around B precisely a factor of -1
can be proved from the Gauss-Bonnet theorem, but we will not attempt
to do so here. Including the two factors of -1, the fermion field ψ^μ is
invariant under parallel transport around the curve B. In other words, it
obeys periodic boundary conditions; this shows that what we are calling
the fermion-emission operator really does create fermions. Conversely, if
the fermion-emission vertex in fig. 7.13 is replaced by a boson-emission
vertex, one would get only a single factor of -1 (from the spin connec-

tion), confirming that what we have been using as boson-emission vertices really do create bosons.

The attempts in the early 1970s to construct a suitable fermion vertex operator enjoyed only partial success. Operators transforming as space-time spinors were indeed constructed, but the required formulas were rather untransparent. Also, the operators that appeared did not have conformal dimension one, the proper value for a vertex operator, but 5/8. The origin of the number 5/8 will appear presently. Because the would-be vertex operators did not have the correct conformal dimension, they could not be manipulated in the conventional ways. Special and rather arduous procedures of projecting at each stage onto physical states had to be invented. It was conjectured that the problem could be cured by correctly incorporating Faddeev–Popov ghosts, but it was not clear how to do this. This problem was finally solved in important work by Friedan, Martinec and Shenker (FMS), and Knizhnik. Some of the highlights of this work will be sketched here, though we will not explain the 'picture-changing' operation of FMS.

The basic tool that makes the construction much simpler than it would at first appear is bosonization of fermions, introduced in §3.2.4. In trying to find operators that transform as spinors of $SO(1,9)$ and create a cut in the RNS field ψ^μ, we may as well concentrate on just, say, the right-moving modes on a closed string. Let us begin with two right-moving RNS fermions ψ^1, ψ^2. Their kinetic energy is of the form

$$\int d^2\sigma \left(\psi^1 \partial_+ \psi^1 + \psi^2 \partial_+ \psi^2 \right) = \int d^2\sigma \psi^{1+i2} \partial_+ \psi^{1-i2}, \qquad (7.3.45)$$

where $\psi^{1\pm i2} = \psi^1 \pm i\psi^2$. We have introduced $\psi^{1\pm i2}$ because the second form of the kinetic energy in (7.3.45) is analogous to the ghost kinetic energy $c^-\partial_+ b_{--}$ that was studied in chapter 3. We can therefore borrow our previous bosonization formulas from §3.2.4. Introducing a right-moving boson ϕ, we write:

$$\begin{aligned} \psi^{1+i2} &= e^{i\phi} \\ \psi^{1-i2} &= e^{-i\phi}. \end{aligned} \qquad (7.3.46)$$

The transformation $\psi^2 \to -\psi^2$ of the RNS model exchanges $\psi^{1\pm i2}$, so the energy–momentum tensor corresponds to $k = 0$ in the language of §3.2.2. The operators

$$D_t = e^{it\phi} \qquad (7.3.47)$$

were shown in §3.2.4 to have conformal dimension

$$d_t = t^2/2 - kt/2. \qquad (7.3.48)$$

Thus, with $k = 0$, the operators $D_{\pm 1/2}$ have conformal dimension 1/8. We claim that the operators $D_{\pm 1/2}$ are the operators that create a cut in the fields ψ^1, ψ^2. This follows indeed from the equation for the correlation functions given in §3.2.4. The three-point correlation function $\langle D_{1/2} D_{1/2} D_{-1} \rangle$ is (writing ψ^{1-i2} for D_{-1})

$$\langle D_{1/2}(\sigma_1^-) D_{1/2}(\sigma_2^-) \psi^{1-i2}(\sigma_3^-) \rangle$$

$$= (\sigma_1^- - \sigma_3^-)^{-1/2} (\sigma_2^- - \sigma_3^-)^{-1/2} (\sigma_1^- - \sigma_2^-)^{1/4}.$$

(7.3.49)

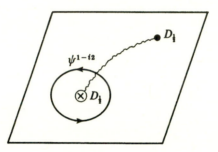

Figure 7.14. The correlation function $\langle D_{1/2} D_{1/2} \psi^{1-i2} \rangle$ changes sign if ψ^{1-i2} is transported in a small loop around one of the D's, indicating that the D's create branch cuts in the ψ field.

Note the square-root branch points at $\sigma_3^- = \sigma_1^-$ and at $\sigma_3^- = \sigma_2^-$. These branch points mean that in the field of the operator $D_{1/2}(\sigma_1^-)$, the elementary fermion field $\psi^{1-i2}(\sigma_3^-)$ is double valued, changing sign if σ_3^- is transported in a small loop around σ_1^- as in fig. 7.14. This sign change means that $D_{1/2}$ creates a branch cut in ψ^{1-i2}. Considering the other correlation functions involving the D's and ψ's in exactly the same way shows that in fact $D_{\pm 1/2}$ creates branch cuts in $\psi^{1\pm i2}$. By analogy with certain phenomena that occur in the two-dimensional Ising model, operators that create cuts in Fermi fields are often called 'spin operators'. Since the $D_{\pm 1/2}$ have dimension 1/8 (for this system of $k = 0$) we have learned that the spin operator of a pair of ordinary Majorana fermions has dimension 1/8.

Now we move on to consider not a single pair of fermions $\psi^{1,2}$ but the ten right-moving fermions of the RNS model. It is convenient to work in Euclidean space so that the $SO(1,9)$ Lorentz symmetry becomes $SO(10)$. With the positive signature, it is natural to label the RNS fermions as

$\psi^\mu, \mu = 1\ldots 10.^*$ To bosonize the ψ^μ we must introduce five right-moving bosons $\phi_i, i = 1,\ldots,5$, defined by the formulas

$$\psi^{1\pm i2} = e^{\pm i\phi_1}, \quad \psi^{3\pm i4} = e^{\pm i\phi_2}, \quad \psi^{5\pm i6} = e^{\pm i\phi_3},$$
$$\psi^{7\pm i8} = e^{\pm i\phi_4}, \quad \psi^{9\pm i10} = e^{\pm i\phi_5}. \tag{7.3.50}$$

The ϕ_k are related (as in various discussions in the text of chapters 5 and 6 and appendix 5.A) to the generators of a Cartan subalgebra of $SO(10)$:

$$J_{12} = i\partial_-\phi_1, \quad J_{34} = i\partial_-\phi_2, \quad J_{56} = i\partial_-\phi_3,$$
$$J_{78} = i\partial_-\phi_4, \quad J_{9,10} = i\partial_-\phi_5. \tag{7.3.51}$$

(As in appendix 5.A, J_{kl} is the operator that generates a rotation in the kl plane.) For each of the ϕ_m we have the 'spin operators'

$$D^m_{\pm 1/2} = e^{\pm i\phi_m/2}, \tag{7.3.52}$$

which create cuts in two of the ψ^μ (in fact, in ψ^{2m-1} and in ψ^{2m}). If we want operators that create cuts in each of the ψ^μ, we must take products of spin operators of all five types. Thus, we define

$$\Theta_\alpha = D^1_{\pm 1/2} D^2_{\pm 1/2} \cdots D^5_{\pm 1/2}. \tag{7.3.53}$$

The subscript 'α' in (7.3.53) refers to the choice of signs $\pm 1/2$. There are five independent choices of sign in (7.3.53), so in all we have defined $2^5 = 32$ 'spin operators' Θ_α. Since $32 = 16 + \overline{16}$ is the dimension of the spinor representation of $SO(10)$, it is natural to suspect that the Θ_α might in fact transform in the spinor representation of $SO(10)$. Indeed, by comparing the form of the Cartan generators in (7.3.51) to the definition of the spin operators in (7.3.52) and (7.3.53), we see that (because of the factors of $1/2$ in the exponents in (7.3.52)) the Θ_α have weights $(\pm 1/2, \pm 1/2, \ldots, \pm 1/2)$. As we know from appendix 5.A, these are the weights of the spinor representation of $SO(10)$. We take this as sufficient evidence that the Θ_α transform as spinors of $SO(10)$.[†]

---

[*] Thus, for purposes of the present discussion, we replace the time-like mode ψ^0 with a tenth space-like mode ψ^{10}.

[†] The outline of a complete proof would be as follows. The $SO(10)$ currents in the spinor description are just $J_{ij} = \psi_i\psi_j$. As we have in (7.3.50) expressions for the ψ_i in terms of the bosons ϕ_k, we can write the $SO(10)$ currents in terms of the bosons; this gives formulas similar to those that appeared in chapter 6 in describing the bosonic representation of E_8. In the bosonic description one could then simply calculate the commutators of the $SO(10)$ charges $Q_{ij} = \oint J_{ij}$ with the Θ_α and show that the Θ_α transform as spinors under $SO(10)$.

Thus, we have come a long way towards finding suitable fermion vertex operators. We have found operators that transform as spinors and generate cuts in the ψ^μ. However, the Θ_α do not have the correct conformal dimension to be fermion vertex operators. As there are five independent factors in the definition (7.3.53) of the Θ_α, and each of these factors has dimension $1/8$, we see that the Θ_α have dimension $5/8$. It is, of course, dimension 1 that is required. We must make up the remaining dimension $3/8$ from some other source.

To motivate a guess about what that source might be, let us recall the nature of the general argument in §1.4.2, which showed that a vertex operator should exist for any external particle type. The key to the argument was conformal invariance – a conformal mapping can project an external string to a finite point; a local operator must then appear at that point. In the strictest sense, conformal invariance only holds when string theory is formulated with the ghosts; only then are the conformal anomalies entirely canceled. Therefore, the argument for the existence of vertex operators suggests that, in general, it is necessary to formulate the vertex operators in a Hilbert space that includes the ghosts.

Of course, we have been able to formulate bosonic vertex operators without worrying about ghosts, so if the ghosts really play an essential role in the fermion vertex operators, this is because of something that is special to the fermions. The essential clue arises by recalling that the convenient gauge-fixed form of the RNS model involves superconformal ghosts $\gamma_{\pm 1/2}$, $\beta_{\pm 3/2}$ (the subscripts denote the world-sheet Lorentz spin and were explained in §4.4.1), which arise upon quantization of an underlying locally supersymmetric theory. In particular, $\beta_{\pm 3/2}$ has the quantum numbers of the supercurrent $J \sim \partial X^\mu \psi_\mu$ of that supergravity model, while the $\gamma_{\pm 1/2}$ have the same quantum numbers except for shifted spin. The fermion vertex operator V_F, which we are trying to construct, should create a cut in ψ^μ but not in X^μ. As a result it must create a cut in the supercurrent J, and hence in the superconformal antighosts $\beta_{\pm 3/2}$ and ghosts $\gamma_{\pm 1/2}$. This fact has no analog for bosonic-emission vertices, and gives a rationale for expecting that the ghosts play a more essential role in the construction of the fermion-emission operator.

Our task is thus to define spin operators that create a cut in the superconformal ghosts γ, β. Their action is rather similar to that of Dirac fermions. The right movers, for instance, are governed by

$$S = -\frac{1}{\pi} \int d^2\sigma (\gamma_{+1/2} \nabla_{+1} \beta_{-3/2}). \qquad (7.3.54)$$

One can therefore expect to define spin operators for the superconformal ghosts by analogy with the way we defined such operators for the ψ^μ.

However, there are quite a few subtleties that arise (and were unraveled by FMS). The subtleties arise because in (7.3.54), γ and β are *commuting* fields, so when we bosonize them we are involved in 'bosonization of bosons', which involves some novelties. We will try to circumvent the more subtle and interesting novelties (which may, however, prove to be very important in the further development of string theory) and bring out just a few of the simpler points.

Figure 7.15. The one-loop diagram that gives rise to an anomaly in the current commutator.

We begin with some observations that could have been used as the starting point in the discussion of ordinary bosonization in chapter 3. The reason for developing this alternative viewpoint is that the treatment in §3.2.4 began by constructing anticommuting fields out of a Bose variable; this is irrelevant in the present case, since γ and β are commuting, rather than anticommuting, and we do not wish to construct anticommuting fields. Accordingly, many of the formulas from chapter 3 do not carry over here. We would like to identify certain of the bosonization formulas that carry over in a relatively straightforward way. Among these are the formulas involving the ghost number current $J_- = \gamma_{+1/2}\beta_{-3/2}$. This current has zero ghost number, and if one uses Poisson brackets, one finds that it commutes with itself. However, as in the case of fermions, one finds at the quantum level a Schwinger anomaly that arises from the one-loop diagram of fig. 7.15, so that it obeys the equal τ commutation relations

$$[J_-(\sigma), J_-(\sigma')] = -\frac{i}{2\pi}\delta'(\sigma - \sigma').\qquad(7.3.55)$$

This anomaly was already discussed in chapter 6 in the case of fermions. For commuting fields γ and β there is one major difference. A minus sign in fig. 7.15 that usually arises because of a closed fermion loop is absent here, so the Schwinger term has the opposite sign from usual.

Except for the minus sign, the commutation relations of (7.3.55) are closely related to the canonical commutation relations of right-moving bosons that were encountered, for instance, in chapter 3. These canonical

commutation relations have a unique irreducible representation, so in fact
we can (just as in the fermion case) write the current J_- in terms of a
right-moving boson ρ^-:

$$J_- = i\partial_-\rho^-. \tag{7.3.56}$$

Because the Schwinger term on the right-hand side of (7.3.55) has the
opposite sign from usual, ρ^- is the right-moving part of a boson ρ that
has the wrong sign of the kinetic energy:

$$S = \frac{1}{2\pi}\int d^2\sigma \partial_\alpha\rho\partial^\alpha\rho = \frac{1}{2\pi}\int d^2\sigma[(\partial_\sigma\rho)^2 - (\partial_\tau\rho)^2]. \tag{7.3.57}$$

It is straightforward to work out the canonical commutators of $\partial_-\rho^-$
from (7.3.57), and show that the current defined in (7.3.56) really has the
commutation relation (7.3.55). Thus, we have shown that the standard
formula (7.3.56) is valid for 'bosonization of bosons' except that ρ in
(7.3.57) has a kinetic energy of the wrong sign. The wrong sign just
reflects the fact that the two-point function of the current J_- computed
from fig. 7.15 was the opposite sign from usual, so that if (7.3.56) is to be
valid, the boson ρ^- must have a propagator of the 'wrong' sign.

We now move on to consider the operators

$$\Sigma_t = e^{it\rho^-}. \tag{7.3.58}$$

The two-point function is

$$\langle\Sigma_t(\sigma^-)\Sigma_{-t}(\sigma'^-)\rangle = (\sigma^- - \sigma'^-)^{t^2}. \tag{7.3.59}$$

The exponent on the right-hand side is t^2 as opposed to $-t^2$ in §3.2.4 since
ρ^- has a propagator of unusual sign. If the dimension of Σ_t is denoted as
d_t, we learn from (7.3.59) that

$$d_t + d_{-t} = -t^2. \tag{7.3.60}$$

If one uses an energy–momentum tensor that treats β and γ symmetrically
($k = 0$ in the language of §3.2.2), then d_t and d_{-t} each have dimension
$t^2/2$. At general k, the argument of §3.2.4 goes through except for a
change of sign to give the shifted formula

$$d_t(k) = -t^2/2 + kt/2. \tag{7.3.61}$$

The superconformal ghosts have $k = 2$, as we saw in §4.4.1.

Just as in conventional bosonization, the spin operators that create cuts in the elementary Fermi fields γ and β are simply $\Sigma_{\pm 1/2}$. This can be demonstrated by considering correlation functions containing the $\Sigma_{\pm 1/2}$ as well as γ or β,[*] but here we accept the statement as plausible. Equation (7.3.61) thus tells us the dimension of the superconformal spin operators; $\Sigma_{+1/2}$ and $\Sigma_{-1/2}$ have dimension $3/8$ and $-5/8$, respectively.

The ghost spin operator $\Sigma_{+1/2}$ of dimension $3/8$ is precisely what we needed to complete the definition of the fermion vertex operator V_F. In fact, for emission of a massless fermion described by a spinor u^α and momentum k^μ we can select the vertex operator

$$V_F(u^\alpha; k^\mu) = \Sigma_{+1/2} \cdot \overline{u}^\alpha \Theta_\alpha \cdot e^{ik \cdot X}. \qquad (7.3.62)$$

More precisely, this is a vertex operator for emission of an open-string massless fermion. In the closed-string case, (7.3.62) must be multiplied by a suitable vertex operator made from modes of opposite handedness. Also, if the GSO projection is made, we must take u^α to be a spinor of one definite handedness only.

One may wonder if the other spin operator $\Sigma_{-1/2}$ of the superconformal ghosts has any utility. In fact, as it has dimension $-5/8$, the combination $\Sigma_{-1/2} \cdot \Theta_\alpha$ is dimensionless; we can make a new dimension-one vertex operator by multiplying this by a suitable dimension-one operator such as the massless boson-emission vertex. We thus define a new fermion-emission vertex as

$$V_F'(u^\alpha; k^\mu) = \Sigma_{-1/2} \cdot \overline{u}\Gamma_\mu\Theta \cdot (\partial_- X^\mu + i\psi_-^\mu \psi_-^\nu k_\nu)e^{ik \cdot X}. \qquad (7.3.63)$$

At first sight it is rather peculiar to find several candidates for a fermion-emission vertex, but on reflection this may not be so surprising; the choice between (7.3.62) and (7.3.63) is analogous to the choice between different 'pictures', which we encountered already for boson emission in the RNS model. In fact, by further study of the superconformal ghosts, FMS greatly clarified the occurrence of different 'pictures' in this model, but this is a matter that we will not delve into here.

In the above presentation, we actually have barely scratched the surface of this subject. One obvious question that must be faced next is to prove that the vertex operators (7.3.62) and (7.3.63), which involve the superconformal ghosts in such an essential way, are compatible with the

---

[*] To calculate such correlation functions, the only information required, in principle, is (7.3.56), which can be integrated to express ρ^- in terms of J_- and then in terms of γ and β.

decoupling of these ghosts as well as the decoupling of the negative norm timelike modes that are unfortunately also conventionally called 'ghosts'. The latter are the only ghosts that we had to worry about in discussing boson emission, since we discussed boson emission in a formalism in which the other sort of ghosts were absent. As we saw in §3.2 and §4.4.2, the physical states in string theory can be neatly characterized by saying that they are annihilated by the BRST charge Q. To show that (7.3.62) and (7.3.63) do not couple physical states to ghosts of any kind, we must show that they commute with Q (up to a total divergence) so that acting on a physical (Q-invariant) state, an integrated vertex operator gives a physical (Q-invariant) state. While it is not difficult to show that the integral of (7.3.62) commutes with the BRST charge, this requires some modest additional machinery which we will not develop here. As for (7.3.63), with a slight modification (addition of a multighost term) it too can be shown to commute with Q. The actual calculations of fermion scattering amplitudes with vertex operators (7.3.62) and (7.3.63) again involve some modest additional machinery and is not presented here. The full development of the subject involves many additional issues for a treatment of which the reader is referred to the literature.

It is interesting to ask what happens to the covariant formalism, some of whose highlights we have sketched, if one eliminates the ghosts and goes to light-cone gauge. In this case, the spin operators $\Sigma_{+1/2}$ of the ghosts can be dropped in (7.3.62). Also, one pair of ψ^μ, say ψ^8 and ψ^9, are dropped in going to light-cone gauge. So in light-cone gauge, the spin operator Θ_α of the elementary fermions is a product of four factors only, not five, and its dimension is $4 \cdot 1/8 = 1/2$. This is a noteworthy number; $1/2$ is the proper canonical dimension for canonical Fermi fields in $1 + 1$ dimensions. What is more, the Θ_α transform, after this truncation, as spinors of $SO(8)$, not $SO(10)$. Of the 16 Θ_α, eight have positive chirality; they are precisely the eight elementary spinor fields of the manifestly supersymmetric formulation of superstrings, in light-cone gauge. Indeed, the $SO(8)$ truncation of (7.3.53) was introduced at the end of §5.2.1 in explaining the relation between the manifestly supersymmetric formulation and the older light-cone formulation.

For another interesting application, consider 16 left-moving fermions λ_i, $i = 1 \ldots 16$. It is, of course, possible to define conserved currents $J_{ij} = \lambda_i \lambda_j$ that generate the group $SO(16)$. They have conformal dimension one, an essential property in order that the conserved charges $Q_{ij} = \oint J_{ij}$ should be conformally invariant. Via bosonization, the 16 λ's can be converted into eight bosons ϕ_m, $m = 1, \ldots, 8$. We can construct corresponding spin operators $D^m_{\pm 1/2}, m = 1, \ldots, 8$, whose products

$$\Theta \sim \prod_m D^m_{\pm 1/2} \qquad (7.3.64)$$

transform as spinors of $SO(16)$. The Θ's have conformal dimension $8 \cdot (1/8) = 1$. Thus, because we started with precisely 16 λ's, not some other number, the Θ's have the conformal dimension appropriate for a conserved current. Indeed, the charges

$$\tilde{Q} = \oint \Theta \qquad (7.3.65)$$

are conserved and (upon making a suitable GSO-like projection) they enlarge the obvious $SO(16)$ symmetry of this system to the larger group E_8. Indeed, the Θ's are the missing ingredient that was needed in §6.3.2 to establish full E_8 current algebra in the fermionic formulation of heterotic strings.

7.4 Superstrings in the Supersymmetric Formulation

The RNS formulation hides the space-time supersymmetry of superstring theory, and in this formulation the construction of fermion vertex operators is a *tour de force* described in the preceding subsection. Calculations based on the space-time supersymmetric formulation, on the other hand, can describe bosons and fermions with equal ease, and enable us to write manifestly supersymmetric amplitudes. The hope that this might be the case was a principal motivation for development of this formalism. There is a price to be paid, however, for manifest supersymmetry and the relatively easy incorporation of fermions. As we explained in chapter 5, quantization of the supersymmetric action is only straightforward in a light-cone gauge. Thus the desired ease of calculation is only achieved in this particular gauge.

7.4.1 Massless Particle Vertices

In this section we derive emission vertices for massless open-string states (vector and spinor). It is only possible to use the formalism based on vertex operators in the light-cone gauge when the momenta of the emitted particles, k^μ_r, satisfy $k^+_r = 0$. With few enough external states of low spin this is no restriction, if Lorentz invariance of the theory is assumed, since this can be achieved by an appropriate choice of Lorentz frame. The general light-cone gauge vertices, coupling strings in arbitrary states with no restrictions on their momenta, will be obtained in chapter 11. The results of this section are important not only for the construction of type I and type II amplitudes, where they are directly applicable, but also to the construction of heterotic string amplitudes, where they are relevant to the right-moving modes.

In chapter 5 we showed that the light-cone gauge quantization of the supersymmetric string action gives two sets of modes denoted α_n^i and S_n^a. The α_n^i occur in the expansion of the transverse spatial coordinates $X^i(\sigma, \tau)$ and satisfy the usual relations

$$[\alpha_m^i, \alpha_n^j] = m\delta_{m+n}\delta^{ij}. \tag{7.4.1}$$

The S_n^a represent modes of the eight surviving components of $\theta^a(\sigma, \tau)$ in the light-cone gauge and satisfy

$$\{S_m^a, S_n^b\} = \delta_{m+n}\delta^{ab}. \tag{7.4.2}$$

The α_n^i transform as an 8_v of the transverse spin(8), whereas the S_n^a transform as an 8_s. The 16 supersymmetry charges are an 8_s and an 8_c given by

$$Q^a = (2p^+)^{1/2}S_0^a \tag{7.4.3}$$

$$Q^{\dot a} = (p^+)^{-1/2}\gamma_{a\dot a}^i \sum_{-\infty}^{\infty} S_{-n}^a \alpha_n^i. \tag{7.4.4}$$

The matrices $\gamma_{a\dot a}^i$, which give the invariant coupling of the three inequivalent 8's of spin(8), were described in appendix 5.B.

Our goal here is to construct vertex operators for emission of the states $|u\rangle$ and $|\zeta\rangle$ that also satisfy appropriate supersymmetry relations. The desired formulas take a particularly simple form if we restrict the kinematics so that $k^+ = 0$. The basic problem with using a general kinematic configuration arises in any string theory in the light-cone gauge from the fact that the expression for the vertex operator contains the factor of $\exp(ik \cdot X)$. In the light-cone gauge X^- is a quadratic form in the X^i's so that the exponential has serious ordering problems. The resolution to the general problem is to use the formalism that introduces a separate Fock space for each string as we shall elucidate in chapter 11. Alternatively, one can use the covariant formalism described earlier. However, for many calculations it is very convenient to use the much simpler vertices derived below in which tree diagrams (as well as one-loop diagrams) are easier to calculate.[*]

Before determining the vertices we need to describe some peculiarities of the kinematics. The ground-state mass-shell condition

$$(k^i)^2 = 2k^+k^-, \qquad (7.4.5)$$

together with $k^+ = 0$ and k^- finite, implies that $(k^i)^2 = 0$, so then we must allow k^i to assume complex values in order to avoid it vanishing identically. In this case only seven of the k^i are independent, but k^- is the eighth independent momentum since its value is not determined by (7.4.5). Similarly, the polarization vector of the vector ground state ζ^μ, satisfying $\zeta^- = k^i\zeta^i/k^+$, has eight independent components. When $k^+ = 0$ it is appropriate to choose $\zeta^i k^i = 0$, which is one relation between the transverse components leaving ζ^- as the eighth independent component. The components of the massless spinor are related by

$$u^a = -\frac{1}{k^+}\gamma^i_{a\dot{a}}k^i u^{\dot{a}}, \qquad (7.4.6)$$

as we saw in appendix 5.B. When $k^+ = 0$ it is convenient to choose

$$\gamma^i_{a\dot{a}}k^i u^{\dot{a}} = 0, \qquad (7.4.7)$$

so that u^a is finite. Equation (7.4.7) implies, in turn, that $u^{\dot{a}}$ only has four independent components (instead of eight). However, there are then four independent components in u^a that are not determined by $u^{\dot{a}}$.

We denote the vertex for the emission of a ground-state boson with polarization ζ^i and momentum k^μ (with $k^+ = 0$) by $V_B(k, \zeta)$. The emission may be from either a fermion or a boson line. The vertex for emitting a massless fermion with wave function $u^{\dot{a}}$ and momentum k^μ is denoted by $V_F(k, u)$. It applies to a process in which the fermion is emitted from a fermion line to the left of the vertex or to the right of the vertex. We will demonstrate that these vertices have the form

$$V_B(\zeta, k) = \zeta \cdot Be^{ik\cdot X} = (\zeta^i B^i - \zeta^- B^+)e^{ik\cdot X} \qquad (7.4.8)$$

$$V_F(u, k) = uFe^{ik\cdot X} = (u^a F^a + u^{\dot{a}}F^{\dot{a}})e^{ik\cdot X}. \qquad (7.4.9)$$

---

[*] The vertex operators in an arbitrary frame can be derived from the ones given here by applying a Lorentz transformation using the light-cone gauge representation of the $J^{\mu\nu}$ given in §5.2.2. Of course, this presupposes that the theory is Lorentz invariant.

The key to the determination of B^i and $F^{\dot{a}}$ is the requirement that V_B and V_F transform into one another under supersymmetry transformations. The vertices are functions of the wave functions so that the transformed vertices should be functions of the transformed wave functions described in §5.3.1. In other words we require that

$$[\eta^a Q^a, V_F(u, k)] \approx V_B(\tilde{\zeta}, k) \tag{7.4.10}$$

$$[\eta^a Q^a, V_B(\zeta, k)] \approx V_F(\tilde{u}, k). \tag{7.4.11}$$

$$[\epsilon^{\dot{a}} Q^{\dot{a}}, V_F(u, k)] \approx V_B(\tilde{\tilde{\zeta}}, k) \tag{7.4.12}$$

$$[\epsilon^{\dot{a}} Q^{\dot{a}}, V_B(\zeta, k)] \approx V_F(\tilde{\tilde{u}}, k). \tag{7.4.13}$$

The expressions for the transformed wave functions $\tilde{\zeta}^i$, $\tilde{u}^{\dot{a}}$, $\tilde{\tilde{\zeta}}^i$ and $\tilde{\tilde{u}}^{\dot{a}}$ were given in §5.3.1 for a general kinematic configuration. The corresponding expressions for the transformations of ζ^- and u^a can be deduced from these before taking the special case with $k^+ = 0$. The $\approx$ means that equality is only required for on-shell matrix elements. In other words, there can be additional terms that are total τ derivatives. They give vanishing contributions to on-shell matrix elements, and do not affect the calculation of tree amplitudes because of the canceled propagator argument. It is remarkable that the requirement of global supersymmetry is sufficient to determine the structure of the vertex operators uniquely.

The solutions to equations (7.4.10)–(7.4.13) are given by

$$B^+ = p^+ \tag{7.4.14}$$

$$B^i = \dot{X}^i - R^{ij} k^j \tag{7.4.15}$$

$$F^{\dot{a}} = (2p^+)^{-1/2}[(\gamma \cdot \dot{X} S)^{\dot{a}} + \tfrac{1}{3} : (\gamma^i S)^{\dot{a}} R^{ij} : k^j] \tag{7.4.16}$$

$$F^a = (p^+/2)^{1/2} S^a, \tag{7.4.17}$$

where R^{ij} is defined by

$$R^{ij}(\tau) = iK^{ij}(\tau) = \frac{1}{4} \gamma^{ij}_{ab} S^a(\tau) S^b(\tau). \tag{7.4.18}$$

This is the generalization of R_0^{ij} (defined in §5.3.1) that has simple angular momentum commutation relations (with an anomalous term) since

$K^{ij}(\tau)$ does. The symbol : : indicates normal ordering of the S_n modes. The verification that these expressions satisfy the commutation relations (7.4.10)–(7.4.13) involves using the supersymmetry transformations of the coordinates S and X, which were given in §5.2.1. Let us start by substituting (7.4.8) and (7.4.15) for V_B in the commutator on the left-hand side of (7.4.13).

$$[\epsilon^{\dot a} Q^{\dot a}, \zeta \cdot B e^{ik \cdot X}] = \frac{1}{\sqrt{p^+}}(\zeta_\mu B^\mu \epsilon \gamma^j k^j S - i\epsilon \gamma^i \zeta^i \dot S$$
$$- \tfrac{1}{2}\epsilon \gamma^k \dot X^k \gamma^{ij} S \zeta^i k^j) e^{ik \cdot X}. \tag{7.4.19}$$

The first term on the left-hand side arises from the variation of $e^{ik \cdot X}$ while the second and third arise from varying $\zeta^i B^i$ (and B^+ does not transform). Using the properties of the γ matrices we have

$$\gamma^k \dot X^k \gamma^{ij} = 2\dot X^i \gamma^j - 2\dot X^j \gamma^i + \gamma^{ij} \gamma^k \dot X^k, \tag{7.4.20}$$

so (7.4.19) simplifies to

$$\frac{1}{\sqrt{p^+}}(\zeta \cdot B \epsilon \gamma^j k^j S - \epsilon \gamma^i k^i S \zeta^j \dot X^j - \tfrac{1}{2}\epsilon \gamma^{ij} \gamma^k \dot X^k S \zeta^i k^j) e^{ik \cdot X}$$
$$- \frac{d}{d\tau}\left(\frac{i}{\sqrt{p^+}}\epsilon \gamma^i \zeta^i S e^{ik \cdot X}\right) + \sqrt{p^+}\epsilon \gamma^i \zeta^i S k^- e^{ik \cdot X}. \tag{7.4.21}$$

Dropping the τ derivative term and substituting for $\zeta \cdot B$ gives

$$V_F(\tilde u, k) = \frac{1}{\sqrt{p^+}}(-\frac{1}{4} : \epsilon \gamma \cdot k S S \gamma^{ij} S : -\tfrac{1}{2}\epsilon \gamma^{ij} \gamma \cdot \dot X S)\zeta^i k^j e^{ik \cdot X}$$
$$- \frac{1}{\sqrt{p^+}}\zeta^- B^+ \epsilon \gamma^i k^i S e^{ik \cdot X} + \sqrt{p^+}\epsilon \gamma^i \zeta^i S k^- e^{ik \cdot X}. \tag{7.4.22}$$

The S^3 term in this expression can now be rearranged by a 'Fierz' transformation, using the formula

$$: S^a S^b := \frac{1}{16} S \gamma^{ij} S \gamma^{ij}_{ab}. \tag{7.4.23}$$

Substituting into (7.4.22) and making the identifications

$$\tilde u^{\dot a} = \frac{1}{\sqrt 2}(\epsilon \gamma^{ij})^{\dot a} k^i \zeta^j + \frac{1}{\sqrt 2}\epsilon^{\dot a} \zeta^i k^i \tag{7.4.24}$$

$$\tilde u^a = \lim_{k^+ \to 0} -\frac{k^i}{k^+}(\gamma^i u)^a, \tag{7.4.25}$$

gives an expression for $V_F(\tilde u, k)$ in the form (7.4.9) with $F^{\dot a}$ and F^a given by (7.4.16) and (7.4.17). The fact that the S^3 term is normal ordered emerges from a more careful discussion of the commutation relations in terms of the mode expansion of $S^a(\tau)$.

The fact that V_F transforms into V_B by the ϵ supersymmetry, (7.4.12), follows in similar fashion.

Turning to the η^a supersymmetry transformations, let us first consider (7.4.10)

$$[\eta^a Q^a, (u^{\dot{a}} F^{\dot{a}} + u^a F^a)e^{ik\cdot X}]$$

$$=u^{\dot{a}}\left((\gamma^i \dot{X}^i \eta)^{\dot{a}} + \tfrac{1}{12}(\gamma^i \eta)^{\dot{a}} S\gamma^{ij} Sk^j \right.$$
$$\left. + \tfrac{1}{6}:(\gamma^i S)^{\dot{a}}\eta\gamma^{ij} S: k^j\right)e^{ik\cdot X} - p^+ u^a \eta^a e^{ik\cdot X} \qquad (7.4.26)$$

$$=(\tilde{\zeta}^i B^i - \tilde{\zeta}^- B^+)e^{ik\cdot X} = V_B(\tilde{\zeta}, k).$$

In the last step we have used the identification $\tilde{\zeta}^i = \gamma_{a\dot{a}}\eta^a u^{\dot{a}}$ from §5.3.1. We have also used $\tilde{\zeta}^- = -\eta^a u^a$, which follows from the expressions for ζ^- and u^a in terms of ζ^i and $u^{\dot{a}}$, respectively. The final commutator to check is (7.4.11), which we shall not consider explicitly.

Even though the vertex operators constructed here are applicable only for a specific noncovariant gauge and restricted kinematics, they allow for an easier determination of certain processes than other formulations. This will become especially evident in chapter 9 when they are used to calculate certain one-loop amplitudes. In order to avoid unnecessarily long expressions we shall choose to consider external states for which the ζ^- polarization vanishes. In this case the B^+ term is absent from V_B. The resulting expressions are sufficiently complete to contain the terms we need to reconstruct the covariant amplitudes.

Let us recalculate the vertex coupling three massless vector states in the supersymmetric formulation. This is given by

$$A_3 = g\langle\zeta_1, k_1|V_B(\zeta_2, k_2)|\zeta_3, k_3\rangle. \qquad (7.4.27)$$

Only the zero modes contribute, which explains why we get no $O(\alpha')$ corrections to the Yang–Mills theory answer. In fact,

$$A_3 = g\,\langle\zeta_1, k_1|\,(\zeta_2 \cdot p - R^{ij}\zeta_2^i\,k_2^j)e^{ik_2\cdot x}\,|\zeta_3, k_3\rangle$$
$$= -g\delta(k_1 + k_2 + k_3)\,\langle\zeta_1|\,(\zeta_2 \cdot k_1 + R_0^{ij}\zeta_2^i\,k_2^j)\,|\zeta_3\rangle, \qquad (7.4.28)$$

where $R_0^{ij} = \tfrac{1}{4}\gamma_{ab}^{ij}S_0^a S_0^b$ is the zero-mode helicity operator defined in §5.3.1.

Using $|\zeta\rangle = |i\rangle \zeta^i$ and the properties of R_0^{ij} in §5.3.1 gives

$$\langle \zeta_1 | \zeta_3 \rangle = \zeta_1 \cdot \zeta_3$$

$$\langle \zeta_1 | R_0^{ij} | \zeta_3 \rangle = \zeta_1^i \zeta_3^j - \zeta_3^i \zeta_1^j. \tag{7.4.29}$$

Together with the physical state conditions $\zeta_r \cdot k_r = 0$ (and dropping the δ function), this gives

$$A_3 = g(\zeta_1 \cdot k_2 \; \zeta_2 \cdot \zeta_3 + \zeta_2 \cdot k_3 \; \zeta_3 \cdot \zeta_1 + \zeta_3 \cdot k_1 \; \zeta_1 \cdot \zeta_2), \tag{7.4.30}$$

in agreement with conventional Yang-Mills theory and our previous result (7.3.40). Note that one of the three terms arises from the $\dot{X}^i$ part of B^i and the other two from the $-R^{ij}k^j$ part. This provides a check on their relative normalization.

The coupling of an emitted massless vector to two massless fermions can be calculated from (7.4.29) by replacing the end states $\langle \zeta_1 |$ and $|\zeta_3\rangle$ by fermion ground states $\langle u_1 |$ and $|u_3\rangle$. In this case we need to use the expression in §5.3.1 for the action of R_0^{ij} on $|u\rangle$. Vertices involving the fermion emission vertex, V_F, can be calculated in a similar manner. This gives two ways of calculating the coupling of a massless vector to two massless fermions, which give consistent answers because

$$g \langle u_1, k_1 | V_F(u_2, k_2) | \zeta_3, k_3 \rangle = - \, g \langle u_2, k_2 | V_B(\zeta, k_3) | u_1, k_1 \rangle$$

$$= - \, 2g \bar{u}_1 \Gamma \cdot \zeta u_2. \tag{7.4.31}$$

Again we have expressed the result in covariant spin$(9, 1)$ notation. It is uniquely determined by the transverse spin(8) result.

There is a striking similarity between the massless vector vertex in (7.3.25) and the one in (7.4.15). In order to make the comparison explicit let us restrict the covariant formula (7.3.25) to transverse momentum and polarization. Then the factor multiplying $\exp(ik \cdot X)$, in each case is

$$B^i = \dot{X}^i - R^{ij}k^j, \tag{7.4.32}$$

where in one formulation

$$R^{ij} = \psi^{[i}\psi^{j]} \tag{7.4.33}$$

and in the other

$$R^{ij} = \tfrac{1}{4}\gamma_{ab}^{ij} S^a S^b. \tag{7.4.34}$$

These two expressions have the same commutation relations including anomalies (affine $SO(8)$). This can be understood in terms of the triality symmetry of $SO(8)$ discussed in appendix 5.A. One representation is based on operators ψ^i in the $\mathbf{8_v}$ and the other on operators S^a in the $\mathbf{8_s}$. The $SO(8)$ triality operation that turns $\mathbf{8_v}$ into $\mathbf{8_s}$ turns (7.4.33) into (7.4.34). It is because the two sets of vertex operators are related by triality that they give the same amplitudes for massless boson scattering.

7.4.2 Open-String Trees

The vertices introduced in the preceding section can be used to calculate tree amplitudes by essentially the same procedure as was used in previous sections. An M-particle tree amplitude is given by

$$A_M = g^{M-2} \langle \phi_1 | V_2 \Delta V_3 \cdots \Delta V_{M-1} | \phi_M \rangle, \qquad (7.4.35)$$

where now the propagator is

$$\Delta = (\tfrac{1}{2} p^2 + N)^{-1} \qquad (7.4.36)$$

and

$$N = \sum_{n=1}^{\infty} (\alpha^i_{-n} \alpha^i_n + n S^a_{-n} S^a_n). \qquad (7.4.37)$$

Each of the vertices can be chosen to be either V_B or V_F to describe emission of a massless vector or spinor. Similarly, each of the states $\langle \phi_1 |$ and $| \phi_M \rangle$ can be chosen to be either a vector $| \zeta \rangle$ or a spinor $| u \rangle$. Since this construction is done in the physical light-cone gauge, tree unitarity is manifest. The Fock space only describes physical degrees of freedom, and there are no spurious states that could appear in pole residues.

Cyclic symmetry is a little more elusive in the present formulation. For one thing we have required that the particles described by the vertices $V_2 \cdots V_{M-1}$ have $k^+ = 0$ in order to keep the formula as simple as possible. On the other hand certain formulas become ill-defined if the k^+ of states $| \phi_M \rangle$ and $\langle \phi_1 |$ vanishes. Therefore more general formulas would be required before comparison with cyclically permuted expressions could be possible. In the light-cone gauge field theory of strings, A_M is actually achieved as a sum of many different diagrams corresponding to different τ orderings. Each contribution is given as an integral with the same integrand and a different integration region. The integration regions piece together to give the complete region required for cyclic symmetry, described in the preceding sections. The boundaries between the different regions are controlled by the k^+ values of the various states. When various $k^+ \to 0$, some regions vanish, and in fact (7.4.35) gives the complete answer since $k_2^+, \ldots, k_{M-1}^+ = 0$.

We now describe the calculation of four-point functions. A convenient way of proceeding is to do all the algebra that involves nonzero modes first without specifying whether the end states are vectors or spinors. Then one attaches specific massless vectors or spinors at the ends of the tree and completes the zero-mode part of the analysis. Consider, for example, a tree in which the two emitted states ($\#\,2$ and $\#\,3$) are massless vectors described by the vertex operators V_B obtained in the preceding section. We begin by calculating the expectation value $\langle k_1|V_B(\zeta_2, k_2 \Delta V_B(\zeta_3, k_3)|k_4\rangle$ where $\langle k_1|$ and $|k_2\rangle$ are understood to be ground states of all the α_n and S_n oscillators with $n \neq 0$. However, the matrix element does not include the S_0 space so that the expression is an operator in this space. The calculation of the matrix element involves the α oscillator techniques described earlier together with the algebra of the fermionic modes and use of the identity

$$\langle 0| \sum_{m=1}^{\infty} S_m^a \sum_{n=1}^{\infty} S_{-n}^b x^n |0\rangle = \frac{x}{1-x}\delta^{ab}. \tag{7.4.38}$$

The result of the calculation is

$$\langle k_1|V_B(\zeta_2, k_2)\Delta V_B(\zeta_3, k_3)|k_4\rangle$$

$$\begin{aligned}
&= \tfrac{1}{2}g^2(1+t/2)\zeta_2 \cdot \zeta_3 B(1-s/2, -1-t/2)\\
&\quad + \tfrac{1}{2}g^2[-k_1 \cdot \zeta_2\, k_4 \cdot \zeta_3 + R_0^{ij}(\zeta_2^i\, k_2^j\, \zeta_3 \cdot k_4 - \zeta_3^i\, k_3^j\, k_1 \cdot \zeta_2)\\
&\quad + R_0^{ij} R_0^{kl}\zeta_2^i\, k_2^j\, \zeta_3^k\, k_3^l]B(-s/2, 1-t/2)\\
&\quad + \tfrac{1}{2}g^2[\zeta_2 \cdot k_3\, \zeta_3 \cdot k_4 + \zeta_2 \cdot k_1\, \zeta_3 \cdot k_2 + \zeta_2 \cdot k_3\, \zeta_3 \cdot k_2\\
&\quad + R_0^{ij}(-\zeta_2^i\, k_2^j\, \zeta_3 \cdot k_2 + \zeta_3^i\, k_3^j\, \zeta_2 \cdot k_3 + \zeta_2 \cdot \zeta_3\, k_2^i\, k_3^j\\
&\quad + \zeta_2^i\, \zeta_3^j\, k_2 \cdot k_3 - k_2^i\, \zeta_3^j\, \zeta_2 \cdot k_3 - \zeta_2^i\, k_3^j\, k_2 \cdot \zeta_3)]B(1-s/2, -t/2).
\end{aligned} \tag{7.4.39}$$

Specific tree amplitudes can be obtained by sandwiching the operator in (7.4.39) between a pair of massless vector states $\langle \zeta_1|$ and $|\zeta_4\rangle$ or a pair of massless spinor states $\langle u_1|$ and $|u_4\rangle$ of the type described in §5.3.1. Using the rule $R_0^{ij}|k\rangle = \delta^{jk}|i\rangle - \delta^{ik}|j\rangle$, derived in §5.3.1, we can derive the requisite identities for the four-vector case

$$\langle \zeta_1|\zeta_4\rangle = \zeta_1 \cdot \zeta_4$$

$$\langle \zeta_1|R_0^{ij}|\zeta_4\rangle = \zeta_4^i\, \zeta_1^j - \zeta_1^i\, \zeta_4^j \tag{7.4.40}$$

$$\langle \zeta_1|R_0^{ij} R_0^{kl}|\zeta_4\rangle = \zeta_1^j\, \zeta_4^k\, \delta^{il} - \zeta_1^i\, \zeta_4^k\, \delta^{jl} - \zeta_1^j\, \zeta_4^l\, \delta^{ik} + \zeta_1^i\, \zeta_4^l\, \delta^{jk}.$$

Using these, some straightforward (but tedious!) algebra gives

$$A_4 = -\tfrac{1}{2}g^2 \frac{\Gamma(-s/2)\Gamma(-t/2)}{\Gamma(1-s/2-t/2)} K(\zeta_1, k_1; \zeta_2, k_2; \zeta_3, k_3; \zeta_4, k_4), \qquad (7.4.41)$$

where the kinematic factor K is given by

$$\begin{aligned}
K = &-\tfrac{1}{4}(st\ \zeta_1 \cdot \zeta_3\ \zeta_2 \cdot \zeta_4 + su\ \zeta_2 \cdot \zeta_3\ \zeta_1 \cdot \zeta_4 + tu\ \zeta_1 \cdot \zeta_2\ \zeta_3 \cdot \zeta_4) \\
&+ \tfrac{1}{2}s(\zeta_1 \cdot k_4\ \zeta_3 \cdot k_2\ \zeta_2 \cdot \zeta_4 + \zeta_2 \cdot k_3\ \zeta_4 \cdot k_1\ \zeta_1 \cdot \zeta_3 \\
&\quad + \zeta_1 \cdot k_3\ \zeta_4 \cdot k_2\ \zeta_2 \cdot \zeta_3 + \zeta_2 \cdot k_4\ \zeta_3 \cdot k_1\ \zeta_1 \cdot \zeta_4) \\
&+ \tfrac{1}{2}t(\zeta_2 \cdot k_1\ \zeta_4 \cdot k_3\ \zeta_3 \cdot \zeta_1 + \zeta_3 \cdot k_4\ \zeta_1 \cdot k_2\ \zeta_2 \cdot \zeta_4 \\
&\quad + \zeta_2 \cdot k_4\ \zeta_1 \cdot k_3\ \zeta_3 \cdot \zeta_4 + \zeta_3 \cdot k_1\ \zeta_4 \cdot k_2\ \zeta_2 \cdot \zeta_1) \\
&+ \tfrac{1}{2}u(\zeta_1 \cdot k_2\ \zeta_4 \cdot k_3\ \zeta_3 \cdot \zeta_2 + \zeta_3 \cdot k_4\ \zeta_2 \cdot k_1\ \zeta_1 \cdot \zeta_4 \\
&\quad + \zeta_1 \cdot k_4\ \zeta_2 \cdot k_3\ \zeta_3 \cdot \zeta_4 + \zeta_3 \cdot k_2\ \zeta_4 \cdot k_1\ \zeta_1 \cdot \zeta_2). \qquad (7.4.42)
\end{aligned}$$

Again, if we assume that the theory is Lorentz invariant, the result can be interpreted as a covariant expression, dropping the kinematic restrictions that went into its derivation. (The extension is unique in this case.) It is then necessary that the factor K satisfy on-shell gauge invariance, i.e., it must vanish when any of the polarization vectors is replaced by the corresponding momentum. This is easily verified to be the case. It is necessary that K have cyclic symmetry in the four external lines, and it does. What may seem surprising, however, is that it actually has total symmetry in the four lines. This feature will be important later when it is appears as a factor in the construction of closed-string amplitudes. We will see in chapter 9 that K can actually be expressed in terms of a trace over the space spanned by the S_0 operators

$$K = \mathrm{tr}_{S_0} \left(R_0^{i_1 i_5} R_0^{i_2 i_6} R_0^{i_3 i_7} R_0^{i_4 i_8} \right) k^{i_1} k^{i_2} k^{i_3} k^{i_4} \zeta^{i_5} \zeta^{i_6} \zeta^{i_7} \zeta^{i_8}, \qquad (7.4.43)$$

which makes its symmetry properties manifest. The fact that this kinematic factor arises as a trace is very natural in the loop calculation although somewhat obscure in the calculation of tree diagrams.

Four-particle amplitudes involving fermion lines can also be calculated from (7.4.35) with essentially the same amount of labor. In all cases one finds that

$$A(1,2,3,4) = -\tfrac{1}{2}g^2 \frac{\Gamma(-s/2)\Gamma(-t/2)}{\Gamma(1-s/2-t/2)} K(1,2,3,4), \qquad (7.4.44)$$

where K is a suitable kinematic factor. K is given by (7.4.42) in the four-vector case. The tree with fermions on the ends and bosons in the middle can be obtained by sandwiching (7.4.39) between $\langle u_1 |$ and $| u_4 \rangle$ using $R_0^{ij} | \dot{a} \rangle = -\frac{1}{2} \gamma_{\dot{a}\dot{b}}^{ij} | \dot{b} \rangle$, and

$$\langle u_1 | u_4 \rangle = (k_4^+)^{-1} u_1 u_4$$

$$\langle u_1 | R_0^{ij} | u_4 \rangle = \frac{1}{2}(k_4^+)^{-1} u_1 \gamma^{ij} u_4 \qquad (7.4.45)$$

$$\langle u_1 | R_0^{ij} R_0^{kl} | u_4 \rangle = \frac{1}{4}(k_4^+)^{-1} u_1 \gamma^{ij} \gamma^{kl} u_4.$$

Expressing the result in covariant form, one finds that

$$\begin{aligned}
K(u_1, \zeta_2, \zeta_3, u_4) = &-\tfrac{1}{2} t \overline{u}_1 \Gamma \cdot \zeta_2 \Gamma \cdot (k_3 + k_4) \Gamma \cdot \zeta_3 u_4 \\
&+ s(\overline{u}_1 \Gamma \cdot \zeta_3 u_4 k_3 \cdot \zeta_2 - \overline{u}_1 \Gamma \cdot \zeta_2 u_4 k_2 \cdot \zeta_3 \qquad (7.4.46) \\
&- \overline{u}_1 \Gamma \cdot k_3 u_4 \zeta_2 \cdot \zeta_3).
\end{aligned}$$

The other possibilities are

$$\begin{aligned}
K(u_1, \zeta_2, u_3, \zeta_4) = &\tfrac{1}{2} t \overline{u}_1 \Gamma \cdot \zeta_2 \Gamma \cdot (k_3 + k_4) \Gamma \cdot \zeta_4 u_3 \\
&+ \tfrac{1}{2} s \overline{u}_1 \Gamma \cdot \zeta_4 \Gamma \cdot (k_2 + k_3) \Gamma \cdot \zeta_2 u_3
\end{aligned} \qquad (7.4.47)$$

and

$$K(u_1, u_2, u_3, u_4) = -\tfrac{1}{2} s \overline{u}_2 \Gamma^\mu u_3 \overline{u}_1 \Gamma_\mu u_4 + \tfrac{1}{2} t \overline{u}_1 \Gamma^\mu u_2 \overline{u}_4 \Gamma_\mu u_3, \qquad (7.4.48)$$

which are obtained in a similar manner by evaluating matrix elements of $V_F V_B$ and $V_F V_F$, respectively. These factors have total symmetry in the boson lines and antisymmetry in the fermion lines. To prove this for (7.4.48) requires Fierz transformation formulas for Majorana–Weyl spinors in ten dimensions. Defining

$$T_1(1234) = \overline{u}_1 \Gamma^\mu u_2 \overline{u}_3 \Gamma_\mu u_4 \qquad (7.4.49)$$

$$T_3(1234) = \overline{u}_1 \Gamma^{\mu\nu\rho} u_2 \overline{u}_3 \Gamma_{\mu\nu\rho} u_4, \qquad (7.4.50)$$

the Fierz identities, which can be proved by the same kind of algebra as in appendix 4.A, are

$$T_1(1432) = -\tfrac{1}{2} T_1(1234) + \tfrac{1}{24} T_3(1234) \qquad (7.4.51)$$

$$T_3(1432) = 18 T_1(1234) + \tfrac{1}{2} T_3(1234). \qquad (7.4.52)$$

Thus the total symmetry of (7.4.42) generalizes to the supermultiplet in the way one would expect.

7.4.3 Closed-String Trees

Closed superstring tree amplitudes can be related to open superstring tree amplitudes in the same way as open and closed bosonic strings are related in §7.2.1 and 7.2.2. For example, let us denote the vertex for three massless open strings by

$$A_3^{op} = g\zeta_1^A \zeta_2^B \zeta_3^C V_{ABC}(k_1, k_2, k_3), \qquad (7.4.53)$$

where A, B, C are superspace indices that can take vector or spinor values. Then this expression describes all the three-point couplings of the previous section. For three massless closed-string states, the three-particle vertex is then given by

$$A_3^{cl} = \kappa \zeta_1^{AA'} \zeta_2^{BB'} \zeta_3^{CC'} V_{ABC}(\tfrac{1}{2}k_1, \tfrac{1}{2}k_2, \tfrac{1}{2}k_3) V_{A'B'C'}(\tfrac{1}{2}k_1, \tfrac{1}{2}k_2, \tfrac{1}{2}k_3). \qquad (7.4.54)$$

This describes all couplings of the 256 massless states since $A, A', \ldots$ each take 16 possible values in the light-cone formalism. In the covariant description they each represent a ten-vector and a 16-component Majorana–Weyl spinor. In type I and IIB theories the same chirality spinor is used for A, B, C as for A', B', C', whereas in the type IIA theory opposite chiralities are used. In the type I theory the spectrum is truncated to the $N = 1$ supergravity multiplet by requiring $\zeta^{AA'}$ to have graded symmetry, i.e., antisymmetry if both indices are spinorial and symmetry otherwise.

Trees with four massless particles can be constructed using (7.2.40). The four-particle open-string trees of the preceding section can be written in the form

$$A_4^{op} = g^2 \zeta_1^A \zeta_2^B \zeta_3^C \zeta_4^D K_{ABCD} \frac{\Gamma(-s/2)\Gamma(-t/2)}{\Gamma(1 + u/2)}, \qquad (7.4.55)$$

where K_{ABCD} represents all the kinematic factors called K in the preceding section. It has total (graded) symmetry under simultaneous interchanges of the indices and the corresponding momenta. To evaluate A_4^{cl} we need the relationship

$$
\begin{aligned}
C(s,t,u) &= \left(\sin\frac{\pi t}{8}\right) \frac{\Gamma(-s/8)\Gamma(-t/8)}{\Gamma(1 - s/8 - t/8)} \times \frac{\Gamma(-t/8)\Gamma(-u/8)}{\Gamma(1 - t/8 - u/8)} \\
&= -\pi \frac{\Gamma(-s/8)\Gamma(-t/8)\Gamma(-u/8)}{\Gamma(1 + s/8)\Gamma(1 + t/8)\Gamma(1 + u/8)},
\end{aligned}
\qquad (7.4.56)
$$

which follows by using $\Gamma(a)\Gamma(1 - a)\sin\pi a = \pi$. In terms of $C(s,t,u)$ we have

$$A_4^{cl} = \kappa^2 \zeta_1^{AA'} \zeta_2^{BB'} \zeta_3^{CC'} \zeta_4^{DD'} K_{ABCD}(k/2) K_{A'B'C'D'}(k/2) C(s,t,u). \tag{7.4.57}$$

Since K and C have total symmetry in the four lines, it follows that A_4^{cl} itself has the requisite total graded symmetry.

7.4.4 Heterotic String Trees

Heterotic strings are closed strings whose right-moving modes are supersymmetric and whose left-moving modes are bosonic. In addition there is the $E_8 \times E_8$ or spin(32)/Z_2 group theory structure described in §6.3. It can be incorporated either by compactifying 16 left-moving dimensions on the appropriate torus or by the use of 32 fermionic degrees of freedom, analogous to those in the RNS formulation of the superstring.

The massless degrees of freedom of the left-moving sector consist of a ten-vector and 496 scalars in the adjoint representation of the gauge group. The vector state tensored with the massless right-moving modes gives the $N = 1$ $D = 10$ supergravity multiplet. Let us start by considering amplitudes that involve these states. The states can be labeled by polarization tensors and spinors $\zeta^{A\mu}$, where A is a supermultiplet label (vector plus spinor) corresponding to right-moving modes (as in §7.4.3.) and μ is a vector label associated with left-moving modes. The cubic couplings of the supergravity multiplet are then given by

$$A_3 = \kappa \zeta_1^{A\mu} \zeta_2^{B\nu} \zeta_3^{C\rho} V_{ABC}(\tfrac{1}{2}k) t_{\mu\nu\rho}(\tfrac{1}{2}k), \tag{7.4.58}$$

where V_{ABC} are the vertices introduced in §7.4.3 and $t_{\mu\nu\rho}$ is given in (7.1.58). Recall that $t_{\mu\nu\rho}$ contains $O(\alpha')$ corrections to the vertices of the minimal point-particle Yang–Mills theory but that V_{ABC} does not. As a result A_3 contains couplings that correspond to $O(\alpha')$ corrections to those of the minimal point-particle $N = 1$ $D = 10$ supergravity theory. In an effective action they are represented by various R^2 and other terms.

The four-particle tree for states of the supergravity multiplet can also be computed using (7.2.40). A needed input is the four-vector tree of the bosonic theory. It is obtained by evaluating (7.1.68) for $M = 4$. One finds that

$$A_4^{bos} = g^2 \zeta_1^\mu \zeta_2^\nu \zeta_3^\rho \zeta_4^\lambda K_{\mu\nu\rho\lambda}^{bos}(k) \frac{\Gamma(-s/2)\Gamma(-t/2)}{\Gamma(1+u/2)}, \tag{7.4.59}$$

where the kinematic factor $K_{\mu\nu\rho\lambda}^{bos}(k)$ has total symmetry in the four lines, just like (7.4.42). It is rather messy to write out, however, involving more than twice as many terms as (7.4.42). In terms of this factor and the factor K_{ABCD} introduced in (7.4.55) the four-particle tree has the form

$$A_4 = \kappa^2 \zeta_1^{A\mu} \zeta_2^{B\nu} \zeta_3^{C\rho} \zeta_4^{D\lambda} K_{ABCD}(k/2) K_{\mu\nu\rho\lambda}^{bos}(k/2) C(s,t,u), \qquad (7.4.60)$$

where $C(s,t,u)$ is defined in (7.4.56).

The massless spectrum of type I superstrings and of heterotic strings is the same. Therefore one can compare the amplitudes (7.4.58) and (7.4.60) with those of (7.4.54) and (7.4.57) for corresponding states. In each case the leading terms in an expansion in powers of α' agree with the vertices and trees generated by the minimal $D = 10$ $N = 1$ supergravity theory (to be described in chapter 13). However, the α' dependent terms in the two cases are quite different. For example, (7.4.54) is independent of α', whereas (7.4.58) is linear in α'. The differences are rather more substantial for M-particle trees with $M > 3$. This implies that the effective field theory description must be quite different in the two cases even though the leading terms are given by the same supergravity theory (described in §13.1) in each case.

Let us now consider some of the vertices and amplitudes for states that carry nontrivial group quantum numbers. The right-moving superstring modes are still described in the same way as above, but now we need to consider the description of left-moving modes that belong to the adjoint representation of the gauge group. In the bosonic description of the group degrees of freedom, the 496 massless modes fall into two distinct categories. Sixteen of them are mutually commuting and belong to a Cartan subalgebra whereas the other 480 carry charges K^I with $K \cdot K = 2$. The 16 in the Cartan subalgebra have $K^I = 0$. Since the method of construction treats the two categories of states so differently the full group symmetry that relates them is not manifest.

Amplitudes involving only states in the Cartan subalgebra are described by replacing the space-time indices by group-theory ones, i.e.,

$$K_{\mu\nu\rho\lambda} \to K_{IJKL}. \qquad (7.4.61)$$

Since the internal momenta are all zero in this case the only nonvanishing terms arise from having the indices I, J, K, L attached to Kronecker δ's rather than momenta. Thus t_{IJK} vanishes and

$$K_{IJKL}(k/2) \propto \frac{tu}{1+s/8}\delta_{IJ}\delta_{KL} + \frac{su}{1+t/8}\delta_{IL}\delta_{JK} + \frac{st}{1+u/8}\delta_{IK}\delta_{JL}$$
$$(7.4.62)$$

may be substituted for $K_{\mu\nu\rho\lambda}^{bos}(k/2)$ in (7.4.60) to describe the scattering of four super Yang–Mills states in the Cartan subalgebra. This amplitude vanishes in the low-energy point-particle limit, in agreement with the fact that ordinary Yang–Mills theory gives vanishing amplitudes for neutral particles in tree approximation.

Let us now consider scattering of charged particles, (*i.e.*, amplitudes involving states associated with the 480 generators that are not in the Cartan subalgebra). These states are generalized left-moving tachyons, in the sense that the tachyon condition $k_L^2 = 2$ for the left-moving 26-momentum $k_L = (\frac{1}{2}k^\mu, K^I)$ is satisifed by $k^2 = 0$, $K^2 = 2$. This state is relevant to the physical spectrum, unlike the ordinary tachyon ($k^2 = 8$, $K^2 = 0$), which cannot satisfy the $L_0 = \tilde{L}_0 - 1$ condition, as explained in §6.3.1. Amplitudes involving these charged states are described by functions in which the left-moving modes are given by the multitachyon amplitudes of the bosonic theory with momenta of the type described above ($k^2 = 0$, $K^2 = 2$). It is also necessary to take account of the cocycles in the vertex operators. For example, the three-particle vertex for three such left-moving modes is given by

$$\epsilon(K) = \langle K_1 | V_L(K_2) | K_3 \rangle, \qquad (7.4.63)$$

where

$$V_L(K_2) = c_{K_2}(P)e^{2iK_2 \cdot X}. \qquad (7.4.64)$$

Using the cocycle choice

$$c_K(P) = (-1)^{P*K} \qquad (7.4.65)$$

of §6.4.5, this factor is simply

$$\epsilon(K) = (-1)^{K_3 * K_2}. \qquad (7.4.66)$$

Since $K_1 + K_2 + K_3 = 0$ and $K_i^2 = 2$, it follows that ϵ is totally antisymmetric under interchanges of the three K coordinates. As a result the vertex operator for three charged particles,

$$A_3 = \kappa \zeta_1^A \zeta_2^B \zeta_3^C \epsilon(K) V_{ABC}(\tfrac{1}{2}k), \qquad (7.4.67)$$

has the appropriate symmetries required by Bose or Fermi statistics. The coefficients $\epsilon(K)$ represent the structure constants for three charged particles in a basis where they are totally antisymmetric.

The amplitude for scattering of four charged particles involves the same basic principles. In analogy with the Mandelstam variables s, t, u it is convenient to introduce lattice invariants

$$S = -(K_1 + K_2)^2, \quad T = -(K_2 + K_3)^2, \quad U = -(K_1 + K_3)^2. \quad (7.4.68)$$

The kinematics $\Sigma K_i = 0$, $K_i^2 = 2$ implies that $S + T + U = -8$. Furthermore the only possible values for S, T, or U are -8, -6, -4, -2, 0. For example if $K_1 = K_2 = -K_3 = -K_4$, then $S = -8$ and $T = U = 0$. The four-tachyon amplitude based on the momenta k_L gives

$$B(-1 - s/8 - S/2, -1 - t/8 - T/2)$$

$$= \frac{\Gamma(-1 - s/8 - S/2)\Gamma(-1 - t/8 - T/2)}{\Gamma(2 + u/8 + U/2)}. \quad (7.4.69)$$

Combining this with the four open superstring tree amplitude, gives the result for four charged heterotic strings

$$A_4 \propto \zeta_1^A \zeta_2^B \zeta_3^C \zeta_4^D K_{ABCD}(k/2)\epsilon(K)D(k, K), \quad (7.4.70)$$

where

$$D(k, K) = \pi \frac{\Gamma(-1 - s/8 - S/2)\Gamma(-1 - t/8 - T/2)\Gamma(-1 - u/8 - U/2)}{\Gamma(1 + s/8)\Gamma(1 + t/8)\Gamma(1 + u/8)}. \quad (7.4.71)$$

The cocycle factor $\epsilon(K)$ is given by

$$\epsilon(K) = c_{K_3}(K_1 + K_2)c_{K_2}(K_1)(-1)^{U/2}$$

$$= (-1)^{K_1 * K_3 + K_3 * K_2 + K_2 * K_1}. \quad (7.4.72)$$

This factor is totally symmetric in the four K_i. The intricate meshing of the dynamical variables and the group-theory coordinates is strikingly subtle and beautiful.

The amplitudes for four charged gauge particles described by (7.4.70) or four neutral ones described by (7.4.60) and (7.4.62) can be unified and expressed in a more conventional form. To do this let us label the external particles by matrices T^a, where the index a takes 496 values corresponding to the generators of $E_8 \times E_8$ or $SO(32)$. In the $SO(32)$ case we use 32×32 matrices (fundamental representation) for the T^a. In the $E_8 \times E_8$ case one can use 32×32 matrices for the $SO(16) \times SO(16)$ subgroup but not for the full group. For the full group 496×496 matrices (adjoint representation) are required. The normalization of traces in this representation can be made to agree with those of the $SO(16) \times SO(16)$ subgroup in the 32-dimensional representation by including a factor of $1/30$ in the definition 'tr'.

In calculating the scattering amplitude we may use the current algebra description of the group-theory degrees of freedom (see §6.2), for the matrix element that enters the calculation

$$\langle T_1 | V(T_2) y^{N-2} V(T_3) | T_4 \rangle = \langle 0 | T_1 \cdot Q_1 T_2 \cdot Q y^{N-2} T_3 \cdot Q T_4 \cdot Q_{-1} | 0 \rangle,$$
(7.4.73)

where $Q = \sum_{-\infty}^{\infty} Q_n$. This can be evaluated using the current algebra relations of §6.2. We shall write the amplitude in the form

$$A_4 = \kappa^2 \zeta_1^A \zeta_2^B \zeta_3^C \zeta_4^D K_{ABCD}(k/2) C(s,t,u) G(k,T),$$
(7.4.74)

(where, for the special case of four charged states, $C(s,t,u) G(k,T) = \epsilon(K) D(k,K)$). Substituting in the formula for the heterotic string four-particle amplitude, one obtains

$$
\begin{aligned}
G(k,T) = & s \, \text{tr}(T_1 T_4 T_2 T_3) + t \, \text{tr}(T_1 T_3 T_4 T_2) + u \, \text{tr}(T_1 T_2 T_3 T_4) \\
& + \tfrac{1}{32} \Big[\frac{tu}{1+s/8} \text{tr}(T_1 T_2) \text{tr}(T_3 T_4) + \frac{su}{1+t/8} \text{tr}(T_1 T_4) \text{tr}(T_2 T_3) \\
& + \frac{st}{1+u/8} \text{tr}(T_1 T_3) \text{tr}(T_2 T_4) \Big].
\end{aligned}
$$
(7.4.75)

If all four states are uncharged, the four T_i are mutually commuting and the first three terms in G cancel using $s + t + u = 0$. The result therefore agrees with (7.4.62) in that case.

One can make contact with the low-energy effective point-particle theory by approximating $C(s,t,u) \sim (stu)^{-1}$ for $\alpha's, \alpha't, \alpha'u \ll 1$. Thus the s-channel poles of $C \cdot G$ are

$$A_4 \sim \kappa^2 \frac{8}{s} \Big[\frac{1}{t} \text{tr}(T_1 T_2 T_3 T_4) + \frac{1}{u} \text{tr}(T_1 T_3 T_4 T_2) + \frac{1}{32} \text{tr}(T_1 T_2) \text{tr}(T_3 T_4) \Big].$$
(7.4.76)

The first two terms correspond to exchanges of adjoint representation states of the super Yang–Mills multiplet, and the last term corresponds to exchange of singlet states of the supergravity multiplet. By comparing the residues and reinstating the powers of α' needed on dimensional grounds, one can deduce the relation $\kappa^2 = \frac{1}{2}\alpha' g_{YM}^2$, valid for heterotic strings.[*]

The same basic techniques can be used to construct amplitudes combining the three categories of particles: gravity sector, neutral and charged. Also, the calculation of amplitudes with more than four external lines is straightforward in principle, although explicit expressions can be quite messy. The real purpose in doing all this algebra is to develop a better sense of the structure of the theory and to convince oneself that formal expressions really correspond to well-defined mathematical quantities. Some of the explicit formulas are also useful for identifying various terms that must appear in an effective action description of the theory.

7.5 Summary

A method for calculating tree-approximation S-matrix elements in each of the known string theories has been described and several examples were presented. The amplitudes were described as correlation functions of products of vertex operators. The formulas represent a straightforward generalization of the Feynman rules of ordinary field theories. A significant difference is that they were not derived from a Lagrangian, although it was argued that they must be correct, because they possess the factorization properties required by unitarity in tree approximation and Regge asymptotic behavior.

The group $SL(2, R)$, representing the one-to-one conformal mappings of the upper half plane onto itself, plays a fundamental role in the case of open-string tree amplitudes. It is the subgroup of the Virasoro algebra that leaves the vacuum invariant. Correspondingly, $SL(2, C)$, the group of one-to-one conformal mappings of the entire complex plane (including the point at infinity) onto itself, is fundamental in the analysis of closed-string tree amplitudes. In order to avoid an infinite volume factor associated with either of these groups, the world-sheet coordinates of three of the emitted particles must take fixed values. The operator construction automatically gives a specific choice ($z_1 = \infty$, $z_2 = 1$, $z_n = 0$).

Open-string diagrams have cyclic symmetry in the external lines, and the complete tree amplitude is given as a sum over cyclically inequivalent terms weighted by suitable group theory factors. Closed-string tree diagrams have total symmetry in the external lines and no group theory

---

[*] When six dimensions compactify into a space of volume V, the effective four-dimensional coupling constants are $\kappa_4 \sim \kappa/\sqrt{V}$ and $g_4 \sim g/\sqrt{V}$. If we suppose that $g_4 \sim 1$ and κ_4 describes gravitational couplings of Newtonian strength, then we see that α' must be about the square of the Planck length ($10^{-33} - 10^{-32}$ cm).

factors can be introduced. The formulas for closed-string amplitudes correspond closely to direct products of open-string formulas with one factor associated with left-moving modes and one factor associated with right-moving modes. In particular, one of the two $SL(2,R)$ factors acts on each sector, and the bosonic vertex operators (for example) have conformal dimension $(1,1)$.

The results obtained in this chapter can be obtained in a variety of other ways – by the path integral methods described in chapter 1, by the functional methods described in chapter 11 and by the light-cone gauge field theory. Hopefully, they will also eventually be obtained from covariant field theory formulations, or some suitable substitute.

Appendix 7.A   Coherent-State
Methods and Correlation Functions

In this appendix we prove (7.1.64) and some related formulas. The reader will note that the identities we consider here are very similar to the ones that entered in our discussion of bosonization in §3.2.4.

We begin with the following identity valid for operators A and B that are linear in harmonic oscillator raising and lower operators

$$\langle : e^A :: e^B : \rangle = e^{\langle AB \rangle}, \tag{7.A.1}$$

where $\langle \ldots \rangle$ is short for the ground state expectation value $\langle 0| \ldots |0\rangle$. If there are many oscillators, the linearity implies that both the left-hand and right-hand side of the equation factorize into separate factors for each of the oscillator modes. Therefore it is sufficient to establish the formula for a single oscillator. In other words, letting $A = c_1 a^\dagger + c_2 a$ and $B = c_3 a^\dagger + c_4 a$, we must prove that

$$\langle 0| e^{c_2 a} e^{c_3 a^\dagger} |0\rangle = e^{c_2 c_3}. \tag{7.A.2}$$

The identity in (7.A.2) is easily established by elementary considerations. However, instead of doing that we describe a technique, based on coherent states, which will prove useful in later chapters as well. We define a coherent state $|\lambda\rangle$ by

$$|\lambda\rangle = \exp(\lambda a^\dagger) |0\rangle = \sum_{n=0}^{\infty} \frac{\lambda^n}{\sqrt{n!}} |n\rangle, \tag{7.A.3}$$

where $|n\rangle$ is the usual normalized number basis state satisfying $a^\dagger a |n\rangle = n |n\rangle$ and $\langle m|n\rangle = \delta_{mn}$. Note that $|0\rangle = |0\rangle$, in particular. Coherent states are eigenstates of the lowering operator a

$$a|\lambda) = \lambda|\lambda),\tag{7.A.4}$$

a fact that is easily established using $[a, a^\dagger] = 1$. It follows that

$$e^{\lambda_1 a}|\lambda_2) = e^{\lambda_1 \lambda_2}|\lambda_2),\tag{7.A.5}$$

and hence that

$$(\mu|\lambda) = (0|e^{\mu^* a}|\lambda) = e^{\mu^* \lambda}.\tag{7.A.6}$$

This formula gives the required result in (7.A.2). Another useful identity is

$$z^{a^\dagger a}|\lambda) = \sum_{n=0}^{\infty} \frac{(\lambda z)^n}{\sqrt{n!}}|n) = |\lambda z).\tag{7.A.7}$$

This is obtained by substituting the definition of $|\lambda)$ from (7.A.3).

Recalling the expansion

$$X^\mu(y) = x^\mu - ip^\mu \log y + i\sum_{n \neq 0} \frac{1}{n}\alpha_n^\mu y^{-n},\tag{7.A.8}$$

the nonzero mode part of the tachyon vertex operator is

$$W_0(k, y) = \exp\left(k \cdot \sum_{n=1}^{\infty} \frac{1}{n}\alpha_{-n}y^n\right)\exp\left(-k \cdot \sum_{n=1}^{\infty} \frac{1}{n}\alpha_n y^{-n}\right).\tag{7.A.9}$$

Thus the nonzero mode part of the correlation function for a product of two vertices is

$$\langle W_0(k_1, y_1)W_0(k_2, y_2)\rangle = (1 - y_2/y_1)^{k_1 \cdot k_2}.\tag{7.A.10}$$

The result is obtained most readily using (7.A.1) and

$$\langle\left(\sum_1^{\infty} \frac{\alpha_n^\mu}{n}y_1^{-n}\right)\left(\sum_1^{\infty} \frac{\alpha_{-n}^\nu}{n}y_2^n\right)\rangle = \eta^{\mu\nu}\sum_1^{\infty} \frac{1}{n}\left(\frac{y_2}{y_1}\right)^n$$

$$= -\eta^{\mu\nu}\log(1 - y_2/y_1),\tag{7.A.11}$$

which is valid for $|y_2| < |y_1|$.

Equation (7.A.1) has a generalization

$$\langle : e^{A_1} :: e^{A_2} : \cdots : e^{A_M} \rangle = \exp[\sum_{i<j} \langle A_i A_j \rangle], \qquad (7.A.12)$$

which is easily established using the same coherent-state methods described above. Therefore the generalization of (7.A.10) is

$$\langle W_0(k_1, y_1) \ldots W_0(k_M, y_M) \rangle = \prod_{i<j} (1 - y_j/y_i)^{k_i \cdot k_j}. \qquad (7.A.13)$$

Next let us consider the contribution of the zero-mode terms. In the case of tachyon matrices these contribute

$$\langle Z_0(k_1, y_1) Z_0(k_2, y_2) \ldots Z_0(k_M, y_M) \rangle$$
$$= (\prod y_i) \langle e^{ik_1 \cdot x} y_1^{k_1 \cdot p} e^{ik_2 \cdot x} y_2^{k_2 \cdot p} \ldots e^{ik_M \cdot x} y_M^{k_M \cdot p} \rangle. \qquad (7.A.14)$$

But

$$e^{ik_M \cdot x} y_M^{k_M \cdot p} |0\rangle = |k_M\rangle,$$
$$e^{ik_{M-1} \cdot x} y_{M-1}^{k_{M-1} \cdot p} |k_M\rangle = y_{M-1}^{k_{M-1} \cdot k_M} |k_{M-1} + k_M\rangle, \qquad (7.A.15)$$

and so forth, so altogether

$$\langle Z_0(1) \ldots Z_0(M) \rangle = (\prod y_i) \prod_{i<j} y_i^{k_i \cdot k_j}. \qquad (7.A.16)$$

Combining this result with (7.A.13) gives

$$\langle \frac{V_0(k_1, y_1)}{y_1} \frac{V_0(k_2, y_2)}{y_2} \ldots \frac{V_0(k_M, y_M)}{y_M} \rangle = \prod_{i<j} (y_i - y_j)^{k_i \cdot k_j}, \qquad (7.A.17)$$

as asserted in (7.1.73).

The fact that the right-hand side of (7.A.17) only involves differences $y_i - y_j$ can be understood as follows. The statement that $V(y)$ has conformal dimension $J = 1$ implies (using $\frac{\partial}{\partial \tau} = iy\frac{\partial}{\partial y}$)

$$[L_m, V(y)] = y^m(y\frac{d}{dy} + m)V(y) = y\frac{d}{dy}[y^m V(y)]. \qquad (7.A.18)$$

In particular, for $m = -1$,

$$[L_{-1}, V(y)/y] = \frac{d}{dy}[V(y)/y]. \qquad (7.A.19)$$

Therefore making common L_{-1} transformations

$$\frac{V(y_i)}{y_i} \to e^{\lambda L_{-1}}\frac{V(y_i)}{y_i}e^{-\lambda L_{-1}} = \frac{V(y_i + \lambda)}{y_i + \lambda}. \qquad (7.A.20)$$

This does not change the correlation function (7.A.17) because the vacuum is $SL(2, R)$ invariant. Hence the result can only depend on the differences $y_i - y_j$ for any $J = 1$ vertices.

Now consider commuting adjacent vertex operators $V(i)$ and $V(i + 1)$ in (7.A.17). Formally, the only change would be that the factor $(y_i - y_{i+1})^{k_i \cdot k_{i+1}}$ on the right-hand side of the equation would be replaced by $(y_{i+1} - y_i)^{k_i \cdot k_{i+1}}$. However, there is a problem, because the series summed above (such as (7.A.11)) only converge for $|y_i| > |y_{i+1}|$ in one case and $|y_{i+1}| > |y_i|$ in the other. To make sense of the formula it is necessary to accompany the interchange with an analytic continuation of $(y_i - y_{i+1})^{k_i \cdot k_{i+1}}$ from $y_i > y_{i+1}$ to $y_i < y_{i+1}$. In doing this we need to decide which side of the branch point at $y_i = y_{i+1}$ to go on. Let us make the choice such that

$$(y_i - y_{i+1})^{k_i \cdot k_{i+1}} \to e^{i\pi k_i \cdot k_{i+1}}(y_i - y_{i+1})^{k_i \cdot k_{i+1}}. \qquad (7.A.21)$$

One sometimes writes

$$\langle X^\mu(y_1)X^\nu(y_2)\rangle = -\eta^{\mu\nu}\log(y_1 - y_2)\lambda, \qquad (7.A.22)$$

as a generalization of (7.A.11) that includes zero modes. Here λ is an infrared cutoff that cancels out of all really well-defined formulas. Equation (7.A.22) is best regarded as a mnemonic for deducing (7.A.17) from (7.A.12). In the same spirit one can write correlation functions for the derivatives of X^μ by differentiating (7.A.22). For example (using a dot to represent a τ derivative and recalling that $\frac{\partial}{\partial \tau} = iy\frac{\partial}{\partial y}$)

$$\langle \frac{\dot{X}^\mu(y_1)}{y_1} X^\nu(y_2) \rangle = -\frac{i\eta^{\mu\nu}}{y_1 - y_2}$$

$$\langle \frac{\dot{X}^\mu(y_1)}{y_1} \frac{\dot{X}^\nu(y_2)}{y_2} \rangle = \frac{\eta^{\mu\nu}}{(y_1 - y_2)^2}.$$

(7.A.23)

The proof of (7.1.49) requires a slight generalization of (7.A.1), namely

$$: e^A :: e^B := : e^{A+B} : e^{\langle AB \rangle}.$$

(7.A.24)

This is easily established using the coherent-state formulas. Since the expression still factorizes into separate pieces for each oscillator, it is sufficient to study a single mode. Using this equation we have

$$V_0(k_1, y_1)V_0(k_2, y_2) =: V_0(k_1, y_1)V_0(k_2, y_2) : e^{-\langle k_1 \cdot X(y_1) k_2 \cdot X(y_2) \rangle}$$

(7.A.25)

and similarly with $1 \leftrightarrow 2$. Then since

$$: V_0(k_1, y_1)V_0(k_2, y_2) :=: V_0(k_2, y_2)V_0(k_1, y_1) :$$

(7.A.26)

it follows that

$$V_0(k_1, y_1)V_0(k_2, y_2) = V_0(k_2, y_2)V_0(k_1, y_1)e^{-\langle [k_1 \cdot X(y_1), k_2 \cdot X(y_2)] \rangle}.$$

(7.A.27)

The exponent in (7.A.27) is given by

$$C(y_1, y_2) = [X(y_1), X(y_2)]$$

$$= \log(y_1/y_2) + \sum_{n=1}^{\infty} \frac{1}{n}[(y_1/y_2)^n - (y_2/y_1)^n]$$

$$= \log(y_1/y_2) - \log(1 - y_1/y_2) + \log(1 - y_2/y_1)$$

$$= \log(-1) = i\pi(1 + 2K),$$

(7.A.28)

where K is an integer. In order to calculate K consider choosing y_1 and y_2 at different values on the unit circle, so that both series converge. All phases can then be calculated once a definite branch of the logarithm is chosen. Choosing the branch for which the imaginary part is between $-i\pi$ and $+i\pi$ gives

$$C(y_1, y_2) = i\pi\epsilon(\arg y_1 - \arg y_2).$$

(7.A.29)

If the points on the circle are projected back onto the real axis this gives the result of (7.1.49)

$$V_0(k_1, y_1)V_0(k_2, y_2) = V_0(k_2, y_2)V_0(k_1, y_1) \exp[\pi i k_1 \cdot k_2 \epsilon(y_1 - y_2)].$$

(7.A.30)

Bibliography

In the following we present a brief chapter by chapter historical discussion of the significant ideas and discoveries in string theory. A few of the most important papers are cited and more complete listings are then given. We apologize in advance for any errors or oversights. The discussion for the introductory chapter is especially brief, so as to minimize duplication with the subsequent chapters.

Chapter 1

String theory originated in the study of strong interactions, as an outgrowth of the bootstrap and Regge-pole theory. Important early developments included saturation of current algebra sum rules with narrow resonances which led to 'finite energy sum rules' [300,264,265] and the duality concept introduced by Dolen, Horn and Schmid [127,128], whose subsequent development had many contributors including Freund [168], Harari [249] and Rosner [380]. This work led to the discovery of the Veneziano amplitude [453], which perhaps constitutes the beginning of string theory. This was quickly followed by the development of multiparticle generalizations [31,457,214,80], which were recast in a particularly elegant form by Koba and Nielsen [290,291]. An alternative set of amplitudes (later recognized as corresponding to closed strings) was invented for four-particle amplitudes by Virasoro [456] and generalized to N particles by Shapiro [418]. The first hint of special properties of the Veneziano model in 26 space-time dimensions came from the study of loop diagrams by Lovelace [303].

The recognition that these dual resonance amplitudes describe the dynamics of a relativistic string originated in independent works of Nambu [341], Nielsen [357] and Susskind [432]. The formula for the action as the area of the world sheet was introduced by Nambu [342], Goto [217] and Hara [247]. The use of the light-cone gauge for quantization of this action in the work of Goddard, Goldstone, Rebbi and Thorn [206] was an important step in clarifying the string picture.

Ramond's discovery of a free wave equation for fermionic strings [372] followed by the discovery of the interacting bosonic sector by Neveu and

Schwarz [351] led to a string theory of interacting fermions and bosons (the RNS model) with critical dimension $D = 10$. It was realized by Gliozzi, Scherk and Olive [202,203] that this model has a supersymmetric spectrum when it is suitably projected. Some years later the projected theory was formulated in a manifestly supersymmetric way by Green and Schwarz [224,225,226]. These and other developments will be discussed in connection with the other chapters.

The fact that the low energy limit of string theory is equivalent to ordinary field theory originates with the work of Scherk [388] who considered the limit of open strings which results in scalar ϕ^3 field theory (a limit which is inconsistent since it does not keep the massless spin 1 particle). Neveu and Scherk [355] showed that a consistent low-energy analysis of open-string tree diagrams leads to the tree diagrams of Yang–Mills theory. Yoneya [472,473] and Scherk and Schwarz [389,391] similarly made the connection between closed strings and the perturbation expansion of Einstein's theory of gravity.

The proposal to use string theory as a unified theory of fundamental forces including gravitation, rather than as a theory of hadrons, was made by Scherk and Schwarz [389,391]. They also noted that in this context it was natural to take seriously the extra dimensions of the known string theories [393,406]. See also [364] for a contemporaneous perspective on string theory.

The Kaluza–Klein idea of extra dimensions [279,286] had probably been the most imaginative of the early suggestions concerning unification of electromagnetism and gravitation. One of the most fascinating of the early papers on this subject, because of the comparatively modern viewpoint expressed, was [139]. See [299] for a perspective on the early work. The non-Abelian extension of Kaluza–Klein theory was first noted in [121] – where it was presented as a homework problem!

General arguments that low energy couplings of a massless spin two particle that does not decouple at zero momentum must mimic general relativity can be found in [461–463].

The path-integral approach to string theory originated in work of Hsue, Sakita and Virasoro [258] and Gervais and Sakita [196]. Once the string action was reformulated using a world-sheet metric tensor [56,120], it became possible for Polyakov [369,370] to introduce path-integral quantization in a form that properly takes account of the conformal symmetry and clarifies the meaning of the critical dimension.

A number of review articles on string theory deserve to be mentioned. They include five *Physics Reports* articles [12,400,454,375,314] which were reprinted together as a book [267]. An excellent review article by Scherk

[392] and a book by Frampton [160] also appeared in the 1970s. More recently, there have been review articles by Schwarz [407,410], Green [228,232] and Brink [62]. The reviews [392,407,228] are reprinted together with many original articles in a two-volume work edited by Schwarz [409]. Many useful survey articles, including some on subjects little developed in this book, can be found in the proceedings of the 1985 Santa Barbara workshop [233].

Chapter 2

As we have also pointed out in connection with chapter 1, the string interpretation of 'dual resonance models' originated in independent work of Nambu [341], Nielsen [357] and Susskind [433]. The formulation of the world-sheet action as an area followed shortly thereafter in work of Nambu [342] and Goto [217]. Reparametrization invariance of the world-sheet action enabled Brink, Di Vecchia and Howe [56] and Deser and Zumino [120] to express string theory in the language of two-dimensional general relativity. Both groups carried out the construction for the RNS model as well as the bosonic string.

The operator description of dual string models was first obtained by studying the factorization properties of dual resonance amplitudes. The factorization was achieved by Fubini and Veneziano, who also developed the operator formalism in a series of important papers [185–188]. They also proposed the idea of ghost cancellation via Ward identities. The factorization was independently obtained by Bardakçi and Mandelstam [33]. The $SL(2, R)$ constraint conditions (generated by L_1, L_0, L_{-1}) were pointed out by Gliozzi [201]. Virasoro noted that when the leading Regge trajectory has intercept $\alpha(0) = 1$ this can be extended to an infinite set of L_n constraints [458]. The so-called 'Virasoro algebra' first appeared in [188]. The occurrence of the anomaly term in the Virasoro algebra was first obtained by J. Weis (unpublished). The gauge conditions that define the physical states were derived by Del Giudice and Di Vecchia [116]. The first hint that $D = 26$ is special appeared in a study of nonplanar loop amplitudes by Lovelace [303]. Brink and Nielsen described a relation between the critical dimension and the mass of the ground state in terms of zero-point fluctuations [53].

The use of light-cone gauge originated in a study of QED by Kogut and Soper [292]. It soon found its home in string theory in the classic work of Goddard, Goldstone, Rebbi and Thorn [206]. They succeeded in putting the string interpretation of dual resonance models on a firm mathematical foundation by quantizing the free bosonic string in the light-cone gauge and showing that the quantum theory is Lorentz invariant only for $D = 26$. Del Giudice, Di Vecchia and Fubini [117] had earlier constructed the transverse physical state operators (described in §2.3.2), which constitute the spectrum generating algebra for $D = 26$. These operators play a central role in two different proofs of the no-ghost theorem given by Brower [66] and Goddard and Thorn [204]. The version of the theorem presented in §2.3.3 draws heavily on recent work of Thorn [441]. The asymptotic density of states described in §2.3.5 provides a specific implementation of a spectrum suggested earlier by Hagedorn [240] based on quite different considerations.

Chapter 3

The necessity of introducing ghost modes in a covariant treatment of nonabelian gauge theories in order to obtain unitary loop amplitudes was first pointed out by Feynman [149]. Faddeev and Popov gave a systematic procedure for deriving the ghost action in the context of path-integral quantization [144]. The modern approach to the path integral treatment of string theory began with the classic papers of Polyakov [369,370] who made use of the formulation of the string action that uses a world-sheet metric [56,120] in order to properly take account of world-sheet symmetries. The Polyakov approach has been developed and elaborated by many authors, notably Friedan [178], Alvarez [15], Durhuus *et al.* [133] and Fujikawa [190]. The developments in string theory have spawned interesting work in two-dimensional conformal field theory, which also has interesting applications in statistical physics. Notable examples include the works of Belavin, Polyakov and Zamolodchikov [38] and Friedan, Qiu and Shenker [179,180].

The BRST symmetry of the quantum action with ghost terms was identified by Becchi, Rouet and Stora [36,37] and (independently) by Tyutin [448]. Its application to string theory first appears in the work of Kato and Ogawa [282]. An important subject which we have not treated is the problem of showing that the BRST cohomology group of appropriate ghost number coincides with the space of physical states. Several treatments have appeared in the last two or three months [164,251,167]. Following observations of Siegel [421,422], the BRST approach is proving to be very important in the construction of a second-quantized string field theory. This is a vast and rapidly developing subject not covered in these volumes. However, references are listed at the end of the bibliographic section of volume 2. It was discovered long ago that in two dimensions two fermions can be replaced by a single boson [426,428,94,

315,307]. Bosonization of fermions and related tools in the theory of affine Lie algebra have surprising applications to the proof of various classical identities [298]. Our derivation of the Jacobi triple product identity in §3.2.4 is a simple example.

Although the rôle of modular invariance originates with the closed-string loop calculation of Shapiro [419] its central importance was only appreciated much later. For example, global aspects of the string world-sheet are discussed in the work of Alvarez [15,16]. The study of string amplitudes associated with higher genus surfaces is a rapidly developing subject. It involves many new techniques and results from algebraic geometry. Unfortunately, this subject goes beyond the scope of our exposition in this book, though some of the background at a very elementary level can be found in chapters 12, 14, and 15.

The relationship between the conformal anomaly and the Virasoro anomaly is due to Polyakov [369]. The derivation we give in §3.2.3 is from [20].

Nonlinear sigma models as an approach to describing string propagation in background fields were developed in [305,72,152–159,69,414,415,268]. There is a vast literature on conformal invariance in general nonlinear models [177,176,163,17,18,19,332,174]. Some more recent references can be found in the bibliography to chapter 16.

Chapter 4

The RNS model, referred to in the bibliography of chapter 1, was developed further by Neveu, Schwarz and Thorn [352], clarifying the role of the super-Virasoro conditions. The world-sheet supersymmetry of that theory was first clearly spelled out in a paper of Gervais and Sakita [195]. Related results were reported by Aharonov, Casher and Susskind [9,10] and by Iwasaki and Kikkawa [266]. At about the same time as world-sheet supersymmetry was being understood, Gol'fand and Likhtman proposed the super-Poincaré algebra as a new symmetry structure for four-dimensional theories [215]. Two years later Wess and Zumino constructed the theory that bears their name [465] and supersymmetry became an active area of research.

The superspace approach to supersymmetry first appeared in the two-dimensional world-sheet context in papers by Montonen [326] and Fairlie and Martin [146]. Later it was developed further by Howe [256] and Martinec [323]. Four-dimensional superspace was first proposed by Salam and Strathdee [383].

The no-ghost theorem for the RNS model was proved in the work of

Goddard and Thorn [204] cited earlier as well as in works by Schwarz [398] and Brower and Friedman [67].

The $N = 1$, $D = 4$ supergravity theory was discovered by Freedman, van Nieuwenhuizen and Ferrara [162] and by Deser and Zumino [119]. For reviews of supergravity see [450,466,192]. The tools of supergravity technology were applied to the construction of a locally supersymmetric superstring action in the previously cited work of Brink, Di Vecchia and Howe [56] and Deser and Zumino [120]. This action was utilized by Polyakov in the second of his pair of important papers [370].

The work of Gliozzi, Scherk and Olive [202,203] made clear for the first time that, after a suitable truncation, a string spectrum with space-time supersymmetry can be obtained. Their work strongly suggested a connection between superstrings and supergravity since the superstring spectrum contains the massless Rarita-Schwinger field associated with local supersymmetry. Their work also introduced the $D = 10$ super Yang–Mills theory, as did a paper by Brink, Schwarz and Scherk [57]. These papers also showed how ordinary field theories with extended supersymmetry, such as the $N = 4$, $D = 4$ super Yang–Mills theory, could be obtained by dimensional reduction.

The models with $N = 2$ and $N = 4$ superconformal symmetry on the world-sheet were formulated by Ademollo et al. [5–7]. The locally supersymmetric world-sheet action for the $N = 2$ theory was constructed by Brink and Schwarz [58]. Other systematic studies of possibilities for new theories by Nahm [336], Ramond and Schwarz [373] and Lovelace [304] showed how limited the possibilities are.

Chapter 5

The supersymmetric form of the superstring action was first described by Green and Schwarz in the light-cone gauge [224,226]. Subsequently, Brink and Schwarz found the covariant action for the superparticle [59] and Siegel [420] pointed out the local fermionic symmetry that it possesses. This was a clue to formulation of the covariant superstring action described in §5.1.2 and §5.1.3 [229]. An interpretation of the S_2 term as a sort of Wess-Zumino term was given by Henneaux and Mezincescu [250]. Various attempts have been made to quantize this action covariantly [230,39,254], but they have not yet led to success. Some suggestions that may prove helpful in future work have been made by Siegel [423,424].

The use of bosonization to relate the RNS and manifestly supersymmetric light cone formulations was given in [468]. The formulation of the covariant superstring action in a curved superspace background geometry was given in [470] for the heterotic string. This was generalized to type II theories in [234] and to include non-vanishing Yang–Mills backgrounds in [27,278].

Chapter 6

In the early days of dual models a scheme for attaching group-theory quantum numbers to open strings was proposed by Chan and Paton [365]. This was supposed to describe a global flavor symmetry, but a few years later Neveu and Scherk noted that it in fact gives a local gauge symmetry [355]. The scheme was originally proposed for unitary groups. The extension to orthogonal or symplectic groups was pointed out in [408]. Marcus and Sagnotti proved that these were the only possibilities for schemes of this type [318].

Following the discovery of anomaly cancellation (described in chapters 10 and 13) by Green and Schwarz [231] the search was on for an $E_8 \times E_8$ superstring theory. Freund suggested that it could be derived by compactification of the $D = 26$ string [173]. This turned out to be partly correct in the heterotic string theory developed by Gross, Harvey, Martinec and Rohm [238,239]. The construction employs a vertex operator representation of affine Lie algebras developed by mathematicians [148,297,413,165]. The subject had been made accessible for physicists in the papers of Goddard and Olive [210,212]. Some aspects of these constructions already appear in [246,30]. The construction of the Virasoro algebra in terms of the currents on the world sheet goes back to the work of Sugawara [429]. A quantum field theory realization for an arbitrary highest weight representation of affine Lie algebras was given in [469], along with a derivation of the quantization of the central charge. The idea of using world-sheet fermions to describe internal symmetry, which is also utilized in the work of Gross et al., was first proposed by Bardakçi and Halpern [34].

More general discussions of toroidal compactification have appeared in works of Englert and Neveu [141], Casher, Englert, Nicolai and Taormina [79] and Narain [343].

Chapter 7

The four-particle Veneziano amplitude [453] was generalized to N-particle tree amplitudes by Bardakçi and Ruegg [32], Virasoro [457], Goebel and Sakita [214] and Chan and Tsou [80,81]. The study of factorization and development of the operator formalism, discussed in connection with chapter 2, followed. In particular, the vertex operator and propagator were introduced by Fubini and Veneziano [187]. Koba and Nielsen re-expressed the amplitudes in a manifestly dual form [290,291]. The twist operator was introduced by Caneschi, Schwimmer and Veneziano [73] and by Amati, Le Bellac and Olive [21]. The general vertex for coupling three open strings was obtained by factorizing tree diagrams by Sciuto [412] and generalized to N strings by Lovelace [302].

The four-particle closed-string amplitude originated with an inspired guess by Virasoro [456], somewhat like the guess Veneziano had made a little earlier for the open strings. (Of course, the string picture was not yet known at this time.) The N-particle generalization of Virasoro's formula was introduced by Shapiro [418]. The relationship between open- and closed-string trees described in §7.2.3 was discovered recently by Kawai, Lewellen and Tye [284].

Bosonic RNS trees were constructed in the original papers of Neveu and Schwarz [351] and Neveu, Schwarz and Thorn [352]. Amplitudes with one fermion line, discussed in §7.3.4, were developed independently by Neveu and Schwarz [353] and by Thorn [437].

The development of the fermion-emission vertex and the calculation of fermion-fermion scattering was a major challenge during the first three years of superstring theory. The vertex operator was developed by Thorn [437], Schwarz [396] and Corrigan and Olive [96]. However, since ghost modes were not included it was still necessary to disentangle the effects of the gauge conditions. This was done by Brink, Olive, Rebbi and Scherk [54,363]. This resulted in the computation of the fermion-fermion scattering amplitude by Schwarz and Wu [399,404]. This work was clarified and expanded upon in subsequent papers by Corrigan, Goddard, Smith and Olive [99] and Bruce, Corrigan and Olive [68]. The result was obtained independently using light-cone gauge methods by Mandelstam [312].

Incorporation of the ghosts has made possible a much more satisfactory understanding of the fermion emission vertex in work by Friedan, Martinec and Shenker [181,184] and Knizhnik [289]. We have sketched some parts of this work in §7.3.5, though we have not discussed the picture changing operation [181,184].

Vertex operators and tree amplitudes in the supersymmetric light-cone formalism of chapter 5 were constructed by Green and Schwarz [225,226]. The tree amplitudes of the type II superstring theories were obtained by Green, Schwarz and Brink [227]. These techniques were applied to the calculation of heterotic string tree amplitudes by Gross, Harvey, Martinec and Rohm [239].

REFERENCES

1. Ademollo, M., Rubinstein, H.R., Veneziano G. and Virasoro, M.A. (1968), 'Bootstrap of meson trajectories from superconvergence', *Phys. Rev.* **176**, 1904.
2. Ademollo, M., Veneziano, G. and Weinberg, S. (1969), 'Quantization conditions for Regge intercepts and hadron masses', *Phys. Rev. Lett.* **22**, 83.
3. Ademollo, M., Del Giudice, E., Di Vecchia, P. and Fubini, S. (1974), 'Couplings of three excited particles in the dual-resonance model', *Nuovo Cim.* **19A**, 181.
4. Ademollo, M., D'Adda, A., D'Auria, R., Napolitano, E., Sciuto, S., Di Vecchia, P., Gliozzi, F., Musto, R. and Nicodemi, F. (1974), 'Theory of an interacting string and dual-resonance model', *Nuovo Cim.* **21A**, 77.
5. Ademollo, M., Brink, L., D'Adda, A., D'Auria, R., Napolitano, E., Sciuto, S., Del Giudice, E., Di Vecchia, P., Ferrara, S., Gliozzi, F., Musto, R., Pettorini, R. and Schwarz, J. (1976), 'Dual string with U(1) colour symmetry', *Nucl. Phys.* **B111**, 77.
6. Ademollo, M., Brink, L., D'Adda, A., D'Auria, R., Napolitano, E., Sciuto, S., Del Giudice, E., Di Vecchia, P., Ferrara, S., Gliozzi, F., Musto, R. and Pettorino, R. (1976), 'Dual string models with non-Abelian colour and flavour symmetries', *Nucl. Phys.* **B114**, 297.
7. Ademollo, M., Brink, L., D'Adda, A., D'Auria, R., Napolitano, E., Sciuto, S., Del Giudice, E., Di Vecchia, P., Ferrara, S., Gliozzi, F., Musto, R., and Pettorino, R. (1976), 'Supersymmetric strings and color confinement', *Phys. Lett.* **62B**, 105.
8. Affleck, Ian. (1985), 'Critical behavior of two-dimensional systems with continuous symmetries', *Phys. Rev. Lett.* **55**, 1355.
9. Aharonov, Y., Casher, A. and Susskind, L. (1971), 'Dual-parton model for mesons and baryons', *Phys. Lett.* **35B**, 512.
10. Aharonov, Y., Casher, A. and Susskind, L. (1972), 'Spin-$\frac{1}{2}$ partons in a dual model of hadrons', *Phys. Rev.* **D5**, 988.
11. Alessandrini, V., Amati, D., Le Bellac, M. and Olive, D. (1970), 'Duality and gauge properties of twisted propagators in multi-Veneziano theory', *Phys. Lett.* **32B**, 285.
12. Alessandrini, V., Amati, D., Le Bellac, M. and Olive, D. (1971), 'The operator approach to dual multiparticle theory', *Phys. Reports* **C1**, 269.
13. Altschüler, D. and Nilles, H.P. (1985), 'String models with lower critical dimension, compactification and nonabelian symmetries', *Phys. Lett.* **154B**, 135.
14. Alvarez, E. (1986), 'Strings at finite temperature', *Nucl. Phys.* **B269**, 596.
15. Alvarez, O. (1983), 'Theory of strings with boundaries: Fluctuations, topology and quantum geometry', *Nucl. Phys.* **B216**, 125.
16. Alvarez, O. (1986), 'Differential geometry in string models', in *Workshop on Unified String Theories, 29 July – 16 August, 1985*, eds. M. Green and D. Gross (World Scientific, Singapore), p. 103.
17. Alvarez-Gaumé, L. and Freedman, D.Z. (1980), 'Kahler geometry and the renormalization of supersymmetric σ models', *Phys. Rev.* **D22**, 846.

18. Alvarez-Gaumé, L. and Freedman, D.Z. (1980), 'Geometrical structure and ultraviolet finiteness in the supersymmetric σ-model', *Commun. Math. Phys.* **80**, 443.

19. Alvarez-Gaumé, L., Freedman, D.Z. and Mukhi, S. (1981), 'The background field method and the ultraviolet structure of the supersymmetric nonlinear σ-model', *Ann. Phys.* **134**, 85.

20. Alvarez-Gaumé, L. and Witten, E. (1983) 'Gravitational anomalies', *Nucl. Phys.* **B234**, 269.

21. Amati, D., Le Bellac, M. and Olive, D. (1970), 'The twisting operator in multi-Veneziano theory', *Nuovo Cim.* **66A**, 831.

22. Ambjørn, J., Durhuus, B., Fröhlich, J. and Orland, P. (1986), 'The appearance of critical dimensions in regulated string theories', *Nucl. Phys.* **B270[FS16]**, 457.

23. Antoniadis, I., Bachas, C., Kounnas, C. and Windey, P. (1986), 'Supersymmetry among free fermions and superstrings', *Phys. Lett.* **171B**, 51.

24. Aoyama, H., Dhar, A. and Namazie, M.A. (1986), 'Covariant amplitudes in Polyakov string theory', *Nucl. Phys.* **B267**, 605.

25. Appelquist, T., Chodos, A. and Freund, P., (1987), *Modern Kaluza-Klein Theory and Applications* (Benjamin/Cummings).

26. Ardalan, F. and Mansouri, F. (1986), 'Interacting parastrings', *Phys. Rev. Lett.* **56**, 2456.

27. Atick, J.J., Dhar, A. and Ratra, B. (1986), 'Superstring propagation in curved superspace in the presence of background super Yang–Mills fields', *Phys. Lett.* **169B**, 54.

28. Atick, J.J., Dhar, A. and Ratra, B. (1986), 'Superspace formulation of ten dimensional supergravity coupled to $N = 1$ super-Yang-Mills theory', *Phys. Rev.* **D33**, 2824.

29. Balázs, L.P. (1986), 'Could there be a Planck-scale unitary bootstrap underlying the superstring?', *Phys. Rev. Lett.* **56**, 1759.

30. Banks, T., Horn, D. and Neuberger, H. (1976), 'Bosonization of the $SU(N)$ Thirring models', *Nucl. Phys.* **B108**, 119.

31. Bardakçi, K. and Ruegg, H. (1968), 'Reggeized resonance model for the production amplitude', *Phys. Lett.* **28B**, 342.

32. Bardakçi, K. and Ruegg, H. (1969), 'Reggeized resonance model for arbitrary production processes', *Phys. Rev.* **181**, 1884.

33. Bardakçi, K. and Mandelstam, S. (1969), 'Analytic solution of the linear-trajectory bootstrap', *Phys. Rev.* **184**, 1640.

34. Bardakçi, K. and Halpern, M.B. (1971), 'New dual quark models', *Phys. Rev.* **D3**, 2493.

35. Batalin, I.A. and Vilkovisky, G.A. (1977), 'Relativistic S-matrix of dynamical systems with boson and fermion constraints', *Phys. Lett.* **69B**, 309.

36. Becchi, C., Rouet, A. and Stora, R. (1974), 'The abelian Higgs Kibble model, unitarity of the S-operator', *Phys. Lett.* **52B**, 344.

37. Becchi, C., Rouet, A. and Stora, R. (1976), 'Renormalization of gauge theories', *Ann. Phys.* **98**, 287.

38. Belavin, A.A., Polyakov, A.M. and Zamolodchikov, A.B. (1984), 'Infinite conformal symmetry in two-dimensional quantum field theory', *Nucl. Phys.* **B241**, 333.

39. Bengtsson, I. and Cederwall, M. (1984), 'Covariant superstrings do not admit covariant gauge fixing', Göteborg preprint 84-21-Rev.
40. Bergshoeff, E., Nishino, H. and Sezgin, E. (1986), 'Heterotic σ-models and conformal supergravity in two dimensions', *Phys. Lett.* **166B**, 141.
41. Bergshoeff, E., Sezgin, E. and Townsend, P.K. (1986), 'Superstring actions in $D = 3,4,6,10$ curved superspace', *Phys. Lett.* **169B**, 191.
42. Bergshoeff, E., Randjbar–Daemi, S., Salam, A., Sarmadi, H. and Sezgin, E. (1986), 'Locally supersymmetric σ-model with Wess–Zumino term in two dimensions and critical dimensions for strings', *Nucl. Phys.* **B269**, 77.
43. Bershadsky, M.A., Knizhnik, V.G. and Teitelman, M.G. (1985), 'Superconformal symmetry in two dimensions', *Phys. Lett.* **151B**, 31.
44. Bershadsky, M. (1986), 'Superconformal algebras in two dimensions with arbitrary N', *Phys. Lett.* **174B**, 285.
45. Bjorken, J.D., Kogut, J.B. and Soper, D.E. (1971), 'Quantum electrodynamics at infinite momentum: Scattering from an external field', *Phys. Rev.* **D3**, 1382.
46. Boucher, W., Friedan, D. and Kent, A. (1986), 'Determinant formulae and unitarity for the $N = 2$ superconformal algebras in two dimensions or exact results on string compactification', *Phys. Lett.* **172B**, 316.
47. Boulware, D.G. and Newman, E.T. (1986), 'The geometry of open bosonic strings', *Phys. Lett.* **174B**, 378.
48. Bouwknegt, P. and Van Nieuwenhuizen, P. (1986), 'Critical dimensions of the $N=1$ and $N=2$ spinning string derived from Fujikawa's approach', *Class. Quant. Grav.* **3**, 207.
49. Bowick, M.J. and Wijewardhana, L.C.R. (1985), 'Superstrings at high temperature', *Phys. Rev. Lett.* **54**, 2485.
50. Bowick, M. and Gürsey, F. (1986), 'The algebraic structure of BRST quantization', *Phys. Lett.* **175B**, 182.
51. Braaten, E., Curtright, T.L. and Zachos, C.K. (1985), 'Torsion and geometrostasis in nonlinear σ models', *Nucl. Phys.* **B260**, 630.
52. Brink, L. and Olive, D. (1973), 'The physical state projection operator in dual resonance models for the critical dimension of space-time', *Nucl. Phys.* **B56**, 253.
53. Brink, L. and Nielsen, H.B. (1973), 'A simple physical interpretation of the critical dimension of space-time in dual models', *Phys. Lett.* **45B**, 332.
54. Brink, L., Olive, D., Rebbi, C. and Scherk, J. (1973), 'The missing gauge conditions for the dual fermion emission vertex and their consequences', *Phys. Lett.* **45B**, 379.
55. Brink, L. and Winnberg, J.O. (1976), 'The superoperator formalism of the Neveu-Schwarz-Ramond model', *Nucl. Phys.* **B103**, 445.
56. Brink, L., Di Vecchia, P. and Howe, P. (1976), 'A locally supersymmetric and reparametrization invariant action for the spinning string', *Phys. Lett.* **65B**, 471.
57. Brink, L., Schwarz, J.H. and Scherk, J. (1977), 'Supersymmetric Yang-Mills theories', *Nucl. Phys.* **B121**, 77.
58. Brink, L. and Schwarz, J.H. (1977), 'Local complex supersymmetry in two dimensions', *Nucl. Phys.* **B121**, 285.

59. Brink, L. and Schwarz, J.H. (1981), 'Quantum superspace', *Phys. Lett.* **100B**, 310.

60. Brink, L. and Green, M.B. (1981), 'Point-like particles and off-shell supersymmetry algebras', *Phys. Lett.* **106B**, 393.

61. Brink, L., Lindgren, O. and Nilsson, B.E.W. (1983), '$N = 4$ Yang–Mills theory on the light cone', *Nucl. Phys.* **B212**, 401.

62. Brink, L. (1985), 'Superstrings', Lectures delivered at the 1985 Les Houches summer school; Göteborg preprint 85-68.

63. Brooks, R., Muhammad, F. and Gates, S.J. (1986), 'Unidexterous $D = 2$ supersymmetry in superspace', *Nucl. Phys.* **B268**, 599.

64. Brower, R.C. and Thorn, C.B. (1971), 'Eliminating spurious states from the dual resonance model', *Nucl. Phys.* **B31**, 163.

65. Brower, R.C. and Goddard, P. (1972), 'Collinear algebra for the dual model', *Nucl. Phys.* **B40**, 437.

66. Brower, R.C. (1972), 'Spectrum-generating algebra and no-ghost theorem for the dual model', *Phys. Rev.* **D6**, 1655.

67. Brower, R.C. and Friedman, K.A. (1973), 'Spectrum-generating algebra and no-ghost theorem for the Neveu-Schwarz model', *Phys. Rev.* **D7**, 535.

68. Bruce, D., Corrigan, E. and Olive, D. (1975), 'Group theoretical calculation of traces and determinants occurring in dual theories', *Nucl. Phys.* **B95**, 427.

69. Callan, C.G., Friedan, D., Martinec, E.J. and Perry, M.J. (1985), 'Strings in background fields', *Nucl. Phys.* **B262**, 593.

70. Callan, C.G. and Gan, Z. (1986), 'Vertex operators in background fields', *Nucl. Phys.* **B272**, 647.

71. Campagna P., Fubini S., Napolitano E. and Sciuto S. (1971), 'Amplitude for N nonspurious excited particles in dual resonance models', *Nuovo Cim.* **2A**, 911.

72. Candelas, P., Horowitz, G., Strominger, A. and Witten, E. (1985), 'Vacuum configurations for superstrings' *Nucl. Phys.* **B258**, 46.

73. Caneschi, L., Schwimmer, A. and Veneziano, G. (1969), 'Twisted propagator in the operatorial duality formalism', *Phys. Lett.* **30B**, 351.

74. Caneschi, L. and Schwimmer, A. (1970), 'Ward identities and vertices in the operatorial duality formalism', *Nuovo Cim. Lett.* **3**, 213.

75. Carbone, G. and Sciuto, S. (1970), 'On amplitudes involving excited particles in dual-resonance models', *Nuovo Cim. Lett.* **3**, 246.

76. Cardy, J.L. (1986), 'Operator content of two-dimensional conformally invariant theories', *Nucl. Phys.* **B270[FS16]**, 186.

77. Casalbuoni, R. (1976), 'Relatively (*sic.*) and supersymmetries', *Phys. Lett.* **62B**, 49.

78. Casalbuoni, R. (1976), 'The classical mechanics for Bose-Fermi systems', *Nuovo Cim.* **33A**, 389.

79. Casher, A., Englert, F., Nicolai, H. and Taormina, A. (1985), 'Consistent superstrings as solutions of the $D = 26$ bosonic string theory', *Phys. Lett.* **162B**, 121.

80. Chan H.M. (1969), 'A generalized Veneziano model for the N - point function', *Phys. Lett.* **28B**, 425.

81. Chan H.M. and Tsou S.T. (1969), 'Explicit construction of the N - point function in the generalized Veneziano model', *Phys. Lett.* **28B**,

485.

82. Chang, L.N. and Mansouri, F. (1972), 'Dynamics underlying duality and gauge invariance in the dual-resonance models', *Phys. Rev.* **D5**, 2535.

83. Chang, L.N., Macrae, K.I. and Mansouri, F. (1976), 'Geometrical approach to local gauge and supergauge invariance: Local gauge theories and supersymmetric strings', *Phys. Rev.* **D13** 235.

84. Chapline, G. (1985), 'Unification of gravity and elementary particle interactions in 26 dimensions?', *Phys. Lett.* **158B**, 393.

85. Chiu, C.B., Matsuda, S. and Rebbi, C. (1969), 'Factorization properties of the dual resonance model: A general treatment of linear dependences', *Phys. Rev. Lett.* **23**, 1526.

86. Chiu, C.B., Matsuda, S. and Rebbi, C. (1970), 'A general approach to the symmetry and the factorization properties of the N-point dual amplitudes', *Nuovo Cim.* **67A**, 437.

87. Chodos, A. and Thorn, C.B. (1974), 'Making the massless string massive', *Nucl. Phys.* **B72**, 509.

88. Christensen, S.M. and Duff, M.J. (1978), 'Quantum gravity in $2 + \epsilon$ dimensions', *Phys. Lett.* **79B**, 213.

89. Clavelli, L. and Ramond, P. (1970), '$SU(1,1)$ analysis of dual resonance models', *Phys. Rev.* **D2**, 973.

90. Clavelli, L. and Ramond, P. (1971), 'Group-theoretical construction of dual amplitudes', *Phys. Rev.* **D3**, 988.

91. Cohen, A., Moore, G., Nelson, P. and Polchinski, J. (1986), 'An off-shell propagator for string theory', *Nucl. Phys.* **B267**, 143.

92. Cohen, E., Gomez, C. and Mansfield, P. (1986), 'BRS invariance of the interacting Polyakov string', *Phys. Lett.* **174B**, 159.

93. Coleman, S., Gross, D. and Jackiw, R. (1969), 'Fermion avatars of the Sugawara model', *Phys. Rev.* **180**, 1359.

94. Coleman, S. (1975), 'Quantum sine-Gordon equation as the massive Thirring model', *Phys. Rev.* **D11**, 2088.

95. Collins, P.A. and Tucker, R.W. (1977), 'An action principle for the Neveu-Schwarz-Ramond string and other systems using supernumerary variables', *Nucl. Phys.* **B121**, 307.

96. Corrigan, E.F. and Olive, D. (1972), 'Fermion-meson vertices in dual theories', *Nuovo Cim.* **11A**, 749.

97. Corrigan, E.F. and Goddard, P. (1973), 'Gauge conditions in the dual fermion model', *Nuovo Cim.* **18A**, 339.

98. Corrigan, E.F. and Goddard, P. (1973), 'The off-mass shell physical state projection operator for the dual resonance model', *Phys. Lett.* **B44**, 502.

99. Corrigan, E.F., Goddard, P., Smith, R.A. and Olive, D.I. (1973), 'Evaluation of the scattering amplitude for four dual fermions', *Nucl. Phys.* **B67**, 477.

100. Corrigan, E.F. and Goddard, P. (1974), 'The absence of ghosts in the dual fermion model', *Nucl. Phys.* **B68**, 189.

101. Corrigan, E.F. (1974), 'The scattering amplitude for four dual fermions', *Nucl. Phys.* **B69**, 325.

102. Corrigan, E.F. and Fairlie, D.B. (1975), 'Off-shell states in dual resonance theory', *Nucl. Phys.* **B91**, 527.

103. Corrigan, E.F. (1986), 'Twisted vertex operators and representations of the Virasoro algebra', *Phys. Lett.* **169B**, 259.
104. Corwin, L., Ne'eman, Y. and Sternberg, S. (1975), 'Graded Lie algebras in mathematics and physics (Bose-Fermi symmetry)', *Rev. Mod. Phys.* **47**, 573.
105. Craigie, N.S., Nahm, W. and Narain, K.S. (1985), 'Realization of the Kac–Moody algebras of 2D QFTs through soliton operators', *Phys. Lett.* **152B**, 203.
106. Cremmer, E. and Scherk, J. (1974), 'Spontaneous dynamical breaking of gauge symmetry in dual models', *Nucl. Phys.* **B72**, 117.
107. Cremmer, E. and Scherk, J. (1976), 'Dual models in four dimensions with internal symmetries', *Nucl. Phys.* **B103**, 399.
108. Cremmer, E. and Scherk, J. (1976), 'Spontaneous compactification of space in an Einstein–Yang–Mills–Higgs model', *Nucl. Phys.* **B108**, 409.
109. Cremmer, E. and Scherk, J. (1977), 'Spontaneous compactification of extra space dimensions', *Nucl. Phys.* **B118**, 61.
110. Crnković, Č. (1986), 'Many pictures of the superparticle', *Phys. Lett.* **173B**, 429.
111. Curtright, T.L. and Zachos, C.K. (1984), 'Geometry, topology and supersymmetry in nonlinear sigma models', *Phys. Rev. Lett.* **53**, 1799.
112. Curtright, T.L., Mezincescu, L. and Zachos, C.K. (1985), 'Geometrostasis and torsion in covariant superstrings', *Phys. Lett.* **161B**, 79.
113. Curtright, T.L., Thorn, C.B. and Goldstone, J. (1986), 'Spin content of the bosonic string', *Phys. Lett.* **175B**, 47.
114. Das, S.R. and Sathiapalan, B. (1986), 'String propagation in a tachyon background', *Phys. Rev. Lett.* **56**, 2664.
115. De Alwis, S.P. (1986), 'The dilaton vertex in the path integral formulation of strings', *Phys. Lett.* **168B**, 59.
116. Del Giudice, E. and Di Vecchia, P. (1971), 'Factorization and operator formalism in the generalized Virasoro model', *Nuovo Cim.* **5A**, 90.
117. Del Giudice, E., Di Vecchia, P. and Fubini, S. (1972), 'General properties of the dual resonance model', *Ann. Phys.* **70**, 378.
118. Della Selva, A. and Saito, S. (1970), 'A simple expression for the Sciuto three-reggeon vertex generating duality', *Nuovo Cim. Lett.* **4**, 689.
119. Deser, S. and Zumino, B. (1976), 'Consistent supergravity', *Phys. Lett.* **62B**, 335.
120. Deser, S. and Zumino, B. (1976), 'A complete action for the spinning string', *Phys. Lett.* **65B**, 369.
121. DeWitt, B., (1964), 'Dynamical theory of groups and fields', in *Relativity, Groups, and Topology*, ed. B. DeWitt and C. DeWitt (New York, Gordon and Breach), p. 587.
122. Di Vecchia, P., Knizhnik, V.G., Petersen, J.L. and Rossi, P. (1985), 'A supersymmetric Wess-Zumino Lagrangian in two dimensions', *Nucl. Phys.* **B253**, 701.
123. Di Vecchia, P., Petersen, J.L. and Zheng, H.B. (1985), '$N = 2$ extended superconformal theories in two dimensions', *Phys. Lett.* **162B**, 327.
124. Di Vecchia, P., Petersen, J.L. and Yu, M. (1986), 'On the unitary representations of $N = 2$ superconformal theory', *Phys. Lett.* **172B**, 211.

125. Di Vecchia, P., Petersen, J.L., Yu, M. and Zheng, H.B. (1986), 'Explicit construction of unitary representations of the $N = 2$ superconformal algebra', *Phys. Lett.* **174B**, 280.

126. Dolan, L. and Slansky, R. (1985), 'Physical spectrum of compactified strings', *Phys. Rev. Lett.* **54**, 2075.

127. Dolen, R., Horn, D. and Schmid, C. (1967), 'Prediction of Regge parameters of ρ poles from low-energy πN data', *Phys. Rev. Lett.* **19**, 402.

128. Dolen, R., Horn, D. and Schmid, C. (1968), 'Finite-energy sum rules and their application to πN charge exchange', *Phys. Rev.* **166**, 1768.

129. Dotsenko, Vl.S. and Fateev, V.A. (1985), 'Operator algebra of two-dimensional conformal theories with central charge $C \leq 1$', *Phys. Lett.* **154B**, 291.

130. Duncan, A. and Moshe, M. (1986), 'First-quantized superparticle action for the vector superfield', *Nucl. Phys.* **B268**, 706.

131. Duncan, A. and Meyer-Ortmanns, H. (1986), 'Lattice formulation of the superstring', *Phys. Rev.* **D33**, 3155.

132. Durhuus, B., Nielsen, H.B., Olesen, P. and Petersen, J.L. (1982), 'Dual models as saddle point approximations to Polyakov's quantized string', *Nucl. Phys.* **B196**, 498.

133. Durhuus, B., Olesen, P. and Petersen, J.L. (1982), 'Polyakov's quantized string with boundary terms', *Nucl. Phys.* **198**, 157.

134. Durhuus, B., Olesen, P. and Petersen, J.L. (1982), 'Polyakov's quantized string with boundary terms (II)', *Nucl. Phys.* **201**, 176.

135. Eastaugh, A., Mezincescu, L., Sezgin, E. and Van Nieuwenhuizen, P. (1986), 'Critical dimensions of spinning strings on group manifolds from Fujikawa's method', *Phys. Rev. Lett.* **57**, 29.

136. Ecker, G. and Honerkamp, J. (1971), 'Application of invariant renormalization to the non-linear chiral invariant pion lagrangian in the one-loop approximation', *Nucl. Phys.* **B35**, 481.

137. Eichenherr, H. (1985), 'Minimal operator algebras in superconformal quantum field theory', *Phys. Lett.* **151B**, 26.

138. Einstein, A. and Mayer, W. (1931), 'Einheitliche Theorie von Bravitation und Elektrizität', *Setz. Preuss. Akad. Wiss.*, 541.

139. Einstein, A. and Bergmann, P. (1938), 'On a generalization of Kaluza's theory of electricity', *Ann. Math.* **39**, 683.

140. Einstein, A., Bargmann, V. and Bergmann, P. (1941), in *Theodore von Kármán Anniversary Volume* (Pasadena) p. 212.

141. Englert, F. and Neveu, A. (1985), 'Non-Abelian compactification of the interacting bosonic string', *Phys. Lett.* **163B**, 349.

142. Evans, M. and Ovrut, B.A. (1986), 'The world sheet supergravity of the heterotic string', *Phys. Lett.* **171B**, 177.

143. Evans, M. and Ovrut, B.A. (1986), 'A two-dimensional superfield formulation of the heterotic string', *Phys. Lett.* **175B**, 145.

144. Faddeev, L.D. and Popov, V.N. (1967), 'Feynman diagrams for the Yang-Mills field', *Phys. Lett.* **25B**, 29.

145. Fairlie, D.B. and Nielsen, H.B. (1970), 'An analogue model for KSV theory', *Nucl. Phys.* **B20**, 637.

146. Fairlie, D.B. and Martin, D. (1973), 'New light on the Neveu-Schwarz model', *Nuovo Cim.* **18A**, 373.

147. Feigin, B.L. and Fuks, D.B. (1982), 'Invariant skew-symmetric differential operators on the line and Verma modules over the Virasoro algebra', *Funct. Analys. Appl.* **16**, 114.
148. Feingold, A. and Lepowsky, J. (1978) 'The Weyl-Kac character formula and power series identities' *Adv. Math.* **29**, 271.
149. Feynman, R.P. (1963), 'Quantum theory of gravitation', *Acta Physica Polonica* **24**, 697.
150. Fradkin, E.S. and Vilkovisky, G.A. (1975), 'Quantization of relativistic systems with constraints', *Phys. Lett.* **55B**, 224.
151. Fradkin, E.S. and Fradkina, T.E. (1978), 'Quantization of relativistic systems with boson and fermion first- and second-class constraints', *Phys. Lett.* **72B**, 343.
152. Fradkin, E.S. and Tseytlin, A.A. (1981), 'Quantization of two-dimensional supergravity and critical dimensions for string models', *Phys. Lett.* **106B**, 63.
153. Fradkin, E.S. and Tseytlin, A.A. (1982), 'Quantized string models', *Ann. Phys.* **143**, 413.
154. Fradkin, E.S. and Tseytlin, A.A. (1985), 'Fields as excitations of quantized coordinates', *JETP Lett.* **41**, 206.
155. Fradkin, E.S. and Tseytlin, A.A. (1985), 'Quantum string theory effective action', *Nucl. Phys.* **B261**, 1.
156. Fradkin, E.S. and Tseytlin, A.A. (1985), 'Effective field theory from quantized strings', *Phys. Lett.* **158B**, 316.
157. Fradkin, E.S. and Tseytlin, A.A. (1985), 'Effective action approach to superstring theory', *Phys. Lett.* **160B**, 69.
158. Fradkin, E.S. and Tseytlin, A.A. (1985), 'Anomaly-free two-dimensional chiral supergravity-matter models and consistent string theories', *Phys. Lett.* **162B**, 295.
159. Fradkin, E.S. and Tseytlin, A.A. (1985), 'Non-linear electrodynamics from quantized strings', *Phys. Lett.* **163B**, 123.
160. Frampton, P. (1974), *Dual Resonance Models*, (Benjamin).
161. Frautschi, S. (1971), 'Statistical bootstrap model of hadrons', *Phys. Rev.* **D3**, 2821.
162. Freedman, D.Z., Van Nieuwenhuizen, P. and Ferrara, S. (1976), 'Progress toward a theory of supergravity', *Phys. Rev.* **D13**, 3214.
163. Freedman, D.Z. and Townsend, P.K. (1981), 'Antisymmetric tensor gauge theories and non-linear σ-models', *Nucl. Phys.* **B177**, 282.
164. Freeman, M.D. and Olive, D.I. (1986), 'BRS cohomology in string theory and the no-ghost theorem,' *Phys. Lett.* **175B**, 151.
165. Frenkel, I.B. and Kac, V.G. (1980), 'Basic representations of affine Lie algebras and dual resonance models', *Inv. Math.* **62**, 23.
166. Frenkel, I.B. (1981), 'Two constructions of affine Lie algebra representations and boson-fermion correspondence in quantum field theory', *J. Funct. Anal.* **44**, 259.
167. Frenkel, I.B., Garland, H. and Zuckerman, G. (1986), 'Semi-infinite cohomology and string theory', (Yale University preprint).
168. Freund, P.G.O. (1968), 'Finite-energy sum rules and bootstraps', *Phys. Rev. Lett.* **20**, 235.
169. Freund, P.G.O. (1969), 'Model for the Pomeranchuk term', *Phys. Rev. Lett.* **22**, 565.

170. Freund, P.G.O. and Rivers, R.J. (1969), 'Duality, unitarity and the Pomeranchuk singularity', *Phys. Lett.* **29B**, 510.

171. Freund, P.G.O. and Kaplansky, I. (1976), 'Simple supersymmetries', *J. Math. Phys.* **17**, 228.

172. Freund, P.G.O. and Nepomechie, R.I. (1982), 'Unified geometry of antisymmetric tensor gauge fields and gravity', *Nucl. Phys.* **B199**, 482.

173. Freund, P.G.O. (1985), 'Superstrings from 26 dimensions', *Phys. Lett.* **151B**, 387.

174. Fridling, B. and van de Ven, A. (1986) 'Renormalization of generalized two dimensional nonlinear σ models', *Nucl. Phys.* **B268**, 719.

175. Fridling, B.E. and Jevicki, A. (1986), 'Nonlinear σ-models as S-matrix generating functionals of strings', *Phys. Lett.* **174B**, 75.

176. Friedan, D. (1980), 'Nonlinear models in $2 + \epsilon$ dimensions,' Ph.D. thesis, published in *Ann. Phys.* **163** (1985) 318.

177. Friedan, D. (1980), 'Nonlinear models in $2 + \epsilon$ dimensions', *Phys. Rev. Lett.* **45**, 1057.

178. Friedan, D. (1984), 'Introduction to Polyakov's string theory', in *Recent Advances in Field Theory and Statistical Mechanics*, eds. J.B. Zuber and R. Stora. Proc. of 1982 Les Houches Summer School (Elsevier), p. 839.

179. Friedan, D., Qiu, Z. and Shenker, S. (1984), 'Conformal invariance, unitarity, and critical exponents in two dimensions', *Phys. Rev. Lett.* **52**, 1575.

180. Friedan, D., Qiu, Z. and Shenker, S. (1985), 'Superconformal invariance in two dimensions and the tricritical Ising model', *Phys. Lett.* **151B**, 37.

181. Friedan, D., Shenker, S. and Martinec, E. (1985), 'Covariant quantization of superstrings', *Phys. Lett.* **160B**, 55.

182. Friedan, D. (1985), 'On two-dimensional conformal invariance and the field theory of strings', *Phys. Lett.* **162B**, 102.

183. Friedan, D. (1986), 'Notes on string theory and two dimensional conformal field theory', in *Workshop on Unified String Theories, 29 July – 16 August, 1985*, eds. M. Green and D. Gross (World Scientific, Singapore), p. 162.

184. Friedan, D., Martinec, E. and Shenker, S. (1986), 'Conformal invariance, supersymmetry and string theory', *Nucl. Phys.* **B271**, 93.

185. Fubini, S., Gordon, D. and Veneziano, G. (1969), 'A general treatment of factorization in dual resonance models', *Phys. Lett.* **29B**, 679.

186. Fubini, S. and Veneziano, G. (1969), 'Level structure of dual-resonance models', *Nuovo Cim.* **64A**, 811.

187. Fubini, S. and Veneziano, G. (1970), 'Duality in operator formalism', *Nuovo Cim.* **67A**, 29.

188. Fubini, S. and Veneziano, G. (1971), 'Algebraic treatment of subsidiary conditions in dual resonance models', *Ann. Phys.* **63**, 12.

189. Fubini, S., Hanson, A.J. and Jackiw, R. (1973), 'New approach to field theory', *Phys. Rev.* **D7**, 1732.

190. Fujikawa, K. (1982), 'Path integral of relativistic strings', *Phys. Rev.* **D25**, 2584.

191. Fujikawa, K. (1983), 'Path integral measure for gravitational interactions', *Nucl. Phys.* **B226**, 437.

192. Gates, S.J., Grisaru, M., Roček, M. and Siegel, W. (1983), *Superspace or One Thousand and One Lessons in Supersymmetry*, (Benjamin/Cummings).

193. Gervais, J.L. (1970), 'Operator expression for the Koba and Nielsen multi-Veneziano formula and gauge identities', *Nucl. Phys.* **B21**, 192.

194. Gervais, J.L. and Sakita, B. (1971), 'Generalizations of dual models', *Nucl. Phys.* **B34**, 477.

195. Gervais, J.L. and Sakita, B. (1971), 'Field theory interpretation of supergauges in dual models', *Nucl. Phys.* **B34**, 632.

196. Gervais, J.L. and Sakita, B. (1971), 'Functional-integral approach to dual-resonance theory', *Phys. Rev.* **D4**, 2291.

197. Gervais, J.L. and Neveu, A. (1972), 'Feynman rules for massive gauge fields with dual diagram topology', *Nucl. Phys.* **B46**, 381.

198. Gervais, J.L. and Sakita, B. (1973), 'Ghost-free string picture of Veneziano model', *Phys. Rev. Lett.* **30**, 716.

199. Gervais, J.L. and Neveu, A. (1986), 'Dimension shifting operators and null states in 2D conformally invariant field theories', *Nucl. Phys.* **B264**, 557.

200. Gleiser, M. and Taylor, J.G. (1985), 'Very hot superstrings', *Phys. Lett.* **164B**, 36.

201. Gliozzi, F. (1969), 'Ward-like identities and twisting operator in dual resonance models', *Nuovo Cim. Lett.* **2**, 846.

202. Gliozzi, F., Scherk, J. and Olive, D. (1976), 'Supergravity and the spinor dual model', *Phys. Lett.* **65B**, 282.

203. Gliozzi, F., Scherk, J. and Olive, D. (1977), 'Supersymmetry, supergravity theories and the dual spinor model', *Nucl. Phys.* **B122**, 253.

204. Goddard, P. and Thorn, C.B. (1972), 'Compatibility of the dual Pomeron with unitarity and the absence of ghosts in the dual resonance model', *Phys. Lett.* **40B**, 235.

205. Goddard, P., Rebbi, C. and Thorn, C.B. (1972), 'Lorentz covariance and the physical states in dual-resonance models', *Nuovo Cim.* **12A**, 425.

206. Goddard, P., Goldstone, J., Rebbi, C. and Thorn, C.B. (1973), 'Quantum dynamics of a massless relativistic string', *Nucl. Phys.* **B56**, 109.

207. Goddard, P., Kent, A. and Olive, D. (1985), 'Virasoro algebras and coset space models', *Phys. Lett.* **152B**, 88.

208. Goddard, P., Olive, D. and Schwimmer, A. (1985), 'The heterotic string and a fermionic construction of the E_8 Kac–Moody algebra', *Phys. Lett.* **157B**, 393.

209. Goddard, P., Nahm, W. and Olive, D. (1985), 'Symmetric spaces, Sugawara's energy momentum tensor in two dimensions and free fermions', *Phys. Lett.* **160B**, 111.

210. Goddard, P. and Olive, D. (1985), 'Algebras, lattices and strings' in *Vertex Operators in Mathematics and Physics, Proceedings of a Conference, November 10 – 17, 1983*, eds. J. Lepowsky, S. Mandelstam, I.M. Singer (Springer–Verlag, New York), p. 51.

211. Goddard, P. and Olive, D. (1985), 'Kac–Moody algebras, conformal symmetry and critical exponents', *Nucl. Phys.* **B257[FS14]**, 226.

212. Goddard, P. and Olive, D. (1986), 'An introduction to Kac–Moody algebras and their physical applications', in *Workshop on Unified String*

Theories, 29 July – 16 August, 1985, eds. M. Green and D. Gross (World Scientific, Singapore), p. 214.

213. Goddard, P., Kent, A. and Olive, D. (1986), 'Unitary representations of the Virasoro and super-Virasoro algebras', *Commun. Math. Phys.* **103**, 105.

214. Goebel, C.J. and Sakita, B. (1969), 'Extension of the Veneziano form to N- particle amplitudes', *Phys. Rev. Lett.* **22**, 257.

215. Gol'fand, Y.A. and Likhtman, E.P. (1971), 'Extension of the algebra of Poincaré group generators and violation of P invariance', *JETP Lett.* **13**, 323.

216. Gomes, J.F. (1986), 'The triviality of representations of the Virasoro algebra with vanishing central element and L_0 positive', *Phys. Lett.* **171B**, 75.

217. Goto, T. (1971), 'Relativistic quantum mechanics of one-dimensional mechanical continuum and subsidiary condition of dual resonance model', *Prog. Theor. Phys.* **46**, 1560.

218. Green, M.B. and Veneziano, G. (1971),'Average properties of dual resonances', *Phys. Lett.* **36B**, 477.

219. Green, M.B. and Shapiro, J.A. (1976), 'Off shell states in the dual model', *Phys. Lett.* **64B**, 454.

220. Green, M.B. (1976), 'Reciprocal space-time and momentum-space singularities in the narrow resonance approximation', *Nucl. Phys.* **B116**, 449.

221. Green, M.B. (1976), 'The structure of dual Green functions', *Phys. Lett.* **65B**, 432.

222. Green, M.B. (1977), 'Point-like structure and off-shell dual strings', *Nucl. Phys.* **B124**, 461.

223. Green, M.B. (1977), 'Dynamical point-like structure and dual strings', *Phys. Lett.* **69B**, 89.

224. Green, M.B. and Schwarz, J.H. (1981), 'Supersymmetrical dual string theory', *Nucl. Phys.* **B181**, 502.

225. Green, M.B. and Schwarz, J.H. (1982), 'Supersymmetric dual string theory (II). Vertices and trees', *Nucl. Phys.* **B198**, 252.

226. Green, M.B. and Schwarz, J.H. (1982), 'Supersymmetrical string theories', *Phys. Lett.* **109B**, 444.

227. Green, M.B., Schwarz, J.H. and Brink, L. (1982), '$N = 4$ Yang–Mills and $N = 8$ supergravity as limits of string theories', *Nucl. Phys.* **B198**, 474.

228. Green, M.B. (1983), 'Supersymmetrical dual string theories and their field theory limits – a review', *Surveys in High Energy Physics* **3**, 127.

229. Green, M.B. and Schwarz, J.H. (1984), 'Covariant description of superstrings', *Phys. Lett.* **136B**, 367.

230. Green, M.B. and Schwarz, J.H. (1984), 'Properties of the covariant formulation of superstring theories', *Nucl. Phys.* **B243**, 285.

231. Green, M.B. and Schwarz, J.H. (1984), 'Anomaly cancellations in supersymmetric $D = 10$ gauge theory and superstring theory', *Phys. Lett.* **149B**, 117.

232. Green, M.B. (1986), 'Lectures on superstrings', in *Workshop on Unified String Theories, 29 July – 16 August, 1985*, eds. M. Green and D. Gross (World Scientific, Singapore), p. 294.

233. Green, M.B., and Gross, D.J. (1986), eds. *Unified String Theories* (World Scientific).
234. Grisaru, M.T., Howe, P., Mezincescu, L., Nilsson, B.E.W. and Townsend, P.K. (1985), '$N = 2$ superstrings in a supergravity background', *Phys. Lett.* **162B**, 116.
235. Gross, D.J., Neveu, A., Scherk, J. and Schwarz, J.H. (1970), 'The primitive graphs of dual-resonance models', *Phys. Lett.* **31B**, 592.
236. Gross, D.J. and Schwarz, J.H. (1970), 'Basic operators of the dual-resonance model', *Nucl. Phys.* **B23**, 333.
237. Gross, D.J., Harvey, J.A., Martinec, E. and Rohm, R. (1985), 'Heterotic string', *Phys. Rev. Lett.* **54**, 502.
238. Gross, D.J., Harvey, J.A., Martinec, E. and Rohm, R. (1985), 'Heterotic string theory (I). The free heterotic string', *Nucl. Phys.* **B256**, 253.
239. Gross, D.J., Harvey, J.A., Martinec, E. and Rohm, R. (1986), 'Heterotic string theory (II). The interacting heterotic string', *Nucl. Phys.* **B267**, 75.
240. Hagedorn, R. (1968), 'Hadronic matter near the boiling point', *Nuovo Cim.* **56A**, 1027.
241. Halpern, M.B., Klein, S.A. and Shapiro, J.A. (1969), 'Spin and internal symmetry in dual Feynman theory', *Phys. Rev.* **188**, 2378.
242. Halpern, M.B. and Thorn, C.B. (1971), 'Dual model of pions with no tachyon', *Phys. Lett.* **35B**, 441.
243. Halpern, M.B. (1971), 'The two faces of a dual pion-quark model', *Phys. Rev.* **D4**, 2398.
244. Halpern, M.B. (1971), 'New dual models of pions with no tachyon', *Phys. Rev.* **D4**, 3082.
245. Halpern, M.B. and Thorn, C.B. (1971), 'Two faces of a dual pion-quark model. II. Fermions and other things', *Phys. Rev.* **D4**, 3084.
246. Halpern, M.B. (1975), 'Quantum "solitons" which are $SU(N)$ fermions', *Phys. Rev.* **D12**, 1684.
247. Hara, O. (1971), 'On origin and physical meaning of Ward-like identity in dual-resonance model', *Prog. Theor. Phys.* **46**, 1549.
248. Harari, H. (1968), 'Pomeranchuk trajectory and its relation to low-energy scattering amplitudes', *Phys. Rev. Lett.* **20**, 1395.
249. Harari, H. (1969), 'Duality diagrams', *Phys. Rev. Lett.* **22**, 562.
250. Henneaux, M. and Mezincescu, L. (1985), 'A σ-model interpretation of Green–Schwarz covariant superstring action', *Phys. Lett.* **152B**, 340.
251. Henneaux, M. (1986), 'Remarks on the cohomology of the BRS operator in string theory', *Phys. Lett.* **177B**, 35.
252. Hlousek, Z. and Yamagishi, K. (1986), 'An approach to BRST formulation of Kac-Moody algebra', *Phys. Lett.* **173B**, 65.
253. Honerkamp, J. (1972), 'Chiral multi-loops', *Nucl. Phys.* **B36**, 130.
254. Hori, T. and Kamimura, K. (1985), 'Canonical formulation of superstring', *Prog. Theor. Phys.* **73**, 476.
255. Hosotani, Y. (1985), 'Hamilton–Jacobi formalism and wave equations for strings', *Phys. Rev. Lett.* **55**, 1719.
256. Howe, P.S. (1977), 'Superspace and the spinning string', *Phys. Lett.* **70B**, 453.
257. Howe, P.S. (1979), 'Super Weyl transformations in two dimensions', *J. Phys.* **A12**, 393.

258. Hsue, C.S., Sakita, B. and Virasoro, M.A. (1970), 'Formulation of dual theory in terms of functional integrations', *Phys. Rev.* **D2**, 2857.

259. Hull, C.M. and Witten, E. (1985), 'Supersymmetric sigma models and the heterotic string', *Phys. Lett.* **160B**, 398.

260. Hull, C.M. (1986), 'Sigma model beta-functions and string compactifications', *Nucl. Phys.* **B267**, 266.

261. Hwang, S. (1983), 'Covariant quantization of the string in dimensions $D \leq 26$ using a Becchi-Rouet-Stora formulation', *Phys. Rev.* **D28**, 2614.

262. Hwang, S. and Marnelius, R. (1986), 'Modified strings in terms of zweibein fields', *Nucl. Phys.* **B271**, 369.

263. Hwang, S. and Marnelius, R. (1986), 'The bosonic string in non-conformal gauges', *Nucl. Phys.* **B272**, 389.

264. Igi, K. and Matsuda, S. (1967), 'New sum rules and singularities in the complex J plane', *Phys. Rev. Lett.* **18**, 625.

265. Igi, K. and Matsuda, S. (1967), 'Some consequences from superconvergence for πN scattering', *Phys. Rev.* **163**, 1622.

266. Iwasaki, Y. and Kikkawa, K. (1973), 'Quantization of a string of spinning material – Hamiltonian and Lagrangian formulations', *Phys. Rev.* **D8**, 440.

267. Jacob, M. editor. (1974), 'Dual theory', *Physics Reports Reprint Volume I*, (North–Holland, Amsterdam).

268. Jain, S., Shankar, R. and Wadia, S. (1985), 'Conformal invariance and string theory in compact space: bosons,' *Phys. Rev.* **D32**, 2713.

269. Jain, S., Mandal, G. and Wadia, S.R. (1987), 'Virasoro conditions, vertex operators, and string dynamics in curved space', *Phys. Rev.* **D35**, 778.

270. Jevicki, A. (1986), 'Covariant string theory Feynman amplitudes', *Phys. Lett.* **169B**, 359.

271. Jimenez, F., Ramirez Mittelbrunn, J. and Sierra, G. (1986), 'Causality on the world-sheet of the string', *Phys. Lett.* **167B**, 178.

272. Jordan, P. (1947), 'Erweiterung der projektiven Relativitätstheorie', *Ann. der Phys.* **1**, 219.

273. Julia, B. (1985), 'Supergeometry and Kac–Moody algebras', in *Vertex Operators in Mathematics and Physics, Proceedings of a Conference, November 10 – 17, 1983*, eds. J. Lepowsky, S. Mandelstam, I.M. Singer (Springer–Verlag, New York), p. 393.

274. Kac, V.G. (1967), 'Simple graded Lie algebras of finite growth', *Funkt. Anali. i ego Prilozhen. 1*, 82. (English translation: *Fuctional Anal. Appl. 1*, 328.)

275. Kac, V.G. (1975), 'Classification of simple Lie superalgebras', *Funct. Analys. Appl.* **9**, 263.

276. Kac, V.G. (1983) *Infinite Dimensional Lie Algebras* (Birkhauser, Boston).

277. Kac, V.G. and Todorov, I.T. (1985), 'Superconformal current algebras and their unitary representations', *Commun. Math. Phys.* **102**, 337; Erratum, *Commun. Math. Phys.* **104**, 175.

278. Kallosh, R. (1986), 'World-sheet symmetries of the heterotic string in $(10 + 496) + 16$-dimensional superspace', *Phys. Lett.* **176B**, 50.

279. Kaluza, Th. (1921), 'On the problem of unity in physics', *Sitz. Preuss. Akad. Wiss.* **K1**, 966.

280. Kantor, I.L. (1968), 'Infinite dimensional simple graded Lie algebras,' *Doklady AN SSR* **179**, 534 (English translation: *Sov. Math. Dokl.* **9** (1968), 409.)

281. Karlhede, A. and Lindström, U. (1986), 'The classical bosonic string in the zero tension limit', *Class. Quant. Grav.* **3**, L73.

282. Kato, M. and Ogawa, K. (1983), 'Covariant quantization of string based on BRS invariance', *Nucl. Phys.* **B212**, 443.

283. Kato, M. and Matsuda, S. (1986), 'Construction of singular vertex operators as degenerate primary conformal fields', *Phys. Lett.* **172B**, 216.

284. Kawai, H., Lewellen, D.C. and Tye, S.-H.H. (1986), 'A relation between tree amplitudes of closed and open strings', *Nucl. Phys.* **B269**, 1.

285. Kawai, T. (1986), 'Remarks on a class of BRST operators', *Phys. Lett.* **168B**, 355.

286. Klein, O. (1926), 'Quantentheorie und fünfdimensionale Relativitätstheorie', *Z. Phys.* **37**, 895.

287. Klein, O. (1955), 'Generalizations of Einstein's theory of gravitation considered from the point of view of quantum field theory', *Helv. Phys. Acta Suppl. IV* (1956) 58.

288. Knizhnik, V.G. and Zamolodchikov, A.B. (1984), 'Current algebra and Wess–Zumino model in two dimensions', *Nucl. Phys.* **B247**, 83.

289. Knizhnik, V.G. (1985), 'Covariant fermionic vertex in superstrings', *Phys. Lett.* **160B**, 403.

290. Koba, Z. and Nielsen, H.B. (1969), 'Reaction amplitude for n-mesons, a generalization of the Veneziano–Bardakçi–Ruegg–Virasoro model', *Nucl. Phys.* **B10**, 633.

291. Koba, Z. and Nielsen, H.B. (1969), 'Manifestly crossing-invariant parametrization of n–meson amplitude', *Nucl. Phys.* **B12**, 517.

292. Kogut, J.B. and Soper, D.E. (1970), 'Quantum electrodynamics in the infinite-momentum frame', *Phys. Rev.* **D1**, 2901.

293. Kosterlitz, J.M. and Wray, D.A. (1970), 'The general N- point vertex in a dual model', *Nuovo Cim. Lett.* **3**, 491.

294. Kraemmer, A.B. and Nielsen, H.B. (1975), 'Quantum description of a twistable string and the Neveu–Schwarz–Ramond model', *Nucl. Phys.* **B98**, 29.

295. Kugo, T. and Ojima, I. (1978), 'Manifestly covariant canonical formulation of Yang–Mills theories physical state subsidiary conditions and physical S-matrix unitarity', *Phys. Lett.* **73B**, 459.

296. Kugo, T. and Ojima, I. (1979), 'Local covariant operator formalism of non-Abelian gauge theories and quark confinement problem', *Suppl. Prog. Theor. Phys.* **66**, 1.

297. Lepowsky, J. and Wilson, R.L. (1978), 'Construction of the affine Lie algebra A_1^1', *Commun. Math. Phys.* **62**, 43.

298. Lepowsky, J. and Wilson, R.L. (1984), 'The structure of standard modules, I: universal algebras and the Rogers–Ramanujan identities', *Inv. Math.* **77**, 199.

299. Lichnerowicz, A. (1955), *Theories Relativistes de La Gravitation et de L'Electromagnetisme* (Masson, Paris).

300. Logunov, A.A., Soloviev, L.D. and Tavkhelidze, A.N. (1967), 'Dispersion sum rules and high energy scattering', *Phys. Lett.* **24**, 181.

301. Lovelace, C. (1968), 'A novel application of Regge trajectories', *Phys. Lett.* **28B**, 264.
302. Lovelace, C. (1970), 'Simple N-Reggeon vertex', *Phys. Lett.* **32B**, 490.
303. Lovelace, C. (1971), 'Pomeron form factors and dual Regge cuts', *Phys. Lett.* **34B**, 500.
304. Lovelace, C. (1979), 'Systematic search for ghost-free string models', *Nucl. Phys.* **B148**, 253.
305. Lovelace, C. (1984), 'Strings in curved space', *Phys. Lett.* **135B**, 75.
306. Lüscher, M., Symanzik, K. and Weisz, P. (1980), 'Anomalies of the free loop wave equation in the WKB approximation', *Nucl. Phys.* **B173**, 365.
307. Luther, A. and Peschel, I. (1975), 'Calculation of critical exponents in two dimension from quantum field theory in one dimension', *Phys. Rev.* **B12**, 3908.
308. Maharana, J. and Veneziano, G. (1986), 'Gauge Ward identities of the compactified bosonic string', *Phys. Lett.* **169B**, 177.
309. Mandelstam, S. (1968), 'Dynamics based on rising Regge trajectories', *Phys. Rev.* **166**, 1539.
310. Mandelstam, S. (1970), 'Dynamical applications of the Veneziano formula', in *Lectures on elementary particles and quantum field theory*, eds. S. Deser, M. Grisaru and H. Pendleton (MIT Press, Cambridge), p. 165.
311. Mandelstam, S. (1973), 'Interacting-string picture of dual-resonance models', *Nucl. Phys.* **B64**, 205.
312. Mandelstam, S. (1973), 'Manifestly dual formulation of the Ramond-model', *Phys. Lett.* **46B**, 447.
313. Mandelstam, S. (1974), 'Interacting-string picture of the Neveu–Schwarz–Ramond model', *Nucl. Phys.* **B69**, 77.
314. Mandelstam, S. (1974), 'Dual-resonance models', *Phys. Reports* **C13**, 259.
315. Mandelstam, S. (1975), 'Soliton operators for the quantized sine-Gordon equation', *Phys. Rev.* **D11**, 3026.
316. Mandelstam, S. (1983), 'Light-cone superspace and the ultraviolet finiteness of the $N=4$ model', *Nucl. Phys.* **B213**, 149.
317. Mansouri, F. and Nambu, Y. (1972), 'Gauge conditions in dual resonance models', *Phys. Lett.* **39B**, 375.
318. Marcus, N. and Sagnotti, A. (1982), 'Tree-level constraints on gauge groups for type I superstrings', *Phys. Lett.* **119B**, 97.
319. Marnelius, R. (1983), 'Canonical quantization of Polyakov's string in arbitrary dimensions', *Nucl. Phys.* **B211**, 14.
320. Marnelius, R. (1983), 'Polyakov's spinning string from a canonical point of view', *Nucl. Phys.* **B221**, 409.
321. Marnelius, R. (1986), 'The bosonic string in $D < 26$ with and without Liouville fields', *Phys. Lett.* **172B**, 337.
322. Martellini, M. (1986), 'Some remarks on the Liouville approach to two-dimensional quantum gravity', *Ann. Phys.* **167**, 437.
323. Martinec, E. (1983), 'Superspace geometry of fermionic strings', *Phys. Rev.* **D28**, 2604.
324. Meetz, K. (1969), 'Realization of chiral symmetry in a curved isospin space', *J. Math. Phys.* **10**, 589.

325. Minami, M. (1972), 'Plateau's problem and the Virasoro conditions in the theory of duality', *Prog. Theor. Phys.* **48**, 1308.
326. Montonen, C. (1974), 'Multiloop amplitudes in additive dual resonance models', *Nuovo Cim.* **19A**, 69.
327. Moody, R.V. (1967), 'Lie algebras associated with generalized Cartan matrices,' *Bull. Am. Math. Soc.* **73**, 217.
328. Moody, R.V. (1968), 'A new class of Lie algebras', *J. Algebra* **10**, 211.
329. Moore, G. and Nelson, P. (1984), 'Anomalies in nonlinear sigma models', *Phys. Rev. Lett.* **53**, 1510.
330. Moore, G. and Nelson, P. (1986), 'Measure for moduli', *Nucl. Phys.* **B266**, 58.
331. Moore, G., Nelson, P. and Polchinski, J. (1986), 'Strings and super-moduli', *Phys. Lett.* **169B**, 47.
332. Morozov, A.Ya., Perelomov, A.M. and Shifman, M.A., (1984), 'Exact Gell-Mann–Low function of supersymmetric Kähler sigma models', *Nucl. Phys.* **B248**, 279.
333. Myung, Y.S. and Cho, B.H. (1986), 'Entropy production in a hot heterotic string', *Mod. Phys. Lett.* **A1**, 37.
334. Myung, Y.S., Cho, B.H., Kim, Y. and Park, Y-J. (1986), 'Entropy production of superstrings in the very early universe', *Phys. Rev.* **D33**, 2944.
335. Nahm, W., Rittenberg, V. and Scheunert, M. (1976), 'The classification of graded Lie algebras', *Phys. Lett.* **61B**, 383.
336. Nahm, W. (1976), 'Mass spectra of dual strings', *Nucl. Phys.* **B114**, 174.
337. Nahm, W. (1977), 'Spin in the spectrum of states of dual models', *Nucl. Phys.* **B120**, 125.
338. Nahm, W. (1978), 'Supersymmetries and their representations', *Nucl. Phys.* **B135**, 149.
339. Nakanishi, N. (1971), 'Crossing-symmetric decomposition of the five-point and six-point Veneziano formulas into tree-graph integrals', *Prog. Theor. Phys.* **45**, 436.
340. Nam, S. (1986), 'The Kac formula for the $N = 1$ and the $N = 2$ super-conformal algebras', *Phys. Lett.* **172B**, 323.
341. Nambu, Y. (1970), 'Quark model and the factorization of the Veneziano amplitude', in *Symmetries and quark models*, ed. R. Chand (Gordon and Breach), p. 269.
342. Nambu, Y. (1970), 'Duality and hydrodynamics', Lectures at the Copenhagen symposium.
343. Narain, K.S. (1986), 'New heterotic string theories in uncompactified dimensions < 10', *Phys. Lett.* **169B**, 41.
344. Ne'eman, Y. (1986), 'Strings reinterpreted as topological elements of space-time', *Phys. Lett.* **173B**, 126.
345. Ne'eman, Y. and Šijački, D. (1986), 'Spinors for superstrings in a generic curved space', *Phys. Lett.* **174B**, 165.
346. Ne'eman, Y. and Šijački, D. (1986), 'Superstrings in a generic supersymmetric curved space', *Phys. Lett.* **174B**, 171.
347. Nemeschansky, D. and Yankielowicz, S. (1985), 'Critical dimension of string theories in curved space', *Phys. Rev. Lett.* **54**, 620.
348. Nepomechie, R.I. (1982), 'Duality and the Polyakov N-point Green's function', *Phys. Rev.* **D25**, 2706.

349. Nepomechie, R.I. (1986), 'Non-Abelian symmetries from higher dimensions in string theories', *Phys. Rev.* **D33**, 3670.

350. Nepomechie, R.I. (1986), 'String models with twisted currents', *Phys. Rev.* **D34**, 1129.

351. Neveu, A. and Schwarz, J.H. (1971), 'Factorizable dual model of pions', *Nucl. Phys.* **B31**, 86.

352. Neveu, A., Schwarz, J.H. and Thorn, C.B. (1971), 'Reformulation of the dual pion model', *Phys. Lett.* **35B**, 529.

353. Neveu, A. and Schwarz, J.H. (1971), 'Quark model of dual pions', *Phys. Rev.* **D4**, 1109.

354. Neveu, A. and Thorn, C.B. (1971), 'Chirality in dual- resonance models', *Phys. Rev. Lett.* **27**, 1758.

355. Neveu, A. and Scherk, J. (1972), 'Connection between Yang-Mills fields and dual models', *Nucl. Phys.* **B36**, 155.

356. Nielsen, H.B. (1969), 'An almost physical interpretation of the dual N point function', Nordita report, (unpublished).

357. Nielsen, H.B. (1970), 'An almost physical interpretation of the integrand of the n-point Veneziano model', submitted to the 15th International Conference on High Energy Physics, (Kiev).

358. Nielsen, H.B. and Olesen, P. (1970), 'A parton view on dual amplitudes', *Phys. Lett.* **32B**, 203.

359. Nielsen, H.B. and Olesen, P. (1973), 'Local field theory of the dual string', *Nucl. Phys.* **B57**, 367.

360. Olesen, P. (1986), 'On the exponentially increasing level density in string models and the tachyon singularity', *Nucl. Phys.* **B267**, 539.

361. Olesen, P. (1986), 'On a possible stabilization of the tachyonic strings', *Phys. Lett.* **168B**, 220.

362. Olive, D. and Scherk, J. (1973), 'No-ghost theorem for the Pomeron sector of the dual model', *Phys. Lett.* **44B**, 296.

363. Olive, D. and Scherk, J. (1973), 'Towards satisfactory scattering amplitudes for dual fermions', *Nucl. Phys.* **B64**, 334.

364. Olive, D. (1974), 'Dual Models', in *Proceedings of the XVII International Conference on High Energy Physics* (Science Research Council, Rutherford Laboratory, Chilton, Didcot, U.K.), p.I-269.

365. Paton, J.E. and Chan H.M. (1969), 'Generalized Veneziano model with isospin', *Nucl. Phys.* **B10**, 516.

366. Patrascioiu, A. (1974), 'Quantum dynamics of a massless relativistic string (II)', *Nucl. Phys.* **B81**, 525.

367. Pauli, W. (1933), 'Über die Formulierung der Naturgesetze mit fünf homogenen Koordinaten', *Ann. der Phys.* **18**, 305, 337.

368. Pernici, M. and Van Nieuwenhuizen, P. (1986), 'A covariant action for the $SU(2)$ spinning string as a hyperkähler or quaternionic nonlinear sigma model', *Phys. Lett.* **169B**, 381.

369. Polyakov, A.M. (1981), 'Quantum geometry of bosonic strings', *Phys. Lett.* **103B**, 207.

370. Polyakov, A.M. (1981), 'Quantum geometry of fermionic strings', *Phys. Lett.* **103B**, 211.

371. Ramond, P. (1971), 'An interpretation of dual theories', *Nuovo Cim.* **4A**, 544.

372. Ramond, P. (1971), 'Dual theory for free fermions', *Phys. Rev.* **D3**, 2415.

373. Ramond, P. and Schwarz, J.H. (1976), 'Classification of dual model gauge algebras', *Phys. Lett.* **64B**, 75.
374. Rayski, J. (1965), 'Unified field theory and modern physics', *Acta Physica Polonica* **27**, 89.
375. Rebbi, C. (1974), 'Dual models and relativistic quantum strings', *Phys. Reports* **C12**, 1.
376. Rebbi, C. (1975), 'On the commutation properties of normal-mode operators and vertices in the theory of the relativistic quantum string', *Nuovo Cim.* **26A**, 105.
377. Redlich, A.N. and Schnitzer, H.J. (1986), 'The Polyakov string in $O(N)$ or $SU(N)$ group space', *Phys. Lett.* **167B**, 315.
378. Redlich, A.N. (1986), 'When is the central charge of the Virasoro algebra in string theories in curved space-time not a numerical constant?', *Phys. Rev.* **D33**, 1094.
379. Rosenzweig, C. (1971), 'Excited vertices in the model of Neveu and Schwarz', *Nuovo Cim. Lett.* **2**, 924.
380. Rosner, J.L. (1969), 'Graphical form of duality', *Phys. Rev. Lett.* **22**, 689.
381. Roy, S.M. amd Singh, V. (1986), 'Quantization of Nambu–Goto strings with new boundary conditions', *Phys. Rev.* **D33**, 3792.
382. Sakita, B. and Virasoro, M.A. (1970), 'Dynamical model of dual amplitudes', *Phys. Rev. Lett.* **24**, 1146.
383. Salam, A. and Strathdee, J. (1974), 'Super-gauge transformations', *Nucl. Phys.* **B76**, 477.
384. Salam, A. and Strathdee, J. (1982), 'On Kaluza–Klein theory', *Ann. Phys.* **141**, 316.
385. Salomonson, P. and Skagerstam, B.S. (1986), 'On superdense superstring gases: A heretic string model approach', *Nucl. Phys.* **B268**, 349.
386. Sasaki, R. and Yamanaka, I. (1985), 'Vertex operators for a bosonic string', *Phys. Lett.* **165B**, 283.
387. Sasaki, R. and Yamanaka, I. (1986), 'Primary fields in a unitary representation of Virasoro algebras', *Prog. Theor. Phys.* **75**, 706.
388. Scherk, J. (1971), 'Zero-slope limit of the dual resonance model', *Nucl. Phys.* **B31**, 222.
389. Scherk, J. and Schwarz, J.H. (1974), 'Dual models for non-hadrons', *Nucl. Phys.* **B81**, 118.
390. Scherk, J. and Schwarz, J.H. (1974), 'Dual models and the geometry of space-time', *Phys. Lett.* **52B**, 347.
391. Scherk, J. and Schwarz, J.H. (1975), 'Dual model approach to a renormalizable theory of gravitation', honorable mention in the 1975 essay competition of the Gravity Research Foundation.
392. Scherk, J. (1975), 'An introduction to the theory of dual models and strings', *Rev. Mod. Phys.* **47**, 123.
393. Scherk, J. and Schwarz, J.H. (1975), 'Dual field theory of quarks and gluons', *Phys. Lett.* **57B**, 463.
394. Scheunert, M. Nahm, W. and Rittenberg, V. (1976), 'Classification of all simple graded Lie algebras whose Lie algebra is reductive. I.', *J. Math. Phys.* **17**, 1626.
395. Schild, A. (1977), 'Classical null strings', *Phys. Rev.* **D16**, 1722.

396. Schwarz, J.H. (1971), 'Dual quark-gluon model of hadrons', *Phys. Lett.* **37B**, 315.
397. Schwarz, J.H. (1972), 'Dual-pion model satisfying current-algebra constraints', *Phys. Rev.* **D5**, 886.
398. Schwarz, J.H. (1972), 'Physical states and Pomeron poles in the dual pion model', *Nucl. Phys.* **B46**, 61.
399. Schwarz, J.H. and Wu, C.C. (1973), 'Evaluation of dual fermion amplitudes', *Phys. Lett.* **47B**, 453.
400. Schwarz, J.H. (1973), 'Dual resonance theory', *Phys. Reports* **C8**, 269.
401. Schwarz, J.H. (1973), 'Off-mass-shell dual amplitudes without ghosts', *Nucl. Phys.* **B65**, 131.
402. Schwarz, J.H. (1974), 'Dual quark-gluon theory with dynamical color', *Nucl. Phys.* **B68**, 221.
403. Schwarz, J.H. and Wu, C.C. (1974), 'Off-mass-shell dual amplitudes (II)', *Nucl. Phys.* **B72**, 397.
404. Schwarz, J.H. and Wu, C.C. (1974), 'Functions occurring in dual fermion amplitudes', *Nucl. Phys.* **B73**, 77.
405. Schwarz, J.H. (1974), 'Off-mass-shell dual amplitudes (III)', *Nucl. Phys.* **B76**, 93.
406. Schwarz, J.H. (1978), 'Spinning string theory from a modern perspective', in *Proc. Orbis Scientiae 1978, New Frontiers in High-Energy Physics*, eds. A. Perlmutter and L.F. Scott (Plenum Press), p. 431.
407. Schwarz, J.H. (1982), 'Superstring theory', *Phys. Reports* **89**, 223.
408. Schwarz, J.H. (1982), 'Gauge groups for type I superstrings', in *Proc. of the Johns Hopkins Workshop on Current Problems in Particle Theory 6*, Florence, 1982, p. 233.
409. Schwarz, J.H., ed. (1985), *Superstrings: The First Fifteen Years of Superstring Theory*, in 2 volumes (World Scientific, Singapore).
410. Schwarz, J.H. (1985), 'Introduction to superstrings', in *Superstrings and Supergravity*, A.T. Davis and D.G. Sutherland, eds. (Edinburgh), p. 301.
411. Schwarz, J.H. (1986), 'Faddeev–Popov ghosts and BRS symmetry in string theories', *Suppl. Prog. Theor. Phys.* **86**, 70.
412. Sciuto, S. (1969), 'The general vertex function in dual resonance models', *Nuovo Cim. Lett.* **2**, 411.
413. Segal, G. (1981), 'Unitary representations of some infinite dimensional groups', *Commun. Math. Phys.* **80**, 301.
414. Sen, A. (1985), 'Heterotic string in an arbitrary background field, ', *Phys. Rev.* **D32**, 2102.
415. Sen, A. (1985), 'Equations of motion for the heterotic string theory from the conformal invariance of the sigma model', *Phys. Rev. Lett.* **55**, 1846.
416. Sen, A. (1986), 'Local gauge and Lorentz invariance of heterotic string theory', *Phys. Lett.* **166B**, 300.
417. Shapiro, J.A. (1969), 'Narrow-resonance model with Regge behavior for $\pi\pi$ scattering', *Phys. Rev.* **179**, 1345.
418. Shapiro, J.A. (1970), 'Electrostatic analogue for the Virasoro model', *Phys. Lett.* **33B**, 361.
419. Shapiro, J.A. (1972), 'Loop graph in the dual-tube model', *Phys. Rev.* **D5**, 1945.

420. Siegel, W. (1983), 'Hidden local supersymmetry in the supersymmetric particle action', *Phys. Lett.* **128B**, 397.
421. Siegel, W. (1984), 'Covariantly second-quantized string II', *Phys. Lett.* **149B**, 157; (1985), 'Covariantly second-quantized string II', *Phys. Lett.* **151B**, 391.
422. Siegel, W. (1984), 'Covariantly second-quantized string III', *Phys. Lett.* **149B**, 162; (1985), 'Covariantly second-quantized string III', *Phys. Lett.* **151B**, 396.
423. Siegel, W. (1985), 'Spacetime-supersymmetric quantum mechanics', *Class. Quant. Grav.* **2**, L95.
424. Siegel, W. (1985), 'Classical superstring mechanics', *Nucl. Phys.* **B263**, 93.
425. Sierra, G. (1986), 'New local bosonic symmetries of the particle, superparticle and string actions', *Class. Quant. Grav.* **3**, L67.
426. Skyrme, T.H.R. (1961), 'Particle states of a quantized meson field', *Proc. Roy. Soc.* **A262**, 237.
427. Slansky, R. (1981), 'Group theory for unified model building', *Phys. Reports* **79**, 1.
428. Streater, R.F. and Wilde, I.F. (1970), 'Fermion states of a boson field', *Nucl. Phys.* **B24**, 561.
429. Sugawara, H. (1968), 'A field theory of currents', *Phys. Rev.* **170**, 1659.
430. Sugawara, H. (1986), 'String in curved space: Use of spinor representation of a noncompact group', *Phys. Rev. Lett.* **56**, 103.
431. Sundborg, B. (1985), 'Thermodynamics of superstrings at high energy densities', *Nucl. Phys.* **B254**, 583.
432. Susskind, L. (1970), 'Dual-symmetric theory of hadrons. – I', *Nuovo Cim.* **69A**, 457.
433. Susskind, L. (1970), 'Structure of hadrons implied by duality', *Phys. Rev.* **D1**, 1182.
434. Teitelboim, C. (1986), 'Gauge invariance for extended objects', *Phys. Lett.* **167B**, 63.
435. 't Hooft, G. (1974), 'A planar diagram theory for strong interactions', *Nucl. Phys.* **B72**, 461.
436. Thorn, C.B. (1970), 'Linear dependences in the operator formalism of Fubini, Veneziano, and Gordon', *Phys. Rev.* **D1**, 1693.
437. Thorn, C.B. (1971), 'Embryonic dual model for pions and fermions', *Phys. Rev.* **D4**, 1112.
438. Thorn, C.B. (1980), 'Dual models and strings: The critical dimension', *Phys. Reports* **67**, 163.
439. Thorn, C.B. (1984), 'Computing the Kac determinant using dual model techniques and more about the no-ghost theorem', *Nucl. Phys.* **B248**, 551.
440. Thorn, C.B. (1985), 'A proof of the no-ghost theorem using the Kac determinant', in *Vertex Operators in Mathematics and Physics, Proceedings of a Conference, November 10 – 17, 1983*, eds. J. Lepowsky, S. Mandelstam, I.M. Singer (Springer–Verlag, New York), p. 411.
441. Thorn, C.B. (1986), 'Introduction to the theory of relativistic strings', in *Workshop on Unified String Theories, 29 July – 16 August, 1985*, eds. M. Green and D. Gross (World Scientific, Singapore), p. 5.

442. Todorov, I.T. (1985), 'Current algebra approach to conformal invariant two-dimensional models', *Phys. Lett.* **153B**, 77.
443. Trautman, A. (1970), 'Fibre bundles associated with space-time', *Rep. Math. Phys.* **1**, 29.
444. Tseytlin, A.A. (1986), 'Covariant string field theory and effective action', *Phys. Lett.* **168B**, 63.
445. Tseytlin, A.A. (1986), 'Effective action for a vector field in the theory of open superstrings', *Pis'ma Zh. Eksp. Teor. Fiz.* **43**, 209.
446. Tye, S.-H.H. (1985), 'The limiting temperature of the universe and superstrings', *Phys. Lett.* **158B**, 388.
447. Tye, S.-H.H. (1985), 'New actions for superstrings', *Phys. Rev. Lett.* **55**, 1347.
448. Tyutin, I.V. (1975), 'Gauge invariance in field theory and in statistical physics in the operator formulation', Lebedev preprint FIAN No. 39 (in Russian), unpublished.
449. Vafa, C. and Witten, E. (1985), 'Bosonic string algebras', *Phys. Lett.* **159B**, 265.
450. Van Nieuwenhuizen, P. (1981), 'Supergravity', *Phys. Reports* **68**, 189.
451. Van Nieuwenhuizen, P. (1986), 'The actions of the $N = 1$ and $N = 2$ spinning strings as conformal supergravities', *Int. J. Mod. Phys.* **A1**, 155.
452. Veblen, O. (1933), *Projektive Relativitäts Theorie* (Springer, Berlin).
453. Veneziano, G. (1968), 'Construction of a crossing-symmetric, Regge-behaved amplitude for linearly rising trajectories', *Nuovo Cim.* **57A**, 190.
454. Veneziano, G. (1974), 'An introduction to dual models of strong interactions and their physical motivations', *Phys. Reports* **C9**, 199.
455. Veneziano, G. (1986), 'Ward identities in dual string theories', *Phys. Lett.* **167B**, 388.
456. Virasoro, M.A. (1969), 'Alternative constructions of crossing-symmetric amplitudes with Regge behavior', *Phys. Rev.* **177**, 2309.
457. Virasoro, M.A. (1969), 'Generalization of Veneziano's formula for the five-point function', *Phys. Rev. Lett.* **22**, 37.
458. Virasoro, M.A. (1970), 'Subsidiary conditions and ghosts in dual-resonance models', *Phys. Rev.* **D1**, 2933.
459. Volovich, I.V. and Katanaev, M.O. (1986), 'Quantum strings with a dynamic geometry', *Pis'ma Zh. Eksp. Teor. Fiz.* **43**, 212.
460. Waterson, G. (1986), 'Bosonic construction of an $N = 2$ extended superconformal theory in two dimensions', *Phys. Lett.* **171B**, 77.
461. Weinberg, S. (1964), 'Derivation of gauge invariance and the equivalence principle from Lorentz invariance of the S-matrix', *Phys. Lett.* **9**, 357.
462. Weinberg, S. (1964), 'Photons and gravitons in S-matrix theory: derivation of charge conservation and equality of gravitational and inertial mass', *Phys. Rev.* **135**, B1049.
463. Weinberg, S. (1965), 'Photons and gravitons in perturbation theory: derivation of Maxwell's and Einstein's equations', *Phys. Rev.* **138**, B988.
464. Weinberg, S. (1985), 'Coupling constants and vertex functions in string theories', *Phys. Lett.* **156B**, 309.

465. Wess, J. and Zumino, B. (1974), 'Supergauge transformations in four dimensions', *Nucl. Phys.* **B70**, 39.
466. Wess, J. and Bagger, J. (1983), *Supersymmetry and Supergravity*, (Princeton Univ. Press).
467. Witten, E. (1983), 'Global aspects of current algebra', *Nucl. Phys.* **B223**, 422.
468. Witten, E. (1983), '$D = 10$ superstring theory', in *Fourth Workshop on Grand Unification*, ed. P. Langacker et al. (Birkhauser), p. 395.
469. Witten, E. (1984), 'Non-Abelian bosonization in two dimensions', *Commun. Math. Phys.* **92**, 455.
470. Witten, E. (1986), 'Twistor-like transform in ten dimensions', *Nucl. Phys.* **B266**, 245.
471. Witten, E. (1986), 'Global anomalies in string theory', in *Symposium on Anomalies, Geometry, Topology, March 28-30, 1985*, eds. W.A. Bardeen and A.R. White (World Scientific, Singapore), p.61.
472. Yoneya, T. (1973), 'Quantum gravity and the zero-slope limit of the generalized Virasoro model', *Nuovo Cim. Lett.* **8**, 951.
473. Yoneya, T. (1974), 'Connection of dual models to electrodynamics and gravidynamics', *Prog. Theor. Phys.* **51**, 1907.
474. Yoneya, T. (1976), 'Geometry, gravity and dual strings', *Prog. Theor. Phys.* **56**, 1310.
475. Yoshimura, M. (1971), 'Operational factorization and symmetry of the Shapiro–Virasoro model', *Phys. Lett.* **34B**, 79.
476. Yu, L.P. (1970), 'Multifactorizations and the four-Reggeon vertex function in the dual resonance models', *Phys. Rev.* **D2**, 1010.
477. Yu, L.P. (1970), 'General treatment of the multiple factorizations in the dual resonance models; the N-Reggeon amplitudes', *Phys. Rev.* **D2**, 2256.
478. Zumino, B. (1974), 'Relativistic strings and supergauges' in *Renormalization and Invariance in Quantum Field Theory*, ed. E. Caianiello (Plenum Press), p. 367.

Index

Printed in the United Kingdom
by Lightning Source UK Ltd.
109039UKS00001B/328-333